APPLIED HYDRO-AEROMECHANICS IN OIL AND GAS DRILLING

APPLIED HYDRO-AEROMECHANICS IN OIL AND GAS DRILLING

EUGENIY G. LEONOV and VALERIY I. ISAEV
Moscow Gubkin State University of Oil and Gas

Translated from Russian into English by Emmanuil G. Sinaiski

A JOHN WILEY & SONS, INC., PUBLICATION

Published by John Wiley & Sons, Inc., Hoboken, New Jersey
Published simultaneously in Canada

Library of Congress Cataloging-in-Publication Data:

Leonov, Eugeniy G., 1935-
Applied hydroaeromechanics in oil and gas drilling / Eugeniy G. Leonov, Valeriy I. Isaev ; translated from Russian into English by Sinaiski E.G.
p. cm.
"Text is based on lectures held by authors in Moscow State Gubkin University of Oil & Gas"–Pref.
Includes index.
ISBN 978-0-470-48756-3 (cloth)
1. Oil well drilling. 2. Gas well drilling. 3. Hydraulics. I. Isaev, Valeriy I., 1940- II. Title.
TN871.2.L42 2010
622′.3381–dc22

2009018426

Printed in the United States of America

10 9 8 7 6 5 4 3 2 1

CONTENTS

PREFACE

At the basis of majority technological operations of oil and gas well drilling lie hydro-aeromechanical processes. Currently, drilling is practically impossible without circulation of drilling and plugging fluids. Actually, removal of cutting from the well, application of turbo drills and jet bits, lowering of casing strings and casing cementing, control of circulation loss and reservoir fluid showings, investigation of lost circulations, and production horizon zones do not make a complete list of operations that go hand in hand with hydro-aeromechanical processes.

Hydro-aeromechanics in drilling is a special branch in oil technology in which flows of circulating and plugging fluids, skeleton, and bed fluid in well drilling are considered and described. Designing, planning, optimization, and realization of drilling schemes would have been impossible without data made available by this branch.

A rapid rise in the number and depth of wells in the last few years has attracted enhanced attention to hydro-aeromechanical problems in drilling. With scientists joining forces, some problems were successfully solved. On the basis of these developments, the hydro-aeromechanical theory underlying the main technological operations of well drilling has been systematically presented in this book. The text is based on lectures given at the Moscow State Gubkin University of Oil and Gas.

EUGENIY G. LEONOV
VALERIY I. ISAEV

Moscow, Russia
September 2009

NOTATION

a	parameter; aeration flow factor; degree of cement slurry aeration
a_{i}	empirical factor
a_r	degree of grouting mortar aeration
a_r^{min}	minimal aeration degree of the grouting mortar
a_s	sound velocity of gas
$\boldsymbol{a}$	acceleration vector
A	pressure drop per pipe unit length
$Ar = \frac{d_p^3 g}{\mu^2} \rho_f(\rho_p - \rho_f)$	Archimedes number
Ar_{cr}	critical Archimedes number
b_{i}	empirical factor
B_i	constant for viscous (viscous-plastic) fluid
c	sound velocity
C	empirical constant
C	filtration resistance factor.
C_i	constant for viscous fluid
C_w	resistance factor of a rigid particle
d	pipe diameter
d_{ax}	diameter of the pipe axis in annular channel
d_{bb}	diameter of the borehole bottom
d_d	diameter of the drill pipe
d_{dc}	external diameter of the drill collar
d_{ex}	external diameter of the well pipe
d_{exdp}	external diameter of the drill pipe

d_{exdc}	external diameter of the drill collar
d_{exj}	external diameter of the joint
d_h	hydraulic diameter
d_{in}	internal diameter of well pipe
d_{indc}	internal diameter of the drill collar
d_{inw}	internal diameter of the well
d_m	maximal diameter of the joint or the coupling; diameter of motor body
d_p	particle diameter; equivalent diameter of a rigid (cutting) particle; maximal diameter of cutting particles to be removed; pipe diameter
d_t	diameter of tubing
d_T	diameter of turbo drill
d_{w}	well diameter
d/dt	total derivative
D_c	contour diameter
e	eccentricity
e_t	gearing eccentricity
E	elastic modulus of fluid (water and mud solution)
E'	elastic modulus of the pipeline material
E'	modulus of elasticity
Ei	integral exponential function
$\boldsymbol{F}$	vector of mass force, gravity force, external force
F_{as}	area of well annular space cross section
$Fr = v^2/(gd_p)$	Froude number
$Fr = \frac{Q_k^2}{S_{as}^2\, g\, d_h}$	Froude number
$Fr = k^2/(1-\beta)^2$	Froude number
Fr_g	Froude number for gas flow
$\boldsymbol{g}$	gravity acceleration
$\overline{g}$	effective gravity acceleration
G	shear modulus; particle weight
H	depth
h_0	height of the cement box
h_1	distance from the bottom to lower boundary of the bench
h_1^*	critical height at which happens outburst of fluid from the well
h(t)	height of the gas bench
H	well depth; cylinder height; slot height
H	gap width
H	formation thickness
H	drilling depth
H^*	height of the clean cement slurry

H_{fc}	fall of foam-cement slurry column level
H_r	occurence depth of bed roof with maximal bed pressure gradient
$He = \frac{\tau_0 d^2 \rho}{\eta^2}$	Hedström number
i	parameter of the equation (9.6.5)
$\boldsymbol{i}, \boldsymbol{j}, \boldsymbol{k}$	basis (unit) vectors of Cartesian coordinate system
k	parameter of the power fluid
k	adiabatic index
k, n	empirical factors characterizing leakage of a turbo drill
k	roughness
k	consistence parameter; permeability factor
κ	Karman universal constant
k	Rittinger constant
k	permeability factor; permeability factor of rigid phase skeleton
k	vector of mixture momentum
k'	correction factor for thixotrope solution
k_a	abnormality factor
k_{av}	average value of permeability factor of rigid phase skeleton
k_c	compressibility factor
k_{eq}	equivalent roughness of the drill pipe in an uncased annular space
k_f	filtration factor
k_f	permeability factor
k_h	hinderness factor
κ_{pc}	factor of piezo-conductivity
k_r	absolute equivalent roughness; reserve factor
k_s	safety factor
k_{sol}	solubility factor
K	safety factor
l	distance from the wall surface.
l	length
l	length of the drill stem
ℓ	section length of drill pipe
ℓ_{cem}	length of the section to be cemented
ℓ_f	see Fig.12.4*b*
ℓ_m	motor length
ℓ_{sh}	depth of the showing formation
l_T	length of the pipe between joints and couplings; length of turbo drill

L_t	depth of reservoir top bedding with maximal reservoir pressure gradient depth of the bottom from the mouth
$\vec{\Delta}l$	vector of phase displacement in time Δt
L	channel length; well depth
L_t	depth of reservoir top bedding with maximal reservoir pressure gradient
m	mass
m	pump operating efficiency
m_n	mass flow rate through a single head
M	moment; torque at the engine shaft
M	point at time t
M'	point at time $t + \Delta t$
$\overline{M}^c$	dimensionless torque
$\overline{M}$	moment of the motor rotor
M_p	down-hole motor moment sufficient to crush rock solid
M_P	turbo-drill moment needed for rock fracture
M_{tr}	reference data of the turbo-drill shaft moment
M_{tr}	turbo-drill shaft moment in the regime of maximal power
n	consistence parameter
n	rotation frequency of the motor rotor
n	number of joints and couplings; number of pumps
n	parameter of the power fluid
n_{joint}	number of joints
$\mathbf{n}$	normal vector
N	power
N	hydraulic power delivered to the bit
N	number of phases, components; number of pump strokes
p	pressure
$p(z)$	current pressure
$p*$	pressure at which changes the flow regime
p_0	normal pressure, initial pressure
p_{ab}	absolute pressure; absorption pressure
p_{al}	allowed pressure of the drill pump
p_{ap}	pressure in the ascending pipe
p_{apcv}	pressure in the ascending pipe of closed well
p_{am}	pressure at the well mouth in the annulus
p_{as}	pressure in annular space
p_{asd}	difference of pressures in the annular space
p_{at}	atmospheric pressure
p_{cp}	counter-pressure

p_{av}	pressure averaged over the depth
p_b	pressures at the bean
p_{bh}	bottom-hole pressure
p_{bean}	pressure at the bean
p_{bit}	pressure in the bit
p_{bitbs}	pressure in the drill string before bit
p_{bitds}	pressure in drillstem before bit
p_{bm}	minimal pressure expected at the bottom
p_{bot}	bottom hole pressure ($p_{bot} = p'_{ap} + \sum \Delta p_{as}$)
p_{bp}	bursting pressure of the casing column
p_c	pressure in the orifice
p_c	circuit pressure
p_{cm}	pressure at the column mouth
p_{cp}	counter-pressure
p_{cr}	critical pressure
p_{dcds}	pressure in the drillstem before weighted drill string
p_{dp}	pressure at the entrance into DP
p_{dptdas}	pressure in the annular space at junction of DC and TD
p_{dpdcas}	pressure in the annular space at junction of DP and DC
p_{dsb}	pressure in the drillstem before the bit ($p_{dsb} = p_{bot} + \Delta p_b$)
p_{dstd}	pressure in the drillstem before the turbo-drill
p_{end}	end pressure
p_{ex}	excess pressure; pressure at the slot counter
p_f	formation pressure; pressure at formation contour
p_{fl}	hydraulic pressures of pure fluid
p_{gtc}	expected pressure indicated in geotechnical conditions
p_h	absorption pressure (hydraulic fracturing pressure)
p_h	hydraulic pressures of solution
p_{hf}	reservoir hydraulic fracture pressure
p_{hf}	hydraulic fracturing pressure
p_{hyd}	hydrostatic pressure of a solution without regard for cutting
p'_{hyd}	hydrostatic pressure of a solution with regard to cutting
p_{ib}	pressure at the ith bean inlet
p_{ib}	pressure at the bean when the top of gas bench reaches the well mouth
p_{in}	pressure in the slot orifice

p_{init}	initial pressure of pumping
p_k	pressure on the slot contour
p_l	pressure in lower part of the drill-stem
p_{lp}	pressure at lower cross section in descending flow
p_m	pressure at the well-head (well mouth)
p_m	manometer pressure
p_{nom}	nominal component of the stress vector
p_{ov}	overpressure
p_{nn}	normal component of the stress vector
$p_{n\tau}$	tangential component of the stress vector
p_p	pressure in a pipe
p_{per}	permissible pressure
p_{pump}	pump pressure
p_{p1}, p_{p2}	absorption or hydrofracturing pressure
p_{pd}	difference of pressures in pipes
p_r	reservoir (pore) pressure
p_s	pressure in sonic flow
p_{sb1}, p_{sb2}	pressures in showing beds; formation pressures
p_{sat}	saturation pressure
p_{st}	pipe strength; pressure opposite formation most disposed to absorption less friction losses
p_t	pressures at the pipe top
p_u	pressures in upper part of the drill stem
p_w	pressure of fluid column in well
p_{wm}	pressure at the well mouth.
p	dimensionless pressure
$\boldsymbol{p}$	stress vector
$\boldsymbol{p}_n$	stress vector to a surface with normal $\boldsymbol{n}$
$p_{rr}, p_{\varphi\varphi}, p_{zz}$	normal stresses
$\boldsymbol{p}_r, \boldsymbol{p}, \boldsymbol{p}_z$	stress vectors of surface force in cylindrical coordinates
$p_{rr} = \tau_{r\varphi}, p_{rz} = \tau_{rz}, p_{\varphi z} = \tau_{\varphi z}$	tangential stress components
Δp	pressure drop
Δp_{an}	pressure drop in annular channel with zero eccentricity
$\Sigma(\Delta p_{as})$	friction along annular space length
Δp_{bit}	losses in water courses of the bit; pressure drop in drill bit
Δp_{bot}	pressure drop on the bottom
Δp_{br}	pressure drop on the motor rotor at braking operation regime
Δp_{cr}	critical pressure drop
Δp_{dhm}	losses in the down-hole motor (turbo drill)
Δp_{fr}	friction loss

Δp_{hp}	difference between hydrostatic pressures of fluid columns in the annular space and pipe
Δp_{hyd}	hydrostatic pressure of fluid column
Δp_{in}	inertial component of pressure drop
Δp_{joint}	pressure drop in the joint
Δp_{las}	pressure drop owing to local resistances in annular space
Δp_{lj}	pressure drop owing to local losses from joints in the annular space
Δp_{lock}	pressure loss from locks in annular space
Δp_{Tlp}	pressure drop owing to losses in local resistances inside pipes
Δp_{lrp}	pressure drop owing to local resistances in pipes
Δp_{lsb}	pressure drop owing to losses in surface binding
Δp_{M}	pressure increase in resistance
$\Sigma(\Delta p_{D})$	pressure drop owing to friction along pipe length
Δp_{Tr}	pressure reserve in the drill bit
Δp_{sb}	pressure loss in the surface bending
Δp_{T}	pressure drop in turbulent flow
Δp_{td}	pressure drop in turbo drill
$\overline{q}$	dimensionless mass flow rate
q_{as}	flow rate in the annulus
q_{fd}	flow rate of the fluid displaced
q_{p}	flow rate in pipe
q_{pd}	pump delivery
Q	flow rate
Q_{0}	volume flow rate of gas (air) at normal conditions
Q_{1}	flow rate of fluid displaced by rigid phase
Q_{as}	flow rate of fluid in the annular space
Q_{c}	volume flow rate of cement slurry
Q_{cr}	critical delivery
Q_{cut}	volumetric flow rate of cuttings
Q_{G}	gas flow rate
Q_{k}	killing fluid delivery
Q_{l}	flow rate (leakage) of washing fluid through the seal of turbo-drill shaft
Q_{p}	pump delivery
Q_{pd}	pump delivery
Q_{pnd}	pump nominal delivery
Q_{td}	fluid flow rate in the turbo drill
Q_{wb}	flow rate of fluid in the wellbore
r	radius; pipe radius; radial coordinate
r_{0}	radius of initial circle

r_c	formation circuit radius; external boundary of reservoir
r_{ex}	external radius
r_{in}	internal radius
r_w	well radius
r, φ, z	cylindrical coordinates
R	gas constant; pipe radius
R	resultant force acting on a particle
Re	Reynolds number
Re^*	Reynolds number of viscous-plastic flow
Re_{as}	Reynolds number in the annular space
Re_{cr}	critical Reynolds number
Re'_{cr}	critical Reynolds number calculated with particle velocity u_p
Re_{cr1}	first critical Reynolds number
Re_{cr2}	second critical Reynolds number
Re_p	Reynolds number of fluid flow in a pipe
Re_s	Reynolds number of the soaring particle
S	surface element; surface area; cylinder lateral surface; pipe area
S	downhole pressure increase (safety margin)
S_{av}	saving of the oil-well cement
S_{as}	annular space cross-section area
S_i	area of annular space; part of pipe area occupied by i-th phase
S_{ki}	areas of sectors
S_n	normal cross section; normal cross-section area
S_p	area of casing
S_{si}	sector area
S_t	area of tubing cross section
ΔS	surface element
$Se = He/Re = \frac{\tau_0 \pi r H^2}{\lvert Q \rvert \eta}$	Saint Venant number
Se_{as}	Saint Venant number for annular space
Se_{cr}	critical Saint Venant number
Se_p	Saint Venant number for pipe
$Sh = \frac{Q_3 t_k}{S_{as} L}$	Strouhal number
t	time
t_{blow}	time counted from the beginning of well blowing
t_c	instant of time of pump disconnection beginning
t_f	inflow time
t_L	time at which the upper boundary of the bench arrives the well mouth

$\bar{t}_p$	time of pump start; time of gate valve closure; time of pumping needed to displace mixture column from the well
t_r	recovery time
Δt	time step
Δt_{pd}	time interval of pump disconnecting
T	temperature
T	absolute temperature
T	time of ball drop
T_{av}	temperature averaged over the depth
T_{bot}	temperature at the well bottom
T_m	temperature at the well mouth
T_k	time of blowout killing
$\overline{T}$	temperature averaged over well depth
$Ta = \frac{1}{2}\sqrt{\frac{1-\delta}{\delta}\frac{v_1 d_h \rho}{\mu}}$	Taylor number
Ta_{cr}	critical Taylor number
u_{ds}	velocity in round trip operation of the drill stem
$u_p(t)$	velocity of pipe motion
v	mean velocity
v_*	dynamic velocity
v_{as}	mean velocity of fluid in hole annulus
v_{av}	average velocity
v_b	velocity of fluid flow in washing holes of the bit
v_c	velocity of particle centre.
v_{cr}	critical velocity
v_{cut}	velocity of cutting
v_d	mechanical rate of drilling
v_{dwf}	velocity of the lower bench boundary displaced by washing fluid
v_f	fluid velocity
v_g	gas velocity
v_k	minimal velocity of fluid lift in hole clearance providing cutting removal
v_M	mechanical drilling rate
v_p	average velocity of a rigid particle
v_r	relative velocity of a rigid particle
v_s	levitate velocity of a particle; velocity of particle start
v_{sed}	sedimentation velocity
V	volume of space, of system part
V	specific volume
V_{as}	annular space volume of the well
V_{cem}	cementing volume
V_{dr}, V_{drf}	volume of driving fluid

V_{ds}	internal volume of the drill stem
V_{flush}	volume of the flushing fluid
V_G	Volume of gas in cement slurry
V_i	volume of ith phase
V_p	volume of casing
V_s	volume of the weighted solution
V_{sh}	volume of the driving fluid
V_w	well volume
δV	specific volume increment
$\boldsymbol{w}$	velocity vector
$\boldsymbol{w}_i$	velocity vector of ith phase in the mixture
w_L	velocity of laminar flow
w_{max}	maximal flow velocity
w_x, w_y	components of the flow velocity in Cartesian coordinates
w_T	velocity of turbulent flow
w_n	normal component of fluid velocity
w_r, w_φ, w_z	velocity components in cylindrical coordinates
w_z	velocity component directed along pipe z axis
W_r	resistance force acting on a particle
$\boldsymbol{w}$	velocity vector
z	coordinate along pipe axis; current well depth with reference point at the well mouth
$\overline{z}$	over-compressibility factor averaged over well depth
$z_{twash}; z_{flush}; z_{driv}; z_{gm}$	coordinates of washing, flushing, driving fluids and grouting mortar
z_t	number of rotor teeth
z_h	coordinate of the absorptive formation
$\alpha = \frac{k_f(1+\varepsilon)^2}{\rho_l g a}$	analog of consolidation factor
α	angle of inclination; zenith angle
α	angle between z axis and the direction of gravity force
α	correction factor for velocity of turbulent flow
α	hydraulic resistance factor of surface element
$\alpha_{ap}, \alpha_{bh}, \alpha_{sw}, \alpha_{ks}$	resistance factors of surface elemets listed in Table 6.1
β_0	compressibility factor
$\beta(Se)$	dimensionless parameter
γ	shear; relative displacement of fluid layer
$\dot{\gamma}$	shear rate
$\dot{\gamma}_c$	mean shear rate
$\dot{\gamma}_w$	shear rate gradient at a wall
$\delta = d_i/d_c$	dimensionless parameter

$\overline{\delta}$	dimensionless tension of rotor in braking operation regime
$\overline{\delta}$	dimensionless tension
δ_0	dimensionless parameter
δ_L	thickness of the laminar sublayer
Δ	discriminant
Δ_{adm}	admissible error
ε	relative roughness
ε	relative accuracy of calculation
$\varepsilon = \varphi_1/\varphi_2$	analog of porosity factor in soil mechanics
$\varepsilon = \varepsilon(p')$	compression curve
$\dot{\varepsilon} = \partial w/\partial z$	deformation rate
$\delta\varepsilon$	relative deformation
ζ	dimensionless parameter; dimensionless variable
η	plastic viscosity factor
η	aeration mass factor; mass factor reflecting the cutting existence; plastic viscosity factor
θ	dead-loss (static) shear stress
θ_1	shear stress; dead-load shear stress
λ	hydraulic resistance factor
λ_{as}	hydraulic friction resistance factor in annular space
λ_c	hydraulic resistance factor
λ_{cr}	critical resistance factor
λ_m	resistance factor of a mixture
λ_t	resistance moment factor
λ_p	hydraulic friction resistance factor in pipes
Λ_i	constant for viscous fluid
μ	dynamic viscosity factor
μ	empirical flow rate factor
μ	discharge factor
μ	correction flow rate factor
μ_{eff}	apparent viscosity factor
μ_{fr}	friction factor
ξ	dimensionless parameter; dimensionless variable; resistance factor
ξ	resistance factor
$\prod = \sqrt{\frac{\delta}{1-\delta}\left(\frac{Ta}{\mathrm{Re}}\right)}$	dimensionless parameter
Π_s	vector of surface force
ρ	density
ρ_0	initial density; gas density at normal pressure
ρ_{ch}	density of the chaser
ρ_{cs}	density of the cement slurry
ρ_{cut}	density of the cutting

ρ_f	density of fluid
ρ_{flush}	density of the flushing fluid
ρ_{wash}	density of the washing fluid
ρ_g	density of gas
ρ_{gtj}	expected density in geological-technical job
ρ_i	density of ith phase
ρ_k	killing fluid density
ρ_m	mixture density
ρ_p	density of a particle; density of cutting particles
ρ_{per}	permissible density of fluid
ρ_s	kill mud density
ρ_w	density of water
ρ_{ws}	density of weighted solution
ρ_{wf}	washing fluid density
$\delta\rho$	density increment
σ	normal stress
σ	root-mean-square deviation
σ	diaphragm ultimate strength
$\Delta\sigma$	surface element
υ	mean velocity
τ	friction stress; tangential stress
τ_0	dynamic shear stress
τ_c	mean stress in fluid
τ_n	normal stress
τ_w	friction stress at a wall
φ	volume concentration
φ_i	volume content of ith phase; volume concentration of ith phase
$\overline{\varphi}_i$	average concentration
Φ	total area of hydromonitor bit heads; area of nozzles; area of the head cross section
Φ_c	total area of all bit head cross sections
$\boldsymbol{\Phi}_i$	force density vector
χ	mass concentration; piezo-conductivity factor
ψ, ψ_0, ψ_i	angles of annular space sectors
ω	angular velocity
$\nabla\cdot$	divergence

ACRONYMS

DC	drill collar
DP	drill pipe
TD	turbodrill
AS	annular space
TJ	tool joint
TJAS	tool joint in annular space
DS	drill string
SL	sludge (cutting)
DM	downhole motor

CHAPTER 1

MAIN RESULTS AND DEVELOPMENT LINES OF HYDRO-AEROMECHANICS OF DRILLING PROCESSES

Intensive investigation of forms and laws of fluid flow in wells began in 1901 when in the United States application of the mechanical rotary drilling with washing, the so-called rotary drilling, was found on Spindletop field in Texas state. In 1911, for the first time in Russia's Suruchan region several wells were bored by rotary method with washing of well bottom by mud solution. After nationalization of the oil industry, the rotary boring began to develop quickly.

With steady increase in well depth and complexity of mine geological conditions, widespread use of jet drilling bit and downhole motors resulted in the washing and plugging back in hydro-aeromechanical well-bed system becoming more costly and power intensive. Since under real hydro-aeromechanical system it is understood that the whole set of well elements and uncovered beds connected with each other in a unified technological set have a complex structure, it is necessary to build a mathematical model of this system. The model was developed in two directions: the description of main hydro-aeromechanical properties of separate elements and the structure of the system as a whole.

Investigation of basic element properties is aimed at finding correlations between pressure, flow rate, and time through relations of theoretical

Applied Hydro-Aeromechanics in Oil and Gas Drilling. By Leonov and Isaev

hydro-aeromechanics and applied hydraulics. Let us point out the most significant results of hydro-aeromechanics in drilling.

Rheological equations formulated for viscous fluids by Newton in 1685 (Krilow, 1936), for viscous-plastic media by Shvedoff in 1889 (Reiner, 1960) and Bingham in 1916 (Bingham, 1922), and for pseudoplastic media by Ostwald in 1924 (Reiner, 1960) are of profound importance in solving problems of drilling hydro-aeromechanics. With the help of these equations, formulas were obtained for pressure distribution in stationary laminar flow of viscous (Poiseuille, 1840, 1841; Stokes, 1845, 1850, 1901), viscous-plastic (Buckingham, 1921), and pseudoplastic (Rabinowitch, 1929; Mooney, 1931) fluids in circular pipes. Solutions have also been obtained for flows in concentric circular channels of viscous (Lamb, 1945), viscous-plastic (Volarovich and Gutkin, 1946), and pseudoplastic (Fredrickson and Bird, 1958) fluids.

On the basis of Bukingham and Volarovich and Gutkin formulas for the flow of viscous-plastic fluids in circular and concentric circular pipes, Grodde (1960) applied convenient graphic method to calculate pressure drop.

Schelkachev (1931) considered laminar stationary flow of viscous fluid in eccentric circular channel and obtained formula for pressure distribution. McLean et al. (1967) gave a general scheme for approximate calculation of pressure distribution in laminar flow of rheologic stationary fluid in concentric circular channel with cross section replaced by conventional sections of concentric channels with independent flows.

The stability of laminar flows of viscous fluid in circular pipes was experimentally investigated by Reynolds during 1876–1883 (Reynolds, 1883). He established transition criterion from laminar to turbulent flow. Hedström (1952) characterized the loss of viscous-plastic fluid laminar flow stability by Reynolds and Saint Venant numbers.

On the basis of boundary layer theory developed by Prandtl during 1904–1925 (Prandtl and Tietjens, 1929, 1931) for turbulent flow of viscous fluid in pipes with smooth and rough walls, Altshul (Altshul and Kiselev, 1975) obtained dependences for hydraulic resistance factors.

In developing the theory of multistage turbine, Shumilov (1943) gave formula for pressure drop in turbo-drill. To derive the pressure change in local resistances of circulation system, Herrick (1932) used the equivalent length method. Shumilov (1943) applied Borda–Karno formula for locks and Torricelli formula for drill bit orifice when determining pressure drop. Laminar flow of viscous fluid around a sphere was considered by Stokes (1845). Experimental investigations of flows around rigid spherical particles in a wide range of Reynolds numbers were generalized in the form of Rayleigh curve. Shischenko and Baklanov (1933) investigated conditions of stability and flow of mud solution around particles.

Targ (1951) found pressure distribution in laminar stationary flow of viscous fluid in an axially symmetric circular channel, one of the wall of which moves with constant velocity. Gukasov (1976) considered laminar flow of viscous-plastic fluid in concentric circular channel with movable internal wall.

Basic hydrodynamic equations for multiphase fluids using empirical relations for concentrations and hydraulic resistance factor were derived by Teletov (1958). On this basis were obtained pressure distributions in pipes and circular channels in well washing by aerated fluid or gas blowdown.

A fundamental contribution to solving the problem of nonstationary flows in hydraulic systems with regard to compressibility of fluids and elasticity of walls was made by Zhukowski (1899–1921), who developed the theory of one-dimensional nonstationary flow of viscous fluid to solve many problems (Zhukowski, 1948).

In connection with problems of oil- and gas-field development in works of Pavlowski (1922), Leibenson (1934), Schelkachev (1990), Charniy (1963), Muskat (1963), and many others, the flow of reservoir fluid in porous medium has been extensively studied to solve problems with opening up of productive buildup and problems with drilling.

Along with the investigation of hydro-aeromechanic properties of system elements, methods to investigate well-bed system as a whole have also been developed. In doing so, there have been established correlations between elements of the system needed to simultaneously solve all equations characterizing separate elements. For example, Herrick (1932) had considered a problem on feed and pressure of drilling pump for circulation of washing fluid and Shazov (1938) devised a scheme of procedure in choosing number and parameters of cementing aggregates for one-step well plugging. Mirzadjanzadeh and his collaborators (Mirzadjanzadeh, 1959) developed a method for analyzing hydro-aerodynamic processes with the help of stochastic and adaptive training models.

Shischenko and Baklanov (1933) were first to systematically outline a number of washing fluid hydraulic problems. Many aspects of hydro-aeromechanics of drilling processes were considered in monographs (Gukasov, 1976; Gukasov and Kochnev, 1991; Goins and Sheffield, 1983; Esman, 1982; Mezshlumov, 1976; Mezshlumov and Makurin, 1967; Mirzadjanzadeh, 1959; Mirzadjanzadeh and Entov, 1985; Shischenko et al., 1976; Macovei, 1982; and others), handbooks (Mittelman, 1963; Filatov, 1973; Gabolde and Nguyen, 1991; and others) and the periodic literature.

At present, there has been a tendency to develop systems approach to drilling hydro-aeromechanics chiefly in building well-bed system models both simplified and more complex ones demanding application of various mathematical methods with regard to designing, building, and operation of wells.

CHAPTER 2

BASIC PROBLEMS OF HYDRO-AEROMECHANICS IN DRILLING PROCESSES

Hydro-aeromechanic processes in drilling occur in the well-bed system consisting in the simplest case of two parts: circulation system of the well along which fluid, gas, or their mixture including rigid particles flow and one or several opened up bed formations.

In general, the hydro-aeromechanic program of well-bed operation would be engineered when distributions of the following parameters are determined and reconciled: (1) flow rates; (2) pressures; (3) densities; (4) stresses; (5) concentrations; (6) temperatures; (7) geometric sizes of system elements (length, diameter, and spatial arrangement of each circulation system including level depth, radius, and thickness of beds); (8) characteristics of compressors and pumps, cementing units, and mixers (deliveries and pumps); (9) strength characteristics of system elements; (10) characteristics of the lifting mechanism of the drilling rig (velocities and accelerations in round trips); (11) characteristics of downhole motors (pressure drops at different flow rates of the flushing fluid); and (12) granulometric compositions of the cutting drilled and carried out from the well.

Distributions 1–6 are connected with each other by common hydro-aeromechanic equations in the region of distributions 7–12 taking place

Applied Hydro-Aeromechanics in Oil and Gas Drilling. By Leonov and Isaev

in drilling. The description of hydro-aeromechanic processes of drilling reduces to finding relations connecting distributions listed in 1–12.

Depending on the goal of technological operation, any distribution from 1 to 12 can be sought or given in the form of technical, technological, economic, or ecological restrictions. In designing and handling hydro-aeromechanic program or its parts, it is required to get distributions or separate values of some of them as functions of flow rate and pressure at given values of the rest.

Figure 2.1 presents a list of main processes 1.1–1.3 and 2.1–2.5 and problems 1.1.1–1.3.3 and 2.1.1–2.5.1 associated with them, which usually happen to be considered in drilling. In order to solve them, it is necessary to investigate distributions 1–12 for stationary and nonstationary flows in well-bed system elements. When solving a concrete problem, one finds one

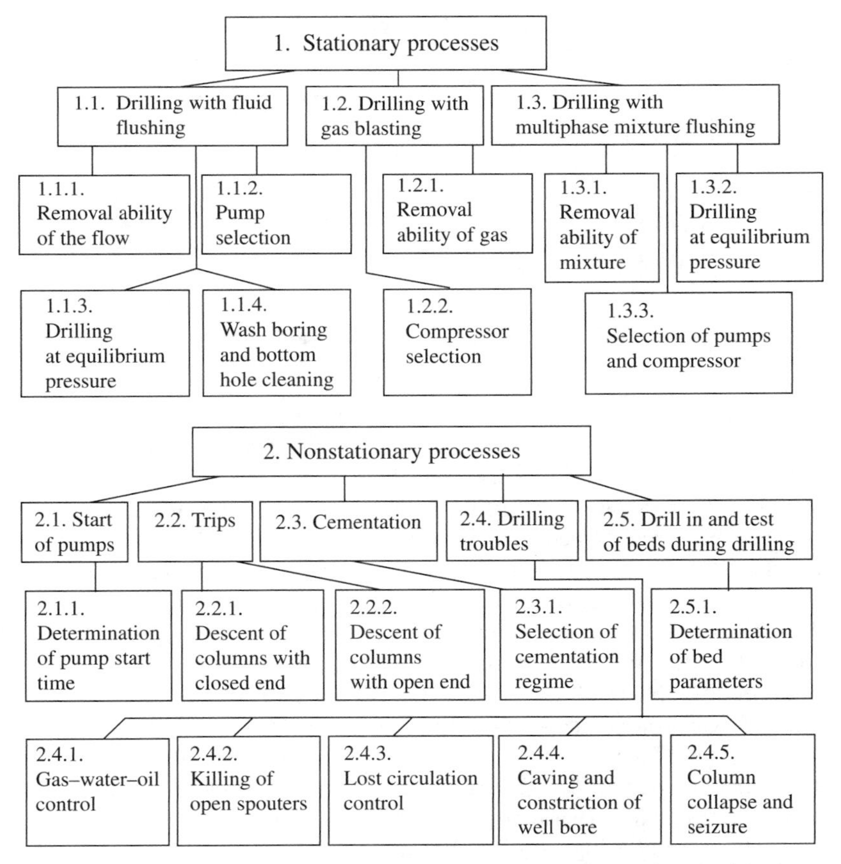

FIGURE 2.1 List of main processes and problems associated with them.

or more distributions among 1–12 so that they would not contradict the rest of them.

For example, let us consider in more detail a distribution of pressure in underground part of the circulation system, which happens to be often determined in carrying out hydro-mechanical process of drilling with fluid washing. Figure 2.2 shows the sought pressure distribution (diagram)

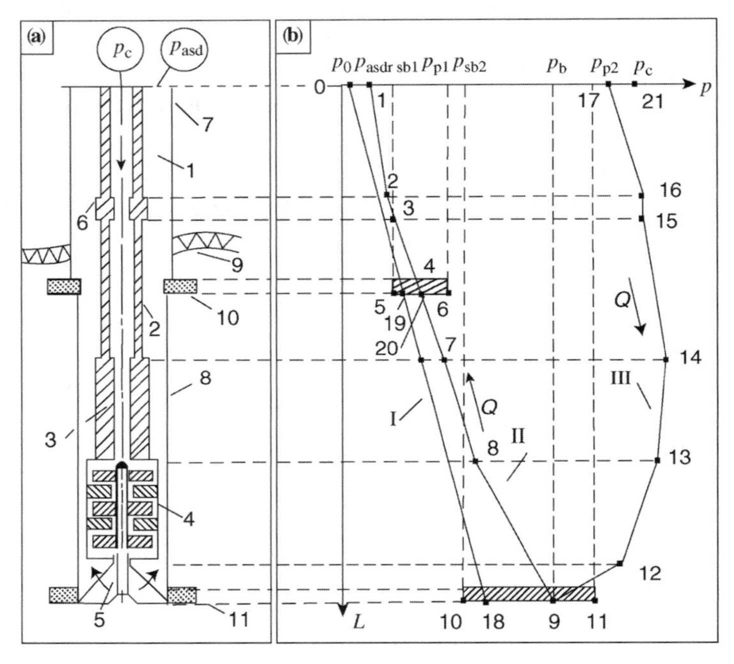

FIGURE 2.2 Diagram of underground part of vertical well circulation system and pressure distribution in bed-well system. (a) Diagram of the underground part of circulation system: 1—annular system; 2—drill pipe; 3—drill collar; 4—downhole motor; 5—drilling bit; 6—joint; 7—the last lowered casing string; 8—opened borehole; 9—covered weak bed; 10—rock under shoe of the last lowered casing string; 11—opening bed. (b) Pressure distribution in system elements (I—hydrostatic; II—at circulation in annular system; III—at circulation in drill stem): 1–2, 3–4, 4–7—after drill pipe; 2–3—after joints; 7–8—after drill collar; 8–9—after motor; 9–12—in drilling bit; 12–13—in downhole motor; 13–14—in drill collar; 14–15, 16–17—in drill pipe; 15–16—in joints. Values of pressure: 1—pressure in annular system at well head; 5, 10—formation pressures p_{sb1} and p_{sb2}; 6, 11—hydro-fracturing (absorption) pressures p_{p1} and p_{p2} in rock and bottom opening bed; 18—bottom-hole hydrostatic pressure; 9—bottom-hole pressure in circulation (washing); 19—hydrostatic pressure in annular system under shoe of casing string; 20—pressure in annular system in washing under shoe of casing string; 21—pressure in ascending pipe.

in circulation system of a vertical well in boring with washing of incompressible fluid at a given arrangement of the drill pipe string (the arrows show directions of the circulation). The pressure in the diagram is determined under the following conditions:

(a) Pressure in the ascending pipe (p_{ap}) does not exceed the allowed pressure of the drill pump (p_{al}); that is, it satisfies the consistency of distributions 2 and 9.

(b) Pressure in uncased parts of the well is higher than pressure in showing beds (p_{sb1} and p_{sb2}) but does not exceed absorption or hydro-fracturing pressures (p_{p1} and p_{p2}):

$$p_{sb1} < p < p_{p1}, \quad p_{sb2} < p < p_{p2}; \tag{2.1}$$

that is, it is a valid consistency of distributions 2 and 9.

(c) Flow rates of fluid in the annular space (Q_{as}) and at the well bore (Q_{wb}) ensure the cutting recovery; these flow rates provide distribution 1.

(d) Difference of pressures in pipes (p_{pd}) and in the annular space (p_{asd}) satisfies condition of the pipe strength (p_{st}):

$$|p_{pd} - p_{asd}| < p_{st}; \tag{2.2}$$

that is, it is a valid consistency of distributions 2 and 9.

In a variety of problems, the expected pressure depends on the characteristics of items 1–12. In performing calculations, it is not necessary to find the whole pressure distribution (diagram). For example, in the absence of weak or showing beds, it is enough to determine the pressure only in the ascending pipe, which should not exceed permissible pressure in the pump. In the process considered, other distributions 1–12 are not mentioned, but it is meant that they satisfy the diagram in Fig. 2.2. And yet the existence of such distributions should be kept in mind, and they must be taken into account when solving concrete problems.

From what has been said, it follows that the basis of all hydrodynamic calculations consists of the facility to find pressure distributions in circulation system elements of the well. In order to calculate pressure distribution and to build pressure diagram, one should be able to determine pressure drop both in concrete circulation system elements and in given cross section of the well element.

CHAPTER 3

MULTIPHASE MEDIA IN DRILLING PROCESSES

Drilling fluid, grouting mortars, special solutions, for example, spacers, reservoir fluids, and skeleton represent complex media consisting of more simple elements. Therefore, such fluids are mixtures of several media with definite properties inherent to each of them.

By setting up hydro-aeromechanic problem, the medium taking part in the process considered happened to be homogeneous or heterogeneous, single phase or multiphase, one component or multicomponent depending on the type of technological operation (Basarov, 1991).

There are macroscopic systems in which ingredients are vastly superior in sizes to molecular sizes. From macroscopic systems are set apart two systems: homogeneous and heterogeneous. *Homogeneous* (*uniform*) systems possess identical properties in any arbitrarily chosen part equal in volume to another part. For example, water and in many cases mud and cement solution may be considered uniform or homogeneous. Rocks can be approximately taken as homogeneous in salt and mud beds. *Heterogeneous* (*nonuniform, multiphase*) systems consist of several different physically homogeneous media. In such systems, one or several physical properties may be safely assumed to undergo a sudden change when going from one

---

Applied Hydro-Aeromechanics in Oil and Gas Drilling. By Leonov and Isaev

point of the volume to another. For example, in gas–liquid (aerated) flushing fluid, it is often assumed that the density instantly changes while going through bubble or air plug boundary.

In mud solution, discontinuity of density is also considered, in particular in passage through boundary between rigid particle of mud weighting material and fluid of water–mud solution. Since it is considered a boundary or a surface of definite thickness between two physically inhomogeneous media, and some properties undergo a great change at this surface, it is called interface and the media are called phases. Thus, for example, the aerated fluid is a two-phase heterogeneous in which one phase is fluid (water, oil) and the other one is gas (air, natural gas). In drilling with flushing, such solution can also contain particles of a slurry when flowing in annular channel.

Hence, in annular channel, the heterogeneous flows three-phase mixture: fluid—phase 1, gas—phase 2, and particles of slurry—phase 3. If it is an aerated mud solution, then in some problems one more phase should be taken into account since mud in water is usually dispersed not up to molecular level, and in some investigations, the system should be considered as a four-phase system.

One should not identify aggregate state with phase state. There are three aggregate states, solid, liquid, and gaseous, but phases may be unbounded in number. For example, many-colored immiscible fluids are in one and the same aggregate state (liquid) but represent separate phases distinguished by a determined property, namely, by color. In water–oil or spacer solution displacement, matters are in one and the same aggregate state, that is, liquid, but it is clear that oil and water and spacer differ essentially in properties; that is, they are different phases.

One should also not identify notions of phases and components. A system has as many components as there are chemical elements or their compounds. Mixture of gases is a single-phase but multicomponent system. In a mixture of chemically nonreacting gases, there are as many components as it has different gases. For example, when the drilling is performed at great depth with the help of aerated water, it may happen that all gas would be dissolved in fluid at molecular level and the resulting solution becomes homogeneous and uniform in properties but multicomponent, containing water and gas. As such solution moves to the well mouth, the pressure reduces, the gas liberates, and the solution transits into heterogeneous two-phase multicomponent state. Other situations are also possible when a homogeneous system is multicomponent and one-component system is multiphase.

A variety of media being used and encountered in well drilling require their properties to be studied. Properties of multiphase systems, in particular of two-phase media, may be different depending on to what degree each phase is dispersed. If one or several phases are dispersed and surrounded by another phase, then such heterogeneous system is sometimes called dispersed system, the crushed phase is called dispersed (discontinuous) phase, and the surrounding phase is called dispersion (continuous) phase. For example, air bubbles in aerated fluid represent dispersed phase and the fluid is continuous phase.

In addition, in some two-phase systems, it is impossible to determine which of the phases is a dispersed phase and which is a continuous phase since it is impossible to find which phase is surrounded by another one. For example, in porous media (rocks) with communicating pores, in gas–liquid and water–oil mixtures with near-equal volume concentration of both phases, they can have continuous distribution.

Classification of heterogeneous systems in dispersivity is presented in physical chemistry. If particles of dispersed phase have sizes 10^{-7} m, the system is called microheterogeneous. The word "micro" denotes dispersivity up to indicated size. If particles of the dispersed phase have sizes from 10^{-9} to 10^{-7} m, the system is called ultraheterogeneous or fine grained. In these systems, particles of dispersed phase are called colloidal particles. One should distinguish colloid systems from true solutions. Recall that true solutions are solutions in which substances are distributed at molecular level and form homogeneous systems, while colloid system is a variant of heterogeneous systems.

True systems can be one component or multicomponent. Heterogeneous systems are suspensions (rigid particles suspended in fluid), emulsions (droplets of one fluid suspended in another one), aerosols (droplets suspended in gas), and so on.

It is required to determine quantitative physical characteristics inherent to homogeneous and heterogeneous systems. On the basis of continuum mechanics (Sedov, 1983; Nigmatullin, 1987), all considered media are taken as macroscopic systems; that is, any volume of medium under consideration is taken as homogeneous or heterogeneous.

Arbitrary macroscopic system or a part of it possesses a mass; that is, it contains a definite amount of substance. In a system let us consider a volume V with mass m. If this system is homogeneous, its density is a continuous function of point location M and can be defined as

$$\rho(M) = \lim_{V \to 0} \frac{m}{V}. \tag{3.1}$$

Thereby, the density of the system is determined at each point.

Density functions of the type (3.1) will be as many as the number of phases since when an arbitrary volume V of such system tends to be zero one gets the density of one or another phase. In doing so in multiphase system with N phases, N densities are obtained. When investigating heterogeneous system motion, it is required to use a notion of density of a volume containing all or several phases. In this connection, introduce a notion of true phase content in the following way. Let V be volume of the system part. Then,

$$V = \sum_{i=1}^{N} V_i = V_1 + V_2 + \cdots + V_N, \tag{3.2}$$

where V_i is the volume of ith phase. If any kth phase does not enter in this volume, then $V_k = 0$.

Relation

$$\varphi_i = V_i / V \tag{3.3}$$

is called true volume content of ith phase or concentration of ith phase in volume V. The sum of all phase concentrations φ_i is equal to

$$\sum_{i=1}^{N} \varphi_i = \varphi_1 + \varphi_2 + \cdots + \varphi_N = \frac{V_1}{V} + \frac{V_2}{V} + \cdots + \frac{V_N}{V} = 1. \tag{3.4}$$

It is evident that

$$\varphi_i \leq 1. \tag{3.5}$$

The true density of the system in the volume V may be determined as follows:

$$\rho = \sum_{i=1}^{N} \varphi_i \rho_i, \tag{3.6}$$

where ρ_i is density of each phase.

Now, find the phase velocity at point M. Let given phase at the instant of time t be at point M and at $t + \Delta t$ shifts to the point M'. The way moved by the phase is $|\Delta \boldsymbol{l}| = MM'$, where $\Delta \boldsymbol{l}$ is vector of phase displacement in time Δt. Then, the velocity is equal to

$$\boldsymbol{w} = \lim_{\Delta t \to 0} \frac{\Delta \boldsymbol{l}}{\Delta t} = \frac{\partial \boldsymbol{l}}{\partial t}. \tag{3.7}$$

The magnitude of the velocity is independent of the frame of reference in which the velocity is considered, but velocity projections on coordinate axes in one coordinate system differ from velocity projections in another coordinate system.

In order to solve problems on motion of multiphase medium, one should know velocities of the system. If at point M, the velocity of phase is $\boldsymbol{w}_i$, the true velocity of the mixture can be represented as

$$\boldsymbol{w} = \sum_{i=1}^{N} \varphi_i \boldsymbol{w}_i. \tag{3.8}$$

The velocity is a vector quantity distinct from density, which is a scalar quantity.

The motion of media will be studied in cylindrical coordinate system because in well drilling the flow takes place chiefly in pipes, annular channels, and beds.

In cylindrical coordinate system, variables are r, φ, z (Fig. 3.1). In accordance with (3.7), velocity projections are

$$w_r = \frac{\partial r}{\partial t}, \quad w_\varphi = r\frac{\partial \varphi}{\partial t}, \quad w_z = \frac{\partial z}{\partial t}. \tag{3.9}$$

Then, phase velocity is

$$\boldsymbol{w} = w_r \boldsymbol{i} + w_\varphi \boldsymbol{j} + w_z \boldsymbol{k}, \tag{3.10}$$

where $\boldsymbol{i}, \boldsymbol{j}, \boldsymbol{k}$ are unit vectors.

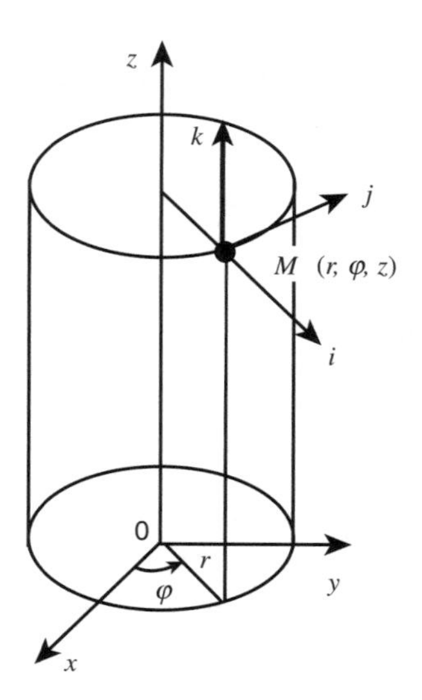

FIGURE 3.1 Cylindrical coordinate system.

Thus, at each point the velocity $\boldsymbol{w}$ is defined as vector quantity, the projections of which depend on point location

$$\boldsymbol{w} = \boldsymbol{w}(M) = \boldsymbol{w}(r, \varphi, z, t);$$
$$w_r = w_r(r, \varphi, z, t), \quad w_\varphi = w_\varphi(r, \varphi, z, t), \quad w_z = w_z(r, \varphi, z, t). \quad (3.11)$$

Take in the medium a surface element S with normal $\boldsymbol{n}$. The flow rate of the medium through this element is

$$Q = \int_S w_n \, \mathrm{d}S, \quad (3.12)$$

where w_n is the velocity projection on the normal $\boldsymbol{n}$.

If the integrand in the expression (3.12) is a projection of an arbitrary vector (it need not be a velocity) on the normal to the surface element, expression (3.12) is called vector flux through the surface S. The flow rate of a phase through normal cross section $Sn(Q = \upsilon S_n)$, where υ is mean velocity of the medium and S_n is the area of the channel cross section, represents a particular case of vector flux.

If the surface is closed, that is, it restricts a volume ΔV, then the relation

$$\nabla \cdot \boldsymbol{w} = \lim_{\Delta V \to 0} \frac{\int_S w_n \, \mathrm{d}S}{\Delta V} \quad (3.13)$$

is called vector divergence, which is vector flux through the surface of infinitely small volume surrounding the considered point.

From (3.13), it ensues that if the flow rate through any closed surface S vanishes, then $\nabla \cdot \boldsymbol{w} = 0$.

It is possible to show that

$$\frac{\mathrm{d}}{\mathrm{d}t}(\Delta V) = \Delta V \cdot \nabla \cdot \boldsymbol{w}. \quad (3.14)$$

This means that the divergence characterizes the relative increase or decrease in medium volume, that is, medium compressibility.

From mathematics, it is known that expression (3.13) in cylindrical coordinates has the form

$$\nabla \cdot \boldsymbol{w} = \frac{1}{r}\frac{\partial r w_r}{\partial r} + \frac{1}{r}\frac{\partial w_\varphi}{\partial \varphi} + \frac{\partial w_z}{\partial z}. \quad (3.15)$$

The motion and internal stress in media are caused by forces that can be classified into internal and external forces. External forces relative to the system are those that are induced by other systems, whereas internal forces are conditioned by another parts of the same system.

If at arbitrarily chosen point M of the medium takes an elementary surface ΔS with normal $\boldsymbol{n}$, on this surface an external force $\Delta \boldsymbol{F}$ will act produced by a part of the medium located, as viewed from the surface, in the normal direction. The surface exhibits a stress equal to the ratio between force and surface area

$$\boldsymbol{p}_n = \frac{\Delta \boldsymbol{F}}{\Delta S}. \tag{3.16}$$

Upon contraction of the area into a point, one obtains stress at point M

$$\boldsymbol{p}_n = \lim_{\Delta S \to 0} \frac{\Delta \boldsymbol{F}}{\Delta S}. \tag{3.17}$$

Designate ideal medium (ideal fluid or ideal gas) as such a medium in which the stress vector $\boldsymbol{p}_n$ at any surface element with normal $\boldsymbol{n}$ is orthogonal to the surface, that is, vector $\boldsymbol{p}_n$ is parallel to $\boldsymbol{n}$.

Figure 3.2 demonstrates decomposition of the vector $\boldsymbol{p}_n$ on normal $\boldsymbol{p}_{nn}$ and tangential $\boldsymbol{p}_{n\tau}$ components. In ideal fluid, there is by definition

$$\boldsymbol{p}_{n\tau} = 0, \tag{3.18}$$

that is, tangential stresses in ideal fluid are absent.

The majority of flows in the drilling practice would be considered one dimensional in the sense that in appropriate coordinate system, Cartesian or cylindrical, only one velocity component plays a significant role. Such suggestion is true in many cases and gives needed accuracy in calculations. For example, flows in pipes and annular channels have only one velocity component w_z directed along the pipe axis z, being dependent on pipe radius r, that is, $w_z = w_z(r)$. The flow (inflow or outflow) of fluid in circular bed may be taken as one dimensional; that is, in cylindrical coordinate

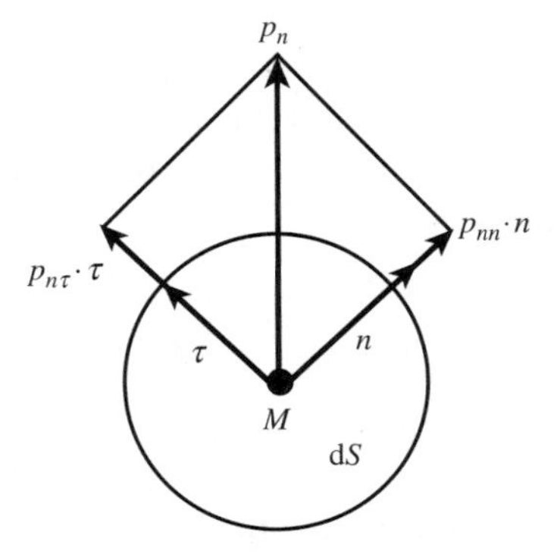

FIGURE 3.2 Decomposition of the stress vector $\boldsymbol{p}_{n\tau} = \boldsymbol{p}_{n\tau}(M, \boldsymbol{n})$ on components.

system, the flow has one velocity component w_r directed along the radius r of the bed and being the function of only z in the limit of formation thickness H and radius r, that is, $w_r = w_r(z, r)$. One dimensionality is of course a matter of convention since radial flow considered in Cartesian coordinates has two components, w_x and w_y.

Depending on the properties of fluids taking part in flows, one can consider them as incompressible or compressible. *Incompressibility* of fluid is defined as invariability of arbitrarily chosen fluid volume in the sense that the volume shape can be deformed but the volume by itself remains constant. One should distinguish between incompressibility and homogeneity notions. If incompressible fluid is *homogeneous*, then everywhere in the flow the density is constant ($\rho = \text{const}$). If the *heterogeneous* fluid is incompressible, then in passage through interface the density changes ($\rho \neq \text{const}$). And yet the heterogeneous gas is compressible and in rare cases it can be taken as incompressible.

The flows, in what follows, will be mainly considered in circular pipes, in annular and concentric channels, and between parallel circular plates. It should be noted that all flows taking place in circulation system of well or in the whole system of well-bed are bounded.

CHAPTER 4

HYDRO-AEROMECHANIC EQUATIONS OF DRILLING PROCESSES

4.1 MASS CONSERVATION EQUATION

Mass conservation law states: the net mass ΔM of a mixture part occupying at time t the space volume ΔV remains constant and at following instants of time if mass change due to internal and external sources is absent (Loitsyansky, 1987)

$$\frac{\mathrm{d}}{\mathrm{d}t}(\Delta M) = 0. \tag{4.1.1}$$

By definition of density $\Delta M \approx \rho \Delta V$, where $\rho = \Sigma \varphi_i \rho_i$, there is

$$\Delta M \cong \sum \varphi_i \rho_i \Delta V. \tag{4.1.2}$$

Differentiation of the left part (4.1.1) gives

$$\frac{\mathrm{d}(\rho \Delta V)}{\mathrm{d}t} = \Delta V \frac{\mathrm{d}\rho}{\mathrm{d}t} + \rho \frac{\mathrm{d}(\Delta V)}{\mathrm{d}t} = 0. \tag{4.1.3}$$

Applied Hydro-Aeromechanics in Oil and Gas Drilling. By Leonov and Isaev

Substituting (3.14) in (4.1.3) and carrying all terms in the left part, one obtains

$$\left(\frac{\mathrm{d}\rho}{\mathrm{d}t}+\rho\nabla\cdot\boldsymbol{w}\right)\Delta V=0.$$

Since $\Delta V \neq 0$, it is

$$\frac{\mathrm{d}\rho}{\mathrm{d}t}+\rho\nabla\cdot\boldsymbol{w}=0$$

or

$$\frac{\mathrm{d}\left(\sum\varphi_i\rho_i\right)}{\mathrm{d}t}+\left(\sum\varphi_i\rho_i\right)\nabla\cdot\left(\sum\varphi_i\boldsymbol{w}_i\right)=0. \tag{4.1.4}$$

In a similar manner, it is possible to derive equations for each phase. In doing so, one gets as many equations of the type (4.1.4) as there are phases. For example, for two-phase mixture, there is

$$\begin{aligned}&\frac{\mathrm{d}\varphi_1\rho_1}{\mathrm{d}t}+\varphi_1\rho_1\nabla\cdot(\varphi_1\boldsymbol{w}_1)=0,\\&\frac{\mathrm{d}\varphi_2\rho_2}{\mathrm{d}t}+\varphi_2\rho_2\nabla\cdot(\varphi_2\boldsymbol{w}_2)=0.\end{aligned} \tag{4.1.5}$$

In accordance with (3.15) for $\nabla\cdot\boldsymbol{w}$ and definition of total derivative, the equation (4.1.4) may be rewritten as

$$\frac{\partial\rho}{\partial t}+\frac{\partial\rho}{\partial r}\cdot\frac{\partial r}{\partial t}+\frac{\partial\rho}{\partial\varphi}\frac{\partial\varphi}{\partial t}+\frac{\partial\rho}{\partial z}\cdot\frac{\partial z}{\partial t}$$

$$+\rho\left(\frac{1}{r}\cdot\frac{\partial rw_r}{\partial r}+\frac{1}{r}\cdot\frac{\partial w_\varphi}{\partial\varphi}+\frac{\partial w_z}{\partial z}\right)=0.$$

As far as $\frac{\partial r}{\partial t}=w_r$, $\frac{\partial\varphi}{\partial t}=\omega=\frac{w_\varphi}{r}$, $\frac{\partial z}{\partial t}=w_z$, where ω is angular velocity, it yields

$$\frac{\partial\rho}{\partial t}+\nabla\cdot(\rho\boldsymbol{w})=0. \tag{4.1.6}$$

From (4.1.6), it follows

$$\frac{\partial\rho}{\partial t}+\frac{1}{r}\cdot\frac{\partial(\rho rw_r)}{\partial r}+\frac{1}{r}\cdot\frac{\partial(\rho w_\varphi)}{\partial\varphi}+\frac{\partial(\rho w_z)}{\partial z}=0. \tag{4.1.7}$$

In the case of stationary flow, that is, when $\partial\rho/\partial t = 0$, (4.1.7) gives

$$\frac{1}{r} \cdot \frac{\partial(\rho r w_r)}{\partial r} + \frac{1}{r} \cdot \frac{\partial(\rho w_\varphi)}{\partial \varphi} + \frac{\partial(\rho w_z)}{\partial z} = 0. \tag{4.1.8}$$

Later on, it will be obtained for one-dimensional symmetric flows: in tubes where only $w_z \neq 0$, ρw_z is a function only of the radial coordinate r; in circular slots where only $w_r \neq 0$, ρw_r is a function only of coordinates r and z; in flows induced by rotation of pipes, where only $w_\varphi \neq 0$, ρw_φ is a function only of r.

For incompressible homogeneous fluid, the density ρ in equation (4.1.8) can be removed from the derivative and thus be canceled.

4.2 MOMENTUM (MOTION) EQUATION

By definition of mixture density using phase volume contents φ_1, φ_2, and so on, the expression

$$\boldsymbol{k} = \varphi_1\rho_1\boldsymbol{w}_1 + \varphi_2\rho_2\boldsymbol{w}_2 + \cdots = \sum \varphi_i\rho_i\boldsymbol{w}_i \tag{4.2.1}$$

may be considered as vector of mixture momentum (Teletov, 1958). Then, the momentum of a mixture filling the volume ΔV will be

$$\sum(\varphi_i\rho_i\boldsymbol{w}_i)\Delta V. \tag{4.2.2}$$

The vector of mass force (gravity force) distributed over the volume ΔV has the form

$$\boldsymbol{F}_{\mathrm{M}} = \boldsymbol{g}\left(\sum \varphi_i\rho_i\right)\Delta V, \tag{4.2.3}$$

where $\mathbf{g}$ is the gravity acceleration.

The surface $\Delta\sigma$ of volume ΔV under the action of external forces is exposed to surface tension. Denote through $\boldsymbol{\Pi}_{\mathrm{s}}$ total vector of surface forces. It will be determined further as applied to drilling problems. Supposing that the theorem of momentum change could be applied to elementary volume of the mixture as a whole, one obtains

$$\frac{\mathrm{d}}{\mathrm{d}t}\left[\left(\sum \varphi_i\rho_i\boldsymbol{w}_i\right)\Delta V\right] = \boldsymbol{g}(\Sigma\varphi_i\rho_i)\Delta V + \boldsymbol{\Pi}_{\mathrm{s}} \tag{4.2.4}$$

and differentiation of the left side yields

$$\sum\left(\varphi_i\rho_i\Delta V\frac{d\boldsymbol{w}_i}{dt}\right)+\sum\left[\boldsymbol{w}_i\frac{d(\varphi_i\rho_i\Delta V)}{dt}\right]=\boldsymbol{g}\left(\sum\varphi_i\rho_i\right)\Delta V+\boldsymbol{\Pi}_s.$$

In the case of the absence of additional mass sources, the second term owing to (4.1.1) vanishes and

$$\sum\left(\varphi_i\rho_i\frac{dw_i}{dt}\right)\Delta V=\boldsymbol{g}(\Sigma\varphi_i\rho_i)\Delta V+\boldsymbol{\Pi}_s. \qquad (4.2.5)$$

For simplicity sake of the following mathematical treatment, let us introduce designations

$$\rho\frac{d\boldsymbol{w}}{dt}=\Sigma\varphi_i\rho_i\frac{d\boldsymbol{w}_i}{dt},\qquad \rho=\Sigma\varphi_i\rho_i. \qquad (4.2.6)$$

Earlier, it has been mentioned that it makes sense to consider main drilling problems related to flows of washing solutions and grouting mortars in pipes, annular channels, circular slots, and beds with the use of cylindrical coordinates. In doing so, let us obtain concrete form of equation (4.2.5).

Take a point with coordinates r, φ, z in moving medium and mark out in the vicinity of this point a moving elementary particle of the mixture with volume ΔV and faces shown in Fig. 4.1. In accordance with notations (4.2.6), the derivation will be done for one-phase continuum. Stress

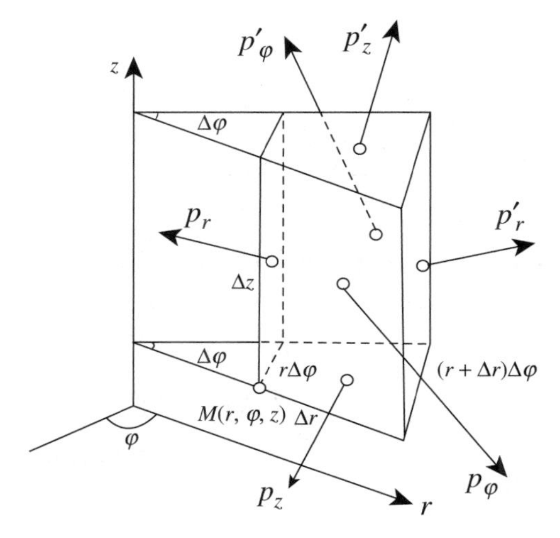

FIGURE 4.1 Components of stress vectors acting on faces of elementary volume.

vectors of surface forces acting on the faces passing through the point M are $\boldsymbol{p}_r, \boldsymbol{p}_\varphi, \boldsymbol{p}_z$, and on the opposite faces $\boldsymbol{p}'_r, \boldsymbol{p}'_\varphi, \boldsymbol{p}'_z$. Then, $\boldsymbol{p}'_r, \boldsymbol{p}'_\varphi, \boldsymbol{p}'_z$ accurate to the second infinitesimal order could be expressed through $\boldsymbol{p}_r, \boldsymbol{p}_\varphi, \boldsymbol{p}_z$. In order to do this, we draw on the first two terms of Taylor expansion

$$\boldsymbol{p}'_r = \boldsymbol{p}_r(r+\Delta r, \varphi, z) = \boldsymbol{p}_r(r, \varphi, z) + \frac{\partial \boldsymbol{p}_r}{\partial r}\Delta r;$$

$$\boldsymbol{p}'_\varphi = \boldsymbol{p}_\varphi(r, \varphi+\Delta\varphi, z) = \boldsymbol{p}_\varphi(r, \varphi, z) + \frac{\partial \boldsymbol{p}_\varphi}{\partial \varphi}\Delta \varphi; \tag{4.2.7}$$

$$\boldsymbol{p}'_z = \boldsymbol{p}_z(r, \varphi, z+\Delta z) = \boldsymbol{p}_z(r, \varphi, z) + \frac{\partial \boldsymbol{p}_z}{\partial z}\Delta z.$$

Due to (4.2.5), we have

$$\rho\frac{\mathrm{d}\boldsymbol{w}}{\mathrm{d}t}\Delta V = \rho\boldsymbol{g}\Delta V + \boldsymbol{\Pi}_\mathrm{s}, \tag{4.2.8}$$

where (see Fig. 4.1) $\Delta V = r\Delta r\Delta\varphi\Delta z$ and $\boldsymbol{\Pi}_\mathrm{s}$ is the sum of all surface forces equal to the product of stress and area of corresponding face. Then,

$$\rho\frac{\mathrm{d}\boldsymbol{w}}{\mathrm{d}t}r\Delta r\Delta\varphi\Delta z = \rho\boldsymbol{g}r\Delta r\Delta\varphi\Delta z - \boldsymbol{p}_r r\Delta\varphi\Delta z + \boldsymbol{p}'_r(r+\Delta r)\Delta\varphi\Delta z$$
$$-\boldsymbol{p}_\varphi\Delta r\Delta z + \boldsymbol{p}'_\varphi\Delta r\Delta z - \boldsymbol{p}_z r\Delta\varphi\Delta r + \boldsymbol{p}'_z r\Delta\varphi\Delta r.$$

Substitute (4.2.7) in this relation and ignore terms of the highest infinitesimal order

$$\rho\frac{\mathrm{d}\boldsymbol{w}}{\mathrm{d}t}r\Delta r\Delta\varphi\Delta z = \rho\boldsymbol{g}r\Delta r\Delta\varphi\Delta z + \boldsymbol{p}_r\Delta r\Delta\varphi\Delta z + \frac{\partial \boldsymbol{p}_r}{\partial r}r\Delta r\Delta\varphi\Delta z$$
$$+\frac{\partial \boldsymbol{p}_\varphi}{\partial \varphi}\Delta r\Delta\varphi\Delta z + \frac{\partial \boldsymbol{p}_z}{\partial z}r\Delta\varphi\Delta r\Delta z.$$

By dividing both sides of the last equality by $r\Delta r\Delta\varphi\Delta z$, one obtains momentum equation in vector form

$$\rho\frac{\mathrm{d}\boldsymbol{w}}{\mathrm{d}t} = \rho\boldsymbol{g} + \frac{\boldsymbol{p}_r}{r} + \frac{\partial \boldsymbol{p}_r}{\partial r} + \frac{1}{r}\cdot\frac{\partial \boldsymbol{p}_\varphi}{\partial \varphi} + \frac{\partial \boldsymbol{p}_z}{\partial z}. \tag{4.2.9}$$

Decompose vectors in this equation on components (projections) at point M

$$\boldsymbol{w} = w_r\boldsymbol{i} + w_\varphi\boldsymbol{j} + w_z\boldsymbol{k}; \tag{4.2.10}$$

$$\boldsymbol{g} = g_r\boldsymbol{i} + g_\varphi\boldsymbol{j} + g_z\boldsymbol{k}; \tag{4.2.11}$$

and

$$\begin{aligned}\boldsymbol{p}_r &= p_{rr}\boldsymbol{i} + p_{r\varphi}\boldsymbol{j} + p_{rz}\boldsymbol{k};\\ \boldsymbol{p}_\varphi &= p_{\varphi r}\boldsymbol{i} + p_{\varphi\varphi}\boldsymbol{j} + p_{\varphi z}\boldsymbol{k};\\ \boldsymbol{p}_z &= p_{zr}\boldsymbol{i} + p_{z\varphi}\boldsymbol{j} + p_{zz}\boldsymbol{k}.\end{aligned} \tag{4.2.12}$$

Find derivatives of vectors since not only vectors but also their derivatives enter in (4.2.9). In accord with differentiation rules, derivatives of vectors $\boldsymbol{p}_r, \boldsymbol{p}_\varphi, \boldsymbol{p}_z$ given by formulas (4.2.12) are

$$\begin{aligned}\frac{\partial \boldsymbol{p}_r}{\partial r} &= \frac{\partial p_{rr}}{\partial r}\boldsymbol{i} + \frac{\partial p_{r\varphi}}{\partial r}\boldsymbol{j} + \frac{\partial p_{rz}}{\partial r}\boldsymbol{k} + p_{rr}\frac{\partial \boldsymbol{i}}{\partial r} + p_{r\varphi}\frac{\partial \boldsymbol{j}}{\partial r} + p_{rz}\frac{\partial \boldsymbol{k}}{\partial r};\\ \frac{\partial \boldsymbol{p}_\varphi}{\partial \varphi} &= \frac{\partial p_{\varphi r}}{\partial \varphi}\boldsymbol{i} + \frac{\partial p_{\varphi\varphi}}{\partial \varphi}\boldsymbol{j} + \frac{\partial p_{\varphi z}}{\partial \varphi}\boldsymbol{k} + p_{\varphi r}\frac{\partial \boldsymbol{i}}{\partial \varphi} + p_{\varphi\varphi}\frac{\partial \boldsymbol{j}}{\partial \varphi} + p_{\varphi z}\frac{\partial \boldsymbol{k}}{\partial \varphi};\\ \frac{\partial \boldsymbol{p}_z}{\partial z} &= \frac{\partial p_{zr}}{\partial z}\boldsymbol{i} + \frac{\partial p_{z\varphi}}{\partial z}\boldsymbol{j} + \frac{\partial p_{zz}}{\partial z}\boldsymbol{k} + p_{zr}\frac{\partial \boldsymbol{i}}{\partial z} + p_{z\varphi}\frac{\partial \boldsymbol{j}}{\partial z} + p_{zz}\frac{\partial \boldsymbol{k}}{\partial z}.\end{aligned} \tag{4.2.13}$$

In these relations, enter partial derivatives of unit vectors. The vector $\boldsymbol{k}$ does not change its direction in going from one point to another. It is always parallel to z-axis and its direction is independent of coordinates r, φ, z, that is, derivatives $\partial\boldsymbol{k}/\partial r$, $\partial\boldsymbol{k}/\partial\varphi$, and $\partial\boldsymbol{k}/\partial z$ vanish. Also, $\partial\boldsymbol{i}/\partial z$ and $\partial\boldsymbol{j}/\partial z$ vanish since at fixed values of coordinates r, φ, and z, change of unit vectors $\boldsymbol{i}$ and $\boldsymbol{j}$ remains parallel, as well as $\partial\boldsymbol{i}/\partial r$ and $\partial\boldsymbol{j}/\partial r$. Nonzero would be $\partial\boldsymbol{i}/\partial\varphi$ and $\partial\boldsymbol{j}/\partial\varphi$. Find them by definition of the derivative

$$\frac{\partial \boldsymbol{i}}{\partial \varphi} = \lim_{\Delta\varphi\to 0}\frac{\Delta\boldsymbol{i}}{\Delta\varphi}.$$

As it is seen from Fig. 4.2, $\Delta\boldsymbol{i} \approx \boldsymbol{j}\Delta\varphi$ or $\Delta\boldsymbol{i}/\Delta\varphi \approx \boldsymbol{j}$ and

$$\lim_{\Delta\varphi\to 0}\frac{\Delta\boldsymbol{i}}{\Delta\varphi} = \frac{\partial \boldsymbol{i}}{\partial \varphi} = \boldsymbol{j}.$$

In the same way, it can be shown that $\partial\boldsymbol{j}/\partial\varphi = -\boldsymbol{i}$. Hence,

$$\frac{\partial \boldsymbol{i}}{\partial \varphi} = \boldsymbol{j}; \qquad \frac{\partial \boldsymbol{j}}{\partial \varphi} = -\boldsymbol{i}. \tag{4.2.14}$$

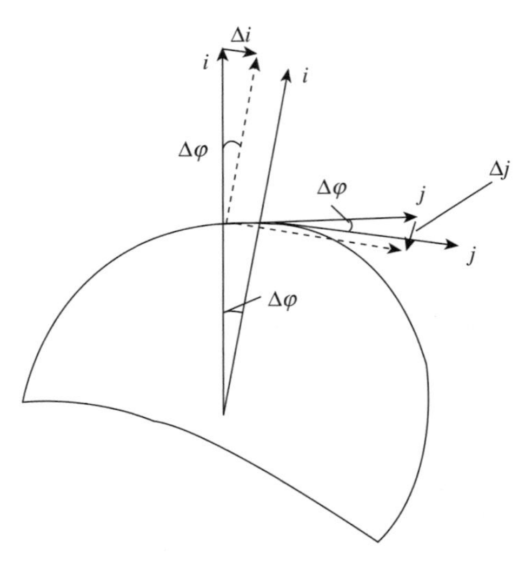

FIGURE 4.2 Derivation of formulas $\partial \boldsymbol{i}/\partial\varphi = \boldsymbol{j}$ and $\partial \boldsymbol{j}/\partial\varphi = -\boldsymbol{i}$.

If the medium is ideal, then in accordance with (3.18) components $\boldsymbol{p}_{n\tau}$ are satisfied equalities

$$p_{\varphi r} = p_{r\varphi} = p_{zr} = p_{rz} = p_{\varphi z} = p_{z\varphi} = 0.$$

Then, $p_{rr}, p_{\varphi\varphi}, p_{zz}$ are only normal stresses and, for example, p_{rr} could be considered as pressure $-p$. It can be shown that other stresses are equal to $-p$; that is, $p_{\varphi\varphi} = -p, p_{zz} = -p$.

With regard to this, in (4.2.13) the common component $-p$ (in what follows it will be considered as only such media) can be separated, so the stress components take the form

$$\begin{aligned} &p_{rr} = -p + \tau_{rr}; \quad p_{\varphi\varphi} = -p + \tau_{\varphi\varphi}; \quad p_{zz} = -p + \tau_{zz}; \\ &p_{r\varphi} = \tau_{r\varphi}; \quad p_{rz} = \tau_{rz}; \quad p_{\varphi z} = \tau_{\varphi z}. \end{aligned} \tag{4.2.15}$$

It is able to prove that $p_{\varphi r} = p_{r\varphi}, p_{zr} = p_{rz}$, and $p_{\varphi z} = p_{z\varphi}$. Then, (4.2.13) $p_{zr} = p_{rz}$ can be rewritten as

$$\begin{aligned} &\frac{\partial \boldsymbol{p}_r}{\partial r} = \frac{\partial(-p + \tau_{rr})}{\partial r}\boldsymbol{i} + \frac{\partial \tau_{r\varphi}}{\partial r}\boldsymbol{j} + \frac{\partial \tau_{rz}}{\partial r}\boldsymbol{k}; \\ &\frac{\partial \boldsymbol{p}_\varphi}{\partial \varphi} = \frac{\partial \tau_{r\varphi}}{\partial \varphi}\boldsymbol{i} + \frac{\partial(-p + \tau_{\varphi\varphi})}{\partial \varphi}\boldsymbol{j} + \frac{\partial \tau_{\varphi z}}{\partial \varphi}\boldsymbol{k} + \tau_{r\varphi}\boldsymbol{j} + (-p + \tau_{\varphi\varphi})(-\boldsymbol{i}); \\ &\frac{\partial \boldsymbol{p}_z}{\partial z} = \frac{\partial \tau_{rz}}{\partial z}\boldsymbol{i} + \frac{\partial \tau_{\varphi z}}{\partial z}\boldsymbol{j} + \frac{\partial(-p + \tau_{zz})}{\partial z}\boldsymbol{k}. \end{aligned} \tag{4.2.16}$$

Determine the total derivative of the velocity with respect to time and its projections. By definition of the total derivative

$$\frac{d\boldsymbol{w}}{dt} = \frac{\partial \boldsymbol{w}}{\partial t} + \frac{\partial \boldsymbol{w}}{\partial r} \cdot \frac{\partial r}{\partial t} + \frac{\partial \boldsymbol{w}}{\partial \varphi} \cdot \frac{\partial \varphi}{\partial t} + \frac{\partial \boldsymbol{w}}{\partial z} \cdot \frac{\partial z}{\partial t}.$$

Since $\partial r/\partial t = w_r$, $\partial \varphi/\partial t = w_\varphi/r$, and $\partial z/\partial t = w_z$ are projections of the velocity **w**, it is

$$\frac{d\boldsymbol{w}}{dt} = \frac{\partial \boldsymbol{w}}{\partial t} + w_r \frac{\partial \boldsymbol{w}}{\partial r} + \frac{w_\varphi}{r} \cdot \frac{\partial \boldsymbol{w}}{\partial \varphi} + w_z \frac{\partial \boldsymbol{w}}{\partial z}. \tag{4.2.17}$$

Derivatives $\partial \boldsymbol{w}/\partial r, \partial \boldsymbol{w}/\partial \varphi,$ and $\partial \boldsymbol{w}/\partial z$ are determined in the same way as $\partial \boldsymbol{p}_r/\partial r$, $\partial \boldsymbol{p}_\varphi/\partial \varphi$, and $\partial \boldsymbol{p}_z/\partial z$ were determined. Consequently, in formulas (4.2.13) $\boldsymbol{p}_r, \boldsymbol{p}_\varphi$, and $\boldsymbol{p}_z$ can be replaced with **w**, as a result of which the following equation is obtained:

$$\begin{aligned}
\frac{\partial \boldsymbol{w}}{\partial r} &= \frac{\partial w_r}{\partial r}\boldsymbol{i} + \frac{\partial w_\varphi}{\partial r}\boldsymbol{j} + \frac{\partial w_z}{\partial r}\boldsymbol{k};\\
\frac{\partial \boldsymbol{w}}{\partial \varphi} &= \frac{\partial w_r}{\partial \varphi}\boldsymbol{i} + \frac{\partial w_\varphi}{\partial \varphi}\boldsymbol{j} + \frac{\partial w_z}{\partial \varphi}\boldsymbol{k} + w_r\boldsymbol{j} + w_\varphi(-\boldsymbol{i});\\
\frac{\partial \boldsymbol{w}}{\partial z} &= \frac{\partial w_r}{\partial z}\boldsymbol{i} + \frac{\partial w_\varphi}{\partial z}\boldsymbol{j} + \frac{\partial w_z}{\partial z}\boldsymbol{k}.
\end{aligned} \tag{4.2.18}$$

Substituting relations (4.2.18) in (4.2.17), then (4.2.17) and (4.2.16) in vector equation of motion (4.2.9), and taking into account that the partial derivative of the velocity with respect to time is equal to

$$\frac{\partial \boldsymbol{w}}{\partial t} = \frac{\partial w_r}{\partial t}\boldsymbol{i} + \frac{\partial w_\varphi}{\partial t}\boldsymbol{j} + \frac{\partial w_z}{\partial t}\boldsymbol{k},$$

it yields

$$\begin{aligned}
&\rho\left[\left(\frac{\partial w_r}{\partial t} + w_r\frac{\partial w_r}{\partial r} + \frac{w_\varphi}{r}\left(\frac{\partial w_r}{\partial \varphi} - w_\varphi\right) + w_z\frac{\partial w_r}{\partial z}\right)\boldsymbol{i}\right.\\
&+\left(\frac{\partial w_\varphi}{\partial t} + w_r\frac{\partial w_\varphi}{\partial r} + \frac{w_\varphi}{r}\cdot\frac{\partial w_\varphi}{\partial \varphi} + \frac{w_r w_\varphi}{r} + w_z\frac{\partial w_\varphi}{\partial z}\right)\boldsymbol{j}\\
&\left.+\left(\frac{\partial w_z}{\partial t} + w_r\frac{\partial w_z}{\partial r} + \frac{w_\varphi}{r}\cdot\frac{\partial w_z}{\partial \varphi} + w_z\frac{\partial w_z}{\partial z}\right)k\right]\\
&=\left[\rho g_r - \frac{\partial p}{\partial r} + \frac{\partial \tau_{rr}}{\partial r} + \frac{1}{r}\cdot\frac{\partial \tau_{r\varphi}}{\partial \varphi} + \frac{\partial \tau_{rz}}{\partial z} + \frac{\tau_{rr} - \tau_{\varphi\varphi}}{r}\right]\boldsymbol{i}
\end{aligned}$$

$$+\left[\rho g_{\varphi}-\frac{1}{r}\cdot\frac{\partial p}{\partial\varphi}+\frac{\partial\tau_{r\varphi}}{\partial r}+\frac{1}{r}\cdot\frac{\partial\tau_{\varphi\varphi}}{\partial\varphi}+\frac{\partial\tau_{\varphi z}}{\partial z}+2\frac{\tau_{r\varphi}}{r}\right]\boldsymbol{j}$$

$$+\left[\rho g_{z}-\frac{\partial p}{\partial z}+\frac{\partial\tau_{rz}}{\partial r}+\frac{1}{r}\cdot\frac{\partial\tau_{\varphi z}}{\partial\varphi}+\frac{\partial\tau_{zz}}{\partial z}+\frac{\tau_{rz}}{r}\right]\boldsymbol{k}. \quad (4.2.19)$$

Equating terms standing by unit vectors $\boldsymbol{i}$, $\boldsymbol{j}$, $\boldsymbol{k}$ in both sides of the last equation, one obtains, instead of one vector equation of motion, three scalar equations of motion in stress projections

$$\begin{aligned}
&\rho\left[\frac{\partial w_r}{\partial t}+w_r\frac{\partial w_r}{\partial r}+\frac{w_\varphi}{r}\cdot\frac{\partial w_r}{\partial\varphi}+w_z\frac{\partial w_r}{\partial z}-\frac{w_\varphi^2}{r}\right]\\
&\qquad=\rho g_r-\frac{\partial p}{\partial r}+\frac{\partial\tau_{rr}}{\partial r}+\frac{1}{r}\cdot\frac{\partial\tau_{r\varphi}}{\partial\varphi}+\frac{\partial\tau_{rz}}{\partial z}+\frac{\tau_{rr}-\tau_{\varphi\varphi}}{r};\\
&\rho\left[\frac{\partial w_\varphi}{\partial t}+w_r\frac{\partial w_\varphi}{\partial r}+\frac{w_\varphi}{r}\cdot\frac{\partial w_\varphi}{\partial\varphi}+w_z\frac{\partial w_\varphi}{\partial z}+\frac{w_r w_\varphi}{r}\right]\\
&\qquad=\rho g_\varphi-\frac{1}{r}\cdot\frac{\partial p}{\partial\varphi}+\frac{\partial\tau_{r\varphi}}{\partial r}+\frac{1}{r}\cdot\frac{\partial\tau_{\varphi\varphi}}{\partial\varphi}+\frac{\partial\tau_{\varphi z}}{\partial z}+2\frac{\tau_{r\varphi}}{r};\\
&\rho\left[\frac{\partial w_z}{\partial t}+w_r\frac{\partial w_z}{\partial r}+\frac{w_\varphi}{r}\cdot\frac{\partial w_z}{\partial\varphi}+w_z\frac{\partial w_z}{\partial z}\right]\\
&\qquad=\rho g_z-\frac{\partial p}{\partial z}+\frac{\partial\tau_{rz}}{\partial r}+\frac{1}{r}\cdot\frac{\partial\tau_{\varphi z}}{\partial\varphi}+\frac{\partial\tau_{zz}}{\partial z}+\frac{\tau_{rz}}{r}.
\end{aligned} \quad (4.2.20)$$

Account of expression (4.2.6) gives momentum (motion) equation written in stresses for multiphase medium

$$\begin{aligned}
&\sum\left[\varphi_i\rho_i\left(\frac{\partial w_{ri}}{\partial t}+w_{ri}\frac{\partial w_{ri}}{\partial r}+\frac{w_{\varphi i}}{r}\cdot\frac{\partial w_{ri}}{\partial\varphi}+w_{zi}\frac{\partial w_{ri}}{\partial z}-\frac{w_{\varphi i}^2}{r}\right)\right]\\
&\quad=\left(\sum\varphi_i\rho_i\right)g_r-\frac{\partial p}{\partial r}+\frac{\partial\tau_{rr}}{\partial r}+\frac{1}{r}\cdot\frac{\partial\tau_{r\varphi}}{\partial\varphi}+\frac{\partial\tau_{rz}}{\partial z}+\frac{\tau_{rr}-\tau_{\varphi\varphi}}{r};\\
&\sum\left[\varphi_i\rho_i\left(\frac{\partial w_{\varphi i}}{\partial t}+w_{ri}\frac{\partial w_{\varphi i}}{\partial r}+\frac{w_{\varphi i}}{r}\cdot\frac{\partial w_{\varphi i}}{\partial\varphi}+w_{zi}\frac{\partial w_{\varphi i}}{\partial z}+\frac{w_{ri}w_{\varphi i}}{r}\right)\right]\\
&\quad=\left(\sum\varphi_i\rho_i\right)g_\varphi-\frac{1}{r}\cdot\frac{\partial p}{\partial\varphi}+\frac{\partial\tau_{r\varphi}}{\partial r}+\frac{1}{r}\cdot\frac{\partial\tau_{\varphi\varphi}}{\partial\varphi}+\frac{\partial\tau_{\varphi z}}{\partial z}+2\frac{\tau_{r\varphi}}{r};\\
&\sum\left[\varphi_i\rho_i\left(\frac{\partial w_{zi}}{\partial t}+w_{ri}\frac{\partial w_{zi}}{\partial r}+\frac{w_{\varphi i}}{r}\cdot\frac{\partial w_{zi}}{\partial\varphi}+w_{zi}\frac{\partial w_{zi}}{\partial z}\right)\right]\\
&\quad=\left(\sum\varphi_i\rho_i\right)g_z-\frac{\partial p}{\partial z}+\frac{\partial\tau_{rz}}{\partial r}+\frac{1}{r}\cdot\frac{\partial\tau_{\varphi z}}{\partial\varphi}+\frac{\partial\tau_{zz}}{\partial z}+\frac{\tau_{rz}}{r}.
\end{aligned} \quad (4.2.21)$$

Since one-phase axial flows in pipes and annular space would be considered further, let us make a separate derivation of motion equation in projection on z-axis.

Apply to an elementary fluid particle of mass Δm the second Newtonian law

$$\Delta m \cdot \boldsymbol{a} = \sum \boldsymbol{F}_i, \tag{4.2.22}$$

where **a** is acceleration of the particle and $\sum \boldsymbol{F}_\mathrm{i}$ is the sum of surface and mass forces distributed over the surface and the volume of the particle.

Dividing (4.2.22) by particle volume ΔV

$$\frac{\Delta m}{\Delta V} \cdot \boldsymbol{a} = \frac{\sum \boldsymbol{F}_i}{\Delta V}, \tag{4.2.23}$$

tending $\Delta V \rightarrow 0$ and using density definition (3.1), one gets

$$\rho \frac{\mathrm{d}\boldsymbol{w}}{\mathrm{d}t} = \sum \Phi_i, \tag{4.2.24}$$

where ρ is the fluid density, **w** is the velocity, and $\sum \Phi_i = \lim_{\Delta V \rightarrow 0} \left(\sum \boldsymbol{F}_i / \Delta V \right)$ is the sum of force densities. Figure 4.3 shows an annular element of the pipe. Z-axis coincides with the direction of the gravity acceleration. Now, obtain the sum of algebraic projections of mass and surface forces on z-axis.

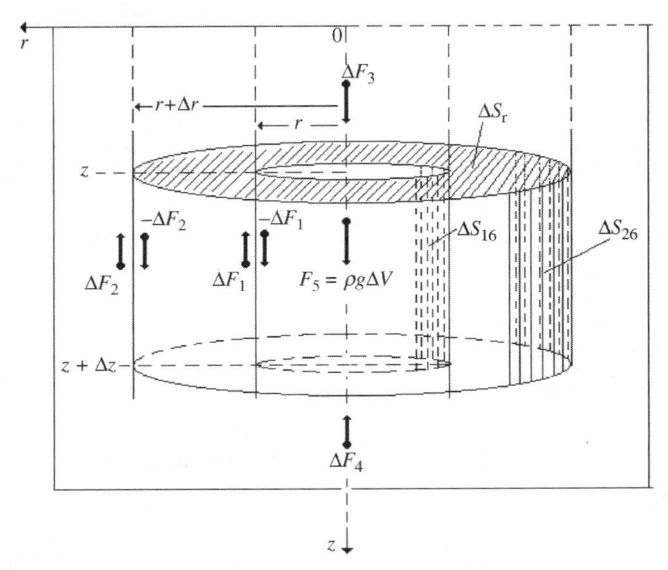

FIGURE 4.3 Derivation of momentum (motion) equation in projections on z-axis.

Projections of surface forces acting at the external and internal surfaces of an elementary particle (Fig. 4.3) are

$$\Delta F_1 = \Delta F_1(r);\;\; \Delta F_2 = \Delta F_2(r+\Delta r);\;\; \Delta F_3 = \Delta F_3(z);\;\; \Delta F_4 = \Delta F_4(z+\Delta z), \tag{4.2.25}$$

where in accordance with (3.16)

$$\Delta F_1(r) = \tau(r)\Delta S_{1б};\quad \Delta F_2(r+\Delta r) = \tau(r+\Delta r)\Delta S_{2б}; \tag{4.2.26}$$

$$\Delta F_3(z) = -p(z)\Delta S_T;\quad \Delta F_4(z+\Delta z) = -p(z+\Delta z)\Delta S_{\mathrm{T}}. \tag{4.2.27}$$

In (4.2.27), the minus sign is especially adapted to make p positive. The mass or volume force is

$$\Delta F_5 = \Delta m \cdot g = \rho \Delta V g. \tag{4.2.28}$$

Corresponding areas and volume are

$$\Delta S_{1б} = 2\pi r \Delta z,\quad \Delta S_{2б} = 2\pi(r+\Delta r)\Delta z, \tag{4.2.29}$$

$$\Delta S_{\mathrm{T}} = \pi(r+\Delta r)^2 - \pi r^2 = 2\pi r \Delta r + \pi \Delta r^2 = \pi \Delta r(2r+\Delta r), \tag{4.2.30}$$

$$\Delta V = \pi(r+\Delta r)^2 \Delta z - \pi r^2 \Delta z = \pi \Delta r \Delta z(2r+\Delta r). \tag{4.2.31}$$

The sum of force projections in the right part of equation (4.2.24) with regard to (4.2.25) – (4.2.31) and direction of force action (force sign) takes the form

$$\rho\frac{\mathrm{d}w}{\mathrm{d}t} = \sum \Phi_i = \lim_{\Delta V \to 0} \frac{\sum F_i}{\Delta V} = \lim_{\Delta V \to 0} \frac{-\Delta F_1 + \Delta F_2 - \Delta F_3 + \Delta F_4 + \Delta F_5}{\pi \Delta r \Delta z(2r+\Delta r)}$$

$$= \lim_{\Delta V \to 0} \frac{-\tau(r)2\pi r\Delta z + \tau(r+\Delta r)2\pi(r+\Delta r)\Delta z + p(z)\pi\Delta r(2r+\Delta r) - p(z+\Delta z)\pi\Delta r(2r+\Delta r) + \rho\Delta V g}{\pi \Delta r \Delta z(2r+\Delta r)}$$

$$= \lim_{\Delta V \to 0, \Delta z \to 0}\left\{\frac{2[(r+\Delta r)\tau(r+\Delta r) - r\tau(r)]}{\Delta r(2r+\Delta r)} - \frac{p(z+\Delta z) - p(z)}{\Delta z} + \rho g\right\}.$$

Passing on to the limit, the following equation is obtained:

$$\sum \Phi_i = \frac{1}{r}\frac{\partial r\tau}{\partial r} - \frac{\partial p}{\partial z} + \rho g. \tag{4.2.32}$$

Substitution of (4.2.32) in (4.2.24) gives motion equation in projections on z-axis

$$\rho\frac{\mathrm{d}w}{\mathrm{d}t} + \frac{\partial p}{\partial z} = \rho g + \frac{1}{r}\frac{\partial r\tau}{\partial r}. \tag{4.2.33}$$

The equation (4.2.33) is appropriate for the third equation (4.2.21) at $i=1$, $\varphi_1=1$, $w_{z1}=w\neq 0$, $w_{r1}=w_{\varphi 1}=0$, $\tau_{zz}=0$, $\partial\tau_{\varphi z}/\partial\varphi=0$, and $\tau_{rz}=\tau$. Last conditions were used when deriving equation (4.2.33). Projections of motion equations on r- and φ-axes are identities when considering flows in pipes and annular channels with motionless walls and that is why they are not exploited.

4.3 THERMODYNAMIC EQUATIONS OF STATE

Equations of state express connection between pressures p_i, densities ρ_i, and temperatures T_i for each of the phases (Basarov, 1991)

$$\Phi(p_i, \rho_i, T_i) = 0, \quad i = 1, 2, \ldots, N. \tag{4.3.1}$$

These equations could be resolved relative pressure

$$p_i = p(\rho_i, T_i).$$

It should be noted that equation (4.3.1) may also be resolved relative density and temperature.

Equation (4.3.1) describes the compressibility degree of each phase. Graphics of some isothermal functions (4.3.1) for one of the phase-resolved relative pressures are shown in Fig. 4.4. These dependences are rather

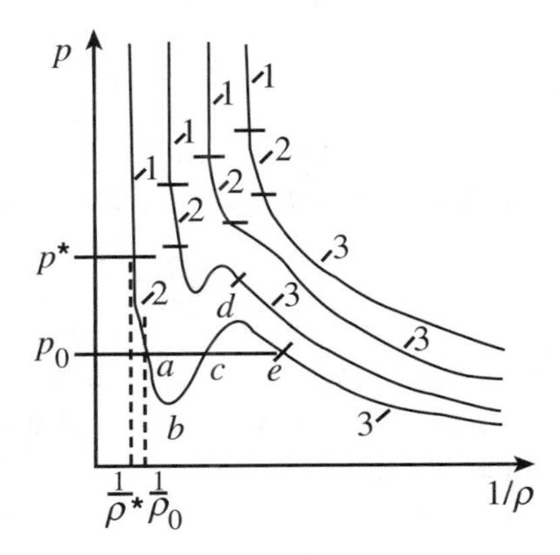

FIGURE 4.4 Dependence of the pressure on specific volume (reverse density): 1—the region of incompressible phase state; 2, 3—regions of barotropic states described by formulas (4.3.3) and (4.3.4).

accurately described by Peng–Robinson thermodynamic state equation

$$p=-\frac{RT}{V-b}-\frac{a}{V^2+2bV-b^2},$$

where $V=1/\rho$ is phase molar volume, factors a and b for n-component mixture are determined by the rule of component mixing.

In Fig. 4.4, the sections 1–3 of curves but with two-phase region *abcd* dependent on conditions (p, $V_i=1/\rho_i$, $T=\text{const}$), at which each of the phases is to be found in practical implementation, can often be approximated by the following relations:

$$\text{in the region 1} \qquad \rho=\text{const}; \tag{4.3.2}$$

$$\text{in the region 2} \qquad \rho=\rho_0[1+\beta_0(p-p_0)]; \tag{4.3.3}$$

$$\text{in the region 3} \qquad p=\rho g\bar{z}RT, \tag{4.3.4}$$

where ρ_0 is the density at initial pressure p_0 at the boundary of the section 2 with two-phase region *abcd*, $\beta_0=-(1/V)/(\partial V/\partial p)$ is the compressibility factor, $\bar{z}$ is the overcompressibility factor, and R is the gas constant.

Among a great number of two-phase mixtures relating to drilling are widely found gas–liquid mixtures in which one phase is gas and another one liquid. In majority of cases, the state of gas can be approximated by Clapeyron formula (4.3.4) and the fluid can be taken as incompressible (4.3.2). Then, (4.3.1) takes the form

$$p=\rho_1 g\bar{z}RT; \tag{4.3.5}$$

$$\rho_2=\text{const}, \tag{4.3.6}$$

where ρ_1 and ρ_2 are densities of gas and fluid.

For two-phase mixtures, one can use the following thermodynamic equation of state:

$$\rho=\rho_1\cdot\varphi+\rho_2\cdot(1-\varphi),$$

where ρ_1 and ρ_2 are densities of the first and second phases determined by one of the equations (4.3.2)–(4.3.4) or by Peng–Robinson, and φ is the concentration of the first phase.

For N-phase mixture, the thermodynamic equation of state has the form

$$\rho=\sum_{i=1}^{N}\rho_i\varphi_i,$$

where ρ_i and φ_i are density and concentration of the ith phase, with $\sum_{i=1}^{N} \varphi_i = 1$.

Below it is accepted that the dependence (4.3.1) is resolved relative density with equal pressures in phases and under isothermal conditions has the form of linear law for each phase (Isaev et al., 2001)

$$\rho_i = a_i + b_i \cdot p, \tag{4.3.7}$$

where a_i and b_i are empirical factors; indices $i = 1$ and 2 denote the first and the second phase. In particular, when the first phase, for example, air, obeys state equation of real gas and the second one, for example, water, is incompressible, $a_i = 0$, $b_i = 1/(\bar{z}RT)$, $a_2 = \text{const}$, and $b_2 = 0$. Here, $\bar{z}$ is the overcompressibility factor, R is the gas constant, and T is the temperature averaged over well depth. When both phases represent slightly compressible fluids with state equations $\rho_i = \rho_{i0}(1 + \beta_i(p - p_{i0}))$, where β_i are compressibility factors and ρ_{i0} and p_{i0} are fixed initial values of densities and pressures, the factors would be $a_i = \rho_{i0}(1 - \beta_i p_{i0})$ and $b_i = \rho_{i0} \cdot \beta_i$. Analysis of state equations of fluids and rigid substances occurring in drilling shows that values of factors a_i and b_i lie in the following ranges: $a_i = 0\text{–}3 \times 10^3\ \text{kg/m}^3$, $b_i = 6 \times 10^{-8}\text{–}1.3 \times 10^{-5}\ \text{kg/(m}^3\text{/Pa)}$.

4.4 RHEOLOGICAL EQUATIONS OF STATE

Enter components of mixture stress τ in momentum equations. Take the simplest suggestion that the stress of the mixture is the sum of phase stress components with coefficients equal to their concentration (Teletov, 1958). Then,

$$\tau = \sum \varphi_i \tau_i. \tag{4.4.1}$$

Examination of τ for a mixture reduces to find the function τ_i for each phase, that is, to determine τ for single-phase media. At this, the phase could be taken as ideal ($\tau_i = 0$) or real ($\tau_i \neq 0$) medium.

Consider a flow of real medium between unbounded parallel surfaces. Let the upper surface of the channel between two plates (Fig. 4.5a) or coaxial cylinders (Fig. 4.5b) move with constant velocity w_0, and the lower one is motionless. Sticking condition at the wall gives rise to stress τ caused by friction resistance owing to sliding of medium layers moving with different velocities w. It has been experimentally shown that in the media, called Newtonian fluid, the stress from friction forces between plates is determined by Newtonian formula

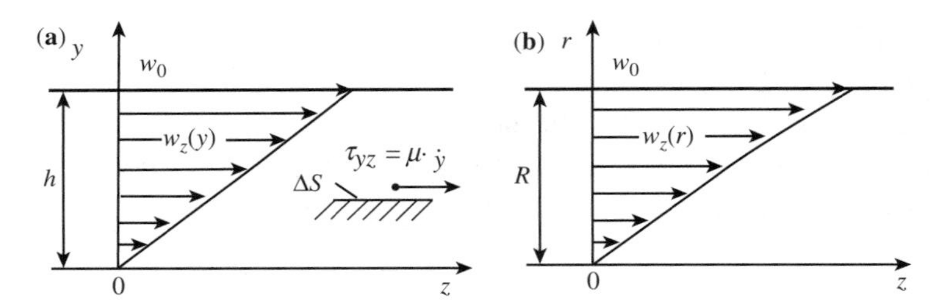

FIGURE 4.5 Scheme of velocity distribution in a gap in laminar flow induced by the motion of upper plate (a) and internal cylinder (b).

$$\tau_{yz} = \mu \frac{w}{y} = \mu \frac{\partial w}{\partial y}. \tag{4.4.2}$$

The factor $\mu = \text{const} > 0$ in formula (4.4.2) is called dynamic viscosity factor. The viscosity favors the fact that faster layers tend to accelerate the adjacent slower ones and inverse slower layers tend to slow down the faster ones. The dimension of viscosity is

$$[\mu] = [\tau][y]/[w] = \text{Pa} \cdot \text{s}. \tag{4.4.3}$$

In this case, the velocity gradient $\partial w/\partial y$ should be equal to relative shear rate of fluid layers $\partial \gamma/\partial t$. Really, due to the velocity definition (3.7)

$$\frac{\partial w}{\partial y} = \frac{\partial(\partial l/\partial t)}{\partial y} = \frac{\partial}{\partial t}\left(\frac{\partial l}{\partial y}\right) = \frac{\partial \gamma}{\partial t} = \dot{\gamma},$$

where $\gamma = \partial l/\partial y$ is relative displacement of fluid layer or simply shear. Thus,

$$\tau_{yz} = \mu\dot{\gamma}. \tag{4.4.4}$$

For a flow between coaxial cylinders, one can write for elementary element of continuum (Fig. 4.5b)

$$\tau_{rz} = \mu \frac{w'_z - w_z}{\Delta r} = \mu \frac{\Delta w_z}{\Delta r}$$

and going to the limit the following equation is obtained

$$\tau_{rz} = \mu \frac{\partial w_z}{\partial r}. \tag{4.4.5}$$

Hence,

$$\tau_{rz} = \mu\dot{\gamma}, \tag{4.4.6}$$

where $\dot{\gamma} = \partial w_z/\partial r$, that is, the shear rate is equal to velocity gradient.

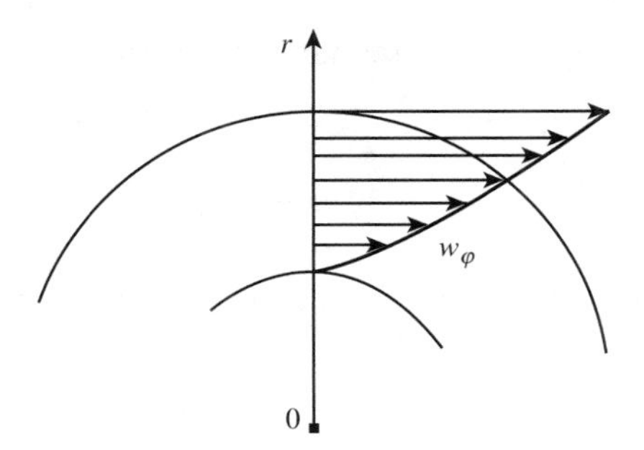

FIGURE 4.6 Scheme of velocity distribution in channel flow between coaxial cylinders under action of outer cylinder rotation.

Consider rotational flow of viscous medium (Fig. 4.6) in a channel between coaxial cylinders under action of one of them, for example, of the outer one. In this case, tangential stresses arising at the outer surface of any rotating cylindrical layer act in the direction of rotation, that is, along φ-axis. Therefore, as in the previous case, it can be written

$$\tau_{r\varphi} = \mu\dot{\gamma}. \tag{4.4.7}$$

In the given case, the shear rate $\dot{\gamma}$ is equal to velocity gradient $\partial w_\varphi/\partial r$ minus the angular velocity of the cylinder rotation ω, namely,

$$\dot{\gamma} = \frac{\partial w_\varphi}{\partial r} - \omega, \tag{4.4.8}$$

where

$$\omega = w_\varphi/r. \tag{4.4.9}$$

Thus, formula (4.4.7) for rotational flow reduces to

$$\tau_{r\varphi} = \mu\left(\frac{\partial w_\varphi}{\partial r} - \frac{w_\varphi}{r}\right). \tag{4.4.10}$$

Consider now radial flow between circular plates (in a circular slot) with an orifice at the center (Fig. 4.7). With the proviso that there is only one

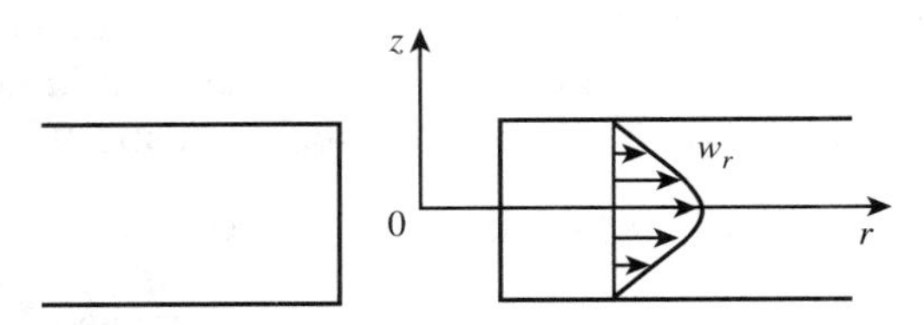

FIGURE 4.7 Scheme of velocity distribution in radial one-dimensional flow between circular plates with an orifice at the center.

nonzero velocity component w_r, the tangential stresses in cross section at a distance r are

$$\tau_{zr} = \mu \frac{\partial w_r}{\partial z}. \tag{4.4.11}$$

So in all four cases, tangential stresses are expressed by one and the same formula

$$\tau = \mu\dot{\gamma}, \tag{4.4.12}$$

in which $\dot{\gamma}$ has its own value.

If in the cases shown in Figs 4.5 and 4.6 to change direction of the outer plate or cylinder motion without changing directions of coordinate axes, the shear rate $\dot{\gamma}$ would be negative. The first case, for example, would be $\dot{\gamma} = w_0/h$ with $w_0 < 0$. At this, the stresses τ_{yz} and τ_{rz} on the surfaces ΔS with normal coinciding in the direction with directions of coordinate axes would also change their sign. Which directions in considered cases coincide with force direction is seen from the definition of stresses (3.16), that is, with velocity direction. Besides, the conservation of laws (4.4.4), (4.4.6), and (4.4.12) at negative $\dot{\gamma}$ and τ demands fulfillment of condition $\mu > 0$.

Thus, function (4.4.12) is defined in the whole range of values $\dot{\gamma}$ and is in odd addition.

In foregoing cases, dependences for tangential stresses of form (4.4.12) were considered for one-dimensional flows in the sense of velocities. Media having such dependence with μ = constant are called Newtonian or viscous fluids. But not all media involved in drilling obey rheological laws (4.4.12). In many cases, even in the above-considered simple one-dimensional flows, the stresses have more general connections (Astarita and Marucci, 1974). At this, the behavior of media is often described by one of the so-called non-Newtonian medium models, namely, viscous-plastic, power, and viscous-elastic models. The model of viscous-plastic fluid has found a widespread application in the majority of drilling solutions (grouting mortars).

Viscous-plastic (Bingham) fluids exhibit fluidity property at the values of tangential stresses τ exceeding a certain value τ_0 called dynamic shear stress. At $|\tau| \leq \tau_0$, the fluid either is at rest or moves as a nondeformable rigid body. The rheological equation of viscous-plastic fluids is

$$\begin{aligned} &\tau = \pm\tau_0 + \eta\dot{\gamma} \quad \text{at} \quad \dot{\gamma} \neq 0; \\ &|\tau| \leq \tau_0 \quad \text{at} \quad \dot{\gamma} = 0. \end{aligned} \tag{4.4.13}$$

The plus sign is taken at $\dot{\gamma} > 0$ and minus at $\dot{\gamma} < 0$.

Some solutions more closely correspond to the model of power fluid for which rheological equation is

$$\tau = k\dot{\gamma}|\dot{\gamma}|^{n-1}, \tag{4.4.14}$$

where factors k and n are called consistency parameters. They characterize the degree of given medium property deviation from the Newtonian law. It is evident that at $n = 1$ relation (4.4.14) coincides with the equation of viscous fluid, factor k taking the value of viscosity factor. At $n < 1$, the power fluid medium is referred to as pseudoplastic fluid and at $n > 1$ as dilatant fluid. Media with $n > 1$ are rarely found.

The main distinction of power fluids from viscous and viscous-plastic ones consists in that the increment of tangential stresses $\Delta\tau$ of the first fluids is not proportional to the shear rate increment $\Delta\dot{\gamma}$. The proportionality of these increments characterizes the linearity of rheological functions $\tau = \tau(\dot{\gamma})$ of viscous-plastic and viscous fluids. Figure 4.8 shows graphs of odd functions (4.4.12)–(4.4.14).

Formulas (4.4.13) and (4.4.14) are sometimes represented by dependences being odd function of $\dot{\gamma}$

$$\tau = \mu_{\text{eff}}(\dot{\gamma})\dot{\gamma}, \tag{4.4.15}$$

where $\mu_{\text{eff}}(\dot{\gamma})$ is the variable characteristic of the medium called effective (apparent) viscosity factor and supposed to be always positive. Since effective viscosity in laminar flows is constant in all shear velocity $\dot{\gamma}$ range only for Newtonian fluids, these fluids are also called linear-viscous fluids. The effective viscosity of viscous-plastic and pseudoplastic fluids varies depending on the shear rate $\dot{\gamma}$; therefore, they are also called nonlinear-viscous fluids. From the above stated, it follows that the linearity of rheological function of the viscous-plastic fluid (4.4.13) could not be identified by its belonging to nonlinear-viscous fluids.

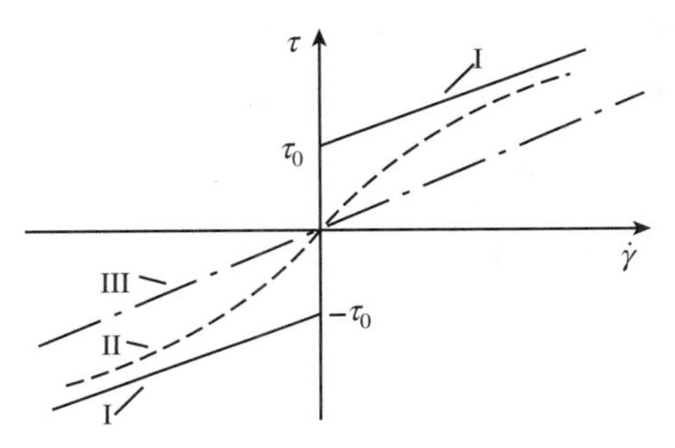

FIGURE 4.8 Forms of dependences $\tau = \tau(\dot{\gamma})$: I—Bingham fluid; II—power fluid ($n < 1$); III—Newtonian fluid.

Sometimes, generalized (Herschel) model with rheological equation is applied

$$\begin{aligned} \tau = \pm\tau_0 + k\dot{\gamma}|\dot{\gamma}|^{n-1} \quad &\text{at} \quad \dot{\gamma} \neq 0; \\ |\tau| \leq \tau_0 \quad &\text{at} \quad \dot{\gamma} = 0. \end{aligned} \tag{4.4.16}$$

In these formulas, τ_0, n, and k enter the consistency parameters. Particular cases of (4.4.16) are equations (4.4.12)–(4.4.14).

In Section 6.5, formula (4.4.14) will also be used to describe stresses in turbulent flows at consistency parameter k depending on the distance from channel walls in accordance with Prandtl–Karman hypothesis.

Equations (4.4.12)–(4.4.15) are obtained through generalization of experimental data for positive values of $\dot{\gamma}$ and using properties of odd function (4.4.15).

EXERCISE 4.4.1

For one-dimensional flows of viscous-plastic fluids, it is required to get rheological equation (4.4.13) for negative $\dot{\gamma}$, if for positive $\dot{\gamma}$ it is experimentally established Bingham dependence $\tau(\dot{\gamma}) = \tau_0 + \eta\dot{\gamma}$.

SOLUTION The function $\tau(\dot{\gamma})$ is odd. The definition of odd function is

$$f(\dot{\gamma}) = -f(-\dot{\gamma}). \tag{4.4.17}$$

In accordance with the given condition for positive $\dot{\gamma}$, we have

$$\tau(\dot{\gamma}) = f(\dot{\gamma}) = \tau_0 + \eta\dot{\gamma}. \tag{4.4.18}$$

For negative $\dot{\gamma}$ the tangential stresses could be found by odd extension of the function (4.4.18), starting from the definition (4.4.17)

$$\tau(\dot{\gamma}) = -f(-\dot{\gamma}) = -[\tau_0 + \eta(-\dot{\gamma})] = -\tau_0 + \eta\dot{\gamma}. \tag{4.4.19}$$

By combining (4.4.18) and (4.4.19), one obtains for all values of $\dot{\gamma}$ formula (4.4.13)

$$\tau = \pm\tau_0 + \eta\dot{\gamma}. \tag{4.4.20}$$

By equating (4.4.13) and (4.4.15), it is possible to get the effective viscosity for all values of $\dot{\gamma}$

$$\mu_{\text{eff}} = \frac{\pm\tau_0 + \eta\dot{\gamma}}{\dot{\gamma}} = \frac{\tau_0}{|\dot{\gamma}|} + \eta. \tag{4.4.21}$$

Then, the rheological law for viscous-plastic fluid (4.4.15) is written in the form

$$\tau = \left(\frac{\tau_0}{|\dot{\gamma}|} + \eta\right)\dot{\gamma}. \tag{4.4.22}$$

There are also media, for example, viscous-elastic dividers, to which the models of viscous-elastic medium suit better. The last along with fluidity possesses the property of elastic form recovery. Various viscous-elastic models have different sets of these properties.

Tangential stresses in viscous-elastic media depend not only on the shear rate $\dot{\gamma}$ but also at least on the shear γ. Therefore, the rheological model of viscous-elastic medium in general could not be characterized by the relation of the form $\tau = f(\dot{\gamma})$. The simplest model of viscous-elastic medium of such a kind is Kelvin–Voigt model of parallel manifestation of viscosity and elasticity properties

$$\tau = f(\gamma, \dot{\gamma}) = G\gamma + \mu\dot{\gamma}, \tag{4.4.23}$$

where μ is the viscosity factor and G is the shear modulus.

The rheological model of the Maxwell viscous-elastic medium (successive manifestation of viscosity and elasticity properties) has the form of differential equation with respect to tangential stress

$$\dot{\tau} + \tau\frac{G}{\mu} = G\dot{\gamma}. \tag{4.4.24}$$

In drilling processes, there are media for which rheological models are not yet built. Among them are the so-called thixotropic fluids in which consistency parameters depend on the history of initial stressed-deformed state formation, velocity, and duration of the shear. It is substantiated that after destruction of the thixotropic medium structure, having been remained still, restores its own properties, for example, the dead-loss shear stress. The stresses at constant shear rate $\dot{\gamma}$ tend with time to be of constant value characterizing equilibrium between destruction and restoration processes of the structure. Figure 4.9 demonstrates typical dependence $\tau(t)$ for nonstationary behavior of thixotropic medium for two constant values of shear rate $\dot{\gamma}_1 = \text{const}_1$ (curves 1 and $1'$), $\dot{\gamma}_2 = \text{const}_2$ (curves 2 and $2'$), and $\dot{\gamma}_1 > \dot{\gamma}_2$. Graphs 1 and $1'$ coincident by superposition are built as a result of two successive tests on the rotary viscometer at $\dot{\gamma}_1 = \text{const}_1$ of one and the same fluid sample being quiescent over a period of time t_0 between tests. Similar results were obtained for curves 2 and $2'$.

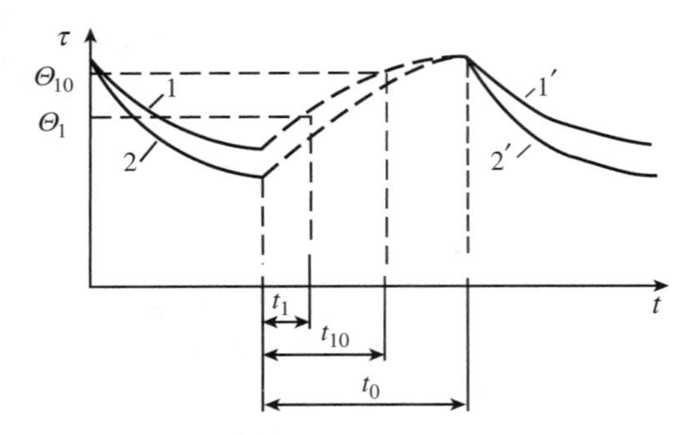

FIGURE 4.9 A graphic of nonstationary behavior of thixotrope medium.

Real drilling solutions commonly possess a certain thixotropic property because of component and phase composition complexity. Therefore, sometimes in the literature thixotropy of viscous-plastic and pseudoplastic fluids is mentioned. For the lack of rheological equations capable of making calculations in practice one would have to characterize the thixotropy of solutions partly by values of dead-loss shear stress θ measured with the special instrument, CHC-2. The procedure to determine θ is arbitrarily chosen. Commonly, periods of rest time are taken equal to $t_1 = 1$ min and $t_{10} = 10$ min after intensive mixing. In Fig. 4.9, through dotted lines is shown the rise of solution strength characterized by stresses τ, measured at different rest periods with CHC-2.

It should be noted that depending on the complexity of rheological curves to describe their isolated sections, different models of concrete form could be used chosen by experimental data handling. The most widespread methods of experimental determination of rheological characteristics used in calculations will be considered in Chapter 16.

Concrete form of the dependence (4.4.1), Newtonian power, and so on is chosen on the ground of phase state idealization.

In practice, the hypothesis on the existence of general function for hydraulic resistance factor of mixture flow λ_c works very well, which permits to write mean absolute stresses of a phase as

$$\tau_i = \frac{\lambda_c}{8} \rho_i v_i^2, \tag{4.4.25}$$

where v_i is mean true phase velocity in the channel cross section.

Dependence of the factor λ_c for laminar flows could be sometimes obtained theoretically in explicit or implicit forms, whereas for turbulent flow it is chiefly obtained experimentally.

4.5 EQUATION OF CONCENTRATIONS

The definition of concentrations φ_1, φ_2, and so on depending in general on all flow parameters was introduced in Chapter 3.

$$\varphi_i = \varphi_i(p, \rho_1, \rho_2, \ldots, \rho_N, w_1, w_2, \ldots, w_N, \tau_1, \tau_2, \ldots, \tau_N, \ldots). \quad (4.5.1)$$

Average concentrations $\bar{\varphi}_i$ in the flow are

$$\bar{\varphi}_i = \frac{1}{\Delta t}\int\limits_{\Delta t}\left(\frac{1}{\Delta V}\int\limits_{\Delta V}\varphi_i \mathrm{d}V\right)\mathrm{d}t. \quad (4.5.2)$$

For average concentration holds equality (3.4)

$$\sum \bar{\varphi}_i = 1. \quad (4.5.3)$$

Expressions for φ_i and $\bar{\varphi}_i$ cannot be obtained theoretically in general. In particular case of equal phase velocities, dependences φ_i and $\bar{\varphi}_i$ have the form

$$\varphi_i = w_i / \sum w_i, \qquad \bar{\varphi}_i = v_i / \sum v_i. \quad (4.5.4)$$

Furthermore, we shall not write the top bar meaning average concentrations. The second quantity in (4.5.4) is conventionally called flow rate concentration and denoted by β_i as distinct from true concentration φ_i.

4.6 FORMULATION OF HYDRO-AEROMECHANICAL PROBLEMS FOR DRILLING PROCESSES

All varieties of multiphase mixtures may be classified by their equations (4.3.1), (4.4.1), and (4.5.1), which completely characterize each concrete mixture and in general should be experimentally determined. Now, write systems of equations for one-dimensional flows.

4.6.1 Axial Flows in Pipes and Annulus

It is accepted that there is only one nonzero velocity component w_{zi}, parallel to z-axis, while other components $w_{\varphi i}$ and w_{ri} vanish. In the flow, the cylindrical surfaces exhibit tangential stresses $\tau_{rzi} \neq 0$. Other stresses vanish. The gravity force is directed parallel to the z-axis, that is, $g_r = 0$, $g_\varphi = 0$, and $g_z = g$, where $g = 9.81$ m/s^2.

From the second equation (4.2.21) at above assumptions, it follows that the pressure is independent of angular coordinate φ. For stationary single-phase flows of viscous, viscous-plastic, and power fluids in pipes and annulus, it is able to show that $\partial \tau_{rz}/\partial z = 0$. Then, from the first equation it is seen that pressure p also does not depend on radius r.

Denote the stress $\tau_{rz} = \tau$ and the velocity $w_{zi} = w_i$. Then, the system of equations describing the given flow takes the form ($i = 1, 2, \ldots, N$)

momentum equation (the third equation of the system (4.2.21))

$$\sum \rho_i \varphi_i \left(\frac{\partial w_i}{\partial t} + w_i \frac{\partial w_i}{\partial z} \right) + \frac{\partial p}{\partial z} = \sum \rho_i \varphi_i g + \frac{\partial \tau}{\partial r} + \frac{\tau}{r}; \qquad (4.6.1)$$

conservation mass equation (4.1.7)

$$\frac{\partial \rho_i \varphi_i}{\partial t} + \frac{\partial \rho_i \varphi_i w_i}{\partial z} = 0; \qquad (4.6.2)$$

thermodynamic equation of state (4.3.1)

$$p = p(\rho_i, T); \qquad (4.6.3)$$

rheological equation of state (4.4.1)

$$\tau = \sum \varphi_i \tau_i, \qquad (4.6.4)$$

where $\tau_i = \tau_i(\dot{\gamma}_i)$;

equation of concentrations (4.5.1)

$$\varphi_i = \varphi_i(p, \rho_1, \rho_2, \ldots, \rho_N, w_1, w_2, \ldots, w_N, \tau_1, \tau_2, \ldots, \tau_N), \qquad (4.6.5)$$

where $\sum \varphi_i = 1$.

When the pipe or annulus is inclined at an angle α to the axis z, one should write in (4.6.1)$\sum \rho_i \varphi_i g \cos \alpha$ instead of $\sum \rho_i \varphi_i g$.

Note that among three momentum equations only one remains. Other equations, besides rheological equation, are as many as there are phases involved in the flow, that is, N. Equation (4.6.3) has different forms, for example, (4.3.2)–(4.3.4). Particular cases of equation (4.6.4) are (4.4.13)–(4.4.14). Equation (4.6.5) could have the form of (4.5.4) or may have the appearance of formulas approximating experimental data. The number of equations should be as many as there are unknown quantities. For example, for single-phase flow ($i = 1$), the number of equations is four, and as many are unknowns, p, w, τ, ρ, since the last equation gives $\varphi = 1$.

4.6.2 Flows Caused by Rotation of Pipes and Walls of Annulus

In the considered case, only angular velocity component $w_{\varphi i} \neq 0$, that is, the flow, occurs along concentric circles centered at the z-axis. By symmetry of the flow, its characteristics are independent of the coordinate φ. Then, at the section perpendicular to the z-axis, $p = p(r, t)$, $w_{\varphi i} = w_{\varphi i}(r, t)$, and only $\tau_{r\varphi} \neq 0$. Take from momentum equations (4.2.21) the first two, since the third one happens to be a solo equation determining only hydraulic pressure independent of pipe rotation. The system of equation (4.2.21) describing the given flow of the mixture is

momentum equation at $w_{\varphi i} = w_i \neq 0$ and $\tau_{r\varphi} = \tau \neq 0$

$$\frac{\partial p}{\partial r} = \sum \rho_i \varphi_i \frac{w_i^2}{r}, \tag{4.6.6}$$

$$\sum \rho_i \varphi_i \frac{\partial w_i}{\partial t} = \frac{\partial \tau}{\partial r} + 2\frac{\tau}{r}; \tag{4.6.7}$$

equation of mass conservation (4.1.7) at $w_{\varphi i} \neq 0$

$$\frac{\partial \rho_i}{\partial t} = 0; \tag{4.6.8}$$

thermodynamic equation of state (4.3.1)

$$p = p(\rho_i, T); \tag{4.6.9}$$

rheological equation of state (4.4.1)

$$\tau = \sum \varphi_i \tau_i, \tag{4.6.10}$$

where $\sum \tau_i = \tau_i(\dot{\gamma}_i)$;
equation of concentrations

$$\varphi_i = \varphi_i(p, \rho_1, \rho_2, \ldots, \rho_N, \; w_1, w_2, \ldots, w_N, \; \tau_1, \tau_2, \ldots, \tau_N), \tag{4.6.11}$$

where $\sum \varphi_i = 1$.

4.6.3 Radial Flow in Circular Slot

In the considered case, only radial velocity component $w_{ri} = w_i \neq 0$. Owing to expansion or contraction of the flow in r-direction, the stresses

$\tau_{rr}, \tau_{\varphi\varphi}, \tau_{zr}$ are nonzero. Nevertheless, only τ_{zr} in momentum equations would be taken into account. For nonstationary flow of viscous fluid, it can be shown that due to rheological law (4.4.15) all terms with τ_{rr} and $\tau_{\varphi\varphi}$ entering in momentum equation (4.2.21) vanish. Non-Newtonian properties would be taken into account only in expression for $\tau_{zr} = \tau$. From the second equation (4.2.21), it follows that the pressure is independent of φ. No account will be taken of the third equation since there are considered flows in which the pressure change along r-coordinate is much more than the pressure change along z-axis. Thus, only the first of the momentum equations (4.2.21) is used.

The system of equations ($i = 1, 2, \ldots, N$) describing given flow takes the form

momentum equation

$$\sum \rho_i \varphi_i \left(\frac{\partial w_i}{\partial t} + w_i \frac{\partial w_i}{\partial r} \right) + \frac{\partial p}{\partial r} = \frac{\partial \tau}{\partial z}; \tag{4.6.12}$$

equation of mass conservation (4.1.7) at $w_i \neq 0$

$$\frac{\partial \rho_i \varphi_i}{\partial t} + \frac{1}{r} \cdot \frac{\partial r \rho_i \varphi_i w_i}{\partial r} = 0; \tag{4.6.13}$$

thermodynamic equation of state (4.3.1)

$$p = p(\rho_i, T); \tag{4.6.14}$$

rheological equation of state (4.4.1)

$$\tau = \sum \varphi_i \tau_i, \tag{4.6.15}$$

where $\sum \tau_i = \tau_i(\dot{\gamma}_i)$;
equation of concentrations

$$\varphi_i = \varphi_i(p, \rho_1, \rho_2, \ldots, \rho_N, w_1, w_2, \ldots, w_N, \tau_1, \tau_2, \ldots, \tau_N), \tag{4.6.16}$$

where $\sum \varphi_i = 1$.

Hence, there are three systems of equations for main flows in circulation in bed-well system.

For example, by solving the system of equations (4.6.1)–(4.6.5), one can find the relation between the pressure p and the distribution of velocities w in pipes in stationary laminar flow of viscous single-phase incompressible fluid. Then, using this relation, it is convenient to get in practice the dependence to calculate the pressure drop Δp over the pipe section with length L and constant cross section in the form of Darcy–Weisbach formula

$$\Delta p = |p_1 - p_2| = \lambda \frac{\rho v^2}{2d} L, \tag{4.6.17}$$

where p_1, p_2 are pressures at the beginning and end of the pipe section, $v = Q/(\pi R^2) = 4Q/\pi d^2$ is the mean velocity in the pipe cross section, Q is the fluid flow rate, d is the pipe diameter, $\lambda = 64/\text{Re}$ is the hydraulic resistance factor, and $\text{Re} = |v| d\rho/\mu$ is the Reynolds number.

Formulas of the type (4.6.17) are obtained by solving the system of equations (4.6.1)–(4.6.5), (4.6.6)–(4.6.11), and (4.6.12)–(4.6.16) for flows of viscous fluid in pipe, annulus, circular slot, and of other types of fluids (power and viscous plastic) at different flow regimes. As the severity of problems increases, for example, for two-phase fluids (slurry–liquid mixture, aerated mixture), it would be worthwhile to average equations over pipe cross section and then to get approximate equations, since obtaining the factor λ in (4.6.17) theoretically for all cases is highly conjectural.

4.6.4 Flows in Pipes and Annulus

Rewrite the equation (4.6.1) in the form

$$\sum \rho_i \varphi_i \frac{\mathrm{d}w_i}{\mathrm{d}t} + \frac{\partial p}{\partial z} = \sum \rho_i \varphi_i g + \frac{1}{r}\frac{\partial r\tau}{\partial r}. \tag{4.6.18}$$

Here, the following notation of total derivative is used:

$$\frac{\mathrm{d}w_i}{\mathrm{d}t} = \frac{\partial w_i}{\partial t} + w_i \frac{\partial w_i}{\partial r}. \tag{4.6.19}$$

Take the integral from both parts of equation (4.6.18) and divide them by pipe area $S = \pi R^2$, that is, perform averaging

$$\frac{2\pi \int_0^R \left[\sum \rho_i \varphi_i \frac{\mathrm{d}w_i}{\mathrm{d}t} + \frac{\partial p}{\partial z}\right] r\,\mathrm{d}r}{\pi R^2} = \frac{2\pi \int_0^R \left[\sum \rho_i \varphi_i g + \frac{1}{r}\frac{\partial r\tau}{\partial r}\right] r\,\mathrm{d}r}{\pi R^2}$$

or

$$\frac{2\pi \int_0^R \left[\sum \rho_i \varphi_i \frac{\mathrm{d}w_i}{\mathrm{d}t}\right] r\,\mathrm{d}r}{\pi R^2} + \frac{2\pi \int_0^R \left[\frac{\partial p}{\partial z}\right] r\,\mathrm{d}r}{\pi R^2} = \frac{2\pi \int_0^R \left[\sum \rho_i \varphi_i g\right] r\,\mathrm{d}r}{\pi R^2} + \frac{2\pi \int_0^R \mathrm{d}(r\tau)}{\pi R^2}. \tag{4.6.20}$$

As far as in considered flow the pressure p, and thus the derivative $\partial p/\partial z$, is constant over the pipe cross section, $\partial p/\partial z$ can be factored out from the integral. Equation (4.6.3) shows that densities ρ_i do not depend on r-coordinate and consequently they could also be factored from the integral. With concentration one could proceed in the same manner since in the cross section it has the meaning of (3.3), that is, in our case

$$\varphi_i = S_i/S, \tag{4.6.21}$$

where S_i is the part of the pipe area occupied by ith phase. The derivative $\mathrm{d}/\mathrm{d}t$ could also be factored from the integral since only velocity depends on the time but not the limits of integrations. Then, (4.6.20) takes the form

$$\frac{\sum\left[\rho_i\varphi_i\frac{\mathrm{d}}{\mathrm{d}t}\left(2\pi\int_0^R w_i r\mathrm{d}r\right)\right]}{\pi R^2}+\frac{2\pi\frac{\partial p}{\partial z}\int_0^R r\,\mathrm{d}r}{\pi R^2}=\frac{2\pi\left[\sum\rho_i\varphi_i g\right]\int_0^R r\,\mathrm{d}r}{\pi R^2}+\frac{2\pi\int_0^R \mathrm{d}(r\tau)}{\pi R^2}. \tag{4.6.22}$$

Since under the sign of derivative is flow rate of ith phase

$$Q_i = 2\pi\int_0^R w_i r\,\mathrm{d}r, \tag{4.6.23}$$

the ratio $Q_i/(\pi R^2)$ would be the average velocity v_i of the ith phase.

Other integrals are

$$2\pi\int_0^R r\,\mathrm{d}r = 2\pi\left.\frac{r^2}{2}\right|_0^R = \pi R^2; \tag{4.6.24}$$

$$\frac{2}{R^2}\int_0^R \mathrm{d}(r\tau) = \frac{2}{R^2}\left.(r\tau)\right|_0^R = \frac{2\tau_\mathrm{w}}{R}, \tag{4.6.25}$$

where τ_w is the friction stress of the mixture at the wall.

With regard to (4.6.24) and (4.6.25), equation (4.6.22) transforms to

$$\sum\rho_i\varphi_i\frac{\mathrm{d}v_i}{\mathrm{d}t}+\frac{\partial p}{\partial z}=\sum\rho_i\varphi_i g+\frac{4\tau_\mathrm{w}}{d}$$

or reminding the notation of total derivative (4.6.19),

$$\sum\rho_i\varphi_i\left(\frac{\partial v_i}{\partial t}+v_i\frac{\partial v_i}{\partial z}\right)+\frac{\partial p}{\partial z}=\sum\rho_i\varphi_i g+\frac{4\tau_\mathrm{w}}{d}, \tag{4.6.26}$$

where $d = 2R$.

Find τ_{w} for the mixture using (4.6.4)

$$\tau_{\mathrm{w}} = \sum \varphi_i \tau_{\mathrm{w}i}. \tag{4.6.27}$$

Consider the equilibrium of ith phase element in the pipe in stationary flow. The equilibrium condition can be written as

$$(p_2 - p_1)\pi R^2 = \tau_{\mathrm{w}i} 2\pi R L. \tag{4.6.28}$$

From here it follows

$$\tau_{\mathrm{w}i} = \frac{(p_2 - p_1)R}{2L}. \tag{4.6.29}$$

Since relation (4.6.17) holds for single-phase flow,

$$\tau_{\mathrm{w}i} = -\frac{\lambda_i \rho_i v_i |v_i|}{2d} L \frac{R}{2L} = -\lambda_i \rho_i \frac{v_i |v_i|}{8}. \tag{4.6.30}$$

Assume that in joint flow of phases $\tau_{\mathrm{w}i}$ has also the form (4.6.30) with $\lambda_i = \lambda_{\mathrm{m}}$ being hydraulic resistance factor of the mixture. Then, substitution of (4.6.30) in (4.6.27) and (4.6.26) leads to the following momentum equation:

$$\sum \rho_i \varphi_i \left(\frac{\partial v_i}{\partial t} + v_i \frac{\partial v_i}{\partial z} \right) + \frac{\partial p}{\partial z} = \sum \rho_i \varphi_i g - \frac{\lambda_{\mathrm{m}}}{2d} \sum \rho_i \varphi_i v_i |v_i|. \tag{4.6.31}$$

Averaging the mass conservation equation (4.6.2), one gets

$$\frac{\partial \rho_i \varphi_i}{\partial t} + \frac{\partial \rho_i \varphi_i v_i}{\partial z} = 0. \tag{4.6.32}$$

The equation of concentrations (4.6.5) looks like (4.6.21).

Media in drilling in many cases could be taken as two phase ($i = 2$). Let us denote $\varphi_1 = \varphi$. Then, $\varphi_2 = 1 - \varphi$, since $\sum \varphi_i = 1$, and the system of equations for two-phase flows in pipes and annulus takes the final form:

momentum equation (Teletov, 1958)

$$\frac{\partial p}{\partial z} = [\varphi \rho_1 + (1-\varphi)\rho_2] g \pm \frac{\lambda_{\mathrm{m}}}{2d} \left[\varphi \rho_1 v_1^2 + (1-\varphi)\rho_2 v_2^2 \right] - \left[\varphi \rho_1 \frac{\mathrm{d}v_1}{\mathrm{d}t} + (1-\varphi)\rho_2 \frac{\mathrm{d}v_2}{\mathrm{d}t} \right]; \tag{4.6.33}$$

equations of mass conservation

$$\frac{\partial \varphi \rho_1}{\partial t} + \frac{\partial \varphi \rho_1 v_1}{\partial z} = 0, \qquad (4.6.34)$$

$$\frac{\partial (1-\varphi)\rho_2}{\partial t} + \frac{\partial (1-\varphi)\rho_2 v_2}{\partial z} = 0; \qquad (4.6.35)$$

thermodynamic equations of state

$$p = p(\rho_1, T), \qquad (4.6.36)$$

$$p = p(\rho_2, T); \qquad (4.6.37)$$

equation of concentrations

$$\varphi = \varphi(v_1, v_2, \rho_1, \rho_2, \mu_1, \mu_2, g, d, \sigma, p); \qquad (4.6.38)$$

equation for hydraulic resistances

$$\lambda_c = \lambda_c(v_1, v_2, \rho_1, \rho_2, \mu_1, \mu_2, g, d, \sigma, \varphi, p). \qquad (4.6.39)$$

The sign before the friction term in (4.6.33) depends on the direction of mixture velocity. If the velocity direction coincides with the direction of gravity force, it should be taken as minus sign. Note that when solving some problems, it is required after averaging to know the hydraulic resistance factor λ_m instead of stresses. As a rule, the function φ in (4.6.38), as well as λ_m in (4.6.39), is obtained by experiments.

4.6.5 Flow in Circular Slot

Average momentum equation (4.6.12) over cylindrical cross section with lateral surface area $S = 2\pi rH$. Integration of (4.6.12) and use of shorthand symbol of total derivative yields

$$\frac{2\cdot 2\pi r \int_0^{H/2} \sum \left(\rho_i \varphi_i \frac{\mathrm{d}w_i}{\mathrm{d}t}\right) \mathrm{d}z}{2\pi rH} + \frac{2\cdot 2\pi r \int_0^{H/2} \frac{\partial p}{\partial r}\, \mathrm{d}z}{2\pi rH} = \frac{2\cdot 2\pi r \int_0^{H/2} \frac{\partial \tau}{\partial r}\, \mathrm{d}z}{2\pi rH},$$

where H is the slot height.

Integrals are taken in the limits from 0 to $H/2$, since the flow is symmetric with respect to z-plane. Since ρ_i, φ_i, and $\partial p/\partial r$ can be factored out of the integral, the following is obtained

$$\sum \rho_i \varphi_i \frac{\mathrm{d}v_i}{\mathrm{d}t} + \frac{\partial p}{\partial r} = \frac{2\tau_w}{H}, \qquad (4.6.40)$$

where τ_w is determined from (4.6.15)

$$\tau_{\mathrm{w}} = \Sigma \varphi_i \tau_{\mathrm{w}i}. \tag{4.6.41}$$

In order to get $\tau_{\mathrm{w}i}$ at the surface of the circulate plate, consider single-phase stationary slow flow when inertial terms could be ignored.

From (4.6.40), ensues

$$\frac{\partial p}{\partial r} = \frac{2\tau_{\mathrm{w}i}}{H}. \tag{4.6.42}$$

For elementary cylindrical element, the last relation gives

$$\frac{\Delta p}{\Delta r} = \frac{2\tau_{\mathrm{w}i}}{H} \quad \text{or} \quad \tau_{\mathrm{w}i} = \frac{\Delta p H}{2\Delta r}.$$

In Chapter 6, it will be shown that for such flow a valid formula is

$$\Delta p = \lambda \frac{\rho v^2}{2H} \Delta r. \tag{4.6.43}$$

Substitution of this expression in the formula for $\tau_{\mathrm{w}i}$ leads to

$$2\tau_{\mathrm{w}i} = \lambda_i \rho_i \frac{v_i |v_i|}{2}. \tag{4.6.44}$$

Suppose $\tau_{\mathrm{w}i}$ in (4.6.41) for multiphase nonstationary flows has the same form as in (4.6.41). Then, substitution of (4.6.44) in (4.6.41) and thereafter the result in (4.6.40) yields the momentum equation

$$\frac{\partial p}{\partial r} = -\frac{1}{H} \sum \left(\varphi_i \lambda_i \rho_i \frac{v_i |v_i|}{2} \right) - \sum \left(\varphi_i \rho_i \left[\frac{\partial v_i}{\partial t} + v_i \frac{\partial v_i}{\partial r} \right] \right). \tag{4.6.45}$$

Thus, for two-phase mixtures ($i = 2$) in circular slot, the averaged system of equations takes the form

momentum equation

$$\frac{\partial p}{\partial r} = -\frac{\lambda_{\mathrm{m}}}{2H} [\varphi \rho_1 v_1 |v_1| + (1-\varphi)\rho_2 v_2 |v_2|] - \left[\varphi \rho_1 \frac{\mathrm{d}v_1}{\mathrm{d}t} + (1-\varphi)\rho_2 \frac{\mathrm{d}v_2}{\mathrm{d}t} \right] \tag{4.6.46}$$

with the proviso that $\lambda_i = \lambda_{\mathrm{m}}$, $\varphi_1 = \varphi$, $\varphi_2 = 1 - \varphi$;

equations of mass conservation

$$\frac{\partial \varphi \rho_1}{\partial t} + \frac{1}{r} \frac{\partial r \varphi \rho_1 v_1}{\partial r} = 0, \tag{4.6.47}$$

$$\frac{\partial (1-\varphi)\rho_2}{\partial t} + \frac{1}{r} \frac{\partial r (1-\varphi)\rho_2 v_2}{\partial r} = 0; \tag{4.6.48}$$

thermodynamic equations of state

$$p = p(\rho_1, T), \tag{4.6.49}$$

$$p = p(\rho_2, T); \tag{4.6.50}$$

equation of concentrations

$$\varphi = \varphi(v_1, v_2, \rho_1, \rho_2, \mu_1, \mu_2, g, d, \sigma, p); \tag{4.6.51}$$

equation for hydraulic resistance factor

$$\lambda_c = \lambda_c(v_1, v_2, \rho_1, \rho_2, \mu_1, \mu_2, g, d, \sigma, \varphi, p). \tag{4.6.52}$$

Note that equation (4.6.46) could be applied not only to the flow in a slot but also to fluid filtration in the bed under condition that φ represents the skeleton concentration of the bed with $v_1 = 0$. If inertial terms are neglected, equation (4.6.46) takes the form of filtration square law

$$\frac{\partial p}{\partial r} = -a\rho v|v|, \tag{4.6.53}$$

where

$$a = \frac{\lambda_m}{2H}(1-\varphi).$$

In particular when $a = \mu/(vk\rho)$, equation (4.6.53) reduces to well-known filtration law

$$\frac{\partial p}{\partial r} = -\frac{\mu}{k}v \quad \text{or} \quad v = -\frac{k}{\mu}\nabla_r p, \tag{4.6.54}$$

where $\nabla_r p = \partial p/\partial r$ and k is permeability factor.

In Chapter 11, several problems related to flow in the formation will be considered.

Flow characteristics in all elements of well-bed circulation system at various drilling processes could be obtained by simultaneous solution of systems of equations (4.6.33)–(4.6.39) and (4.6.46) and (4.6.52) with regard to conditions of their conjugation at the boundaries between elements of well-bed circulation system.

Thus, the solution of technological problems is based on the above-stated systems of equations at given values of parameters at outer and inner boundaries of the well-bed circulation system and initial values of flow parameters, initial and boundary conditions.

CHAPTER 5

HYDROSTATICS OF SINGLE-PHASE FLUIDS AND TWO-PHASE MIXTURES IN GRAVITY FIELD

5.1 HYDROSTATICS OF SINGLE-PHASE FLUIDS

From the equation (4.6.26) for single-phase fluid ($i = 1$, $v_i = 0$, $\varphi_i = \varphi_1 = 1$, $\rho_i = \rho$) follows the basic equation of hydrostatic ($v_i = 0$) for Newtonian and non-Newtonian fluids in gravity field written in differential form (Isaev, 2006)

$$\frac{\mathrm{d}p}{\mathrm{d}z} = \rho g + \frac{4\tau_{wz}}{d}, \tag{5.1.1}$$

where τ_{wz} is tangential stress at the wall; $d = d(z)$ variable diameter of the channel depending on z-coordinate.

The second term in the right part (5.1.1) does not always vanish at equilibrium state of fluid. It presents if there are surface forces of different nature, for example, forces of surface tension between channel wall and fluid and/or free surface, as well as forces due to stresses caused by pressure gradient applied to the fluid. The magnitude of arisen stress depends on applied pressure gradient and can reach its maximal value in limiting equilibrium state of the fluid. It should be noted that owing to the smallness of the stress τ_w in pipes of great diameter (as distinct from capillaries), the magnitude of this term is insignificant and in unbounded fluid vanishes even at finite but enough great τ_w, that is

$$\lim_{d \to \infty} \frac{\tau_{wz}}{d} = 0. \tag{5.1.2}$$

Thus, for unbounded fluid or smallness of the ratio τ_w/d, we have fundamental equation of hydrostatic in differential form commonly cited in textbooks of hydrodynamics

$$\frac{\mathrm{d}\,p}{\mathrm{d}\,z} = \rho g \tag{5.1.3}$$

or

$$\mathrm{d}\,p = \rho g\,\mathrm{d}\,z. \tag{5.1.4}$$

From this it follows that isobaric surfaces, that is surfaces at which the pressure is everywhere constant, are horizontal planes. Therefore, on isobaric surfaces $p = \text{const}$ and owing to (5.1.4) it should be $\mathrm{d}\,z = 0$ or

$$z = \text{const}. \tag{5.1.5}$$

This equation defines a family of horizontal planes for fluids of any density. Thus, to know the pressure inside the channel (in the cross section) it is enough to measure the pressure at its wall.

5.2 HYDROSTATICS OF INCOMPRESSIBLE FLUID AT $\tau_W = 0$

Integration of the equation (5.1.4) for homogeneous incompressible fluid ($\rho = \text{const}$) gives the following basic equation of hydrostatics (Fig. 5.1)

$$p_1 = p_2 + \rho g h, \tag{5.2.1}$$

that is read as follows: the pressure p_1 (at the bottom or in the lower section of the channel) is equal to the pressure p_2 (at the top or in the upper section of the channel) plus the product of fluid density into gravity acceleration and vertical distance between horizontal (isobaric) planes.

If in the well there are several nonmixed fluids with densities ρ_i in form of benches with heights h_i, the hydrostatic pressure would be determined in accord with the formula

$$p_1 = p_2 + \sum \rho_i g h_i. \tag{5.2.2}$$

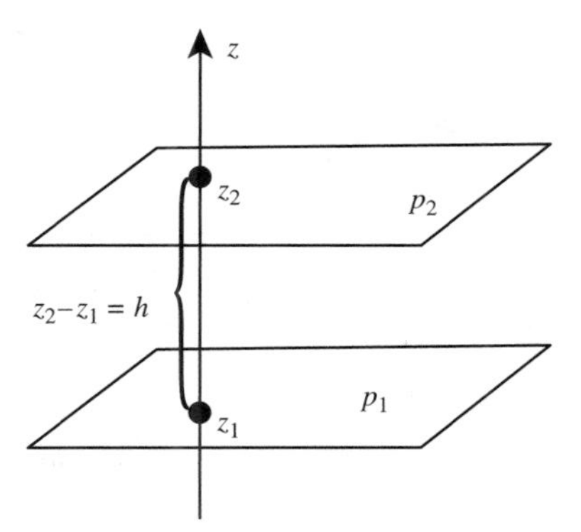

FIGURE 5.1 Derivation of the equation (5.2.1).

Sizes of the vessel do not enter in the equation (5.2.1); therefore, hydrostatic pressure is independent of the vessel shape (Fig. 5.2). Since the pressure does not depend on the shape of the vessel, pressures in each horizontal cross section of communicating vessels filled by one and the same fluid are equal. Pipe and annular spaces of the well may be represented as communicated vessels (Fig. 5.3). Using equation (5.2.1), one can calculate the hydrostatic pressure in any channel cross section with given pressure in one of the channel cross sections. It should be remembered that in basic equation absolute pressures enter though it is valid also for excess (manometer) pressure that is measured by manometers since absolute p_{ab} and excess p_{ex} pressures are related through the formula

$$p_{ab} = p_{ex} + p_{at}. \tag{5.2.3}$$

EXERCISE 5.2.1

It is required to calculate the bottom-hole pressure in accordance with Fig. 5.3 in an open well filled with quiescent washing fluid at given data: fluid density

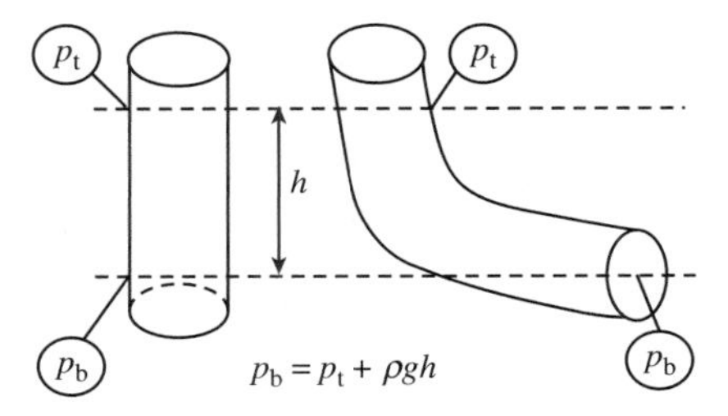

FIGURE 5.2 Hydrostatic pressures at the bottom p_b in different vessels with equal fluid density and equal pressures at the top p_t do not depend on vessel shape.

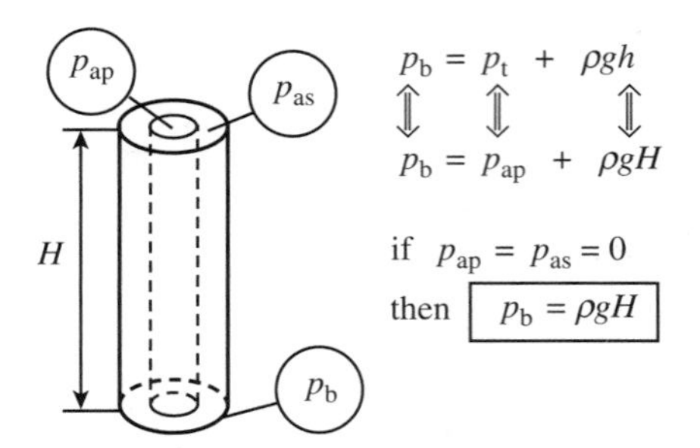

FIGURE 5.3 Pipe and annular spaces of the well as communicating vessels and formulas to calculate down-hole hydrostatic pressure.

$\rho = 1100\,\mathrm{kg/m^3}$ and well vertical depth $H = 1500\,\mathrm{m}$. Manometers at the well mouth show the excess pressure beyond atmospheric one.

SOLUTION In order to calculate the bottom-hole pressure $p_1 = p_b$ we apply the formula (5.2.1). Since, the well is open the manometer pressures in ascending pipe and annular space at the mouth are equal to zero ($p_2 = p_{ap} = p_{as} = 0$). The distance from the mouth to the bottom hole in vertical is $H = 1500\,\mathrm{m}$. Performing calculations in basic SI system of units and transforming results to another units one obtains

$$p_b = 0 + \rho gh = \rho gH = 1100 \times 9.81 \times 1500 = 16\,186\,500\,\mathrm{Pa} = 16.19\,\mathrm{MPa}$$
$$= 162\,\mathrm{bar} = 165\,\mathrm{atm} = 165\,\mathrm{kg/cm^2} = 2347\,\mathrm{psi}.$$

5.3 HYDROSTATICS OF SINGLE-PHASE COMPRESSIBLE FLUID (GAS) AT $\tau_W = 0$

In this case the density throughout the height of the well is not constant and the thermodynamic equation of state has the form of (4.3.5)

$$p = \rho_1 g \bar{z} RT. \tag{5.3.1}$$

Solving this equation with respect to ρ_1, substituting the result in (5.1.3), and performing integration (see Fig. 5.1) we get the basic equation of hydrostatics for compressible fluid (gas)

$$p_1 = p_2\, \mathrm{e}^{h/\bar{z}RT}. \tag{5.3.2}$$

In this equation enter the absolute pressures, and the calculation with this equation should be carried out in absolute pressures. Going in this formula to manometer pressures with regard to relation (5.2.3) one gives

$$p_{1M} + p_{at} = (p_{2M} + p_{at})\, e^{h/\bar{z}RT}$$

or

$$p_{1M} = (p_{2M} + p_{at})\, e^{h/\bar{z}RT} - p_{at}. \tag{5.3.3}$$

At $p_{2M} = 0$, the following is obtained

$$p_{1M} = p_{at}(e^{h/\bar{z}RT} - 1). \tag{5.3.4}$$

EXERCISE 5.3.1

It is required to calculate hydrostatic manometer pressure with regard to Fig. 5.3 at the bottom hole of a well that is drilled with bubble aeration at given data: air density at normal conditions $\rho = 1.29\,\mathrm{kg/m^3}$, vertical height of the well $H = 1500\,\mathrm{m}$, $p_{at} = 9.81 \times 10^4\,\mathrm{Pa}$, mean temperature in the well $T = 293\,\mathrm{K}$, overcompressibility factor is $\bar{z} = 1$, gas constant $R = 29.27\,\mathrm{m/K}$.

SOLUTION To calculate the bottom-hole pressure $p_{1M} = p_b$ take advantage of the formula (5.3.4), since the mouth in the annular space is open and manometer pressure at the annular space mouth is zero ($p_{2M} = p_{as} = 0$). The distance from the mouth to the bottom in vertical is $H = 1500\,\mathrm{m}$. There is

$$p_{1M} = p_{at}(e^{H/\bar{z}RT} - 1) = 9.81 \times 10^4 \left(e^{\frac{1500}{1\times 29.27\times 293}} - 1\right)\,\mathrm{Pa} = 1.874$$
$$\times 10^4\,\mathrm{Pa} = 0.191\,\mathrm{atm} = 0.195\,\mathrm{bar} = 2.823\,\mathrm{psi}.$$

Calculated pressure shows the manometer at the bottom.

5.4 HYDROSTATICS OF SLIGHTLY COMPRESSIBLE FLUID AT $\tau_W = 0$

In this case the density depends on pressure and is determined by the equation (4.3.3)

$$\rho = \rho_0[1 + \beta_0(p - p_0)]. \tag{5.4.1}$$

Integration of equation (5.1.4) with regard to Fig. 5.1 gives

$$p_1 = p_2 + (1/\beta_0 + p_2 - p_0)(e^{\rho_0 g \beta_0 h} - 1). \tag{5.4.2}$$

In this formula one can substitute both absolute and manometer pressures. At $\beta_0 \to 0$ (incompressible fluid), the expression (5.4.2) goes into standard equation of incompressible fluid hydrostatics (5.2.1)

$$p_1 = p_2 + \rho g h. \tag{5.4.3}$$

EXERCISE 5.4.1

Calculate the hydrostatic manometer pressure in accordance with Fig. 5.3 at the well bottom that is drilled with the help of slightly compressible fluid at given data: fluid density $\rho_0 = 1030\,\text{kg/m}^3$, pressure $p_0 = p_{at}$, mean compressibility factor $\beta_0 = 43.1 \times 10^{-11}\,\text{Pa}^{-1}$, vertical depth of the well $H = 1500\,\text{m}$, $p_{at} = 9.81 \times 10^4\,\text{Pa}$. Mean temperature in the well $T = 293\,\text{K}$.

SOLUTION To calculate the bottom pressure $p_1 = p_b$, apply the formula (5.4.2) since the mouth in annular space is open and manometer pressure at the mouth is zero ($p_2 = p_{as} = 0$). The distance from the mouth to the bottom in vertical is $H = 1500\,\text{m}$. We have

$$p_1 = p_2 + (1/\beta_0 + p_2 - p_0)(e^{\rho_0 g \beta_0 h} - 1) = 0 + (1/(4.31 \times 10^{-11})$$
$$+ (0 - 9.81 \times 10^4)) \times (e^{1030 \times 9.81 \times 43.1 \times 10^{-11} \times 1500} - 1) = 15.21\,\text{MPa}.$$

The calculated pressure shown by manometer is a little less than the bottom pressure that would show the manometer as if the fluid were incompressible. Therefore, calculation of the bottom pressure with formula (5.4.3) yields

$$p_1 = p_2 + \rho g h = 0 + \rho g H = 1030 \times 9.81 \times 1500 = 15.156\,\text{MPa}.$$

5.5 HYDROSTATICS OF A FLUID WITH DYNAMIC SHEAR STRESS ($\tau_0 \neq 0$)

This is a sufficiently large class of fluids. Among them are Bingham or viscous-plastic fluids whose rheological equations are given by dependence (4.4.13)

$$\begin{aligned} \tau &= \pm\tau_0 + \eta\dot{\gamma} && \text{at} \quad \dot{\gamma} \neq 0; \\ |\tau| &\leq \tau_0 && \text{at} \quad \dot{\gamma} = 0. \end{aligned} \tag{5.5.1}$$

The sign plus is taken at $\dot{\gamma} > 0$ and minus at $\dot{\gamma} < 0$.

Since there is considered hydrostatic state of fluid at $\dot{\gamma} = 0$, the fluid state does not depend on momentum (motion) equation and is governed by the second (hydrostatic) equation (5.5.1) in which the stress τ is the stress at the channel walls, namely

$$|\tau_w| \leq \tau_0 \quad \text{at} \quad \dot{\gamma} = 0. \tag{5.5.2}$$

To determine basic hydrostatic equation for such fluids, it is necessary to use equation (5.1.1)

$$\frac{dp}{dz} = \rho g + \frac{4\tau_{wz}}{d}. \tag{5.5.3}$$

If the fluid is only under action of gravity force, the stress at the wall vanishes and the hydrostatic pressure is determined by basic equation (5.2.1). In attempt to displace the fluid increasing pressure gradient in the left side of (5.5.3), there came into existence stresses at the wall τ_w, which during further enhancement of pressure gradient can achieve the value τ_0. At $|\tau_w| \leq \tau_0$ the fluid is still in hydrostatic equilibrium called limiting equilibrium because by additional increase of pressure gradient the stress in near-wall region would exceed τ_0, the fluid takes fluidity property owing to destruction of physical–chemical structure at $\tau > \tau_0$ and the fluid begins to flow. The further rise of pressure gradient already in flow-state would be accompanied by enlargement of the fluid destruction region in the direction to the channel axis. The nondestructive part of the fluid is called flow core.

Integrate equation (5.5.3) with regard to notations in Figs. 5.1 and 5.2

$$\int_{p_1}^{p_2} dp = \int_{z_1}^{z_2} \rho g \, dz + \int_{z_1}^{z_2} \frac{4\tau_{wz}}{d} \, dz.$$

Since $\tau_{wz} = \tau_w / \cos\alpha$ and $dl = dz/\cos\alpha$, it is

$$\int_{p_1}^{p_2} dp = \int_{z_1}^{z_2} \rho g \, dz + \int_{l_1}^{l_2} \frac{4\tau_w}{d} \, dl$$

or

$$p_1 = p_2 + \rho g h + \frac{4\tau_w}{d} l, \tag{5.5.4}$$

where

$$|\tau_w| \leq \tau_0 \quad \text{or} \quad \tau_w = \pm\tau_0. \tag{5.5.5}$$

The sign of the stress τ_w depends on the fluid displacement direction, namely minus is taken when the projection of shear coincides with the direction of gravity force and plus when the shear is opposite in direction.

The substitution of τ_w in the right part of (5.5.4) with regard to its sign gives the following inequality for hydrostatic pressure of fluids having dynamic shear stress τ_0

$$p_2 + \rho gh - \frac{4\tau_0}{d} l \leq p_1 \leq p_2 + \rho gh + \frac{4\tau_0}{d} l. \tag{5.5.6}$$

This inequality is true for vertical and inclined circular pipes. At $\tau_0 \to 0$, it turns into common hydrostatic equation for incompressible fluid (5.2.1)

$$p_1 = p_2 + \rho gh.$$

Now, the hydrostatic pressure depends on the shape of the vessel, since in (5.5.6) pipe diameter and two characteristic lengths enter the height of the channel in vertical h and total length of the channel l. This property should be taken into account when employing such fluids in technique. For example, water-level gauge glass cannot be used to determine the level indication of such fluid in the vessel because in communicating vessels the equilibrium would be set at different levels depending on the side from where the fluid was poured into the vessel. This fact is shown in Fig. 5.4.

If pipes of a different diameter d_i has series connection, the relations (5.5.4) and (5.5.6) take form

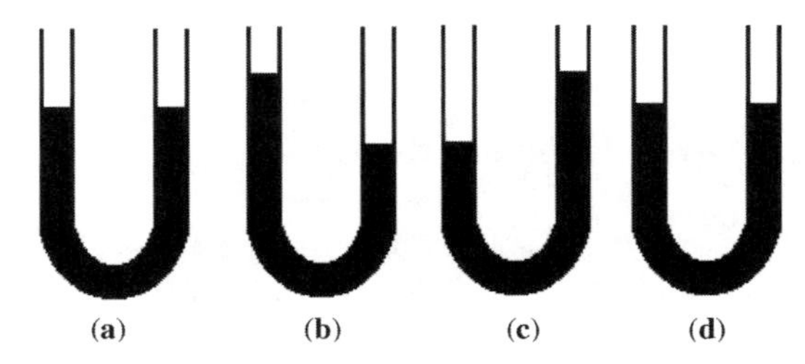

FIGURE 5.4 Hydrostatic equilibrium of fluids in communicating vessels: (a) fluid with $\tau_0 = 0$; (b) the vessel is filled by fluid with $\tau_0 \neq 0$ from left, only under the action of gravity force; (c) the vessel is filled by fluid with $\tau_0 \neq 0$ from right, only under the action of gravity force; (d) at fluid forcing by piston from the right from position (c) up to equal levels in bends (by continuing forcing, the position (b) may be obtained).

$$p_1 = p_2 + \rho g h + \sum \frac{4\tau_w}{d_i} l_i \tag{5.5.7}$$

or

$$p_2 + \rho g h - \sum \frac{4\tau_0}{d_i} l_i \le p_1 \le p_2 + \rho g h + \sum \frac{4\tau_0}{d_i} l_i. \tag{5.5.8}$$

For annular channel, the equilibrium condition develops form

$$p_1 = p_2 + \rho g h + \sum \frac{4\tau_w}{d_{hi}} l_i \tag{5.5.9}$$

or

$$p_2 + \rho g h - \sum \frac{4\tau_0}{d_{hi}} l_i \le p_1 \le p_2 + \rho g h + \sum \frac{4\tau_0}{d_{hi}} l_i, \tag{5.5.10}$$

where d_{hi} is hydraulic diameter of the ith annular space.

The formulas (5.5.7) and (5.5.10) are also true for fluids having static shear stress θ.

For channels with another cross sections, there are their own equations similar to (5.5.7)–(5.5.10). If channels have additional local resistances, the stresses should increase up to dynamic shear stress in order to make the fluid move. It is often difficult to theoretically get the pressure increase Δp_M in local resistances; therefore, they should be determined experimentally. In the considered case, formulas (5.5.19) and (5.5.10) take the form

$$p_1 = p_2 + \rho g h + \sum \frac{4\tau_w}{d_{hi}} l_i + \Delta p_M \tag{5.5.11}$$

or

$$p_2 + \rho g h - \sum \frac{4\tau_0}{d_{hi}} l_i - \Delta p_M \le p_1 \le p_2 + \rho g h + \sum \frac{4\tau_0}{d_{hi}} l_i + \Delta p_M. \tag{5.5.12}$$

EXERCISE 5.5.1

It is required to determine the change of the bottom-hole hydrostatic manometer pressure in the well with regard to Fig. 5.3 when starting pump before beginning of fluid flow. The well is filled by incompressible viscous-plastic fluid. Initial data are density $\rho_0 = 1100\,\text{kg/m}^3$; dynamic shear stress $\tau_0 = 10\,\text{Pa}$; vertical depth of the well $H = 1500$ m; total depth of the well $L = 1900$ m; internal diameter of the well $d_{in} = 0.214$ m; external diameter of drill pipe $d_{ex} = 0.114$ m; outer diameter of drill collar $d_{dc} = 0.178$ m; the length of drill collar 180 m. The mouth in the annular space is open, therefore absolute pressure at the mouth is equal to atmospheric and manometer pressure vanishes.

SOLUTION The bottom-hole pressure in starting of the pump before beginning of fluid flow including annular space is accompanied by increase of stresses in the fluid from zero to dynamic shear stress. The last is realized in drill pipes in the direction of gravity force and in opposite direction in annular space. Thus, since the pressure at the top (at the mouth of annular space) is known, one can use the right part of equation (5.5.7)

$$p_1 = p_{bot} \leq p_2 + \rho g h + \sum \frac{4\tau_0}{d_{hi}} l_i = 0 + 1100 \times 9.81 \times 1500$$

$$+ \frac{4 \times 10}{0.214 - 0.114}(1900 - 180) + \frac{4 \times 10}{0.214 - 0.178} 180 = 17.1 \text{ MPa}.$$

So the pressure in static state begins to rise from the value determined by formula (5.4.3) $p_1 = p_b = p_2 + \rho g h = 0 + \rho g H = 1100 \times 9.81 \times 1500 = 16.2 \times 10^6$ Pa $= 16.2$ MPa. Increment of the pressure through growth of the stress before fluid shear is

$$\Delta p_{bot} = \sum \frac{4\tau_0}{d_{hi}} l_i = \frac{4 \times 10}{0.214 - 0.114}(1900 - 180) + \frac{4 \times 10}{0.214 - 0.178} 180 = 0.888 \text{ MPa}.$$

Thus, the bottom-hole pressure just before the beginning of shear enhances from 16.2 to 17.1 MPa.

EXERCISE 5.5.2

It is required to determine up to what value rises the pressure in ascending pipe as pump starts from the beginning of fluid flow in accordance with Fig. 5.3. The well is filled by viscous-plastic fluid. Initial data are density $\rho_0 = 1100$ kg/m^3; dynamic shear stress $\tau_0 = 10$ Pa; vertical depth of the well $H = 1500$ m; total length of the well $L = 1900$ m; internal diameter of the well $d_{inw} = 0.214$ m; external diameter of drill pipe $d_{exdp} = 0.114$ m; internal diameter of the drill pipe $d_{indp} = 0.100$ m; external diameter of the drill collar $d_{exdc} = 0.178$ m; internal diameter of the drill collar $d_{indc} = 0.09$ m. Just before the shear, the pressure drop in the drilling bit Δp_{bit} rises up to 0.8 MPa and in joints Δp_{joint} up to 1 MPa. The mouth in the annular space is open, therefore absolute pressure at the mouth in the annular space is equal to the atmospheric one and manometer pressure is equal to zero.

SOLUTION Pressure in the pump p_{pump} from its start up to fluid flow beginning elevates in accordance with stress rise in fluid along the whole well from zero up to the value of dynamic shear stress. The shear is performing along gravity force in drill pipes and in opposite direction in annular space. The bottom-hole pressure was calculated in Exercise 5.5.1 with use of the right part of the equation (5.5.8) or of the right part of the equation (5.5.12) for annular space in which local resistances were supposed to be absent or negligible small. The bottom-hole pressure could be also calculated by use of the left part of the equation (5.5.12) (along pipe space) in which there are local resistances (bit and joints) and the pressure p_1 is equal to the pressure in the pump p_{pump}. We have

$$p_{\text{pump}} + \rho g h - \sum \frac{4\tau_0}{d_i} l_i - \Delta p_{\text{M2}} \leq p_{\text{bot}} \leq p_{\text{as}} + \rho g h + \sum \frac{4\tau_0}{d_i} l_i + 0$$

or omitting p_{bot}

$$p_{\text{pump}} + \rho g h - \sum \frac{4\tau_0}{d_i} l_i - \Delta p_{\text{M2}} \leq p_{\text{as}} + \rho g h + \sum \frac{4\tau_0}{d_i} l_i.$$

After some algebra, we obtain the rise of the pressure in the pump p_{pump} to the beginning of fluid flow

$$p_{\text{pump}} \leq 0 + \sum \frac{4\tau_0}{d_i} l_i + \sum \frac{4\tau_0}{d_{\text{h}i}} l_i + \Delta p_{\text{bit}} + \Delta p_{\text{joint}}. \tag{5.5.13}$$

Substitution of numerical values gives pressure change in the pump at its start

$$\begin{aligned} p_{\text{pump}} &\leq 0 + \sum \frac{4\tau_0}{d_i} l_i + \sum \frac{4\tau_0}{d_{\text{h}i}} l_i + \Delta p_{\text{bit}} + \Delta p_{\text{joint}} \\ &= 0 + \frac{4 \times 10}{0.1}(1900-180) + \frac{4 \times 10}{0.09} 180 + 1 \times 10^6 + 0.8 \times 10^6 \\ &\quad + \frac{4 \times 10}{0.214-0.114}(1900-180) + \frac{4 \times 10}{0.214-0.178} 180 \\ &= 3.46 \times 10^6 \text{ Pa} = 3.46 \text{ MPa} \end{aligned}$$

Consequently, the pressure in the pump just before the shear of flushing fluid having dynamic shear stress growth from 0 up to 3.46 MPa.

5.6 HYDROSTATICS OF TWO-PHASE FLUIDS

To solve many technological problems in oil–gas-field industry, it is important to calculate correctly the hydrostatic pressure of media consisting of two phases differing in density and compressibility. For example, gas–liquid aerated mixture contains strongly compressible light gas and lightly compressible fluid of significantly greater density; water–oil emulsion consists of well-compressible light oil and heavier worse compressible water; bed is commonly represented by compressible light fluid and less compressible heavy skeleton. In drilling with foam washing the washing fluid has dynamic shear stress. Let us get formulas to enable to calculate hydrostatic pressure of two-phase media with different compressibilities of phases and availability of dynamic shear stress.

Hydrostatic equilibrium of two-phase fluid is in general case described by equation (5.5.3)

$$\frac{\mathrm{d}p}{\mathrm{d}z} = \rho g + \frac{4\tau_{wz}}{d}, \tag{5.6.1}$$

where density ρ is given by relation (3.6) at $N=2$

$$\rho = \rho_1\varphi + \rho_2(1-\varphi) \tag{5.6.2}$$

and φ is volume concentration of the first phase.

Substitution of (5.6.2) in (5.6.1) leads to

$$\frac{\mathrm{d}p}{\mathrm{d}z} = (\rho_1\varphi + \rho_2(1-\varphi))g + \frac{4\tau_{wz}}{d}. \tag{5.6.3}$$

The density may be expressed through mass content of the first phase χ or aeration mass factor η as follows

$$\rho_m = \frac{\rho_1\rho_2}{\rho_2\chi + \rho_1(1-\chi)} \quad \text{or} \quad \rho_m = \frac{(\eta+1)\rho_1\rho_2}{\rho_2\eta + \rho_1}. \tag{5.6.4}$$

Mass concentration χ is related to aeration mass factor η through the formula $\chi = \frac{\eta}{1+\eta}$. The aeration mass factor η is in its turn connected with aeration flow factor a by formula $\eta = a\rho_0/\rho_2$, where factor $a = Q_0/Q_2$. Here, ρ_0 and Q_0 are gas density and compressor delivery at atmospheric conditions; ρ_2 and Q_2 are density of washing fluid and pump delivery, respectively. The quantities χ and η in the absence of phase solubility as distinct from a and φ do not depend on pressure and thus they are convenient to use. They are also lightly interconverted. It should be noted that it is considered hydrostatics of two-phase fluids, though in relations for densities enter dynamic variables. But at hydrostatic conditions at the moment of startup or slowdown of the pump, when stable foam is absent and (or) has not yet appeared mechanisms of sedimentation or floating-up, such estimations would be true. Relations (5.6.2) and (5.6.4) could be also interconverted.

To account for solubility or evaporation of phases, it may be taken that solubility of one phase (gas) in another one (fluid) or evaporation of fluid phase in gas phase is described in the first approximation by linear dependence on pressure, that is an amount of matter transferred from one phase into another one is proportional to the pressure $\chi = k_{\text{sol}} p$. Then the density of mixture can be written as

$$\rho_{\text{m}} = \frac{\rho_1 \rho_2}{\rho_2(\chi - k_{\text{sol}} p) + \rho_1(1 - (\chi - k_{\text{sol}} p))}. \tag{5.6.5}$$

With regard to this formula, the equation (5.6.1) takes form

$$\frac{\mathrm{d}p}{\mathrm{d}z} = \frac{\rho_1 \rho_2}{\rho_2(\chi - k_{\text{sol}} p) + \rho_1(1 - (\chi - k_{\text{sol}} p))} g + \frac{4\tau_{\text{wz}}}{d}. \tag{5.6.6}$$

In order to integrate equation (5.6.6) at isothermal conditions and by doing so to get equation of two-phase hydrostatic, one should specify thermodynamic equations of state for each phase, that is dependences of densities on pressure. The phases could be incompressible, compressible, slightly compressible, and possessing shear stress.

Use dependence (4.3.7)

$$\rho_i = a_i + b_i \cdot p \tag{5.6.7}$$

and substitute it in (5.6.6) from which by method of separation of variables, the following equation is obtained

$$\mathrm{d}z = \frac{1}{g} \left\{ \frac{Ap^2 + Bp + C}{A_1 p^2 + B_1 p + C_1} \right\} \mathrm{d}p, \tag{5.6.8}$$

where

$$A = k_{\text{sol}}(b_1 - b_2); \qquad B = b_2 \chi + b_1(1 - \chi) + k_{\text{sol}}(a_1 - a_2);$$
$$C = a_2 \chi + a_1(1 - \chi); \qquad A_1 = b_1 b_2 + D k_{\text{sol}}(b_1 - b_2);$$
$$B_1 = a_1 b_2 + a_2 b_1 + DB; \quad C_1 = a_1 a_2 + DC; \qquad D = 4\tau_0/(\mathrm{d}g). \tag{5.6.9}$$

The form of the solution of equation (5.6.8) depends on the sign of denominator discriminant Δ

$$\begin{aligned} \Delta = 4A_1 C_1 - B_1^2 &= 4[b_1 b_2 + D\, k_{\text{sol}}(b_1 - b_2)](a_1 a_2 + DC) \\ &\quad - [a_2 b_1 + a_1 b_2 + DB]^2. \end{aligned} \tag{5.6.10}$$

Let us take boundary condition as

$$p = p_0 \quad \text{at} \quad z = z_0. \tag{5.6.11}$$

Then, the general solution of the equation (5.6.8) takes form for $\Delta < 0$

$$z - z_0 = f_1(p) - f_1(p_0) + f_2(p) - f_2(p_0),$$

where

$$f_1(p) = \frac{A}{gA_1}\left\{p + 0.5\left(\frac{B}{A} - \frac{B_1}{A_1}\right)\ln(A_1p^2 + B_1p + C_1)\right\}; \tag{5.6.12}$$

$$f_2(p) = \frac{A}{gA_1}\left(0.5B_1\left(\frac{B_1}{A_1} - \frac{B}{A}\right) - A_1\left(\frac{C_1}{A_1} - \frac{C}{A}\right)\right) \times \frac{1}{\sqrt{-\Delta}}\ln\frac{(2A_1p + B_1 - \sqrt{-\Delta})}{(2A_1p + B_1 + \sqrt{-\Delta})};$$

for $\Delta > 0$

$$z - z_0 = f_3(p) - f_3(p_0) + f_4(p) - f_4(p_0),$$

where

$$f_3(p) = f_1(p) = \frac{A}{gA_1}\left\{p + 0.5\left(\frac{B}{A} - \frac{B_1}{A_1}\right)\ln(A_1p^2 + B_1p + C_1)\right\}; \tag{5.6.13}$$

$$f_4(p) = \frac{A}{g\sqrt{\Delta}A_1}\left(B_1\left(\frac{B_1}{A_1} - \frac{B}{A}\right) - 2A_1\left(\frac{C_1}{A_1} - \frac{C}{A}\right)\right)\arctan\frac{(2A_1p + B_1)}{\sqrt{\Delta}}.$$

If to consider the hydrostatic of a mixture of gas and incompressible fluid at $k_{sol} = 0$, or a mixture of gas and slurry at the instant of deadlock, then, $a_1 = b_2 = 0$ and the discriminant would be

$$\Delta = -[a_2b_1 + Db_1(1-\chi)]^2 < 0. \tag{5.6.14}$$

In this case (5.6.8) gives

$$z-z_0=\frac{1}{g}\left\{\frac{B(p-p_0)}{B_1}+\frac{CB_1-BC_1}{B_1^2}\ln\frac{B_1p+C_1}{B_1p_0+C_1}\right\}. \qquad (5.6.15)$$

If both phases are incompressible media, for example, a mixture of flushing fluid and a slurry or two incompressible fluids, then $b_1=b_2=0$, the discriminant would be equal to zero ($\Delta=0$), and the solution is

$$z-z_0=\frac{C}{gC_1}(p-p_0). \qquad (5.6.16)$$

The case $a_1=a_2=0$ represents a mixture of two gases, that is a single-phase medium, since gases do not form interface. But even in this case the discriminant vanishes and

$$z-z_0=\frac{B}{gA_1}\ln\left(\frac{p+B_1/A_1}{p_0+B_1/A_1}\right). \qquad (5.6.17)$$

Thus, equations (5.6.12) and (5.6.13) describe the hydrostatics of two-phase mixtures of Newtonian and non-Newtonian fluids including fluids possessing dynamic shear stress. From (5.6.12) and (5.6.13) as limiting cases ensue above considered equations of hydrostatics of Newtonian and non-Newtonian fluids. From (5.6.16) it follows

$$p-p_0=g\frac{C_1}{C}(z-z_0) \qquad (5.6.18)$$

or

$$p-p_0=g\frac{\rho_1\rho_2}{\rho_2\chi+\rho_1(1-\chi)}(z-z_0)+\frac{4\tau_0(z-z_0)}{d\cos\alpha}. \qquad (5.6.19)$$

Relation (5.2.2) gives

$$p=(p_0+B_1/A_1)(\exp(z-z_0)gA_1/B)-B_1/A_1 \qquad (5.6.20)$$

or

$$p=\left(p_0+D\frac{b_2\chi+b_1(1-\chi)}{b_1b_2}\right)e^{\frac{(z-z_0)gb_1b_2)}{b_2\chi+b_1(1-\chi)}}-\frac{b_2\chi+b_1(1-\chi)}{b_1b_2}. \quad (5.6.21)$$

For gases in enough large channels, one can believe that $D=0$. That is, gases in great volumes do not have dynamic shear stress, and therefore, they behave themselves as unified gas in accordance with Dalton law. That is why the equation of state of gas mixture (5.3.1) having been transformed to (5.6.7) is represented as follows

$$\rho=\frac{p}{\bar{z}RT}=bp=\frac{b_1b_2}{b_2\chi+b_1(1-\chi)}p \quad (5.6.22)$$

or

$$\frac{b_1b_2}{b_2\chi+b_1(1-\chi)}=b=\frac{1}{\bar{z}RT} \quad (5.6.23)$$

From (5.6.21), the main hydrostatic law of gas in the gravity field, the so-called barometric height formula follows

$$p_1=p_2\exp(h/zRT), \quad (5.6.24)$$

where p_1 and p_2 are pressures at the lower and upper plates, respectively and $h=z-z_0$ is the distance between two horizontal plates under consideration.

Table 5.1 represents classification of two-phase mixture together with limiting cases depending on the form of thermodynamic equations of state of considered phases that are characterized by, wherever possible, zeroth or nonzeroth coefficients a_i and b_i $(i=1, 2)$.

In the Table 5.1 each number No1 in the second column represents one case of phase state. Numbering in the first column No2 reflects all possible variants of combinations of coefficients a_i and b_i occurring in practice.

In Table 5.2, there are given basic equations of hydrostatics most demanded for calculations in drilling and well exploitation.

Equation (5.6.12) as well as in limiting cases (see in Table 5.2 rows 4 and 5) are implicit algebraic equations with respect to pressure p. To get pressure at given depth p, the indicated equations are solved numerically

TABLE 5.1

No1	No2	a_1	b_1	a_2	b_2	Type of Medium
1	1	0	0	0	0	Vacuum
2	2	0	0	$\neq 0$	0	Single-phase incompressible fluid or bed
	3	$\neq 0$	0	0	0	
3	4	0	0	0	$\neq 0$	Single-phase strongly compressible fluid or gas
	5	0	$\neq 0$	0	0	
4	6	$\neq 0$	$\neq 0$	0	0	Single-phase slightly compressible fluid
	7	0	0	$\neq 0$	$\neq 0$	
5	8	$\neq 0$	0	0	$\neq 0$	Gas–liquid mixture, aerated fluid or gas with slurry
	9	0	$\neq 0$	$\neq 0$	0	
6	10	0	$\neq 0$	0	$\neq 0$	Single-phase mixture of gases
7	11	$\neq 0$	0	$\neq 0$	0	Two-phase mixture of two incompressible fluids or mixture of incompressible fluid with slurry
8	12	$\neq 0$	$\neq 0$	$\neq 0$	0	Two-phase mixture of immiscible incompressible and slightly compressible fluids
	13	$\neq 0$	0	$\neq 0$	$\neq 0$	
9	14	0	$\neq 0$	$\neq 0$	$\neq 0$	Two-phase mixture of gas and slightly compressible fluid
	15	$\neq 0$	0	$\neq 0$	$\neq 0$	
10	16	$\neq 0$	$\neq 0$	$\neq 0$	$\neq 0$	Mixture of two immiscible slightly compressible fluids

or graphically using these equations for fluid under consideration. The graphics for several cases are plotted below.

EXERCISE 5.6.1

In Table 5.3, initial data for two two-phase mixtures and single-phase incompressible fluid (water) are presented at $D=0$ ($\tau_0=0$). The first is gas–liquid mixture, one phase of which is gas and the other slightly compressible fluid with its own coefficients a_i and b_i corresponding the formula (5.6.7). The second mixture consists of slightly compressible phases with different coefficients a_i and b_i.

TABLE 5.2

No.	Basic Equation	Properties of Fluids	Type of Fluid
1	$p_1 = p_2 + \rho g h$ (5.2.1)	$\rho = \text{const}, \tau_0 = 0$	Incompressible fluid
2	$p_{1M} = p_{atm}(e^{h/\bar{z}RT} - 1)$ (5.3.4)	$\rho \neq \text{const}, \tau_0 = 0, p = \rho_1 g \bar{z} RT$	Strongly compressible fluid, gas
3	$p_2 + \rho g h - \frac{4\tau_0}{d} l \leq p_1 \leq p_2 + \rho g h + \frac{4\tau_0}{d} l$ (5.5.8)	$\rho = \text{const}, \tau_0 \neq 0$	Fluid possessing of dynamic shear stress, for example, viscous-plastic fluid
4	$(1 + \eta)\rho_2 g h = \eta \ln(p_1/p_2) + p_1 - p_2$. The limit of equation (5.6.12) for case 5 from Table 5.1 at $D = 0$	$\rho_2 = \text{const}, \tau_0 = 0, p = \rho_1 g \bar{z} RT$	Gas–liquid mixture, for example, aerated fluid

TABLE 5.3

Number of Mixture	First Phase a_1, (kg/m^3)	First Phase b_1, (kg/(m^3 Pa))	First Phase χ	Second Phase a_2, (kg/m^3)	Second Phase b_2, (kg/(m^3 Pa))	Pressure p (MPa) at the depth $z = 1000$ m
1	0	1.28×10^{-5}	1.024×10^{-4}	1703	8.515×10^{-7}	17.33
2	334	3.34×10^{-7}	0.0301	1682	8.415×10^{-7}	14.88
3	1000	0	1	0	0	10.00

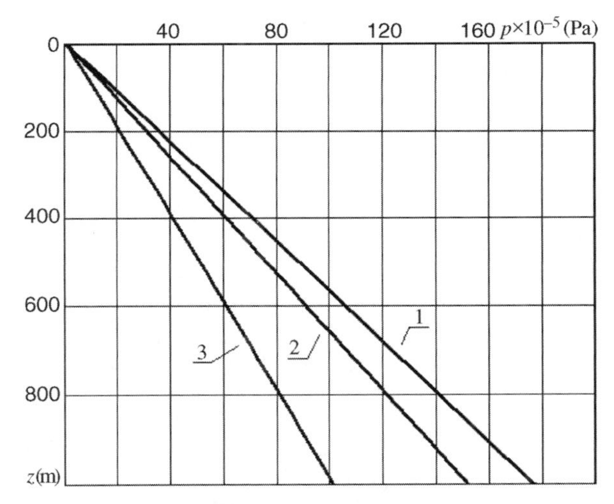

FIGURE 5.5 Distribution of pressure over well depth.

In Table 5.3, the values of a_i, b_i, and χ are chosen so that both mixtures at atmospheric pressure $p_0 = p_{at} = p_{10} = p_{20} = 9.81 \times 10^5$ Pa would have the same initial density $\rho_{0m} = 1500$ kg/m^3.

SOLUTION In Fig. 5.5 are plotted curves of pressure $p = p(z)$ in the depth of the well at data taken from the table with the help of mathematical package MathCad2000. When plotting curves, in equation (5.6.12) were given pressures p and got depth h or in other notation z. Despite insignificant density of the first phase because of its great compressibility, curve 1 is located above the curve 2. At one of the same depth ($z = 1000$ m), the hydrostatic pressure of the second mixture is less on 2.45 MPa than that of the pressure of the first one.

Hence, with formulas given above, it is possible to calculate distribution of hydrostatic pressure of two-phase mixtures having different compressibility and concentration of phases. It was shown that at equal initial densities of different two-phase media, mixtures of incompressible and slightly compressible phases could produce lesser hydrostatic pressure as compared to gas–liquid mixture through specially chosen concentrations and factors for thermodynamic equations of state.

CHAPTER 6

STATIONARY FLOW OF FLUIDS IN ELEMENTS OF THE WELL CIRCULATION SYSTEM

6.1 EQUATIONS FOR STATIONARY FLOWS OF HOMOGENEOUS INCOMPRESSIBLE FLUIDS

Consider stationary flows, that is, flows whose parameters do not depend on time. Equations for flows of homogeneous incompressible fluids result from general equations for multiphase media (4.6.1)–(4.6.5), (4.6.6)–(4.6.11), and (4.6.12)–(4.6.16) if in them to accept $N = 1$. The pressure is a function only of z for flows in circular and annular pipes and only of r for flows in circular slots and annular channels with rotation of walls.

6.1.1 Flows in Pipes and Annular Channels ($w_z = w \neq 0$)

Such flows are described by

momentum (motion) equation

$$\rho w \frac{\partial w}{\partial z} + \frac{\partial p}{\partial z} = \rho g + \frac{1}{r}\frac{\partial r\tau}{\partial r}; \tag{6.1.1}$$

Applied Hydro-Aeromechanics in Oil and Gas Drilling. By Leonov and Isaev

equation of mass conservation

$$\frac{\partial w}{\partial z} = 0; \tag{6.1.2}$$

equation of state

$$\rho = \text{const}; \tag{6.1.3}$$

rheological equation of state

$$\tau = \tau(\dot{\gamma}), \tag{6.1.4}$$

where $\dot{\gamma} = \partial w / \partial r$.

The equation (6.1.4) is determined by the type of fluid used.

6.1.2 Flows in Rotation of Pipes and Annulus Walls ($w_\varphi = w \neq 0$)

The governing equations are

momentum equations

$$\frac{\partial p}{\partial r} = \rho \frac{w^2}{r}, \tag{6.1.5}$$

$$\frac{\partial \tau}{\partial r} + 2\frac{\tau}{r} = 0; \tag{6.1.6}$$

equation of mass conservation

$$\frac{\partial w}{\partial \varphi} = 0; \tag{6.1.7}$$

equation of state

$$\rho = \text{const}; \tag{6.1.8}$$

rheological equation of state

$$\tau = \tau(\dot{\gamma}), \tag{6.1.9}$$

where according to (4.4.8)

$$\dot{\gamma} = \frac{\partial w}{\partial r} - \frac{w}{r}.$$

6.1.3 Radial Flow in a Circular Slot ($w_r = w \neq 0$)

Basic equations are

momentum equation

$$\rho w \frac{\partial w}{\partial r} + \frac{\partial p}{\partial r} = \frac{\partial \tau}{\partial z}; \qquad (6.1.10)$$

equation of mass conservation

$$\frac{1}{r}\frac{\partial rw}{\partial r} = 0; \qquad (6.1.11)$$

equation of state

$$\rho = \text{const}; \qquad (6.1.12)$$

rheological equation of state

$$\tau = \tau(\dot{\gamma}), \qquad (6.1.13)$$

where $\dot{\gamma} = \partial w / \partial r$.

6.2 CALCULATION OF PRESSURE IN LAMINAR FLOWS OF VISCOUS INCOMPRESSIBLE FLUIDS IN CIRCULAR SLOTS, PIPES, AND ANNULAR CHANNELS

6.2.1 Flow in a Circular Slot

Get a formula relating pressure drop $\Delta p = |p_{\text{in}} - p_{\text{ex}}|$ with flow rate $Q = v \cdot S$, where p_{in} and p_{ex} are pressures in the orifice and on the slot contour; v is the mean fluid velocity through any cylindrical surface $S = 2\pi r H$ (Fig. 6.1). To do this it is required to solve the system of equations (6.1.10)–(6.1.13) with concrete form of relation (6.1.13).

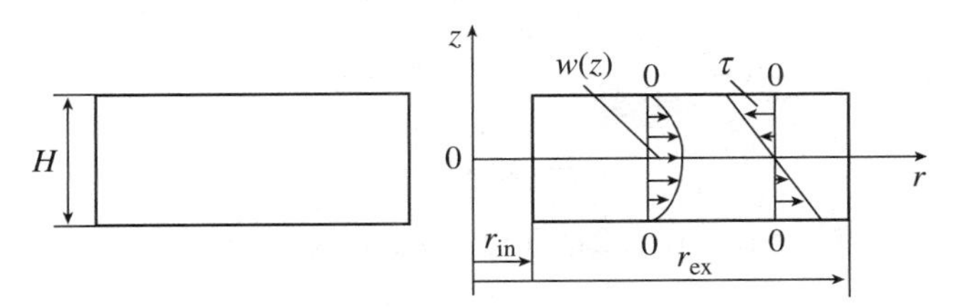

FIGURE 6.1 Distribution of velocity and stress in laminar radial flow of viscous fluid in a circular slot.

Take the following boundary conditions

$$w = 0 \quad \text{at} \quad z = \pm H/2; \tag{6.2.1}$$

$$\begin{aligned} p &= p_{\text{in}} \quad \text{at} \quad r = r_{\text{in}}; \\ p &= p_{\text{ex}} \quad \text{at} \quad r = r_{\text{ex}}. \end{aligned} \tag{6.2.2}$$

The rheological equation (6.1.13) for viscous fluid is

$$\tau = \mu \frac{\partial w}{\partial z}. \tag{6.2.3}$$

From equation of mass conservation (6.1.11) at $\partial v/\partial \varphi = 0$, it follows that the product $r{\cdot}w$ is a function only of z

$$r \cdot w = f_1(z). \tag{6.2.4}$$

Substituting the velocity from (6.2.4) into (6.2.3), one obtains that the product $r{\cdot}\tau$ is also a function only of z

$$r \cdot \tau = f(z). \tag{6.2.5}$$

Find at first the solution of the above formulated system of equations when the product $w(\partial w/\partial r)$ in (6.1.10) is small compared to $\partial p/\partial r$. Such cases could be in slow flows or flows of high-viscosity fluid. Then, after substitution of (6.2.5) in (6.1.10) the following is obtained

$$r \frac{\partial p}{\partial r} = \frac{\partial f(z)}{\partial z}. \tag{6.2.6}$$

The left side of (6.2.6) depends only on r whereas the right part on z. Therefore, the equality (6.2.6) could be true only when both sides of the equation are equal to a constant A to be determined

$$r \frac{\partial p}{\partial r} = \text{const} = A. \tag{6.2.7}$$

Integration of (6.2.7) gives

$$p = A \ln r + B. \tag{6.2.8}$$

Boundary conditions (6.2.2) yield constants A and B

$$A = \frac{p_{\text{in}} - p_{\text{ex}}}{-\ln(r_{\text{ex}}/r_{\text{in}})}, \qquad B = p_{\text{in}} - A \ln r_{\text{in}}. \tag{6.2.9}$$

Then

$$\frac{p - p_{\text{in}}}{p_{\text{ex}} - p_{\text{in}}} = \frac{\ln(r/r_{\text{in}})}{\ln(r_{\text{ex}}/r_{\text{in}})}. \tag{6.2.10}$$

The relation (6.2.10) gives pressure distribution in slow flow of viscous incompressible fluid between the circular plates.

Turn now to the right part of (6.2.6)

$$\frac{\partial f(z)}{\partial z} = A = \text{const} \tag{6.2.11}$$

or after integration

$$f(z) = Az + B_1. \tag{6.2.12}$$

Note that the flow symmetry leads to condition $\partial w/\partial z = 0$ at $z = 0$, and consequently from (6.2.3) ensues that the stress $\tau = 0$. From (6.2.5) also follows that $f(z) = 0$ at $z = 0$. Then (6.2.12) gives $B_1 = 0$.

Insertion of $f(z)$ from (6.2.12) and τ from (6.2.3) into (6.2.5) and integration of the result with regard to conditions (6.2.1) gives velocity distribution

$$w = \frac{A}{2r\mu}\left[z^2 - \left(\frac{H}{2}\right)^2\right]. \tag{6.2.13}$$

Determine the flow rate Q through a cylindrical surface with height H and radius r

$$Q = 4\pi r \int_0^{H/2} w\,\mathrm{d}z = -\frac{\pi A}{6\mu}H^3. \tag{6.2.14}$$

Mean velocity through different cylindrical surfaces is

$$v = \frac{Q}{2\pi r H} = -\frac{A}{12\mu r}H^2. \tag{6.2.15}$$

Substituting here A from (6.2.9) and solving resulting equation relative $(p = |p_{\text{in}} - p|$, we find the sought formula (Pihachev and Isaev, 1973)

$$\Delta p = \frac{6\mu|Q|}{\pi H^3}\ln\frac{r}{r_{\text{in}}}. \tag{6.2.16}$$

This formula is true at $Q > 0$, $A < 0$, as well as at $Q < 0$, $A > 0$.

The last equation could be transformed into Darcy–Weisbach formula

$$\Delta p_1 = \lambda_1 \frac{\rho v^2}{2H}(r - r_{in}). \tag{6.2.17}$$

At $r = r_{ex}$, there is

$$\lambda_1 = \frac{24}{Re} f(\delta), \quad Re = \frac{|v| H \rho}{\mu}, \quad f(\delta) = \frac{\ln \delta}{1-\delta}, \quad \delta = \frac{r_{in}}{r_{ex}}.$$

When deriving the dependence (6.2.17), the inertial term $w(\partial w/\partial r)$ in momentum equation was not taken into account; that is, it was considered pressure drop caused by the action only of viscous forces. Values of Δp calculated in fluid flow between circular plates with formula (6.2.17) ignoring inertia effect could be strongly understated or overstated by a relative total pressure drop.

Consider now the flow in which the chief role plays inertial term. Then the momentum equation (4.6.40) at $i = 1$, $\varphi = 1$, and $\tau_w = 0$ will be

$$v \frac{\partial v}{\partial r} = -\frac{1}{\rho} \frac{\partial p}{\partial r}.$$

Integration of this equation yields

$$\frac{v^2}{2} + \frac{p}{\rho} = \text{const}.$$

Replace the velocity v by its value from (6.2.15)

$$\frac{1}{2} \cdot \frac{Q^2}{4\pi^2 r^2 H^2} + \frac{p}{\rho} = \text{const}.$$

Determine the constant through accounting the second condition (6.2.2). Then

$$\text{const} = \frac{1}{2} \frac{Q^2}{4\pi^2 r_{in}^2 H^2} + \frac{p_{in}}{\rho} = \frac{1}{2} \frac{Q^2}{4\pi^2 r^2 H^2} + \frac{p}{\rho}.$$

From here it follows

$$\Delta p_2 = |p_{in} - p| = \frac{\rho Q^2}{8\pi^2 H^2} \left(\frac{1}{r_{in}^2} - \frac{1}{r^2} \right). \tag{6.2.18}$$

The pressure drop due to inertia (6.2.18) does not depend on the type of fluid rheological law (6.1.13) and is applicable to any incompressible fluid.

The dependence (6.2.18) could be also transformed into Darcy–Weisbach formula

$$\Delta p_2 = \lambda_2 \frac{\rho v^2}{2H}(r - r_{\text{in}}),$$

where

$$\lambda_2 = \frac{H}{r_{\text{in}}}\left(1 + \frac{r}{r_{\text{in}}}\right).$$

From here it follows that the resistance factor when taking into account only inertia force is independent of the fluid velocity.

The total pressure drop is

$$\Delta p = \lambda \frac{\rho v^2}{2H}(r - r_{\text{in}}) = |\Delta p_1 \pm \Delta p_2|,$$

where $\lambda = |\lambda_1 \pm \lambda_2|$.

Note that λ depends significantly on the flow direction, that is, on the sign of Q. At outflow of the fluid one should take minus and at inflow plus should be taken.

EXERCISE 6.2.1

It is required to determine pressure drop in fluid inflow from circular slot under condition that pressure drops caused by friction and inertia forces could be summarized. Given data are $Q = -0.0116\,\text{m}^3/\text{s}$, $\rho = 10^3\,\text{kg/m}^3$, $\mu = 0.01$ Pa s, $r = 100$ m, $r_{\text{in}} = 0.1$ m, and $H = 0.01$ m.

SOLUTION In accordance with formula (6.2.16)

$$\Delta p_1 = \frac{6\mu|Q|}{\pi H^3}\ln\frac{r}{r_{\text{in}}} = \frac{6 \times 0.01 \times 0.0116}{3.14 \times (0.01)^3}\ln\frac{100}{0.1} = 1.53 \times 10^3\ \text{Pa}.$$

and (6.2.18)

$$\Delta p_2 = \frac{\rho Q^2}{8\pi^2 H^2}\left(\frac{1}{r_{\text{in}}^2} - \frac{1}{r^2}\right) = \frac{10^3 \times (0.0116)^2}{8(3.14)^2(0.01)^2}\left(\frac{1}{0.1^2} - \frac{1}{100^2}\right) = 1.7 \times 10^3\ \text{Pa}.$$

The total pressure drop is

$$\Delta p = \Delta p_1 + \Delta p_2 = 1.53 \times 10^3 + 1.7 \times 10^3\ \text{Pa} = 2.23 \times 10^3\ \text{Pa}.$$

Calculation shows that at given conditions the friction and the inertia forces act much the same on pressure drop. If to elevate the viscosity up to

1 Pa s, the influence of friction force markedly enhances while the effect of inertia force does not change. As the distance between plates increases, inertia forces would play the chief role, whereas as the well diameter increases friction forces would do so.

6.2.2 Flows in Pipes and Annular Channels

Determine relation between pressure drop $\Delta p = |p_2 - p_1|$ and flow rate $Q = v \cdot S$, where v is mean fluid velocity through channel cross section, $S = \pi R^2 = \pi d_p^2/4$ is the area of pipe cross section, and $S = \pi(d_{ex}^2 - d_{in}^2)/4$ is the area of annular space cross section.

To determine required function $\Delta p = \Delta p(Q)$, it is necessary to solve the system of equations (6.1.1)–(6.1.4) at the following boundary conditions

$$\begin{aligned} w &= 0 \quad \text{at} \quad r = R_1 = d_{in}/2, \\ w &= 0 \quad \text{at} \quad r = R_2 = d_{ex}/2; \end{aligned} \tag{6.2.19}$$

$$\begin{aligned} p &= p_1 \quad \text{at} \quad z = 0, \\ p &= p_2 \quad \text{at} \quad z = L. \end{aligned} \tag{6.2.20}$$

The rheological equation (6.1.4) in considered case is

$$\tau = \mu \frac{\partial w}{\partial r}. \tag{6.2.21}$$

The equation (6.1.1) with regard to (6.1.2) has form

$$\frac{\partial p}{\partial z} = \rho g + \frac{1}{r} \frac{\partial r\tau}{\partial r}. \tag{6.2.22}$$

From equation (6.1.2), it follows that in axially symmetric and stationary flow, the velocity w is a function only of radial coordinate r, and thus from (6.2.21) ensues that τ is also function only of r. So the right part of (6.2.22) depends only on r, whereas the left part only on z, since the pressure p is a function only of z. Consequently, both sides of equation (6.2.22) have to be equal to a constant A. In order to estimate $\Delta p(Q)$ caused only by friction force, let us omit gravity term ρg in (6.2.22). As a result the following is obtained

$$\frac{\partial p}{\partial z} = A = \text{const} \quad \text{and} \quad \frac{1}{r} \frac{\partial r\tau}{\partial r} = A. \tag{6.2.23}$$

Integration of the first equation (6.2.23) with regard to (6.2.20) gives

$$p = Az + B, \tag{6.2.24}$$

where

$$A = \frac{p_2 - p_1}{L}, \qquad B = p_1.$$

Substitution of (6.2.21) into the second equation (6.2.23) and further integration with regard to (6.2.19) yields velocity distribution in the annular channel

$$w = \frac{A}{4\mu}\left[r^2 - \left(\frac{d_{\rm ex}}{2}\right)^2 + \frac{d_{\rm ex}^2 - d_{\rm in}^2}{4}\frac{\ln(2r/d_{\rm ex})}{\ln(d_{\rm in}/d_{\rm ex})}\right]. \tag{6.2.25}$$

In the limiting case $d_{\rm in} \to 0$ from (6.2.25), velocity distribution in the pipe with diameter $d_{\rm p}$ is obtained (Fig. 6.2)

$$w = \frac{A}{4\mu}\left(r^2 - \frac{d_{\rm p}^2}{4}\right). \tag{6.2.26}$$

The flow rate through channel cross section could be obtained using (6.2.25)

$$Q = \int_0^{2\pi}\int_{d_i/2}^{d_{\rm p}/2} wr\,{\rm d}r\,{\rm d}\varphi = -\frac{\pi A}{128\mu}\left[(d_{\rm ex}^4 - d_{\rm in}^4) + \frac{(d_{\rm ex}^2 - d_{\rm in}^2)^2}{\ln(d_{\rm in}/d_{\rm ex})}\right]. \tag{6.2.27}$$

The dependence (6.2.27) is called Boussinesq formula. At $d_{\rm in} \to 0$, one gets the flow rate in a pipe called Hagen–Poiseuille formula

$$Q = -\frac{\pi A}{128\mu}d_{\rm p}^4. \tag{6.2.28}$$

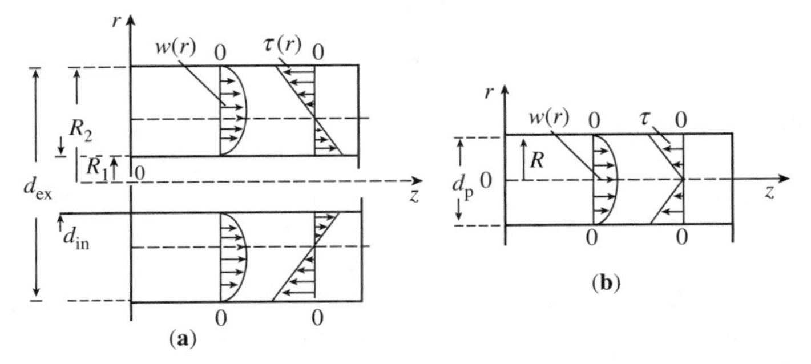

FIGURE 6.2 Distribution of velocity and stress in laminar flow of viscous fluid in the annular channel (a) and in the pipe (b).

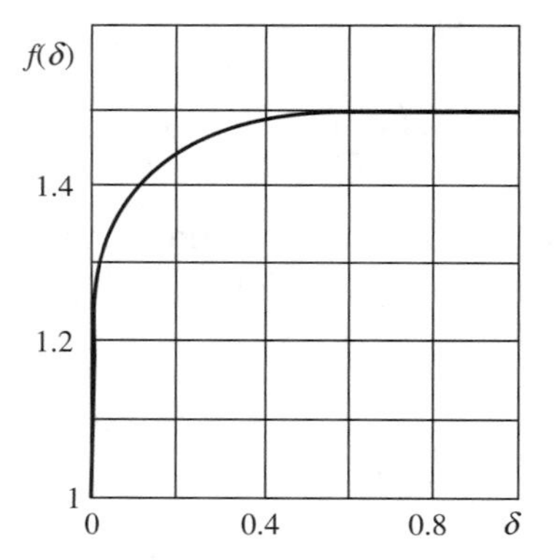

FIGURE 6.3 Graphic of $f(\delta)$.

The formula (6.2.27) can be transformed into Darcy–Weisbach formula

$$\Delta p = |p_2 - p_1| = \lambda \frac{\rho v^2}{2d_h} L, \tag{6.2.29}$$

where $v = Q/S$, $d_h = d_{ex} - d_{in}$ is the hydraulic diameter, $\lambda = (64/Re)f(\delta)$ is the hydraulic resistance factor, $\delta = d_{in}/d_{ex}$, $Re = |v|d_h\rho/\mu$ is the Reynolds number, and $f(\delta) = (1-\delta)^2/(1+\delta^2+(1-\delta^2/\ln\delta))$.

The function $f(\delta)$ is plotted in Fig. 6.3.

At $d_{in} \to 0$, there are $d_h \to d_c$ and $f(\delta) \to 1$, that is the formula (6.2.29) can be used to calculate pressure drop in pipes by taking in it $d_h = d_c$. For the sake of convenience, the formula (2.6.29) may be written by untangling λ and Re as follows:

for annular channels

$$\Delta p = \frac{32\mu|v|}{d_h^2} f(\delta)L = \frac{128\mu|Q|}{\pi d_h^3(d_{ex}+d_{in})} f(\delta)L; \tag{6.2.30}$$

for pipes

$$\Delta p = \frac{128\mu|Q|}{\pi d_p^4} L. \tag{6.2.31}$$

Formulas (6.2.30) and (6.2.31) are true for flows along z-axis ($Q > 0$) as well as for flows in opposite direction, that is, opposite the direction of z-axis ($Q < 0$).

The total pressure drop is a sum of (6.2.31) and $\rho g L$.

EXERCISE 6.2.2

It is required to get pressure drop in the annular space after the column of drill pipes in flushing by viscous fluid with flow rate $Q = 0.018\,\mathrm{m^3/s}$ at given data: $d_{ex} = 0.214\,\mathrm{m}$, $d_{in} = 0.114\,\mathrm{m}$, $\mu = 0.15\,\mathrm{Pa\,s}$, and $L = 1000\,\mathrm{m}$.

SOLUTION Determine the hydraulic diameter $d_h = d_{ex} - d_{in} = 0.214 - 0.114 = 0.1\,\mathrm{m}$. Calculate the ratio $\delta = d_{in}/d_{ex} = 0.114/0.214 = 0.533$. From Fig. 6.3, one gets $f(\delta)$: $f(\delta = 0.533) = 1.49$. From the formula (6.2.30), the following is obtained

$$\Delta p = \frac{128\mu|Q|}{\pi d_h^3 (d_{ex} + d_{in})} f(\delta) L = \frac{128 \times 0.15 \times 0.018}{3.14 \times 0.1^3 (0.214 + 0.114)} 1.49 \times 1000$$
$$= 0.5 \times 10^4\ \mathrm{Pa}.$$

EXERCISE 6.2.3

It is required to get pressure drop in drill pipes at the conditions of Exercise . The internal diameter of pipes is $d_p = 0.094\,\mathrm{m}$.

SOLUTION With formula (6.2.31), the following is obtained

$$\Delta p = \frac{128\mu|Q|}{\pi d_p^4} L = \frac{128 \times 0.15 \times 0.018}{3.14(0.094)^4} 1000 = 1.41 \times 10^4\ \mathrm{Pa}.$$

6.3 CALCULATION OF PRESSURE IN LAMINAR FLOWS OF VISCOUS-PLASTIC FLUIDS IN CIRCULAR SLOTS, PIPES, AND ANNULAR CHANNELS

6.3.1 Flows in a Circular Slot

Find relation between pressure drop $\Delta p = |p_{in} - p_{ex}|$ and flow rate $Q = v \cdot S$, where v is mean flow velocity through cylindrical surface $S = 2\pi r H$ (Fig. 6.4) in laminar flow of viscous-plastic fluid. The laminar flow of viscous-plastic fluid is also called structure flow because of a peculiar kind of velocity due to existence of flow core moving with constant velocity w_0. To get the sought relation, it is required to solve the system of equations (6.1.10)–(6.1.13) in the region between walls of the slot and the flow core with diameter H_0 at the following boundary conditions

$$w = 0 \quad \text{at} \quad z = \pm H/2; \qquad (6.3.1)$$

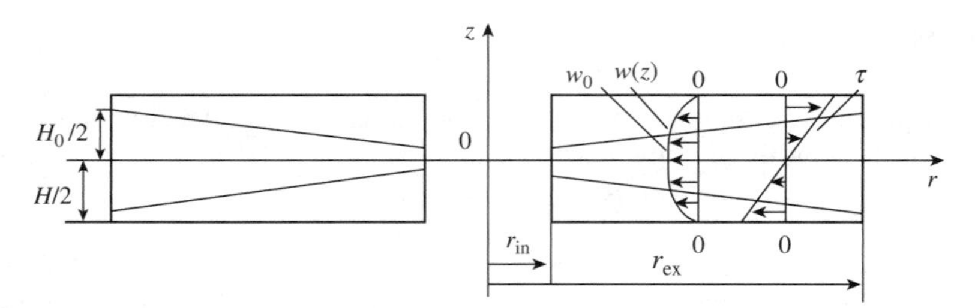

FIGURE 6.4 Distribution of velocity and stress in laminar radial flow (inflow) of viscous-plastic fluid in circular slot.

$$\frac{\partial w}{\partial z} = 0 \quad \text{at} \quad z = \pm H_0/2; \tag{6.3.2}$$

$$w = w_0 \quad \text{at} \quad -H_0/2 \le z \le H_0/2. \tag{6.3.3}$$

At first solve the problem without regard for inertial term in (6.1.10), that is, without $w\partial w/\partial z$. The rheological equation (6.1.13) for viscous-plastic fluid with regard to (4.4.16) is

$$\tau = \pm\tau_0 + \eta \frac{\partial w}{\partial z}. \tag{6.3.4}$$

It could be shown that such a flow, as for viscous fluid, holds relations (6.2.5) and (6.2.12). Then

$$r\tau = Az + B_1$$

and (6.3.4) goes into

$$r\left(\pm\tau_0 + \eta \frac{\partial w}{\partial z}\right) = Az + B_1.$$

Solving this equation with respect to $\partial w/\partial z$

$$\frac{\partial w}{\partial z} = \frac{A}{r\eta} z + \frac{B_1}{r\eta} \mp \frac{\tau_0}{\eta} \tag{6.3.5}$$

and having integrated over z one obtains

$$w = \frac{A}{r\eta}\frac{z^2}{2} + \left(\frac{B_1}{r\eta} \mp \frac{\tau_0}{\eta}\right) z + C_1. \tag{6.3.6}$$

Using the condition $\partial w/\partial z = 0$ at $z = H_0/2$ in (6.3.5), we find B_1 and substituting it in (6.3.6) with condition $w = 0$ at $z = H/2$ we get C_1.

Obtained values of B_1 and C_1 enable to get from (6.3.6) the velocity distribution in the upper half of the flow, in which the minus sign should be taken.

$$w = \frac{A}{r\eta}\left[\frac{z+H/2}{2} - \frac{H_0}{2}\right]\left(z - \frac{H}{2}\right) \quad \text{at} \quad \frac{H_0}{2} \le z \le \frac{H}{2}; \tag{6.3.7}$$

$$w = w_0 \quad \text{at} \quad 0 \le z \le H_0/2. \tag{6.3.8}$$

In a similar way, using in (6.3.5) and (6.3.6) expressions for w and $\partial w/\partial z$ taken at negative z and repeating all foregoing for the lower half of the flow, the following is obtained

$$w = \frac{A}{r\eta}\left[\frac{z-H/2}{2} + \frac{H_0}{2}\right]\left(z + \frac{H}{2}\right) \quad \text{at} \quad -\frac{H}{2} \le z \le -\frac{H_0}{2}; \tag{6.3.9}$$

$$w = w_0 \quad \text{at} \quad -H_0/2 \le z \le 0. \tag{6.3.10}$$

The velocity of the flow core w_0 could be found from (6.3.7) or (6.3.9), if $z = H_0/2$ or $z = -H_0/2$ is substituted. But in (6.3.7)–(6.3.10), the quantity H_0 of the flow core remains unknown. To get $\partial w/\partial z$ substituted with known B_1 from (6.3.5) in (6.3.4)

$$r(\tau - \tau_0) = A(z - H_0/2). \tag{6.3.11}$$

Since at $z = 0$ (at the symmetry axis of the flow), it is taken as the absence of the stress, there is

$$H_0 = 2r\tau_0/A. \tag{6.3.12}$$

The flow rate in given cross section with regard to flow symmetry is determined as follows:

$$Q = 2\left\{2\pi r\frac{H_0}{2}v_0 + 2\pi r\int_{H_0/2}^{H/2} w\,\mathrm{d}z\right\}, \tag{6.3.13}$$

where w should be taken from (6.3.7) and $v_0 = w_0$.

After integration one obtains (Volarovich and Gutkin, 1946)

$$v = \frac{Q}{2\pi rH} = -\frac{(H/2)^2}{3\eta}\left(\frac{\partial p}{\partial r} - \frac{3}{2}\frac{\tau_0}{(H/2)} + \frac{1}{2}\frac{\tau_0^3}{(H/2)^3}\frac{1}{(\partial p/\partial r)^2}\right). \tag{6.3.14}$$

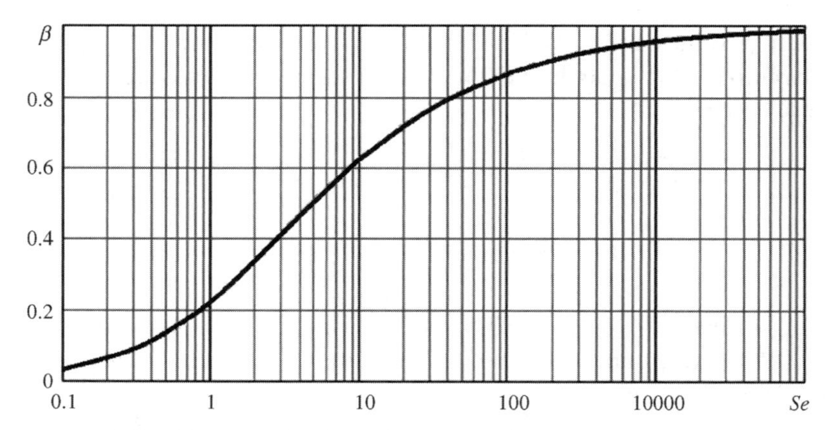

FIGURE 6.5 The dependence $\beta = \beta(Se)$ for circular slot.

At $\tau_0 = 0$, the formula (6.3.14) with regard to (6.2.7) goes into (6.2.15) for viscous fluid.

The relation (6.3.14) can be transformed into dimensionless form

$$Se = \frac{3\beta}{1 - \frac{3}{2}\beta + \frac{1}{2}\beta^3}, \tag{6.3.15}$$

where $\beta = 2\tau_0/H(\partial p/\partial r) > 0$ and $Se = \tau_0 \pi r H^2/|Q|\eta$ is Saint Venant number for the flow in circular slot. The graphic of dependence (6.3.15) is shown in Fig. 6.5. It should be noted that if $\beta > 0$, then $Q < 0$. This case is consistent with considered flow to the center of the circular slot (inflow). If $\beta < 0$ and $Q > 0$, the flow happens to the slot periphery (outflow) and absolute value $|\beta|$ should be inserted in (6.3.15). To get the pressure drop with use of (6.3.15) it is necessary at first to calculate the number Se at $r = r_{ex}$, then to find $|\beta|$ from Fig. 6.5, and at last to determine the pressure drop $\Delta p = |p_{in} - p_{ex}|$ using formula

$$\Delta p = \frac{r_{ex} 2\tau_0}{H|\beta|} \ln \frac{r_{ex}}{r_{in}}. \tag{6.3.16}$$

The formula (6.3.16) follows directly from (6.2.9). Since $\Delta p = |p_{in} - p_{ex}|$, one gets $p_{in} - p_{ex}$ from (6.2.9) and A from (6.2.7)

$$\Delta p = r\left|\frac{\partial p}{\partial r}\right| \ln \frac{r_{ex}}{r_{in}}.$$

At $r = r_{ex}$, owing to definition of β in (6.3.15), it is found

$$\left.\frac{\partial p}{\partial r}\right|_{r=r_{ex}} = \frac{2\tau_0}{H\beta}.$$

Using this value in the foregoing formula at $r = r_{ex}$ one obtains relation (6.3.16). It is able to account for pressure drop due to inertial forces in the total pressure drop if to assume that the latter is approximately equal to the sum of pressure drops (6.3.16) and (6.2.18)

$$\Delta p = \Delta p_{fr} \pm \Delta p_{in},$$

where Δp_{fr} is friction loss determined by (6.3.16); Δp_{in} is inertial loss determined by (6.2.18). In Δp_{in} the sign plus is taken at outflow and the sign minus at inflow.

EXERCISE 6.3.1

It is required to determine pressure loss at inflow in circular slot at given data: $Q = 0.018\,\text{m}^3/\text{s}$, $\rho = 1200\,\text{kg/m}^3$, $\eta = 0.015\,\text{Pa s}$, $\tau_0 = 5\,\text{Pa}$, $r_{ex} = 100\,\text{m}$, $r_c = 0.214\,\text{m}$, and $H = 0.001\,\text{m}$.

SOLUTION Assume that is laminar flow. Determine Saint Venant number at $r = r_{ex}$

$$Se = \frac{\pi r_{ex} H^2 \tau_0}{Q\eta} = \frac{3.14 \times 100 \times 0.001^2 \times 5}{0.018 \times 0.015} = 5.81.$$

Determine $|\beta|_{r=r_{ex}}$ from the graphic of Fig. 6.5

$$|\beta|_{r=r_{ex}} = 0.534.$$

In accordance with (6.3.16), the friction loss is

$$\Delta p_{fr} = \frac{r_{ex} 2\tau_0}{H|\beta|_{r=r_{ex}}} \ln \frac{r_{ex}}{r_{in}} = \frac{100 \times 2 \cdot 5}{0.001 \times 0.534} \ln \frac{100}{0.214} = 115 \times 10^5\ \text{Pa}.$$

In accordance with (6.2.18), inertia loss is

$$\Delta p_{in} = \frac{\rho Q^2}{8\pi^2 H^2} \left(\frac{1}{r_{in}^2} - \frac{1}{r_{ex}^2} \right) = \frac{1200 \times 0.018^2}{8 \times 3.14 (0.001)^2} \left(\frac{1}{0.214^2} - \frac{1}{100^2} \right) = 1.08 \times 10^5\ \text{Pa}.$$

Total pressure drop is

$$\Delta p = \Delta p_{fr} - \Delta p_{in} = 115 \times 10^5 - 1.08 \times 10^5 \approx 114 \times 10^5\ \text{Pa}.$$

6.3.2 Flows in Pipes

Deduce relation between pressure drop $\Delta p = |p_2 - p_1|$ and flow rate $Q = v \cdot S$ for viscous-plastic fluid in pipe section with length L (Fig. 6.6). As in the

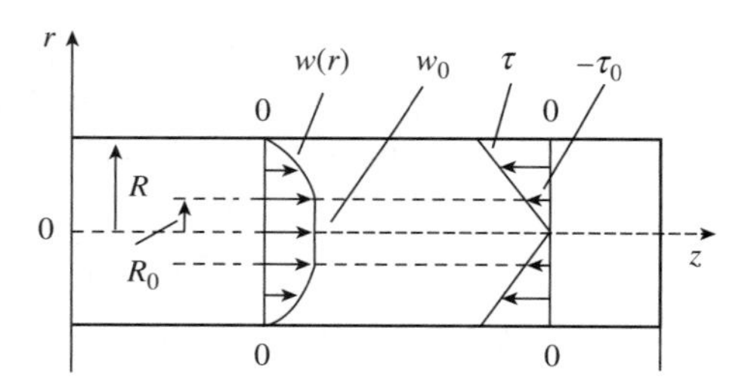

FIGURE 6.6 Distribution of velocity and stress in laminar flow of viscous--plastic fluid in a pipe.

previous case, the flow has a core of radius R_0 moving with velocity v_0. To get pressure drop, it is needed to solve the system of equations (6.1.1)–(6.1.4) without inertia and gravity forces in the region between pipe and flow core at the following boundary conditions

$$w = 0 \quad \text{at} \quad r = R = d_{\text{in}}/2, \qquad w = v_0 = \text{const} \quad \text{at} \quad 0 \leq r \leq R_0; \tag{6.3.17}$$

$$\frac{\partial w}{\partial r} = 0 \quad \text{at} \quad r = R_0. \tag{6.3.18}$$

Rheological equation (6.1.4) with regard to (4.4.13) for viscous-plastic fluid in a pipe at $\partial w/\partial r < 0$ is

$$\tau = -\tau_0 + \eta \frac{\partial w}{\partial r} \quad \text{at} \quad r \geq R_0. \tag{6.3.19}$$

Performing for viscous-plastic fluid the same reasoning as for viscous fluid in Section 6.2, we find the validity of relations (6.2.23) so that pressure distribution along the pipe would be expressed through (6.2.24)

$$p = Az + B_1, \tag{6.3.20}$$

where $A = (p_2 - p_1)/L = -\Delta p/L$ and $B_1 = p_1$.

Substitution of (6.3.19) in the second equation (6.2.23) and further integration gives

$$w = \frac{A}{4\eta} r^2 + \frac{B}{\eta} \ln r + \frac{\tau_0}{\eta} r + C. \tag{6.3.21}$$

In this equation, the constant $B = 0$, in which one can make sure by jointly considering core equilibrium equation, equation (6.3.21), and boundary condition (6.3.18). Another constant C would be determined if in (6.3.21) at $B=0$ the first condition (6.3.17) is used

$$C = -\frac{A}{4\eta}R^2 - \frac{\tau_0 R}{\eta}. \tag{6.3.22}$$

Then (6.3.21) takes the following form

$$w = -\frac{A}{4\eta}(R^2 - r^2) - \frac{\tau_0}{\eta}(R - r). \tag{6.3.23}$$

Since the flow at $r \leq R_0$ represents a motion of continuum cylinder with undisturbed structure, the equilibrium condition of forces acting on the flow core would be

$$\pi R_0^2 \Delta p = 2\pi \times R_0 L \tau_0. \tag{6.3.24}$$

This equation yields core radius

$$R_0 = 2L\tau_0/\Delta p. \tag{6.3.25}$$

Applying to (6.3.23) the second condition (6.3.17) one gets the core velocity

$$w_0 = -\frac{A}{4\eta}(R^2 - R_0^2) - \frac{\tau_0}{\eta}(R - R_0). \tag{6.3.26}$$

Use of relations (6.3.23) and (6.3.26) gives the fluid flow rate

$$Q = 2\pi \int_0^{R_0} w_0 r \, \mathrm{d}r + 2\pi \int_{R_0}^{R} wr \, dr.$$

Integration yields the Buckingham formula

$$Q = \frac{\pi R^4 \Delta p}{8\eta \cdot L}\left[1 - \frac{4}{3}\frac{2\tau_0 L}{R\Delta p} + \frac{1}{3}\left(\frac{2\tau_0 L}{R\Delta p}\right)^4\right]. \tag{6.3.27}$$

This formula could be written in dimensionless form

$$Se = \frac{8\beta}{1 - \frac{4}{3}\beta + \frac{1}{3}\beta^4}, \tag{6.3.28}$$

where

$$\beta = \frac{2\tau_0 L}{R\Delta p}; \qquad Se = \frac{\tau_0 d \cdot S}{\eta Q}. \tag{6.3.29}$$

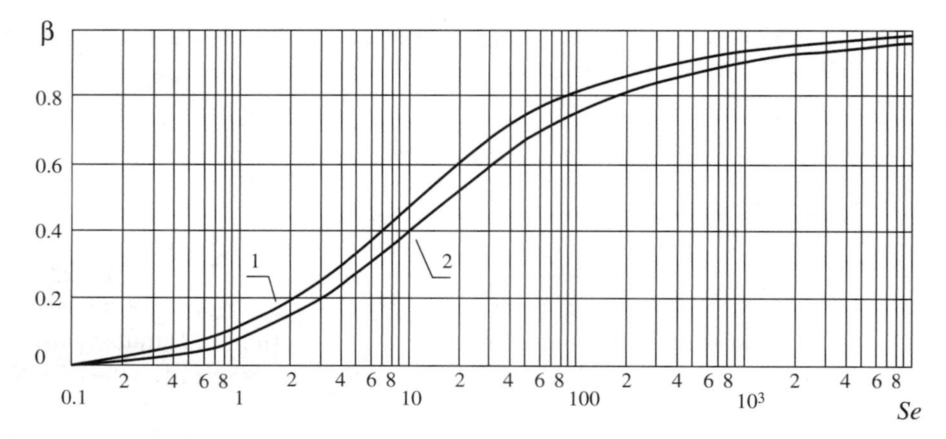

FIGURE 6.7 Dependences $\beta = \beta(Se)$ for pipes of circular (1) and annular (2) cross sections.

The graphic of dependence (6.3.28) (curve 1) in semilogarithmic scale is shown in Fig. 6.7 (Grodde, 1960). Resolving (6.3.29) with respect to Δp

$$\Delta p = \frac{2\tau_0 L}{R\beta}$$

and introducing diameter $d_h = 2R$ it is obtained

$$\Delta p = \frac{4\tau_0 L}{d_h \beta}. \tag{6.3.30}$$

Thus, the formula (6.3.30) gives the pressure drop (friction loss) Δp in the flow of viscous-plastic fluid in pipes. To do this, it is required to calculate from (6.3.29) the number *Se*, to determine from the graphic β versus *Se* (Fig. 6.7, curve 1) the value β for the obtained value of *Se*, and then from the formula (6.3.30) to get the sought pressure drop Δp.

EXERCISE 6.3.2

Determine the pressure drop in pipes in washing with viscous-plastic fluid at given data: $Q = 0.015\ \mathrm{m^3/s}$, $\eta = 0.02\ \mathrm{Pa\,s}$, $\tau_0 = 10\ \mathrm{Pa}$, $d_{in} = 0.094\ \mathrm{m}$, and $L = 1000\ \mathrm{m}$.

SOLUTION Calculate from (6.3.29) Saint Venant number

$$Se = \frac{\tau_0 d_{in}(\pi d_{in}^2/4)}{\eta Q} = \frac{10 \times 0.094}{0.02 \times 0.015} \cdot \frac{3.14}{4} 0.094^2 = 21.7.$$

From Fig. 6.7, curve 1 find $\beta = 0.62$. In accordance with formula (6.3.30), calculate the pressure drop

$$\Delta p = \frac{4\tau_0 L}{d_{in}\beta} = \frac{4 \times 10 \times 1000}{0.094 \times 0.62} = 6.86 \times 10^5 \text{ Pa}.$$

6.3.3 Flows in Annular Concentric Channels

Obtain the relation between pressure drop $\Delta p = |p_2 - p_1|$ and flow rate $Q = v \cdot S_{as}$, where v is mean velocity in the annular space with cross section area $S_{as} = \pi(d_{ex}^2 - d_{in}^2)/4$. In flows of viscous-plastic fluid in annular space as well as in pipes, a core having a form of hollow cylinder with cross section area $\pi(b^2 - a^2)$, lateral surface $2\pi(a + b)L$, and moving with velocity w_0 is formed (Fig. 6.8). This core divides the flow into two gradient layers: (I) in which $\partial w/\partial r < 0$ and (II) in which $\partial w/\partial r > 0$. In this connection when determining velocity profile w, the system (6.1.1)–(6.1.4) should be solved in each layer separately since the rheological equation (6.1.4) with regard to (4.4.13) has its own form in each layer:

$$\tau = -\tau_0 + \eta \frac{\partial w}{\partial r}, \quad \text{for the first layer} \tag{6.3.31}$$

and

$$\tau = \tau_0 + \eta \frac{\partial w}{\partial r} \quad \text{for the second layer.} \tag{6.3.32}$$

The condition of equilibrium of forces acting on the core is written as follows:

$$\pi(b^2 - a^2)\Delta p = 2\pi\tau_0(a + b)L. \tag{6.3.33}$$

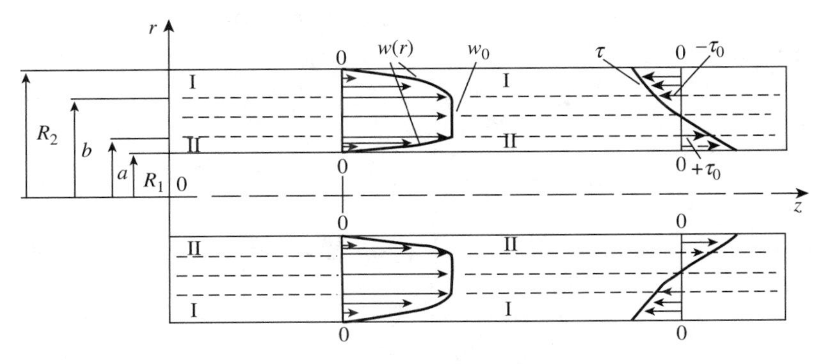

FIGURE 6.8 Distribution of velocity and stress in laminar flow of viscous--plastic fluid in annular channel.

Boundary conditions in the absence of slip along channel walls are

$$\begin{aligned} w = 0 \quad &\text{at} \quad r = R_1; \\ w = 0 \quad &\text{at} \quad r = R_2. \end{aligned} \tag{6.3.34}$$

Since the core moves with constant velocity w_0, the velocity should obey the following conditions:

$$w = w_0 \quad \text{at} \quad a \le r \le b \tag{6.3.35}$$

and

$$\frac{\partial w}{\partial r} = 0 \quad \text{at} \quad r = a \quad \text{and} \quad r = b. \tag{6.3.36}$$

Thus, similar to (6.3.21) the velocity profile in each gradient layer is

for layer I ($b \le r \le R_2$)

$$w = \frac{A}{4\eta} r^2 + \frac{B}{\eta} \ln r + \frac{\tau_0}{\eta} r + C; \tag{6.3.37}$$

for layer II ($R_1 \le r \le a$)

$$w = \frac{A}{4\eta} r^2 + \frac{B_1}{\eta} \ln r - \frac{\tau_0}{\eta} r + C_1. \tag{6.3.38}$$

Unknown constants B, C, B_1, C_1 as well as sizes a, b, and velocity of the core w_0 are determined from seven Volarovitch–Gutkin equations (Volarovich and Gutkin, 1946) obtained by substitution of boundary conditions (6.3.34), (6.3.35), and (6.3.36) into (6.3.37) and (6.3.38) with addition of condition (6.3.33)

$$\begin{aligned}
&\frac{A}{4\eta} R_2^2 + \frac{B}{\eta} \ln R_2 + \frac{\tau_0}{\eta} R_2 + C = 0; \\
&\frac{A}{4\eta} R_1^2 + \frac{B_1}{\eta} \ln R_1 - \frac{\tau_0}{\eta} R_1 + C_1 = 0; \\
&w_0 = \frac{A}{4\eta} b^2 + \frac{B}{\eta} \ln b + \frac{\tau_0}{\eta} b + C; \\
&w_0 = \frac{A}{4\eta} a^2 + \frac{B_1}{\eta} \ln a - \frac{\tau_0}{\eta} a + C_1; \\
&\frac{A}{2\eta} a + \frac{B_1}{\eta a} - \frac{\tau_0}{\eta} = 0;
\end{aligned}$$

$$\frac{A}{2\eta}b + \frac{B}{\eta b} + \frac{\tau_0}{\eta} = 0;$$
$$(b-a)\Delta p = 2\tau_0 L. \tag{6.3.39}$$

The constants in (6.3.39) could not be obtained in explicit form due to transcendence of the system of equations. Therefore, we proceed as follows. Taking B, C, B_1, and C_1 in (6.3.37) and (6.3.38) as well as w_0 as known quantities, we get the flow rate in annular space

$$Q = 2\pi \int_{R_1}^{R_2} wr\,\mathrm{d}r = 2\pi \left\{ \int_{R_1}^{a} \left[\frac{A}{4\eta} r^2 + \frac{B_1}{\eta} \ln r - \frac{\tau_0}{\eta} r + C_1 \right] r\,\mathrm{d}r \right.$$
$$\left. + \int_a^b w_0 r\,\mathrm{d}r + \int_b^{R_2} \left[\frac{A}{4\eta} r^2 + \frac{B}{\eta} \ln r + \frac{\tau_0}{\eta} r + C \right] r\,\mathrm{d}r \right\}, \tag{6.3.40}$$

where A from (6.2.23) is

$$A = \frac{p_2 - p_1}{L} = -\frac{p_1 - p_2}{L} = -\frac{\Delta p}{L}. \tag{6.3.41}$$

The system of equations (6.3.39) and (6.3.40) could be reduced into two equations. Integration and elimination of unknowns B, C, B_1, C_1, and a from (6.3.40) and (6.3.39) gives (Fredrickson and Bird, 1958)

$$Q = \frac{\pi R_2^4 \Delta p}{8\eta L} \left[1 - \delta^4 - 2\frac{b}{R_2}\left(\frac{b}{R_2} - \frac{2\tau_0 L}{\Delta p R_2}\right)(1-\delta^2) \right.$$
$$\left. - \frac{4}{3}(1+\delta^3)\frac{2\tau_0 L}{R_2\,\Delta p} + \frac{1}{3}\left(2\frac{b}{R_2} - \frac{2\tau_0 L}{R_2\,\Delta p}\right)^3 \frac{2\tau_0 L}{R_2\,\Delta p} \right], \tag{6.3.42}$$

where b is determined by Volarovitch–Gutkin equation (Volarovich and Gutkin, 1946)

$$b\left(-\tau_0 + b\frac{\Delta p}{L}\right) \ln \frac{R_1 b}{R_2(b - (2\tau_0 L/\Delta p))} - \tau_0(R_2 + R_1)$$
$$+ \frac{\tau_0}{2}\left(2b - \frac{2\tau_0 L}{\Delta p}\right) - \frac{1}{4}\frac{\Delta p}{L}(-R_2^2 + R_1^2) = 0. \tag{6.3.43}$$

These two equations are transformed into

$$Se = \frac{8\beta}{\frac{1+\delta^2}{(1-\delta)^2} - \frac{2\xi}{1-\delta}\left(\frac{\xi}{1-\delta}-\beta\right) - \frac{4}{3}\frac{1-\delta+\delta^2}{(1-\delta)^2}\beta + \frac{1}{3}\left(\frac{2\xi}{1-\delta}-\beta\right)^3 \beta \frac{1-\delta}{1+\delta}}, \quad (6.3.44)$$

$$\xi\beta(1-\delta)+\xi(\xi-\beta(1-\delta))\ln\frac{\xi\delta}{\xi-\beta(1-\delta)}+\frac{1-2\beta(1-\delta)-(\beta(1-\delta)+\delta)^2}{2}=0, \quad (6.3.45)$$

where $\delta = 2R_1/2R_2 = d_{ex}/d_{in}$ is the ratio of external diameter of interior pipe and internal diameter of exterior pipe; $\xi = 2b/d_{in}$ is the ratio of doubled distance from the pipe axis up to external boundary of annular flow core and internal diameter of exterior pipe with $\delta+\beta(1-\delta)\le\xi\le 1$; Se is the Saint Venant parameter equal to

$$Se = \frac{\tau_0(d_{ex}-d_{in})S_{as}}{\eta Q} \quad (6.3.46)$$

and

$$\beta = \frac{4\tau_0 L}{(d_{ex}-d_{in})\Delta p}. \quad (6.3.47)$$

From (6.3.44) with regard to (6.3.45), the graphic $Se = Se(\beta, \delta)$ could be plotted. In Fig. 6.7, it is shown that only one curve 2 averaged over δ in the range met in the practice.

Hence, with the help of (6.3.47) it may be determined the pressure drop in annular space

$$\Delta p = \frac{4\tau_0 L}{\beta(d_{ex}-d_{in})} \quad (6.3.48)$$

by factor β obtained from Fig. 6.7 (curve 2) and number Se previously calculated for given flow rate and rheological τ_0, η and geometric d_{ex}, d_{in} data.

Graphical method of manual calculation of pressure loss in the flow of viscous-plastic fluid in pipes and annual channels was developed by Grodde (Grodde, 1960).

This method is convenient and its accuracy is determined only by the error of finding β with Fig. 6.7. It can be shown that at $\delta = 0$ the expression (6.3.45) gives $\xi = \beta$ and (6.3.44) goes into solution for the flow of viscous-plastic fluid in a pipe. Denote also that (6.3.48) is the most general solution covering both viscous and viscous-plastic flows in pipes and annular space, converting, respectively, into (6.2.31), (6.3.30), and (6.2.30).

Really, as it follows from (6.3.47) at $\tau_0 \to 0$ the factor $\beta \to 0$. In order to get (6.2.29) from (6.3.48), lim τ_0/β at $\tau_0 \to 0$ should be calculated. Substitution of the expression for *Se* from (6.3.46) into (6.3.44) gives

$$\frac{\tau_0}{\beta} = \frac{8 \times \eta Q/S}{\frac{1+\delta^2}{(1-\delta)^2} + \frac{2\xi}{1-\delta}\left(\frac{\xi}{1-\delta} - \beta\right) - \frac{4}{3}\frac{(1-\delta+\delta^2)}{(1-\delta^2)}\beta + \frac{1}{3}\left(\frac{2\xi}{1-\delta} - \beta\right)^3 \beta \frac{1-\delta}{1+\delta}},$$

where ξ, β, and δ are related by

$$\xi\beta(1-\delta) + \xi(\xi-\beta(1-\delta))\ln\frac{\xi\delta}{\xi-\beta(1-\delta)} + \frac{1-2\beta(1-\delta)-(\beta(1-\delta)+\delta)^2}{2} = 0.$$

Tending in this relations $\tau_0 \to 0$, we get

$$\lim_{\tau_0 \to 0} \frac{\tau_0}{\beta} = \frac{8\eta \times Q/S}{\frac{1+\delta^2}{(1-\delta)^2} - \frac{2\xi}{1-\delta}\frac{\xi}{(1-\delta)}}; \quad \xi^2 \ln\delta + \frac{1-\delta^2}{2} = 0.$$

The last equality gives ξ^2 and after substituting it in the above equation yields

$$\lim_{\tau_0 \to 0} \frac{\tau_0}{\beta} = \frac{8\eta \cdot Q/S}{\frac{(1+\delta^2)}{(1-\delta)^2} + \frac{1-\delta^2}{(1-\delta)^2 \ln\delta}} = \frac{8(1-\delta)^2}{1+\delta^2 + \frac{1-\delta^2}{\ln\delta}} \cdot \eta \frac{Q}{S}.$$

Using in (6.3.48) the last relation instead of τ_0/β, taking into account that $\eta \to \mu$ and performing needed transformations, the following is gained

$$\Delta p = \lim_{\tau_0 \to 0} \frac{4\tau_0 L}{(d_{\text{ex}} - d_{\text{in}})\beta} = \frac{4L8(1-\delta)^2}{\left[1+\delta^2 + \frac{1-\delta^2}{\ln\delta}\right](d_{\text{ex}} - d_{\text{in}})} \mu \frac{Q}{S} = \lambda \frac{\rho v^2}{2d_{\text{h}}} L,$$

where factor λ has the same meaning as in (6.2.29).

Consequently, the formula (6.3.48) at $\tau_0 \to 0$ goes continually into (6.2.29). Thus, the formula is theoretically justified. For more accurate calculations, there are PC programs to solve equations (6.3.44)–(6.3.45) for any $\tau_0 \neq 0$ (Leonov et al., 1980).

Note that (6.3.48) and thus (6.3.30) could be transformed into (6.2.29) with its own hydraulic resistance factors λ. Therefore,

$$\Delta p = \frac{4\tau_0 L}{\beta d_{\text{h}}} = \frac{8\tau_0}{\beta\rho v^2} \frac{\rho v^2}{2d_{\text{h}}} L = \lambda \frac{\rho v^2}{2d_{\text{h}}} L, \tag{6.3.49}$$

where

$$\lambda = \frac{8\tau_0}{\beta\rho v^2} = \frac{8\tau_0 d_h}{\beta \frac{\rho v}{\eta} v \eta d_h} = \frac{64}{Re}\frac{Se}{8\beta}.$$

Since in accordance with (6.3.44)–(6.3.45) $\beta = \beta(Se, \delta)$, there is $Se/(8\beta) = f(Se, \delta)$, that is

$$\lambda = \frac{64}{Re} f(Se, \delta). \tag{6.3.50}$$

This expression in compliance with above made algebra at $\tau_0 \to 0$, $Se \to 0$ leads to relation for viscous fluid

$$\lambda = \frac{64}{Re} f(\delta),$$

where $f(\delta)$ was given in elucidations to the formula (6.2.29).

It should be noted that the formula (6.3.48) is true for calculation of the pressure drop Δp in viscous-plastic fluid starting when $w = 0$. At this $Se \to \infty$ and (6.3.44) shows that $\beta \to 1$. Hence, the relation (6.3.48) at $\beta = 1$ takes form

$$\Delta p = 4\tau_0 L / d_h. \tag{6.3.51}$$

The formula (6.3.51) is used at $w = 0$ to calculate pressure drop needed to break adhesion with channel wall not only for viscous-plastic fluid but also for any fluid having initial shear stress θ

$$\Delta p = 4\theta L / d_h. \tag{6.3.52}$$

EXERCISE 6.3.3

Determine pressure drop in annular space of a well at given data: $Q = 0.015\,m^3/s$, $\eta = 0.02$ Pa s, $\tau_0 = 10$ Pa, $\rho = 1200\,kg/m^3$, $d_{in} = 0.214$ m, $d_{ex} = 0.114$ m, and $L = 1000$ m.

SOLUTION Calculate from (6.3.46) Saint Venan number

$$Se = \frac{\tau_0 (d_{ex} - d_{in}) S_{as}}{\eta Q} = \frac{10(0.214 - 0.114)(3.14/4)(0.214^2 - 0.114^2)}{0.02 \times 0.015} = 85.8.$$

From Fig. 6.7, curve 2 determines $\beta = 0.72$. In accordance with (6.3.48), the pressure drop is

$$\Delta p = \frac{4\tau_0 L}{\beta(d_{ex} - d_{in})} = \frac{4 \times 10 \times 1000}{0.72 \cdot (0.214 - 0.114)} = 5.56 \times 10^5 \text{ Pa}.$$

EXERCISE 6.3.4

Determine pressure drop at flow initiation of viscous-plastic fluid with $\theta = 15$ Pa in a pipe with diameter $d = 0.094$ m and length $L = 500$ m.

SOLUTION From (6.3.52) we determine

$$\Delta p = \frac{4\theta L}{d} = \frac{4 \times 15 \times 500}{0.094} = 3.19 \times 10^5 \text{ Pa}.$$

6.4 CALCULATION OF PRESSURE IN LAMINAR FLOWS OF POWER INCOMPRESSIBLE FLUIDS IN SLOTS, PIPES, AND ANNULAR CHANNELS

6.4.1 Circular Slot

Determine the relation between pressure drop $\Delta p = |p_{in} - p_{ex}|$ and flow rate $Q = v \cdot S$, where v is mean velocity of the flow through cylindrical surface $S = 2\pi r H$.

In order to get the dependence sought, it is required to solve the system of equations (6.1.10)–(6.1.13) with concrete form of rheological equation of state (3.1.13) and the following boundary conditions

$$w = 0 \quad \text{at} \quad z = \pm H/2; \tag{6.4.1}$$

$$\begin{aligned} p &= p_{in} \quad \text{at} \quad r = r_{in}; \\ p &= p_{ex} \quad \text{at} \quad r = r_{ex}. \end{aligned} \tag{6.4.2}$$

The rheological equation (6.1.13) for power fluid is

$$\tau = k \frac{\partial w}{\partial z} \left| \frac{\partial w}{\partial z} \right|^{n-1}. \tag{6.4.3}$$

As well as for viscous fluid from mass conservation equation ensues condition (6.2.4). Substitution of the velocity from (6.2.4) into (6.4.3) leads to the conclusion that the product $r^n\tau$ would be a function only of z-coordinate

$$r^n\tau = f(z). \tag{6.4.4}$$

Let us first seek the velocity distribution in given flow as well as in Section 6.2 assuming that the term $w\partial w/\partial r$ in (6.1.10) could be ignored. Then substituting (6.4.4) in (6.1.10), we obtain

$$r^n \frac{\partial p}{\partial r} = \frac{\partial f(z)}{\partial z}. \tag{6.4.5}$$

Since the left part of this equation depends only on r and the right part only on z, it should be satisfied equalities

$$r^n \frac{\partial p}{\partial r} = A = \text{const}, \tag{6.4.6}$$

$$\frac{\partial f(z)}{\partial z} = A. \tag{6.4.7}$$

Integration of (6.4.6) gives pressure distribution for power fluid in circular slot

$$p = \frac{A}{(1-n)r^{n-1}} + B. \tag{6.4.8}$$

The constants A and B could be found with use of boundary conditions (6.4.2)

$$\begin{aligned} A &= \frac{(p_{\text{in}}-p_{\text{ex}})(1-n)}{(1/r_{\text{in}})^{n-1}-(1/r_{\text{ex}})^{n-1}}; \\ B &= p_{\text{ex}} - \frac{A}{(1-n)r_{\text{ex}}^{n-1}}. \end{aligned} \tag{6.4.9}$$

Then (6.4.8) is

$$\frac{p-p_{\text{ex}}}{p_{\text{ex}}-p_{\text{in}}} = \frac{1-(r_{\text{ex}}/r)^{n-1}}{(r_{\text{ex}}/r_{\text{in}})^{n-1}-1}. \tag{6.4.10}$$

Integration of (6.4.7) gives

$$f(z) = Az + B_1. \tag{6.4.11}$$

As well as in (6.2.12) one gets $B_1 = 0$ and substitution of (6.4.11) and (6.4.3) in (6.4.4) yields

$$k\frac{\partial w}{\partial z}\left|\frac{\partial w}{\partial z}\right|^{n-1} = \frac{Az}{r^n}.$$

Then in the inflow ($A < 0$) for upper half of the flow ($z \geq 0$) takes place

$$\frac{\partial w}{\partial z} = -\frac{1}{r}\left(\frac{-Az}{k}\right)^{1/n}.$$

Integration of this equation gives

$$w = \frac{1}{r\left(\frac{1}{n}+1\right)}\left(\frac{-A}{k}\right)^{1/n} z^{\frac{1}{n}+1} + f_2(r). \tag{6.4.12}$$

The function $f_2(r)$ could be determined by use of the condition (6.4.1) at $z = H/2$, that is, for upper half of the flow ($z \geq 0$)

$$f_2(r) = -\frac{1}{r\left(\frac{1}{n}+1\right)}\left(\frac{-A}{k}\right)^{1/n}\left(\frac{H}{2}\right)^{\frac{1}{n}+1}.$$

Substitution of obtained $f_2(r)$ in (6.4.12) gives velocity distribution of the power fluid in the upper part of the flow

$$w = \frac{A}{r\left(\frac{1}{n}+1\right)k^{1/n}}(-A)^{\frac{1}{n}-1}\left[z^{\frac{1}{n}+1} - \left(\frac{H}{2}\right)^{\frac{1}{n}+1}\right]. \tag{6.4.13}$$

In the lower part of the flow ($z \leq 0$), the velocity profile will be symmetric to the profile (6.4.13). The fluid flow rate at the inflow ($Q > 0$) through cylindrical section is determined by the use of the formula (6.4.13)

$$Q = 2 \times 2\pi r \int_0^{H/2} w\,\mathrm{d}z = \frac{-A}{k^{1/n}}|A|^{\frac{1}{n}-1}\left(\frac{H}{2}\right)^{\frac{1}{n}+2}\frac{4\pi n}{1+2n}. \tag{6.4.14}$$

The expression (6.4.14) is true for inflow ($A < 0$, $Q > 0$) as well as for outflow ($A > 0$, $Q < 0$). It should be noted that at $n = 1$ and $k = \mu$ from (6.4.13) and (6.4.14), the velocity profile (6.2.13) and flow rate (6.2.14) for viscous fluid are resulted. From (6.4.14) it follows

$$|A| = k\left[\frac{|Q|(1+2n)}{4\pi n}\right]^n\left(\frac{2}{H}\right)^{2n+1}. \tag{6.4.15}$$

To get connection between Δp and Q substitute A from (6.4.9) into (6.4.15). As a result, the following is obtained

$$\Delta p = \frac{k}{1-n}\left[\frac{|Q|(1+2n)}{4\pi n}\right]^n\left(\frac{2}{H}\right)^{2n+1}\left[\left(\frac{1}{r_{\text{in}}}\right)^{n-1}-\left(\frac{1}{r_{\text{ex}}}\right)^{n-1}\right]. \quad (6.4.16)$$

This equation could be transformed to the form of Darcy–Weisbach relation

$$\Delta p = \lambda\frac{\rho v^2}{2H}(r-r_{\text{in}}), \quad (6.4.17)$$

where $v = Q/(2\pi rH)$ is mean velocity at distance r.

At $r = r_{\text{ex}}$

$$\lambda = \frac{24}{Re}f(\delta,n); \quad Re = \frac{|v|^{2-n}H^n\rho}{k};$$

$$f(\delta,n) = \left(\frac{1+2n}{n}\right)^n\left[\frac{1}{\delta^{n-1}}-1\right]\frac{2^{n-1}}{3(1-n)(1-\delta)};$$

$$\delta = r_{\text{in}}/r_{\text{ex}}; \quad v = \frac{Q}{2\pi r_{\text{ex}}H}.$$

At $n \to 1$, $k \to \mu$ the formula (6.4.17) goes into (6.2.17).

The effect of inertia on the flow of power fluid could be taken into account using corresponding formulas of Section 6.2.

6.4.2 Flow of Power Fluid in Pipes

Here, our interest is the dependence of pressure drop $\Delta p = |p_2 - p_1|$ on flow rate $Q = v{\cdot}S$, where v is mean velocity and $S = \pi R^2 = \pi d_{\text{in}}^2/4$ is the area of pipe section. To get the sought formula, one should solve the system of equations (6.1.1)–(6.1.4) at boundary conditions

$$w = 0 \quad \text{at} \quad r = R = d_{\text{in}}/2; \quad (6.4.18)$$

$$p = p_1 \quad \text{at} \quad z = 0; \qquad p = p_2 \quad \text{at} \quad z = L. \quad (6.4.19)$$

The rheological equation (6.1.4) in considered case is

$$\tau = k\frac{\partial w}{\partial r}\left|\frac{\partial w}{\partial r}\right|^{n-1}. \quad (6.4.20)$$

There are also true relations (6.2.23) and (6.2.24) since they are not influenced by the form of the rheological equation (6.4.20). Thus, the pressure distribution in the pipe is expressed by relation (6.2.24)

$$p = Az + B, \tag{6.4.21}$$

where

$$A = \frac{p_2 - p_1}{L}; \qquad B = p_1.$$

Integration of the second equation (6.2.23) gives

$$\tau = \frac{Ar}{2} + \frac{B}{r}. \tag{6.4.22}$$

Since at $r = 0$, the stress τ should be finite, and it is needed to take $B = 0$. Consider a flow with $Q > 0$, that is, at $A < 0$ and $\partial w/\partial r < 0$. Substitution of relation (6.4.20) in (6.4.22) yields

$$\frac{\partial w}{\partial r} = -\left(\frac{-A}{2k}\right)^{1/n} r^{1/n}.$$

After integration following is obtained

$$w = -\left(\frac{n}{n+1}\right)\left(\frac{-A}{2k}\right)^{1/n} r^{\frac{1}{n}+1} + C. \tag{6.4.23}$$

The constant C could be found by use of sticking boundary condition (6.4.18)

$$C = \left(\frac{n}{n+1}\right)\left(\frac{-A}{2k}\right)^{1/n}\left(\frac{d_{\text{in}}}{2}\right)^{\frac{1}{n}+1}. \tag{6.4.24}$$

The resulting velocity profile in laminar flow of power fluid in a pipe is

$$w = \frac{A}{2k}\left(\frac{n}{n+1}\right)\left(\frac{|A|}{2k}\right)^{\frac{1}{n}-1}\left[r^{\frac{1}{n}+1} - \left(\frac{d_{\text{in}}}{2}\right)^{\frac{1}{n}+1}\right]. \tag{6.4.25}$$

Calculate the flow rate taking into account relation (6.4.25)

$$Q = 2\pi \int_0^{d_{\text{in}}/2} wr\,\mathrm{d}r = -\frac{\pi n}{3n+1}\left(\frac{d_{\text{in}}}{2}\right)^{\frac{1}{n}+3}\frac{A}{2k}\left(\frac{|A|}{2k}\right)^{\frac{1}{n}-1}. \tag{6.4.26}$$

This formula is true for $A < 0$, $w > 0$, as well as for $A > 0$ and $w < 0$. At $n = 1$ and $k = \mu$, the velocity profile (6.4.25) and the flow rate (6.4.26) go into corresponding quantities of viscous fluid (6.2.26) and (6.2.28).

From (6.4.26) it follows

$$\frac{A}{2k}\left(\frac{|A|}{2k}\right)^{\frac{1}{n}-1} = -\frac{Q(3n+1)}{\pi n\left(\frac{d_{\text{in}}}{2}\right)^{\frac{1}{n}+3}}.$$

Substitute this relation in (6.4.25) and reduce the result to the form (Fredrickson and Bird, 1958)

$$\frac{w}{v} = \frac{3n+1}{n+1}\left[1-\left(\frac{r}{R}\right)^{\frac{n+1}{n}}\right],$$

where $v = Q/(\pi R^2)$.

Typical velocity profiles plotted by this formula are shown in Fig. 6.9 (Wilkenson, 1960).

The expression (6.4.26) may be transformed to the form of Darcy–Weisbach formula using relation $A = (p_2 - p_1)/L$

$$\Delta p = |p_2 - p_1| = \lambda \frac{\rho v^2}{2d_{\text{in}}} L, \tag{6.4.27}$$

where

$$\lambda = \frac{64}{Re} f(n); \quad v = \frac{4Q}{\pi d_{\text{in}}^2}; \quad Re = \frac{|v|^{2-n} d_{\text{in}}^n \rho}{k}; \quad f(n) = 2^{n-3}\left(\frac{3n+1}{n}\right)^n.$$

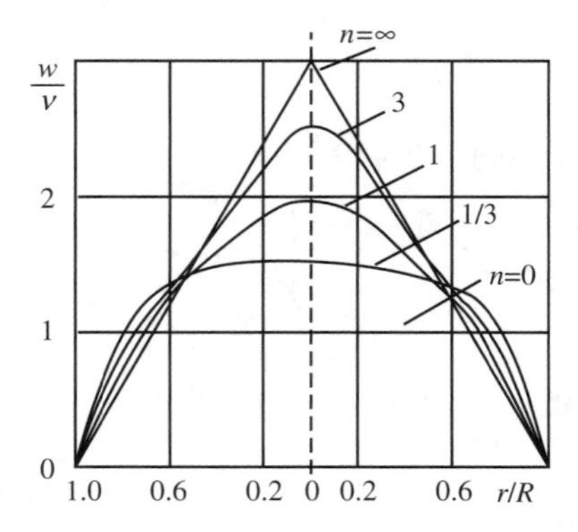

FIGURE 6.9 Typical velocity profiles in the flow of power fluid in pipes at different n.

6.4.3 Flow in Annular Channels

To get the relation between pressure drop $\Delta p = |p_2 - p_1|$ and flow rate $Q = v \cdot S$, where v is mean velocity, $S = \pi(d_{ex}^2 - d_{in}^2)/4$ is the cross section area of annular space, it is required to solve the system of equations (6.1.1)–(6.1.4) in two flow regions, since in each of them the rheological equation for power fluid would have its own sign (see Fig. 6.10):

for region I

$$\tau = k\frac{\partial w}{\partial r}\left(-\frac{\partial w}{\partial r}\right)^{n-1}, \quad \frac{d_{ax}}{2} \le r \le \frac{d_{ex}}{2}; \tag{6.4.28}$$

for region II

$$\tau = k\left(\frac{\partial w}{\partial r}\right)^{n}, \quad \frac{d_{in}}{2} \le r \le \frac{d_{ax}}{2}; \tag{6.4.29}$$

boundary conditions

$$\begin{aligned} w &= 0 \quad \text{at} \quad r = d_{in}/2, \\ w &= 0 \quad \text{at} \quad r = d_{ex}/2, \\ \frac{\partial w}{\partial r} &= 0 \quad \text{at} \quad r = d_{ax}/2. \end{aligned} \tag{6.4.30}$$

$$\begin{aligned} p &= p_1 \quad \text{at} \quad z = 0, \\ p &= p_2 \quad \text{at} \quad z = L. \end{aligned} \tag{6.4.31}$$

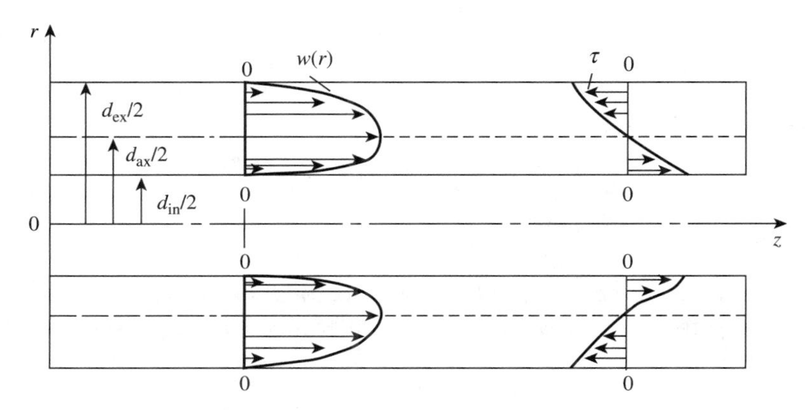

FIGURE 6.10 Profiles of velocity and stress in laminar flow of power fluid in annular channel.

To get velocity profile in each region, employ the same reasoning as was used before for the flow of power fluid in pipe. Therefore, in annular space would be true relation between (6.4.21) and (6.4.22) for both regions except that now $B \neq 0$.

Alternately, substitution of (6.4.28) and (6.4.29) in (6.4.22) gives for region I

$$\frac{\partial w}{\partial r} = -\left(\frac{-A}{2k}\right)^{1/n}\left[r + \frac{2B}{Ar}\right]^{1/n}; \tag{6.4.32}$$

for region II

$$\frac{\partial w}{\partial r} = \left(\frac{-A}{2k}\right)^{1/n}\left[-\frac{2B}{Ar} - r\right]^{1/n}. \tag{6.4.33}$$

The constant B is got from (6.4.32) or (6.4.33) through the third boundary condition (6.4.30)

$$\left.\frac{\partial w}{\partial r}\right|_{r=d_{ax}/2} = -\frac{2B}{A(d_{ax}/2)} - \frac{d_{ax}}{2} = 0,$$

from which

$$B = -\frac{A}{2}\left(\frac{d_{ax}}{2}\right)^2. \tag{6.4.34}$$

Insertion of (6.4.34) in (6.4.32) and (6.4.33) gives velocity distributions at fulfillment of the first two conditions (6.4.30):

for region I

$$w = -\frac{A}{2k}\left(\frac{|A|}{2k}\right)^{\frac{1}{n}-1}\int_{r}^{d_c/2}\left[r - \left(\frac{d_{ax}}{2}\right)^2\frac{1}{r}\right]^{1/n} \mathrm{d}r; \tag{6.4.35}$$

for region II

$$w = -\frac{A}{2k}\left(\frac{|A|}{2k}\right)^{\frac{1}{n}-1}\int_{d_{in}/2}^{r}\left[\frac{1}{r}\left(\frac{d_{ax}}{2}\right)^2 - r\right]^{1/n} \mathrm{d}r. \tag{6.4.36}$$

Unknown diameter d_{ax}, at which the velocities are equal and that divides the flow into two regions, could be determined by equating (6.4.35) and (6.4.36) at $r = d_{ax}/2$

$$\int_{d_{ax}/2}^{d_{ex}/2}\left[r - \left(\frac{d_{ax}}{2}\right)^2\frac{1}{r}\right]^{1/n} \mathrm{d}r = \int_{d_{in}/2}^{d_{ax}/2}\left[\left(\frac{d_{ax}}{2}\right)^2\frac{1}{r} - r\right]^{1/n} \mathrm{d}r.$$

Introduce notations $\delta = d_{in}/d_{ex}$, $\zeta = d_{ax}/d_{ex}$, and $y = 2r/d_{ex}$. Then the foregoing equality would be written as

$$\int_{\delta}^{\zeta} \left[\frac{\zeta^2}{y} - y\right]^{1/n} dy = \int_{\zeta}^{1} \left[y - \frac{\zeta^2}{y}\right]^{1/n} dy. \tag{6.4.37}$$

In accordance with equation (6.4.37) in Fig. 6.11 are plotted graphics of the function $\zeta = \zeta(1/n, \delta)$. Fluid flow rate may be found by using (6.4.35) and (6.4.36)

$$Q = 2\pi \int_{d_{in}/2}^{d_{ex}/2} wr\, dr = \pi \left(\frac{d_{ex}}{2}\right)^{\frac{1}{n}+3} \left(-\frac{A}{2k}\right) \left(\frac{|A|}{2k}\right)^{\frac{1}{n}-1} \int_{\delta}^{1} \left|\frac{\zeta^2}{y} - y\right|^{\frac{1}{n}+1} y\, dy$$

or

$$Q = \pi \left(\frac{d_{ex}}{2}\right)^{\frac{1}{n}+3} \left(-\frac{A}{2k}\right) \left(\frac{|A|}{2k}\right)^{\frac{1}{n}-1} I, \tag{6.4.38}$$

where

$$I = \int_{\delta}^{1} \left|\frac{\zeta^2}{y} - y\right|^{\frac{1}{n}+1} y\, dy.$$

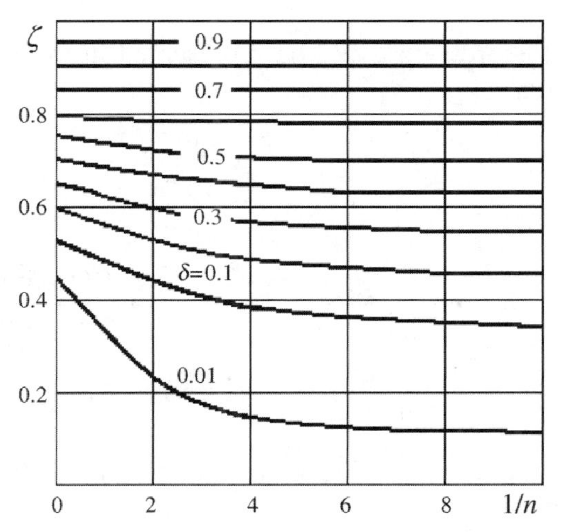

FIGURE 6.11 Graphics of the function $\zeta\ (1/n, \delta)$.

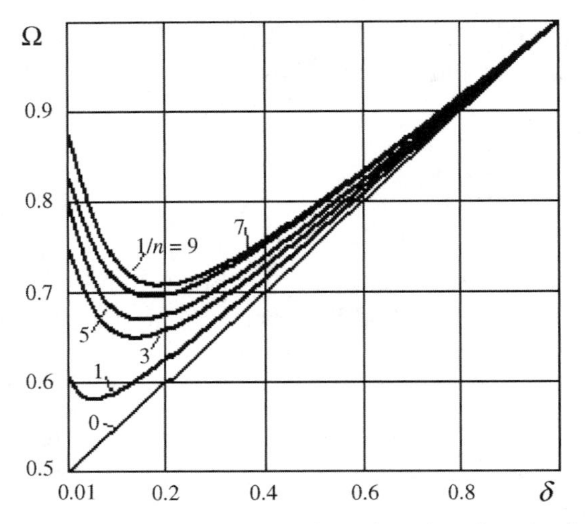

FIGURE 6.12 Graphics of the function Ω (1/*n*, δ).

Figure 6.12 shows the dependence Ω (1/*n*, δ) (Fredrickson and Bird, 1958), which allows to get integral

$$I = \frac{\Omega(1-\delta)^{\frac{1}{n}+2}}{\left(\frac{1}{n}+2\right)}. \tag{6.4.39}$$

Resolve (6.4.38) with respect to Δp

$$\Delta p = \left(\frac{|Q|}{\pi I}\right)^{n} \frac{2kL}{(d_c/2)^{3n+1}}. \tag{6.4.40}$$

The obtained expression could be represented in the form of Darcy–Weisbach formula

$$\Delta p = |p_2 - p_1| = \lambda \frac{\rho v^2}{2d_h} L, \tag{6.4.41}$$

where

$$\lambda = \frac{64}{Re} f(\delta, n); \quad Re = \frac{|v|^{2-n} d_h^n \rho}{k}; \quad f(\delta, n) = \frac{2^{n-3}(1-\delta^2)^n (1-\delta)^{n+1}}{I^n};$$

$$v = \frac{4Q}{\pi(d_{ex}^2 - d_{in}^2)}; \quad d_h = d_{ex} - d_{in}.$$

At $n \to 1$ the relation (6.4.41) goes into the formula (6.2.29) for viscous fluid with $f(\delta, n) \to f(\delta)$.

EXERCISE 6.4.1

It is required to determine pressure drop in annular channel of a well at the following given data: $Q = 0.0189\,m^3/s$, $d_{ex} = 0.214\,m$, $d_{in} = 0.114\,m$, $n = 0.2$, $k = 0.1\,Pa{\cdot}s^{0.2}$, and $L = 1000\,m$.

SOLUTION Calculate

$$\delta = 0.114/0.214 = 0.533; \quad 1/n = 1/0.2 = 5.$$

From Fig. 6.12 it is found $\Omega = 0.78$ and with formula (6.4.39) the following is obtained

$$I = \frac{\Omega(1-\delta)^{\frac{1}{n}+2}}{\frac{1}{n}+2} = \frac{0.78(1-0.533)^{5+2}}{5+2} = 5.39 \times 10^{-4}.$$

In accordance with (6.4.40)

$$\Delta p = \left[\frac{|Q|}{\pi I}\right]^n \frac{2k}{(d_{ex}/2)^{3n+1}} L = \left[\frac{0.0189}{3.14 \times 5.39 \times 10^{-4}}\right]^{0.2} \times \frac{2 \times 0.1 \times 1000}{(0.214/2)^{3\times 0.2+1}} = 0.116 \times 10^5\ \text{Pa}.$$

6.5 CALCULATION OF PRESSURE IN TURBULENT FLOWS IN PIPES AND ANNULAR CHANNELS

To calculate pressure in turbulent flows in pipes and annular channels with length L, Darcy–Weisbach formula is commonly used.

$$\Delta p = \lambda \cdot \frac{\rho v^2}{2 \cdot d_h} L, \tag{6.5.1}$$

where λ is the factor of hydraulic resistance for turbulent flow; d_h is the hydraulic diameter equal to pipe diameter d for flows in pipes or to find difference $d_h = d_{ex} - d_{in}$ of diameters of pipes restricting the annular space.

The relation (6.5.1) was obtained in preceding sections by solving the system of equations for different forms of rheological equations for viscous, viscous-plastic, and power fluids used in actual practice. General form of λ for viscous fluids demonstrates Fig. 6.13 (Schlichting, 1964).

Laminar flow of fluids with a different rheology ceases at certain Reynolds number called critical (Re_{cr}). On reaching the first critical Reynolds number ($Re = Re_{cr1}$) begins the so-called transient regime when the first evidences of turbulent (chaotic, eddy) flow regime are coming into

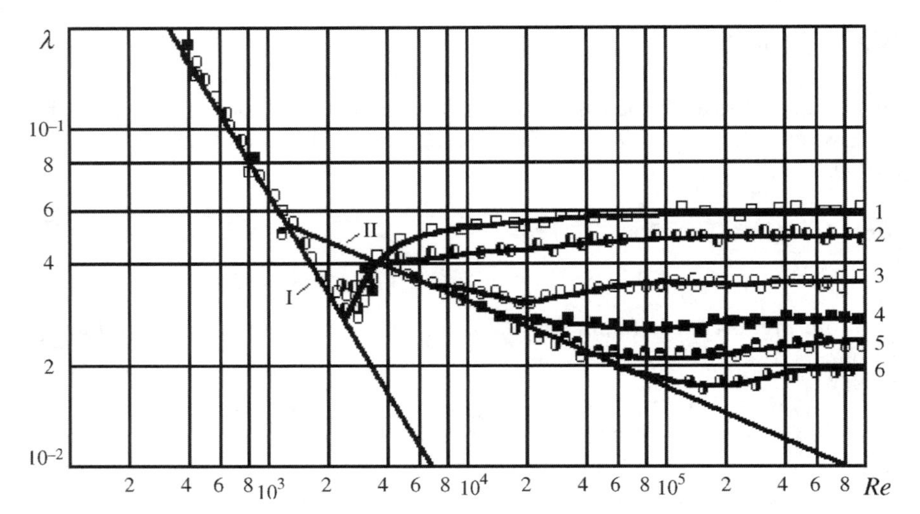

FIGURE 6.13 Resistance factor of viscous fluids: I, II—Poiseuille and Blasius dependences; Nikuradse empirical data at different values of relative roughness $\varepsilon \cdot 10^3$: 1—66.7; 2—32.7; 3—16.7, 4—7.94; 5—3.97; 6—1.97.

view. Parallel flow of fluid jets is disturbed, occur fluctuations of longitudinal velocity, and comes into view lateral motion of produced eddies that move to the pipe axis with fluctuated velocity. Experiments performed by Prandtl showed that maximal velocity fluctuations occur between wall and axis of the pipe decreasing down-flow along the flow axis and directly at the wall. At this the hydraulic resistance increases because on formation and motion of eddies additional energy is expended. As showed by Nikuradse experiments, the hydraulic resistance factor λ in the vicinity of transition regime is ambiguous function of Reynolds number *Re*. Therefore, formulas for λ, although they exist, are not given here. Go directly to consideration of fully developed turbulent flow using it as upper estimation for factor λ in transition region.

At $Re \geq Re_{cr2}$, where Re_{cr2} is the second critical Reynolds number (for Newtonian fluids $Re_{cr2} = 2320$), the flow is turbulized up to the pipe axis resulting in the production of turbulent flow core consisting of a multitude of eddies fluctuating across the flow in average motion along the pipe axis (Fig. 6.14) and laminar sublayer of depth δ_L in which is true rheological equation of laminar flow. In the transition zone between laminar sublayer and turbulent core (not to be confused with the transition zone (Fig. 6.14) in the turbulent flow structure with transition regime from laminar to turbulent flow), the fluid can be taken as having rheological equation of mixed type. The fluid in the turbulent core has its own rheological equation.

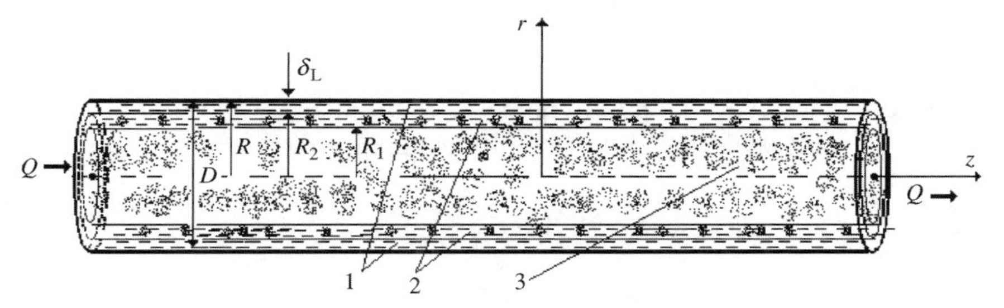

FIGURE 6.14 The structure of turbulent flow in a pipe: 1—laminar sublayer; 2—transition layer (zone); 3—developed turbulent flow (core).

For each structure layer 1–3 (Fig. 6.14) of the turbulent flow in a pipe should be solved the system of equation (6.1.1)–(6.1.4) with its own rheological equations, sticking boundary conditions at the wall and additional boundary conditions reflecting equality of velocities at the boundaries dividing structure layers.

Consider a flow directed opposite z-coordinate. At this $\partial w/\partial z > 0$, $\partial p/\partial z = A > 0$, and longitudinal velocity $w(r)$ to be determined would be negative. Single out near the pipe wall a laminar sublayer with thickness $\delta_L = R - R_2$ (Fig. 6.14) in which rheological properties are defined by one of the above considered fluid models (viscous, viscous-plastic, power, respectively)

$$\tau = \mu \frac{\partial w}{\partial r}; \tag{6.5.2}$$

$$\tau = \tau_0 + \mu \frac{\partial w}{\partial r}; \tag{6.5.3}$$

$$\tau = k \left(\frac{\partial w}{\partial r} \right)^n. \tag{6.5.4}$$

The thickness δ_L for viscous fluid is a function of Reynolds number Re, for viscous-plastic fluid—of Re and He numbers and for power fluid—of Re and exponent n:

$$\delta_L = \delta_L(Re); \tag{6.5.5}$$

$$\delta_L = \delta_L(Re, He); \tag{6.5.6}$$

$$\delta_L = \delta_L(Re, n). \tag{6.5.7}$$

In the sublayer δ_L owing to great value of the derivative $\partial w/\partial r$, the stress τ can also reach significant value. Consequently in (6.5.3) $\tau_0 \ll \eta \partial w/\partial r$, that is non-Newtonian properties of viscous-plastic fluid do not play significant role and (6.5.6) could be replaced by (6.5.5), what could not be said about power fluid (6.5.4) in which the influence of n may be significant in (6.5.7).

In accordance with Prandtl hypothesis in turbulent flow arise additional tangential stresses as a result of energy expenditure on motion of eddies

$$\tau = \rho(\kappa l)^2 \left(\frac{\partial w}{\partial r}\right)^2, \tag{6.5.8}$$

where κ is Karman universal constant; $l = R - r$ is the distance from the wall surface.

Then the flow in the transition layer (Fig. 6.14) could be described by rheological equation of one of the used fluid models (viscous, viscous-plastic, power) extending Prandtl hypothesis up to the pipe axis

$$\tau = \mu \frac{\partial w}{\partial r} + \rho(\kappa l)^2 \left(\frac{\partial w}{\partial r}\right)^2; \tag{6.5.9}$$

$$\tau = \tau_0 + \mu \frac{\partial w}{\partial r} + \rho(\kappa l)^2 \left(\frac{\partial w}{\partial r}\right)^2; \tag{6.5.10}$$

$$\tau = k \left(\frac{\partial w}{\partial r}\right)^n + \rho(\kappa l)^2 \left(\frac{\partial w}{\partial r}\right)^2. \tag{6.5.11}$$

The account for transition layer permits to conserve the hypothesis of laminar layer sticking up to the wall and to build a model explaining lowering of the resistance factor with the help of additives.

For developed turbulent flow a dominant role in formulas plays turbulent terms, and the rheological equation for flow core is

$$\tau = \rho(\kappa l)^2 \left(\frac{\partial w}{\partial r}\right)^2. \tag{6.5.12}$$

Consider formulation of the problem permitting to determine flow characteristics and consequently the dependence (6.5.1) for turbulent flow. It is required to solve the system of equations (6.1.1)–(6.1.4) representing equation (6.1.4), for example, for viscous-plastic fluid, through (6.5.3), (6.5.10), or (6.5.12):

$$\begin{aligned}
&\tau = \rho(\kappa l)^2 \left(\frac{\partial w}{\partial r}\right)^2 \quad \text{at} \quad 0 \le r < R_1; \\
&\tau = \tau_0 + \mu \frac{\partial w}{\partial r} + \rho(\kappa l)^2 \left(\frac{\partial w}{\partial r}\right)^2 \quad \text{at} \quad R_1 \le r \le R_2; \\
&\tau = \tau_0 + \mu \frac{\partial w}{\partial r} \quad \text{at} \quad R_2 < r \le R.
\end{aligned} \tag{6.5.13}$$

The solution of (6.1.1)–(6.1.3) together with (6.5.13) may be obtained, but it has too cumbersome form. Therefore, let us give a solution for viscous-fluid flow ($\tau_0 = 0$) at $R_1 \approx R_2$, that is, neglecting transition layer (two-layer model). The system of equations (6.1.1)–(6.1.4) with (6.5.12) and $l = R - r$ reduces to two equations

$$\frac{1}{r} \cdot \frac{\partial r\tau}{\partial r} = \frac{\partial p}{\partial z}, \tag{6.5.14}$$

$$\tau = \rho\kappa^2 (R-r)^2 \cdot \left(\frac{\partial w}{\partial r}\right)^2 \tag{6.5.15}$$

under condition $0 < r < R - \delta$.

In considered case as well as for laminar flow, $p = p(z)$ and $w = w(r)$ would be valid relations. Then both parts of the equation (6.5.14) have to be constant

$$\frac{\partial p}{\partial z} = A; \tag{6.5.16}$$

$$\frac{1}{r} \cdot \frac{\partial r\tau}{\partial r} = A. \tag{6.5.17}$$

Substitution of (6.5.15) into (6.5.17) after some algebra gives

$$\frac{\partial w}{\partial r} = \frac{1}{\kappa}\sqrt{\frac{A}{2\rho}\frac{\sqrt{r}}{R-r}}, \tag{6.5.18}$$

where $0 \le r \le R_2$; $\delta_L = R - R_2$ is the thickness of laminar sublayer.

Integration of this equation yields logarithmic velocity profile of the pipe flow

$$w(r) = \frac{1}{\kappa}\sqrt{\frac{Ad}{4\rho}}\left(-2\sqrt{\frac{r}{R}} + \ln\left(\frac{1+\sqrt{r/R}}{1-\sqrt{r/R}}\right)\right) + C. \tag{6.5.19}$$

The constant C is equal to maximal flow velocity w_{max}, since at the flow axis $r = 0$ should be $w = w_{max}$. Transform the formula (6.5.19) introducing into it the distance from the pipe wall $l = R - r$ and replacing velocity with dimensionless one

$$r = R - l, \quad \frac{w_{max} - w}{v_*} = -\frac{2}{\kappa}\left(\ln\left(\frac{1+\sqrt{1-l/R}}{\sqrt{l/R}}\right) - \sqrt{1-l/R}\right), \quad v_* = \sqrt{\frac{Ad}{4\rho}}.$$

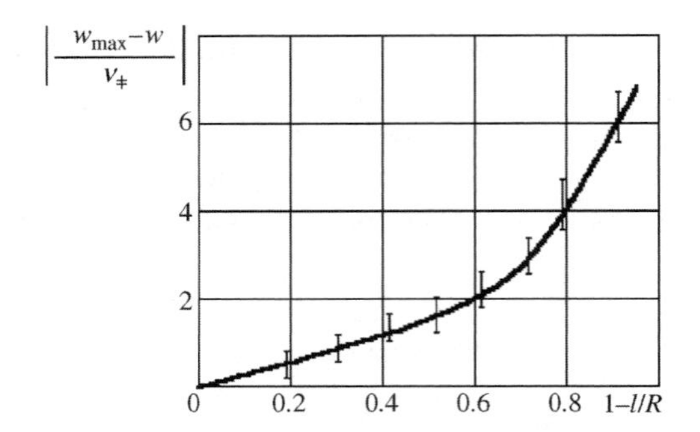

FIGURE 6.15 Dependence of relative velocity on relative distance from the wall.

At small but nonzero l/R, the following asymptotic formula is obtained (Loitsyansky, 1987; Schlichting, 1964):

$$\frac{w_{\max} - w}{v_*} = -\frac{1}{\kappa}\ln\frac{R}{l}. \tag{6.5.20}$$

Figure 6.15 represents Nikuradse empirical data (Loitsyansky, 1987) for Reynolds number $4 \times 10^4 < Re < 3.24 \times 10^6$ (vertical segments) and the curve plotted by formula (6.5.20) at $\kappa = 0.4$. As it is seen the formula (6.5.20) satisfactorily describes real velocity distribution.

Obtain $w_{\max}$ from the matching condition $w_T = \alpha \cdot w_L$ of turbulent and laminar velocity profiles at $r = R_1$. The correction factor α is introduced due to ignoring the transition layer. The velocity of the laminar flow (6.2.26) directed opposite the z-axis is

$$\begin{aligned} w_L(r) &= -\frac{AR^2}{4\mu}\left(1 - \frac{r^2}{R^2}\right) = -v_*^2\frac{\rho}{\mu}\frac{R}{2}\left(1 - \frac{r^2}{R^2}\right) \\ &= -v_*^2\frac{\rho}{\mu}\frac{R}{2}\left(1 - \left(\frac{R-l}{R}\right)^2\right) = -v_*^2\frac{\rho}{\mu}\frac{l}{2}\left(2 - \frac{l}{R}\right). \end{aligned} \tag{6.5.21}$$

From (6.5.19) at $C = w_{\max}$ ensues

$$w_T = w_{\max} + \frac{2v_*}{\kappa}\left(\ln\left(\frac{1+\sqrt{1-l/R}}{\sqrt{l/R}}\right) - \sqrt{1-l/R}\right).$$

As far as $w_T = \alpha \cdot w_L$ at $l = \delta_L$ we get

$$w_{\max} = -\frac{2v_*}{\kappa}\left(\ln\left(\frac{1+\sqrt{1-\delta_L/R}}{\sqrt{\delta_L/R}}\right) - \sqrt{1-\delta_L/R}\right) - \alpha v_*^2\frac{\rho\delta}{2\mu}(2-\delta_L/R).$$

Substitution of $w_{\max}$ in (6.5.20) gives velocity distribution

$$\frac{w}{v_*} = -\frac{1}{\kappa}\ln\frac{\delta_L}{l} + \frac{2}{\kappa}\left(\sqrt{1-\frac{\delta_L}{R}} - \sqrt{1-\frac{l}{R}}\right) + \frac{2}{\kappa}\ln\left(\frac{1+\sqrt{1-\frac{l}{R}}}{1+\sqrt{1-\frac{\delta_L}{R}}}\right)$$

$$-\alpha v_*\frac{\rho\delta_L}{2\mu}\left(2-\frac{\delta_L}{R}\right).$$

Since $\delta_L/R \ll 1$, it is

$$\frac{w}{v_*} = -\frac{1}{\kappa}\ln\frac{\delta_L}{l} + \frac{2}{\kappa}\left(1-\sqrt{1-\frac{l}{R}}\right) + \frac{2}{\kappa}\ln\left(\frac{1+\sqrt{1-\frac{l}{R}}}{2}\right) - \alpha v_*\frac{\rho\delta_L}{2\mu}. \tag{6.5.22}$$

In order to determine δ_L, we exploit the condition of velocity derivative matching at $l = \delta_L$. Use of relations (6.5.18) and (6.2.26) at $l = \delta_L$ yields

$$\left.\frac{\partial w_L}{\partial r}\right|_{l=\delta_L} = \frac{A}{2\mu}(R-\delta_L) = \frac{1}{\alpha}\left.\frac{\partial w_T}{\partial r}\right|_{l=\delta_L} = \frac{1}{\alpha\kappa}\sqrt{\frac{A}{2\rho}}\frac{\sqrt{R-\delta_L}}{\delta}.$$

Since $\delta_L/R \ll 1$, one gets

$$\delta_L = \frac{\mu}{\alpha\cdot\kappa\cdot v_*\rho}; \qquad \frac{\delta_L}{R} = \frac{\mu}{\alpha R\cdot\kappa\cdot v_*\rho}$$

and

$$\frac{\delta_L}{l} = \frac{\mu}{\alpha R\cdot\kappa\cdot v_*\cdot\rho l}. \tag{6.5.23}$$

Substitution of δ_L from (6.5.23) in (6.5.22) gives

$$\frac{w}{v_*} = -\frac{1}{\kappa}\ln\frac{v_*\cdot\rho l}{\mu} + \frac{1}{\kappa}\ln\frac{1}{\alpha\kappa} + \frac{2}{\kappa}\left(1-\sqrt{1-\frac{l}{R}}\right) + \frac{2}{\kappa}\ln\left(\frac{1+\sqrt{1-\frac{l}{R}}}{2}\right) - \frac{1}{\kappa}. \tag{6.5.24}$$

Replace the third and fourth terms through their mean values. Then

$$\frac{w}{v_*} = -\frac{1}{\kappa}\ln\frac{v_*\cdot\rho l}{\mu} + \frac{1}{\kappa}\ln\frac{1}{\alpha\kappa} + \frac{1}{\kappa}\ln\left(\frac{1}{2}\right).$$

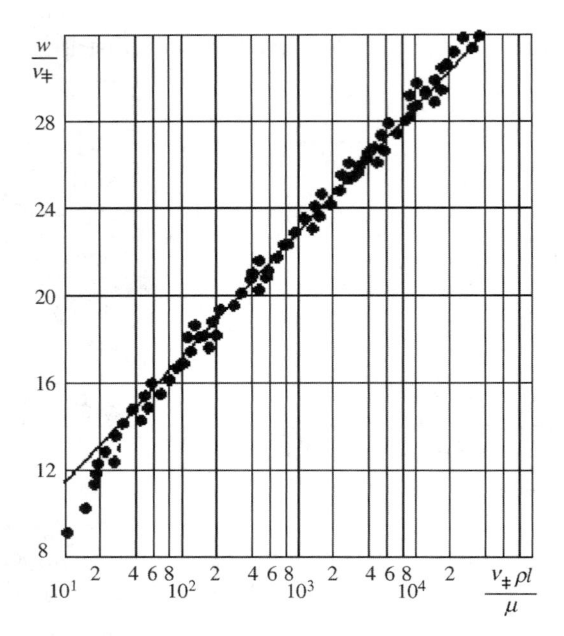

FIGURE 6.16 Dependence (6.5.25).

At $\kappa = 0.4$ and $\alpha = 11.28$, this relation goes into Prandtl universal law of absolute velocity distribution depicted in Fig. 6.16 (points represent Nikuradse empirical data (Loitsyansky, 1987) for Reynolds numbers $4 \times 10^4 < Re < 3.24 \times 10^6$)

$$\frac{w}{v_*} = 5.75 \lg \frac{v_* \times \rho l}{\mu} + 5.5. \tag{6.5.25}$$

The formula (6.5.25) describes universal velocity profile for all Reynolds numbers. The flow rate Q may be received by

$$Q = 2\pi \int_0^R wr \, \mathrm{d}r = 2\pi \int_0^{R-\delta_L} w_T r \, \mathrm{d}r + 2\pi \int_{R-\delta_L}^{R} w_L r \, \mathrm{d}r = 2\pi w_T \frac{r^2}{2}\bigg|_0^{R-\delta_L}$$

$$+ 2\pi w_L \frac{r^2}{2}\bigg|_{R-\delta_L}^{R} - 2\pi \int_0^{R-\delta_L} \frac{\partial w_T}{\partial r}\frac{r^2}{2}\,\mathrm{d}r - 2\pi \int_{R-\delta_L}^{R} \frac{\partial w_L}{\partial r}\frac{r^2}{2}\,\mathrm{d}r. \tag{6.5.26}$$

Inserting w_L from (6.5.21), $w_T = \alpha \cdot w_L$ at $r = R - \delta_L$, $\partial w_T/\partial r$ from (6.5.18) in (6.5.26) and accepting $\delta_L/R \ll 1$, we obtain

$$Q = -\alpha\pi(v_*)^2 \frac{\rho\delta_L}{\mu} R^2 - \frac{\pi v_* R^2}{\kappa}\left(-\frac{46}{15} + 2\ln 2 - \ln\left(\frac{\delta_L}{R}\right)\right).$$

Transform the velocity v_* into

$$v_* = \sqrt{\frac{|A|R}{2\rho}} = \sqrt{\frac{\Delta p \cdot R}{2\rho L}} = \sqrt{\frac{\lambda(\rho v^2/2d)Ld}{4\rho L}} = \sqrt{\frac{\lambda}{8}}|v| = \sqrt{\frac{\lambda}{8}}\frac{|Q|}{\pi R^2}. \tag{6.5.27}$$

Then in expression for Q substitute found value v_*, δ_L/R from (6.5.23), $\kappa = 0.4$, and earlier accepted value $\alpha = 11.28$. Finally, we get Prandtl formula (Schlichting, 1964; Loitsyansky, 1987)

$$\frac{1}{\sqrt{\lambda}} = 2.04\log(Re \cdot \sqrt{\lambda}) - 0.8, \tag{6.5.28}$$

which is well confirmed by experiments. The pressure drop in turbulent flow in smooth pipes could be received by Darcy–Weisbach formula

$$\Delta p = \lambda\frac{\rho v^2}{2d}L = \lambda\frac{\rho Q^2}{2d \cdot S^2}L, \tag{6.5.29}$$

where λ is determined by (6.5.28).

For convenience sake the equation (6.5.28) for various Re could be approximated as (Loitsyansky, 1987)

$$\lambda = \frac{C}{Re^m}. \tag{6.5.30}$$

At $C = 0.316$ and $m = 0.25$ for $Re < 10^5$, the expression (6.5.30) represents well-known Blasius formula

$$\lambda = \frac{0.316}{Re^{0.25}}. \tag{6.5.31}$$

Karman has shown that the use of (6.5.31) leads to power velocity profile (one-seventh law) being the approximation of universal profile (6.5.25) for $Re < 10^5$

$$\frac{w}{w_{\max}} = \left(\frac{l}{R}\right)^{1/7}. \tag{6.5.32}$$

At other *Re* numbers, values of C and m in formula (6.5.30) will give another power law approximating the profile (6.5.25)

$$\frac{w}{w_{\max}} = \left(\frac{l}{R}\right)^{1/N}. \tag{6.5.33}$$

Between exponents m and N, there is a simple relation

$$m = \frac{2}{N+1}. \tag{6.5.34}$$

The velocity profile could be represented in dimensionless coordinates w/v_*, $v_* l\rho/\mu$ as follows:

$$\frac{w}{v_*} = C(N)\left(\frac{v_* l\rho}{\mu}\right)^{1/N}. \tag{6.5.35}$$

Below are exhibited values of C and N depending on *Re* (Schlichting, 1964)

Re	1.1×10^5	8.0×10^5	2.0×10^6	3.2×10^6
N	7	8	9	10
$C(N)$	8.74	9.71	10.6	11.5

Formulas given above are true when the absolute equivalent roughness k_r (height of surface wall asperities) in the pipe is lesser than the layer thickness δ_L. In this case, the wall roughness does not influence the factor λ, that is the pressure drop would be the same as in pipes with $k_r = 0$ (hydraulic smooth pipes).

At $k_r > \delta_L$, the laminar sublayer plays lesser role than the pipe roughness, and at a certain value of roughness the resistance λ becomes constant, that is, it is set as the so-called self-similar turbulence (complete rough pipes). Thus, since $\delta_L = f(Re)$ for one fluid flow rates, the pipe behaves as hydraulic smooth whereas for another as rough pipe. The relative roughness ε is defined as ratio between absolute equivalent roughness and pipe radius

$$\varepsilon = k_r/R. \tag{6.5.36}$$

The roughness of seamless steel pipes is $k_r = (1 \div 2) \cdot 10^{-5}$ m. After several years of exploitation, it reaches the value $k_r = (15 \div 30) \cdot 10^{-5}$ m. Hence, the factor λ in turbulent flow depends not only on *Re* but also on ε (Fig. 6.13)

$$\lambda = \lambda(Re,\ \varepsilon). \tag{6.5.37}$$

For viscous-plastic and power fluids curves $\lambda = \lambda(Re, \varepsilon)$ may be considered as similar going with a different roughness on one and the same self-similar limiting curve.

Different authors have developed dependences λ for viscous, viscous-plastic, and power fluids by approximate methods. The most successful and tested in the practice is Altshul formula for viscous fluids (Altshul and Kiselev, 1975). For $Re < 10^5$ it has form

$$\lambda = 0.1\left(\frac{1.46k_r}{d} + \frac{100}{Re}\right)^{0.25}. \tag{6.5.38}$$

This formula is a generalization of the well-known Blasius formula following from (6.5.38) at $k_r = 0$

$$\lambda = \frac{0.316}{Re^{0.25}}. \tag{6.5.39}$$

For pipes with high roughness for which the second term in (6.5.38) is much less than the first one, there is Shifrinson formula (see Altshul and Kiselev, 1975)

$$\lambda = 0.11\left(\frac{k_r}{d}\right)^{0.25}. \tag{6.5.40}$$

In self-similar flow in hydraulic smooth pipes, it may be taken λ 0.0128. Since in elements of well-circulation systems the roughness is commonly unknown, it is often accepted that $\lambda = 0.02$.

6.5.1 Turbulent Flow in Annular Channel

To get relation between pressure drop and flow rate in the case under consideration, it is required to solve the equation (6.5.14) together with rheological law (6.5.15) for turbulent core and then to match the previously obtained solution with the solution for laminar layers.

Simplify the problem. Accept that the velocity power distribution (6.5.35) takes place in each half of the annular channel flow (see Fig. 6.17). Describe both branches and then match them using equality of tangential stresses at $r = R_a$ (Gukasov, 1976). Thus, the velocity profile is represented as

$$w_1(r) = C(N)v_{*1}\left(v_{*1}\frac{(r-R_1)\rho}{\mu}\right)^{1/N}, \quad R_1 \le r \le R_a, \tag{6.5.41}$$

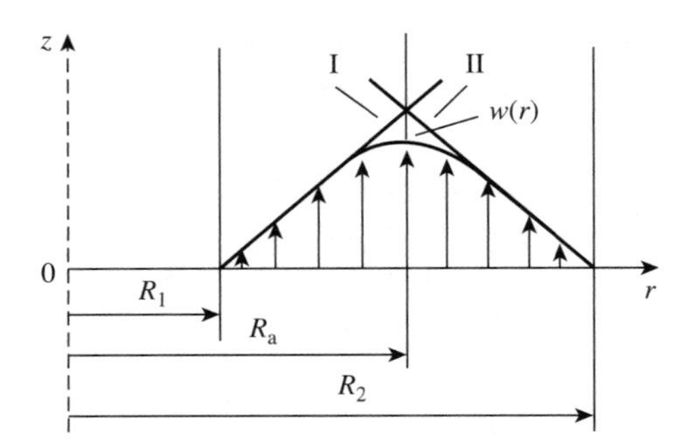

FIGURE 6.17 A scheme of velocity profile approximation in turbulent flow in annular channel: I—velocity profile branch in accord to formula (6.5.41); II—the same in accord to (6.5.42); $w(r)$—real velocity profile.

$$w_2(r) = C(N)v_{*2}\left(v_{*2}\frac{(R_2-r)\rho}{\mu}\right)^{1/N}, \quad R_a \le r \le R_2, \tag{6.5.42}$$

where $v_{*1} = \sqrt{\tau_1/\rho}$, $v_{*2} = \sqrt{\tau_2/\rho}$, τ_1, and τ_2 are stresses at inner and outer surfaces of the annular space; R_a is the radius to be determined at which the velocities are required to be equal.

Equating (6.5.41) and (6.5.42) at $r = R_a$, we get the relation between τ_1 and τ_2

$$\tau_1 = \tau_2\left(\frac{R_2-R_a}{R_a-R_1}\right)^{2(N+1)}. \tag{6.5.43}$$

It remains to find relation between stresses τ_1, τ_2, and pressure drop Δp. To do this, it is sufficient to write out the condition of fluid dynamic equilibrium in the whole annular space and in one of its part (for example, in branch II of the velocity profile) spaced between $r = R_a$ and $r = R_1$ or $r = R_2$:

$$\pi\Delta p(R_2^2-R_1^2) = 2\pi \cdot R_1 L\tau_1 + 2\pi \cdot R_2 L\tau_2; \tag{6.5.44}$$

$$\pi\Delta p(R_2^2-R_a^2) = 2\pi \cdot R_2 L\tau_2. \tag{6.5.45}$$

Determining τ_1 and τ_2 from (6.5.44) and (6.5.45) and substituting them in (6.5.43), we obtain the equation for radius R_a where velocities of both profiles would be equal

$$\frac{R_2}{R_1}\frac{R_a^2-R_1^2}{R_2^2-R_a^2} = \left(\frac{R_2-R_a}{R_a-R_1}\right)^{2(N+1)}. \tag{6.5.46}$$

Eliminating from (6.5.41) and (6.5.42) τ_1 and τ_2 with the help of (6.5.44)–(6.5.46) we receive velocity profile

$$w_1(r) = \varphi_1(r-R_1)^{1/N}, \quad R_1 \le r \le R_a; \tag{6.5.47}$$

$$w_2(r) = \varphi_2(R_2-r)^{1/N}, \quad R_a \le r \le R_2, \tag{6.5.48}$$

where

$$\varphi_1 = 0.98C(N)\left[\frac{\Delta p(R_2^2-R_a^2)}{2\rho L R_2}\right]^{(N+1)/2N}\left[\left(\frac{R_2-R_a}{R_a-R_1}\right)\frac{\rho}{\mu}\right]^{1/N}; \tag{6.5.49}$$

$$\varphi_2 = 0.98C(N)\left[\frac{\Delta p(R_2^2-R_a^2)}{2\rho L R_2}\right]^{(N+1)/2N}\left[\frac{\rho}{\mu}\right]^{1/N}. \tag{6.5.50}$$

The factor 0.98 is introduced to compensate the difference between real values of velocities at $r = R_a$ and those calculated by (6.5.47) and (6.5.48).

Dividing (6.5.49) by (6.5.50), one obtains

$$\frac{\varphi_1}{\varphi_2} = \left(\frac{R_2-R_a}{R_a-R_1}\right)^{1/N} = \left(\frac{1-\delta_a}{\delta_a-\delta}\right)^{1/N}, \tag{6.5.51}$$

where $\delta = R_1/R_2$; $\delta_a = R_a/R_2$.

Find now the flow rate in annular channel employing (6.5.47) and (6.5.48)

$$Q = 2\pi\int_{R_1}^{R_2} wr\,\mathrm{d}r = 2\pi\varphi_2\left[\left(\frac{1-\delta_a}{\delta_a-\delta}\right)^{1/N}\int_{R_1}^{R_a}(r-R_1)^{1/N} r\,\mathrm{d}r + \int_{R_a}^{R_2}(R_2-r)^{1/N} r\,\mathrm{d}r\right]$$

$$= 2\pi\varphi_2 {R_2}^{1/N+2}\frac{N}{(N+1)(2N+1)}(1-\delta_a)^{1/N}(1-\delta)[\delta_a + N(\delta+1)]. \tag{6.5.52}$$

Now from (6.5.52) with (6.5.50) ensues

$$\Delta p = \lambda\frac{\rho v^2}{2d_h}L, \tag{6.5.53}$$

where $\lambda = f(\delta, N)/Re^{2/(N+1)}$;

$$f(\delta,N) = \frac{2^{(N+5)/(N+1)}(1+\delta)^{2N/(N+1)}\left(\frac{1-\delta}{1-\delta_a}\right)^{(N+3)/(N+1)}}{0.98C(N)^{2N/(N+1)}(1-\delta_a)\left[\frac{N}{(N+1)(2N+1)}[\delta_a + N(\delta+1)]\right]^{2N/(N+1)}}. \tag{6.5.54}$$

The quantity δ_a is determined from equation

$$\frac{1}{\delta}\left(\frac{\delta_a^2-\delta^2}{1-\delta_a^2}\right)^{1/N} = \left(\frac{1-\delta_a}{\delta_a-1}\right)^{2/(N+1)} \tag{6.5.55}$$

following from (6.5.46).

The value of δ_a at $N=7$ may be calculated by the expression

$$\delta_a = \delta + 0.5(1-\delta)\delta^{0.225} \tag{6.5.56}$$

giving good approximation to the formula (6.5.55).

Then (6.5.54) at $N=7$ becomes

$$f(\delta) = \frac{9.54}{1+\delta_a}\left(\frac{1-\delta}{1-\delta_a}\right)^{1.25}\left[\frac{(1+\delta)}{\delta_a+7(\delta+1)}\right]^{1.75}. \tag{6.5.57}$$

At $\delta \to 0$, that is at $\delta_a \to 0$, Blasius formula for λ is obtained as

$$\lambda = 0.316/Re^{0.25}.$$

Calculations showed that $f(\delta) \approx \text{const} = 0.334$ at $N=7$. The graphic of the function (6.5.54) at $N=7$ is plotted in Fig. 6.18.

With (6.5.53) friction loss having calculated the velocity could be determined as $v = Q/[\pi(R_2^2-R_1^2)]$ and reading values of the function (6.5.54) from the graphic. At $N=7$ for $Re < 10^5$ the factor λ could be written as $\lambda = 0.334/Re^{0.25}$. The formula for λ in annular space with regard to the roughness similar to (6.5.38) can be represented as follows:

$$\lambda = 0.106\left(\frac{1.46k_r}{d_{ex}-d_{in}} + \frac{100}{\text{Re}}\right)^{0.25}. \tag{6.5.58}$$

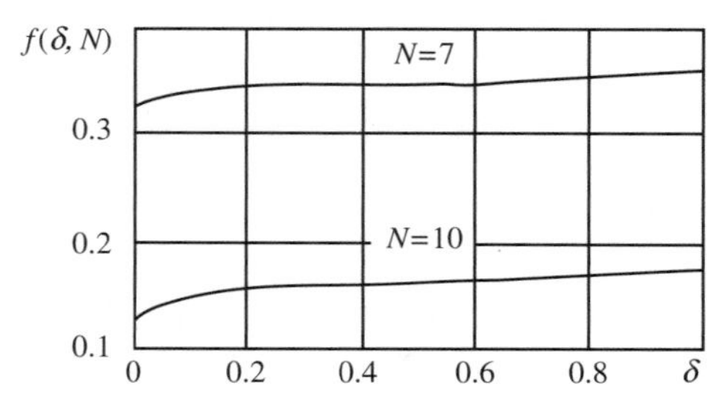

FIGURE 6.18 Graphic of the function $f(\delta, N)$ at $N=7$ and $N=10$.

6.5.2 Turbulent Flow of Viscous-Plastic Fluid in Pipes

In this case, it is required to solve the system of equations (6.1.1)–(6.1.4) together with (6.5.13). The line of reasoning is the same when considering the flow of viscous fluid. A formula similar to (6.5.21) for laminar flow of viscous-plastic fluid would be obtained by transforming the velocity profile in (6.3.23) for the flow in pipes at $w<0$ and $A>0$ as follows:

$$w_L(r) = -\frac{AR^2}{4\eta}\left(1-\frac{r^2}{R^2}\right)+\frac{\tau_0 R}{\eta}\left(1-\frac{r}{R}\right) = -v_*^2\frac{\rho}{\eta}\frac{l}{2}\left(2-\frac{l}{R}\right)+\frac{\tau_0 l}{\eta}. \tag{6.5.59}$$

The velocity distribution would be obtained in the form of (6.5.22) with additional term $(\tau_0/\eta)l$

$$\frac{w}{v_*} = -\frac{1}{\kappa}\ln\frac{\delta_L}{l}+\frac{2}{\kappa}\left(1-\sqrt{1-\frac{l}{R}}\right)+\frac{2}{\kappa}\ln\left(\frac{1+\sqrt{1-\dfrac{l}{R}}}{2}\right)-\alpha v_*\frac{\rho\delta_L}{\eta}+\alpha\frac{\tau_0}{\eta}\frac{\delta_L}{v_*}. \tag{6.5.60}$$

As well as in deriving (6.5.23), the thickness of laminar layer by equating velocity derivatives is found

$$\left.\frac{dw}{dr}\right|_{r=\delta_L} = \frac{A}{2\eta}(R-\delta)-\frac{\tau_0}{\eta} = \frac{1}{\alpha\kappa}\sqrt{\frac{A}{2\rho}}\frac{\sqrt{R-\delta_L}}{\delta_L},$$

from which it follows

$$v_*\frac{\rho}{\eta}-\frac{\tau_0}{\eta v_*} = \frac{1}{\alpha\kappa\delta_L}$$

or

$$\delta_L = \frac{1}{\alpha\kappa\left(v_*\dfrac{\rho}{\eta}-\dfrac{\tau_0}{\eta v_*}\right)}, \tag{6.5.61}$$

where the correction factor α is introduced. The substitution of (6.5.61) into (6.5.60) gives velocity distribution

$$\frac{w}{v_*} = -\frac{1}{\kappa}\ln\left(\frac{v_*\rho l}{\mu}-\frac{\tau_0 l}{\eta v_*}\right)+\frac{1}{\kappa}\ln\frac{1}{\alpha\kappa}+\frac{1}{\kappa}\ln\left(\frac{1}{2}\right).$$

The velocity (6.5.25) and obtained velocity should coincide at $\tau_0 = 0$. Thus, let us accept $\kappa = 0.4$ and $\alpha = 11.28$. Then for absolute velocities the following formula would be true

$$\frac{w}{v_*} = 5.75 \log\left(\frac{v_* \rho l}{\eta} - \frac{\tau_0 l}{\eta v_*}\right) + 5.5. \tag{6.5.62}$$

The fluid flow rate through pipe cross section follows from (6.5.26) with regard to (6.5.59) and (6.5.18)

$$Q = -\alpha\pi v_*^2 \frac{\rho\delta_{\mathrm{L}}}{\eta} R^2 + \alpha\pi R^2 \delta_{\mathrm{L}} \frac{\tau_0}{\eta} - \frac{\pi v_* R^2}{\kappa}\left(-\frac{46}{15} + 2\ln 2 - \ln\left(\frac{\delta_{\mathrm{L}}}{R}\right)\right). \tag{6.5.63}$$

Using now (6.5.61) and (6.5.27), we get resistance factor λ for turbulent flow

$$\frac{1}{\sqrt{\lambda}} = 2.04 \log\left(Re\sqrt{\lambda} - \frac{8He}{Re\sqrt{\lambda}}\right) - 0.8. \tag{6.5.64}$$

In the range $2.3 \times 10^3 \leq Re \leq 3.7 \times 10^5$, it could be represented approximately as

$$\lambda = \frac{0.316}{Re^{0.25}} + \frac{10He}{Re^2}. \tag{6.5.65}$$

In Fig. 6.19, curves 7 are plotted from the formula (6.5.65). For turbulent flow of viscous-plastic fluid, the pressure drop in hydraulic smooth pipes can be calculated with Darcy–Weisbach formula

$$\Delta p = \lambda \frac{\rho v^2}{2 \cdot d_{\mathrm{h}}} L, \tag{6.5.66}$$

where λ is expressed through relation (6.5.64) or (6.5.65).

At $\tau_0 = 0$, the relation (6.5.65) goes into Blasius formula (6.5.31) for viscous fluid. When deriving (6.5.64) wall roughness was not taken into account. In the flow of viscous-plastic fluid, the resistance factor λ is

$$\lambda = 0.106\left(\frac{1.46k_{\mathrm{r}}}{d_{\mathrm{h}}} + \frac{100}{Re}\left(1 + \frac{He}{Re}\right)\right)^{0.25}. \tag{6.5.67}$$

At great Reynolds numbers ($Re > 6 \times 10^5$) for pressure drop Δp of non-Newtonian fluids can be used in formula (6.5.38) or (6.5.58).

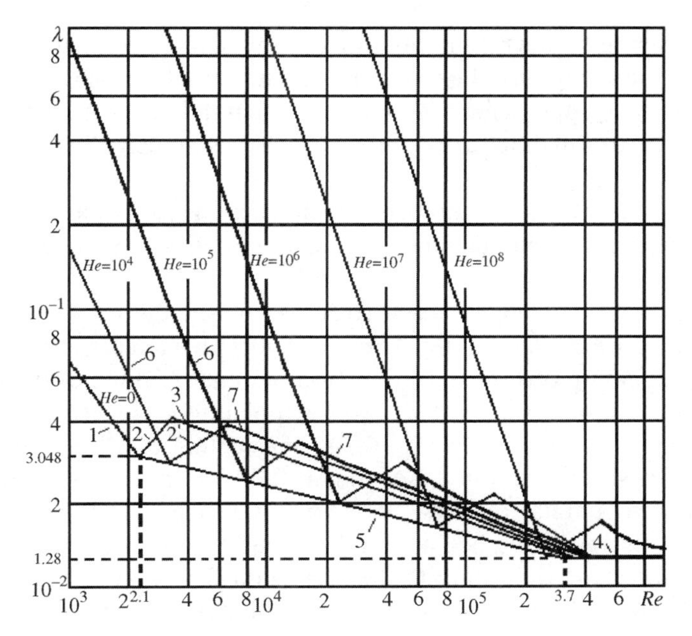

FIGURE 6.19 Dependences of resistance factor in hydraulic smooth pipes: 1—Poiseuille dependence; 2, 2′—transition regime for viscous and viscous-plastic fluids; 3, 4—turbulent flow of viscous fluid ($He = 0$); 4, 5—critical regimes; 6—Buckingham dependence; 4, 7—turbulent flow of viscous-plastic fluid depending on *He*.

The roughness k_r of circulation system elements is commonly unknown. Inside pipe space and in cased sections of annular space, it can noticeably differ from noncased part of the annular space. Therefore, when making upper estimation one should in calculations of pressure drop with (6.5.38) take for pipes and cased sections exaggerated roughness $k_r = 3 \times 10^{-4}$ m as if after several years of pipe exploitation. For noncased sections of annular channels as an estimation, the roughness of a pipe made from unwrought concrete with $k_r = 3 \times 10^{-3}$ m is taken.

6.6 TRANSITION OF LAMINAR FLOW OF VISCOUS, VISCOUS-PLASTIC, AND POWER FLUIDS INTO TURBULENT ONE

As it was mentioned above, the transition regime for viscous fluids determined experimentally begins at critical Reynolds number $Re_{cr} = 2100$ (see Fig. 6.19). The basis to determine Re_{cr} for viscous-plastic and power fluid is a hypothesis, which implies that a flow of any non-Newtonian fluid has four main regions: laminar, transition, turbulent, and self-similar turbulent. With approaching of non-Newtonian fluid properties to viscous ones the curve of hydraulic resistance approaches to such curve for viscous fluid.

At very high Reynolds numbers, the fluids with any physical properties have in hydraulic smooth pipes one self-similarity region of hydraulic resistance.

Blasius empirical dependence (6.5.31) crosses the curve of laminar regime $\lambda = 64/Re$ at $Re = 1187 < Re_{cr} = 2100$ with intersection point being not the beginning of flow deviation from laminar regime. Thus, in what follows the curve 5 is taken for non-Newtonian fluids as critical curve in the first approximation (see Fig. 6.19), connecting two limiting regions common to viscous and non-Newtonian fluids. Hence, this straight line connects two points: the first with coordinates $Re_{cr} = 2100$ and $\lambda_{cr} = 0.03048$ referring to the beginning of viscous fluid transition regime and being at the same time limiting for transition flow regime beginning of non-Newtonian fluids when approaching their properties to the viscous ones; the second with coordinates $Re_{cr} = 3.7 \times 10^5$ and $\lambda_{cr} = 0.0128$ defines the curve of self-similar viscous fluid regime beginning in accord with used hypothesis to this point asymptotically approach resistances of non-Newtonian fluids at degeneracy of their properties. Now the equation for the curve 5 (Fig. 6.19) connecting critical Reynolds numbers in the range $2.1 \times 10^3 < Re_{cr} < 3.7 \times 10^5$ and critical resistances may be written as

$$\lambda_{cr} = \frac{0.11}{Re_{cr}^{0.168}}. \tag{6.6.1}$$

At $Re_{cr} \geq 3.7 \times 10^5$, it can be taken

$$\lambda_{cr} = 0.0128. \tag{6.6.2}$$

For viscous-plastic and power fluids in laminar flow dependences λ on Re are known (see (6.3.50) and (6.4.27)). Substitution in them $Re = Re_{cr}$ gives expressions relating λ_{cr}, Re_{cr}, and parameters characterizing non-Newtonian fluid, namely He or Se for viscous-plastic fluid and n for power fluid

$$\lambda_{cr} = \frac{64}{Re_{cr}} \frac{Se}{8\beta}; \tag{6.6.3}$$

$$\lambda_{\text{cr}} = \frac{64}{Re_{\text{cr}}} 2^{n-3} \left(\frac{3n+1}{n} \right)^n. \tag{6.6.4}$$

Consider the system of equations (6.6.1)–(6.6.3) for flows of viscous-plastic fluids at $\delta = 0$, that is, in pipes. Elimination of λ_{cr} from equations (6.6.1) and (6.6.3) gives

$$\frac{0.11}{Re_{\text{cr}}^{0.168}} = \frac{64}{Re_{\text{cr}}} \frac{Se}{8\beta} = \frac{64}{Re_{\text{cr}}} \frac{He}{Re_{\text{cr}} 8\beta}. \tag{6.6.5}$$

Parameters $Se = He/Re_{\text{cr}}$ and β are related by equation (6.3.28)

$$\frac{He}{Re_{\text{cr}} 8\beta} = \frac{1}{1 - \frac{4}{3}\beta + \frac{1}{3}\beta^4}. \tag{6.6.6}$$

As a result of numerical calculation of the system (6.6.5), the dependence $Re_{\text{cr}} = f(He)$ represented by Fig. 6.20 is obtained (6.6.6). The latter is well consistent with Solov'ev formula received by handling of a large body of experimental data for non-Newtonian fluid flows including drill solutions (Filatov, 1973)

$$Re_{\text{cr}} = 2100 + 7.3 (He)^{0.58}, \tag{6.6.7}$$

where He is Hedström number.

Hence when calculating Re_{cr} for viscous-plastic fluids, one can use the graphic plotted in Fig. 6.20 (Hanks, 1963) or the formula (6.6.7).

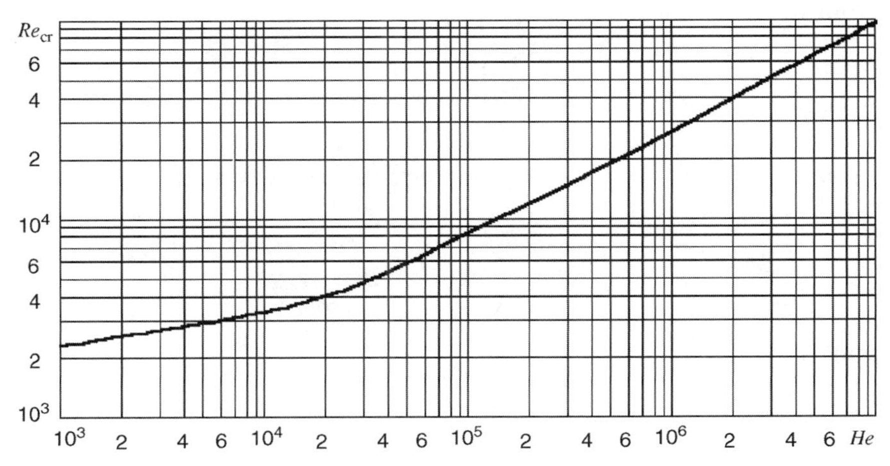

FIGURE 6.20 Critical Reynolds number of transition to turbulent flow regime.

In drilling practice has found a wide utility the formula for turbulent regime beginning being satisfactory applicable at great values of Re_{cr} and Se:

$$v_{cr} = \sqrt{\frac{\tau_0}{\rho}}, \tag{6.6.8}$$

where v_{cr} is critical velocity and $C = 25$ empirical constant obtained by (Filatov, 1973). This formula is true for self-similar turbulent flow in hydraulic smooth pipes. Insertion of $\lambda_{cr} = 0.0128$ in (6.6.3) and replacement of $f = Se/8\beta$ at great values of He/Re on its approximation $f = 0.125\, He/Re$ gives

$$\frac{64}{Re_{cr}} 0.125 \frac{He}{Re_{cr}} = 0.0128$$

or

$$Re_{cr} = 25\sqrt{He}. \tag{6.6.9}$$

Passage to dimensional variables yields

$$v_{cr} = 25\sqrt{\tau_0/\rho}. \tag{6.6.10}$$

This formula becomes inaccurate at $He < 4 \times 10^4$.

Reasoning for power fluid at $\delta = 0$ similar to that made above brings to

$$\frac{64}{Re_{cr}} 2^{n-3} \left(\frac{3n+1}{n}\right)^n = \frac{0.11}{Re_{cr}^{0.168}}$$

from which it follows

$$Re_{cr} = 2100\left[2^{n-3}\left(\frac{3n+1}{n}\right)^n\right]^{1.2}. \tag{6.6.11}$$

In self-similar flow regime taking in (6.6.4) when $\lambda_{cr} = 0.0128$, we get

$$\frac{64}{Re_{cr}} 2^{n-3} \left(\frac{3n+1}{n}\right)^n = 0.0128$$

and

$$Re_{cr} = 5000 \times 2^{n-3}\left(\frac{3n+1}{n}\right)^n. \tag{6.6.12}$$

Critical velocity is

in non-self-similar flow regime

$$v_{cr} = \left\{2100 \frac{k}{\rho d_h^n} \left[2^{n-3} \left(\frac{3n+1}{n}\right)^n\right]^{1.2}\right\}^{\frac{1}{2-n}}, \tag{6.6.13}$$

in self-similar flow regime

$$v_{cr} = \left\{5000 \frac{k}{\rho d_h^n} 2^{n-3} \left(\frac{3n+1}{n}\right)^n\right\}^{\frac{1}{2-n}}. \tag{6.6.14}$$

Analogous formulas for viscous-plastic and power fluids can be also got for annular channel flows at $\delta \neq 0$, but appropriate experimental data are insufficient to validate them. In this case to determine Re_{cr} one can use the formula (6.6.7) or (6.6.11), inserting in it the appropriate values of hydraulic diameter $d_h = d_{ex} - d_{in}$.

6.7 CALCULATION OF PRESSURE IN FLOWS IN ECCENTRIC ANNULUS: FORMATION OF STAGNATION ZONES[1]

Consider flows of different fluids in eccentric annular channel. The eccentricity e is defined as a distance between axes of exterior and interior pipes (Fig. 6.21)

$$e = 00_1. \tag{6.7.1}$$

There are two possible cases of pipes arrangement in eccentricity magnitude

$$2e < d_{in} \quad \text{and} \quad 2e > d_{in}, \tag{6.7.2}$$

the case $e = 0$ corresponding to concentric arranged pipes. If the interior pipe is tangent to exterior one, they are arranged with limiting eccentricity. Consider the case $2e < d_{in}$ (Fig. 6.21a) at $\delta = d_{in}/d_{ex} > 0.5$. The latter inequality is typical for arrangement of casing strings in well setting.

When calculating flows in eccentric channel it is applied method (McLean et al., 1967), which permits without solving the system of

[1] The Section 6.7 is written in collaboration with V.G. Broon.

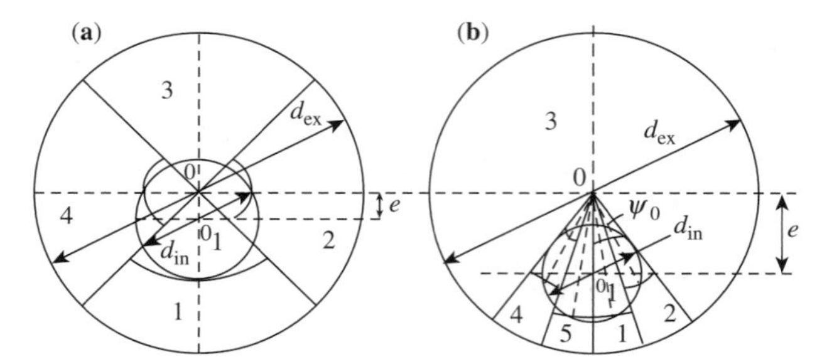

FIGURE 6.21 To flow calculation in eccentric channel: 1–5—sectors by which is divided eccentric annulus.

equations (6.1.1)–(6.1.4) to use results obtained in previous sections. Suppose that the flow in eccentric circular channel could be replaced by flows in a finite number of eccentric circular channels and pipe sectors happening independently of one another under action of common pressure drop Δp. Calculation of mean velocity v_i and flow rate Q_i for each sector could be performed singly. In doing so in neighboring sectors may exist different flow regimes (laminar, turbulent) and in some sectors the flow can be absent (stagnation zones). An eccentric channel formed by exterior and interior circles 0 and 0_1 is replaced by sectors of conventional concentric channels as follows. The eccentric channel is divided by sectors of the circle 0 in $2k$ equal parts, which in its turn are replaced by sectors of conventional concentric channels. Such segmentation is shown in Fig. 6.21a for $k=2$. Let the axis 00_1 correspond to the angle $\psi=0$ (counterclockwise angle reading is taken as positive). The index i refers to a sector number numerated in positive direction beginning from the sector containing bisector 00_1. As conventional diameter of interior circle of ith sector it is taken doubled distance from the center 0 up to the intersection point of sector bisector with interior circle of real annular channel, which in notation of Fig. 6.21a is described by

$$d_i = 2\left(e\cos\psi_i + \sqrt{\frac{d_{in}^2}{4} - e^2\sin^2\psi_i}\right); \tag{6.7.3}$$

$$\psi_i = \pi(i-1)/k, \quad i = 1, 2, \ldots, 2k. \tag{6.7.4}$$

Here, ψ_i is the angle calculated in rad referring to bisector of the ith real channel sector.

The case ($2e > d_{in}$) when the circle 0 is divided into five sectors is shown in Fig. 6.21b. In this case, the circle is divided into odd ($2k + 1$) sectors representing annular spaces and pipes, the sector ($-\psi_0$, ψ_0) with

$$\psi_0 = \arcsin(d_{in}/2e) \tag{6.7.5}$$

having been divided into $2k$ equal parts.

The conventional interior diameters of annular space sectors are determined by (6.7.3). For flow to the right of the axis 00_1 in k sectors there is

$$\psi_i = \psi_0(i-1)/k + \psi_0/(2k), \quad i = 1, 2, \ldots, k. \tag{6.7.6}$$

In the left half of the flow conventional sectors are symmetric located. Conventional diameters are determined by

$$d_i = 2e \cos \psi_i + \sqrt{d_{in}^2 - 4e^2 \sin^2 \psi_i}, \quad i = 1, 2, \ldots, k. \tag{6.7.7}$$

Quantities ψ_i are found from (6.7.6) as for annular space sectors. The conventional diameters of pipes d_{k+1} are taken equal to d_{ex}

$$d_{k+1} = d_{ex}. \tag{6.7.8}$$

In previous sections were determined formulas to calculate pressure drop at given flow rate Q

for laminar flow of viscous fluid

$$\Delta p = \frac{64}{Re} f(\delta) \frac{\rho Q^2 L}{2 d_h S_k^2} = f(\delta) \frac{8 \eta Q}{d_h S_k} \cdot \frac{4L}{d_h}; \tag{6.7.9}$$

for laminar flow of viscous-plastic fluid

$$\Delta p = \frac{4 \tau_0 L}{\beta d_h}, \tag{6.7.10}$$

$$\beta = \beta(Se) \quad \text{and} \quad Se = \frac{\tau_0 d_h S_k}{\eta Q}. \tag{6.7.11}$$

For laminar flow of power fluid Δp, it is determined by (6.4.40).

For turbulent flow in hydraulic smooth pipes ($Re > Re_{cr}$)

$$\Delta p = \lambda \frac{\rho Q^2}{2 d_h S_k^2} L, \tag{6.7.12}$$

where $\lambda = 0.334/\sqrt[4]{Re}$ at $\lambda > 0.025$.

For other *Re* ($Re > 3.2 \times 10^4$), it is taken

$$\lambda = 0.025. \tag{6.7.13}$$

A peculiarity of flow in eccentric channel consists in that different sectors could at one time exist in different flow regimes, and it would be true for one of the formulas for laminar or turbulent flow regimes. The transition regime at which the resistance is somewhat lower, the turbulent one will be included in turbulent regime. In doing so, we overstate a little the pressure drop.

Consider now a flow of viscous-plastic fluid in an eccentric channel. In sectors where the flow is absent the forces caused by tangential stresses are more or equal to pressure forces. Therefore, the condition of flow absence forms

$$\tau_0 \pi (d_{ex} + d_i) L \geq \Delta p \pi \frac{(d_{ex}^2 - d_i^2)}{4} \tag{6.7.14}$$

or

$$\frac{4\tau_0}{d_{ex} - d_i} \geq \Delta p / L. \tag{6.7.15}$$

From (6.7.15) it follows that in sectors with

$$d_i \geq d_{ex} - 4\tau_0 L / \Delta p \tag{6.7.16}$$

would be stagnation zones.

Thus, at given Δp the sizes of stagnation zones as well as flow rates Q_i in correspondent sectors could be determined. The flow rate in each of these sectors would be $Q_i S_{si}/S_i$, where S_{si} is sector area and S_i is the area of corresponding annular space. The total flow rate in all sectors is $\Sigma Q_i S_{si}/S_i$. Since in considered case $S_{si}/S_i = 1/2k$, the total flow rate is $\Sigma Q_i/2k$.

Inasmuch as flow rates in sectors are not known, the method of successive approximations is used. The procedure of successive approximations is as follows. The total flow rate Q through all sectors is known. As the first approximation let us take Δp_k relevant to $e = 0$, that is, to concentric channel.

For each sector it is found

Hedström number

$$He = \frac{\tau_0 d^2 \rho}{\eta^2},$$

critical Reynolds number

$$Re_{\text{cr}} = 2100 + 7.3He^{0.58}, \tag{6.7.17}$$

critical Saint Venant number

$$Se_{\text{cr}} = He/Re_{\text{cr}}.$$

From the graphic of Fig. 6.7 critical value of the factor β_{cr} is obtained. Calculate critical pressure drop

$$\Delta p_{\text{cr}} = \frac{4\tau_0 L}{\beta_{\text{cr}} d}. \tag{6.7.18}$$

Determine search interval of required pressure drop taking as extreme limits of Δp values

$$(\Delta p)_1 = \Delta p_k/2; \qquad (\Delta p)_2 = \Delta p_k. \tag{6.7.19}$$

If $(\Delta p)_1$ or $(\Delta p)_2$ does not exceed Δp_{cr}, (6.7.9) and (6.7.10) are the formulas taken for laminar flow. In the opposite case, (6.7.12) or (6.7.13) is the formula taken for turbulent flow. For the sake of calculation convenience, resolve (6.7.9)–(6.7.13) with respect to Q_i. As a result from (6.7.9) for laminar flow yields

$$Q_i = \frac{\Delta p S_i (d_{\text{ex}} - d_{\text{in}})^2}{32 f(\delta) \eta L} = \Delta p \Lambda_i, \tag{6.7.20}$$

where

$$\Lambda_i = \frac{S_i (d_{\text{ex}} - d_i)^2}{32 f(\delta) \eta L}$$

is a constant for viscous fluid in given sector with area S_i.

From (6.7.11) for viscous-plastic fluid ensues

$$Q_i = \frac{\tau_0 (d_{\text{ex}} - d_i) S_i}{Se_i \eta} = \frac{1}{Se_i} C_i, \tag{6.7.21}$$

where $C_i = \tau_0 (d_{\text{ex}} - d_i) S_i / \eta$ is a constant for viscous-plastic fluid in given sector; Se_i is found from Fig. 6.7 at predetermined calculated parameter

$$\beta_i = \frac{\tau_0}{\Delta p} \cdot \frac{4L}{(d_{\text{ex}} - d_i)} = \frac{B_i}{\Delta p}, \tag{6.7.22}$$

where $B_i = 4\tau_0 L/(d_{ex} - d_i)$ is a constant for viscous-plastic fluid in given sector.

Resolving (6.7.12) for turbulent flow of viscous fluid following is obtained

$$Q_i = \Delta p^{4/7} T_i, \tag{6.7.23}$$

where

$$T_i = 2.78\left[\frac{(d_{ex} - d_i)^5}{\rho^3 \eta L^4}\right]^{1/7} S_i.$$

Now (6.7.13) gives

$$Q_i = \Delta p^{1/2} A_i, \tag{6.7.24}$$

where

$$A_i = 7.02(d_{ex} - d_i)^{3/2}(d_{ex} + d_i)/(\rho L)^{1/2}.$$

The region of the formula (6.7.23) applicability is $\Delta p_{cr} < \Delta p < \Delta p_\pi$, where Δp_π for each sector is

$$\Delta p_\pi = \lambda \frac{\rho Q^2 L}{2 d_h S^2} = \lambda Re^2 \frac{\eta^2 L}{2 \rho d_h} = 12.8 \cdot 10^6 \frac{\eta^2 L}{\rho d_h^3}, \tag{6.7.25}$$

and $Re = \rho Q d_h/(\eta S) = 3.2 \times 10^4$; $\lambda = 0.025$.

The region of formula (6.7.24) applicability is $\Delta p \geq \Delta p_\pi$.

Hence, except for stagnation zones, there are found flow regimes and flow rates Q_i in annular spaces appropriate to each sector. Then compare $\Sigma Q_i/2k$ with given Q. If the accuracy of ΣQ_i is unsatisfactory, one should take the next approximation

$$(\Delta p)_3 = \frac{(\Delta p)_1 + (\Delta p)_2}{2}. \tag{6.7.26}$$

Depending on relation between ΣQ_i and $2kQ$, the following approximation is used

$$(\Delta p)_4 = \frac{(\Delta p)_3 + (\Delta p)_2}{2} \tag{6.7.27}$$

or

$$(\Delta p)_4 = \frac{(\Delta p)_1 + (\Delta p)_3}{2}. \tag{6.7.28}$$

Such calculations are being continued as long as the obtained approximation would reach the approximation given in advance, that is when

$$\left| \frac{\Sigma Q_i - 2kQ}{2kQ} \right| 100\% \leq \varepsilon, \tag{6.7.29}$$

where ε is relative accuracy given in percents.

This method is convenient by handy and computer operations, since the exact solution is absent. Note that relations (6.7.9) and (6.7.10) in previous sections were presented in the form of Darcy–Weisbach formula

$$\Delta p = \frac{\lambda \rho Q^2}{2dS^2} L.$$

Figure 6.22 shows dependence $\lambda = \lambda(Re)$ calculated with above stated procedure and plotted at different values of eccentricity at $k = 2$, $d_{\text{in}}/d_{\text{ex}} = 0.542$, and $He = 4.03 \times 10^5$.

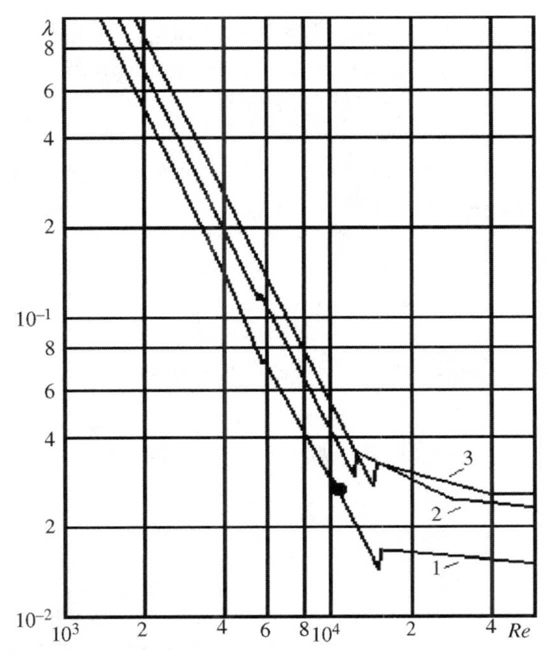

FIGURE 6.22 Dependence $\lambda = \lambda(Re, \bar{e})$. Curves 1, 2, 3 correspond to $\bar{e} = 2e/d_{\text{h}}$ equal to 1; 0.5; 0.

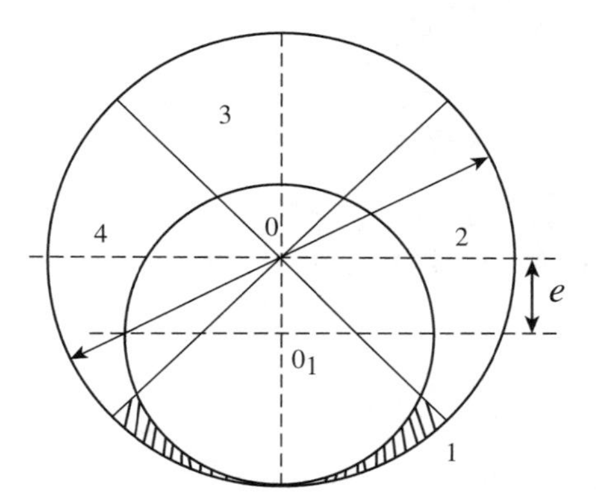

FIGURE 6.23 Scheme for Exercise .

Consider an example of concrete calculation of pressure drop in eccentric channel using method of successive approximations with channel partition on sectors.

EXERCISE 6.7.1

It is required to determine pressure drop in flow of washing fluid in annular channel with limiting eccentricity at given data: $L = 200$ m, $Q = 0.05$ m^3/s, $d_{\text{in}} = 0.310$ m, $d_{\text{ex}} = 0.168$ m, $\tau_0 = 5$ Pa, $\eta = 0.02$ Pa·s, $\rho = 1600$ kg/m^3, and $\varepsilon \leq 3\%$.

SOLUTION Divide annular channel into four sectors (see Fig. 6.23), that is $k = 2$. The limiting eccentricity is

$$e = \frac{d_{\text{ex}} - d_{\text{in}}}{2} = \frac{0.310 - 0.168}{2} = 0.071 \text{ m}.$$

Determine through (6.7.3) conventional diameters

$$d_i = 2\left(e \cos \psi_i + \sqrt{\frac{d_{\text{in}}^2}{4} - e^2 \sin^2 \psi_i}\right);$$

$$\psi_i = \pi(i - 1)/k \quad (i = 1, 2, 3, 4)$$

and other conventional parameters of sectors: hydraulic diameters $d_{\text{h}} = d_{\text{ex}} - d_{\text{in}}$, areas $S_{ki} = \pi(d_{\text{ex}}^2 - d_{\text{in}}^2)/4$, numbers He_i, Re_{cri}, Se_{cri}, and pressure drops $\Delta p_{\text{cri}}/L = 4\tau_0/(d_{\text{h}i}\beta_i)$. β_i are determined by graphic of Fig. 6.7.

For the sector 1 ($i=1$) we have

$$\psi_1 = \pi(1-1)/2 = 0;$$

$$d_1 = 2\left(0.071\cos 0 + \sqrt{\frac{0.168^2}{4} - 0.071^2\sin^2 0}\right) = 0.31\ \mathrm{m};$$

$$d_{h1} = d_{ex} - d_1 = 0.31 - 0.31 = 0; \quad S_{k1} = \pi(d_{ex}^2 - d_1^2)/4 = 0.$$

It is taken that the flow in the first sector is absent, that is $Q_1 = 0$; therefore, we begin calculations with sector 2. From the formula (6.7.3) we find conventional diameter of the second sector ($i=2$) having previously calculated from the formula (6.7.4) the angle

$$\psi_2 = \pi(2-1)/2 = \pi/2;$$

$$d_2 = 2\left(0.071\cos\frac{\pi}{2} + \sqrt{\frac{0.168^2}{4} - 0.071^2\sin^2\frac{\pi}{2}}\right) = 0.0898\ \mathrm{m}.$$

Determine the hydraulic diameter d_h and conventional area S_{k2}

$$d_h = d_{ex} - d_2 = 0.31 - 0.0898 = 0.22\ \mathrm{m};$$

$$S_{k2} = \frac{\pi(d_{ex}^2 - d_2^2)}{4} = \frac{3.14}{4}(0.31^2 - 0.0898^2) = 0.0691\ \mathrm{m}^2.$$

Determine

$$He = \frac{\tau_0 d_h^2 \rho}{\eta^2} = \frac{5 \times 0.22^2 \times 1600}{0.02^2} = 9.68 \times 10^5$$

and Re_{cr} for given sector

$$Re_{cr} = 2100 + 7.3\,He^{0.58} = 2100 + 7.3(9.68 \times 10^5)^{0.58} = 2.37 \times 10^4.$$

Obtain

$$Se_{cr} = \frac{He}{Re_{cr}} = \frac{9.68 \times 10^5}{2.37 \times 10^4} = 40.8.$$

From the curve 2 of Fig. 6.7 at $Se_{cr} = 40.8$ find $\beta_{cr} = 0.63$ and then quantities Δp_{cr}, Δp_{π}, A, and T:

$$\Delta p_{cr} = \frac{4\tau_0 L}{d_h \beta_{cr}} = \frac{4 \times 5 \times 200}{0.22 \times 0.63} = 2.89 \times 10^4 \text{ Pa};$$

$$\Delta p_{\pi} = 12.8 \times 10^6 \frac{L\eta^2}{\rho d_h^3} = 12.8 \times 10^6 \frac{200 \times 0.02^2}{1600 \times 0.22^3} = 6 \times 10^4 \text{ Pa};$$

$$A = 7.02(d_{ex} - d_2)^{3/2}(d_{ex} + d_2)/(\rho L)^{1/2}$$

$$= \frac{7.02(0.31 - 0.0898)^{3/2}(0.31 + 0.0898)}{(1600 \times 200)^{1/2}} = 5.13 \times 10^{-4} \text{ m}^3 \text{ Pa}^{-0.5}/\text{s};$$

$$T = 2.78\left[\frac{(d_{ex} - d_2)^5}{\rho^3 \eta L^4}\right]^{1/7} S_{k2} = 2.78\left[\frac{(0.31 - 0.0898)^5}{1600^3 \times 0.02 \times 200^4}\right]^{1/7} 0.0691$$

$$= 2.34 \times 10^{-4} \text{ m}^3 \text{ Pa}^{-4/7}/\text{s}.$$

In the fourth sector the conventional parameters $d_h S_k$, He, Re_{cr}, Se_{cr}, β_{cr}, Δp_{cr}, and consequently the flow rate Q because of the symmetry are the same as in the sector 2.

Calculate parameters of the sector 3 from the formulas given above: $\psi = \pi$, $d_3 = 0.026$ m; $S_{k3} = 0.0749$ m^2; $He = 1.61 \times 10^6$; $d_h = 0.284$ m; $Re_{cr} = 3.12 \times 10^4$; $S_{cr} = 51.6$; $\beta_{cr} = 0.67$; $\Delta p_{cr} = 2.1 \times 10^4$ Pa; $\Delta p_{\pi} = 0.279 \times 10^5$ Pa; $A = 6.31 \times 10^{-4}$ m^3 Pa$^{-0.5}$/s; $T = 3.04 \times 10^{-4}$ m^3 Pa$^{-4/7}$/s.

Since each sector is a quarter of imagined annular space, the total flow rate Q in eccentric channel would be equal to $Q = Q_1/4 + Q_2/4 + Q_3/4 + Q_4/4$. As far as $Q_1 = 0$, $Q_2 = Q_4$, it is $4Q = 2Q_2 + Q_3$.

In order to get Δp in the eccentric channel determine the range of its search $\Delta p_{as}/2 < \Delta p < \Delta p_{as}$, where Δp_{as} is pressure drop in the annular channel with zero eccentricity.

Find Δp_{as} at $e = 0$. But before this determine the hydraulic diameter d_h, the area S_k, numbers He, Re_{cr}, Se, and Re

$$d_h = d_{ex} - d_{in} = 0.31 - 0.168 = 0.142 \text{ m};$$

$$S_k = \pi(d_{ex}^2 - d_{in}^2)/4 = 3.14(0.31^2 - 0.168^2)/4 = 0.0533 \text{ m}^2;$$

$$He = \frac{\tau_0 d_h^2 \rho}{\eta^2} = \frac{5 \times 0.142^2 \times 1600}{0.02^2} = 4.03 \times 10^5;$$

$$Re_{cr} = 2100 + 7.3\, He^{0.58} = 2100 + 7.3(4.03 \times 10^5)^{0.58} = 1.51 \times 10^4;$$

$$Se = \frac{\tau_0 d_h S_k}{\eta Q} = \frac{5 \times 0.142 \times 0.0533}{0.02 \times 0.05} = 37.8;$$

$$Re = He/Se = 4.03 \times 10^5/37.8 = 1.07 \times 10^4.$$

Since $Re = 1.07 \times 10^4 < Re_{cr} = 1.51 \times 10^4$, the flow in concentric channel is laminar.

The pressure drop is

$$\Delta p_{\rm as} = \frac{4\tau_0 L}{d_{\rm h}\beta} = \frac{4 \times 5 \times 200}{0.142 \times 0.61} = 0.462 \times 10^5 \text{ Pa}.$$

Thus, the pressure drop $\Delta p_{\rm as}$ in eccentric channel should be sought in the range (one can be certain that at $\Delta p = \Delta p_{\rm as}/2$, there is $\Sigma Q_i < 4Q$, whereas at $\Delta p = \Delta p_{\rm as}$, it is $\Sigma Q_i > 4Q$)

$$0.231 \times 10^5 \text{ Pa} = \Delta p_{\rm as}/2 < \Delta p < \Delta p_{\rm as} = 0.462 \times 10^5 \text{ Pa}.$$

If to accept $\Delta p_1 = 0.231 \times 10^5 \Pi a$, the flow regime in the Section 6.3 would be turbulent since

$$\Delta p_{\rm cr3} = 0.21 \times 10^5 \text{ Pa} < \Delta p_{\rm as}/2 = 0.231 \times 10^5 \text{ Pa} < \Delta p_{n3} = 0.279 \times 10^5 \text{ Pa},$$

while the regime in sectors 2 and 4 is laminar since

$$\Delta p_{\rm cr2} = 0.289 \times 10^5 \text{ Pa} > \Delta p_{\rm as}/2 = 0.231 \times 10^5 \text{ Pa}.$$

Flow rates in sectors 2 and 4 can be obtained from (6.7.21) having prior calculated β from (6.7.22)

$$\beta = \frac{\tau_0 4L}{(\Delta p_{\rm as}/2) d_{\rm h}} = \frac{5 \times 4 \times 200}{0.231 \times 10^5 \times 0.22} = 0.787.$$

From the graphic of Fig. 6.7 one gets $Se = 150$. Then (6.7.21) gives

$$Q_2 = Q_4 = \frac{\tau_0 d_{\rm h} S}{\eta Se} = \frac{5 \times 0.22 \times 0.0691}{0.02 \times 150} = 0.0253 \text{ m}^3/\text{s}.$$

Determine the flow rate in the sector 3 from (6.7.23)

$$Q_3 = \Delta p^{4/7} T_3 = (0.231 \times 10^5)^{4/7} \times 3.04 \times 10^{-4} = 0.0947 \text{ m}^3/\text{s}.$$

Calculate

$$\Sigma Q_i = 0.0253 + 0.0947 + 0.0253 = 0.1453 \text{ m}^3/\text{s} < 4Q = 0.2 \text{ m}^3/\text{s};$$

$$\varepsilon = \left| \frac{0.2 - 0.1453}{0.2} \right| 100\% = 27.4\% > 3\%.$$

Assume further $\Delta p = \Delta p_{\rm as} = 0.462 \times 10^5$ Pa.

Since $2.89 \times 10^4 < 4.62 \times 10^4 < 6 \times 10^4$ and $2.1 \times 10^4 < 2.79 \times 10^4 < 4.62 \times 10^4$, the regime in all sections is turbulent. The calculation in sectors 2 and 4 should be carried out by formula (6.7.23) and in sector 3 by formula (6.7.24)

$$Q_2 = Q_4 = (0.462 \times 10^5)^{4/7} \times 2.34 \times 10^{-4} = 0.108\,\mathrm{m^3/s};$$

$$Q_3 = (0.462 \times 10^5)^{1/2} \times 6.31 \times 10^{-4} = 0.136\,\mathrm{m^3/s};$$

$$\Sigma Q_i = 2 \times 0.108 + 0.136 = 0.352 > 4Q = 0.2\,\mathrm{m^3/s};$$

$$\varepsilon = \left|\frac{0.2 - 0.352}{0.2}\right| 100\% = 76\% > 3\%.$$

As the next approximation, it is taken

$$\Delta p = \frac{\Delta p_{\mathrm{as}} + \Delta p_{\mathrm{as}}/2}{2} = \frac{0.462 \times 10^5 + 0.231 \times 10^5}{2} = 0.347 \times 10^5\ \mathrm{Pa}.$$

Since $2.89 \times 10^4 < 3.47 \times 10^4 < 6 \times 10^4$ and $2.79 \times 10^4 < 3.47 \times 10^4$, the regime in all sectors is turbulent. The calculation in sectors 2 and 4 should be made with the formula (6.7.23) and in sector 3 with (6.7.24)

$$Q_2 = Q_4 = (0.347 \times 10^5)^{4/7} \times 2.34 \times 10^{-4} = 0.092\,\mathrm{m^3/s};$$

$$Q_3 = (0.347 \times 10^5)^{0.5} \times 6.31 \times 10^{-4} = 0.118\,\mathrm{m^3/s};$$

$$\Sigma Q_i = 2 \times 0.092 + 0.118 = 0.302\,\mathrm{m^3/s} > 0.2\,\mathrm{m^3/s};$$

$$\varepsilon = \left|\frac{0.2 - 0.302}{0.2}\right| 100\% = 51\% > 3\%.$$

Make the fourth approximation

$$\Delta p = \frac{0.347 \times 10^5 + 0.231 \times 10^5}{2} = 0.289 \times 10^5\ \mathrm{Pa}.$$

Comparison with Δp_{cr} shows that in sectors 2 and 4 the regime is laminar whereas in the sector 3 it is turbulent, and the calculation in the third sector has to be performed with the formula (6.7.24).

Find now flow rates Q_2, Q_3, Q_4:

$$Q_2 = Q_4 = \frac{\tau_0 d_{\mathrm{h}} S}{\eta Se} = \frac{5 \times 0.22 \times 0.0691}{0.02 \times 40} = 0.095\,\mathrm{m^3/s},$$

where $Se = 40$ is determined from the graphic of Fig. 6.7 with given number β

$$\beta = \frac{\tau_0 4L}{\Delta p d_{\mathrm{h}}} = \frac{5 \times 4 \times 200}{0.289 \times 10^5 \times 0.22} = 0.63,$$

$$Q_3 = \Delta p^{1/2} A_3 = (0.289 \times 10^5)^{0.5} \cdot 6.31 \times 10^{-4} = 0.107\,\mathrm{m^3/s}.$$

Calculate

$$\Sigma Q_i = 0.095 + 0.107 + 0.095 = 0.297\ \text{m}^3/\text{s} > 4Q = 0.2\ \text{m}^3/\text{s};$$

$$\varepsilon = \left|\frac{0.297-0.2}{0.2}\right| 100\% = 48.5\% > 3\%.$$

The fifth approximation gives

$$\Delta p = \frac{0.289 \times 10^5 + 0.231 \times 10^5}{2} = 0.26 \times 10^5\ \text{Pa}.$$

Comparison of this value with (p_{cr} shows that the regime in sectors 2 and 4 is laminar and in 3 turbulent.

Flow rates are

$$Q_2 = Q_4 = \frac{\tau_0 d_{\text{h}} S}{\eta Se} = \frac{5 \times 0.22 \times 0.0691}{0.02 \times 60} = 0.063\ \text{m}^3/\text{s}.$$

The number $Se = 60$ is found from the graphic of Fig. 6.7 for *a priori* calculated number β

$$\beta = \frac{\tau_0 4L}{\Delta p d_{\text{h}}} = \frac{5 \times 4 \times 200}{0.26 \times 10^5 \times 0.22} = 0.7.$$

Since $2.1 \times 10^5 < 0.26 \times 10^5 < 0.279 \times 10^5$, it is

$$Q_3 = \Delta p^{4/7} T_3 = (0.26 \times 10^5)^{4/7} \times 3.04 \times 10^{-4} = 0.101\ \text{m}^3/\text{s};$$

$$\Sigma Q_i = 0.063 + 0.101 + 0.063 = 0.227\ \text{m}^3/\text{s} > 4Q = 0.2\ \text{m}^3/\text{s};$$

$$\varepsilon = \left|\frac{0.227-0.2}{0.2}\right| 100\% = 13.5\% > 3\%.$$

The sixth approximation is

$$\Delta p = \frac{0.26 \times 10^5 + 0.231 \times 10^5}{2} = 0.246 \times 10^5\ \text{Pa}.$$

Now find Q_i. Here, the flow in sectors 2 and 4 is also laminar and in 3 turbulent. Calculate

$$\beta = \frac{\tau_0 4L}{\Delta p d_{\text{h}}} = \frac{5 \times 4 \times 200}{0.246 \times 10^5 \times 0.22} = 0.74.$$

From the graphic of Fig. 6.7 follows $Se = 80$. Then

$$Q_2 = Q_4 = \frac{\tau_0 d_{\text{h}} S}{\eta Se} = \frac{5 \times 0.22 \times 0.0691}{0.02 \times 80} = 0.048\ \text{m}^3/\text{s}.$$

Since $0.246 \times 10^5 < \Delta p_\pi = 0.279 \times 10^5$, the calculation in the sector 3 is made with the help of formula (6.7.23)

$$Q_3 = \Delta p^{4/7} T_3 = (0.246 \times 10^5)^{4/7} \times 3.04 \times 10^{-4} = 0.098\ \text{m}^3/\text{s};$$
$$\sum Q_i = 0.048 + 0.098 + 0.048 = 0.194\ \text{m}^3/\text{s} < 4Q = 0.2\ \text{m}^3/\text{s};$$
$$\varepsilon = \left| \frac{0.194 - 0.2}{0.2} \right| 100\% = 3\%.$$

Come to a stop on this approximation. As a result, the pressure drop at given conditions is $\Delta p = 0.246 \times 10^5$ Pa. The pressure drop as compared to the flow in annular channel without eccentricity is lowered in 1.88, that is nearly twice

$$n = \frac{\Delta p_{\text{as}}}{\Delta p} = \frac{0.462 \times 10^5}{0.246 \times 10^5} = 1.88.$$

Determine quantities $\lambda, Re, \bar{e}$ for the whole annular space

$$\lambda = \frac{\Delta p S^2 2 d_{\text{h}}}{L \rho Q^2} = \frac{0.246 \times 10^5 (0.0533)^2 2 \times 0.142}{200 \times 1600 \times 0.05^2} = 0.0248;$$
$$Re = \frac{Q d_{\text{h}} \rho}{S \eta} = \frac{0.05 \times 0.142 \times 1600}{0.0533 \times 0.02} = 1.07 \times 10^4;$$
$$\bar{e} = \frac{2e}{d_{\text{h}}} = \frac{2 \times 0.071}{0.142} = 1.$$

In Fig. 6.22 the obtained factor $\lambda = 0.0248$ is marked by thick point.

6.8 EFFECT OF INTERNAL PIPE ROTATION ON PRESSURE IN ANNULUS

Pipe columns in drilling or viscometer cylinders in rheological researches rotate. In this connection determine the influence of interior pipe (cylinder) rotation on pressure drop in annular channel. Such a flow in the absence of pump delivery is shown in Fig. 6.24.

At first consider a flow in annular space when both the cylinders rotate with angular velocities ω_1 and ω_2 (Loitsyansky, 1987; Ustimenko, 1977; Schlichting, 1964). Determine the relation between stresses and angular velocities of cylinders. In order to get the sought expression it is required to solve the system of equations (6.1.5)–(6.1.9):

momentum equations

$$\frac{\partial p}{\partial r} = \rho \frac{w^2}{r},$$

$$\frac{\partial \tau}{\partial r} + 2\frac{\tau}{r} = 0;$$

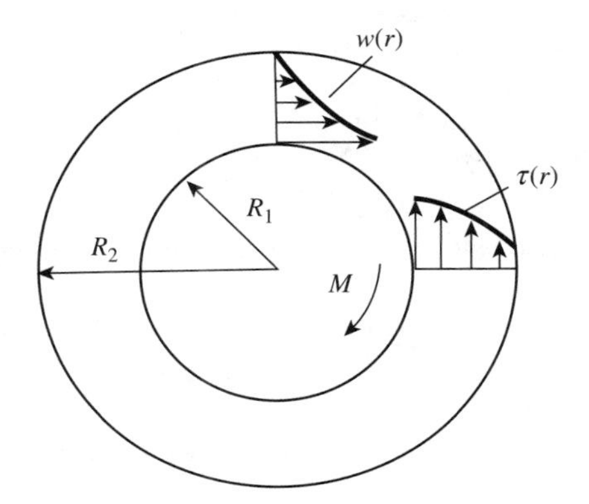

FIGURE 6.24 Distribution of velocity and tangential stress in the gap between annular space in the flow of viscous fluid caused by interior cylinder rotation.

thermodynamic equation of state

$$\rho = \text{const};$$

rheological equation of state

$$\tau = \tau(\dot{\gamma}),$$

in which

$$\dot{\gamma} = \frac{\partial w}{\partial r} - \frac{w}{r}.$$

Boundary conditions answering fluid sticking at the surfaces of rotating cylinders are

$$\begin{aligned} w &= \omega_1 R_1 \quad \text{at} \quad r = R_1 = d_{\text{in}}/2; \\ w &= \omega_2 R_2 \quad \text{at} \quad r = R_2 = d_{\text{ex}}/2. \end{aligned} \tag{6.8.1}$$

Rheological equation (6.1.9) for viscous fluid is

$$\tau = \mu\left(\frac{\partial w}{\partial r} - \frac{w}{r}\right). \tag{6.8.2}$$

Momentum equations (6.1.5) and (6.1.6) with regard to (6.8.2) take form

$$\frac{\partial p}{\partial r} = \rho\frac{w^2}{r}; \quad \frac{\partial}{\partial r}\left[\frac{\partial w}{\partial r} + \frac{w}{r}\right] = 0. \tag{6.8.3}$$

The second equation (6.8.3) gives

$$\frac{\partial w}{\partial r} + \frac{w}{r} = \text{const} = 2C_1. \qquad (6.8.4)$$

If in this equation to change the variable with $w = ur$, it reduces to

$$r\frac{\partial u}{\partial r} = 2(C_1 - u).$$

The solution could be found by method of separation of variables

$$u = C_1 - C_2\frac{1}{r^2}.$$

Returning to old variable $u = w/r$, the following is obtained

$$w = C_1 r - C_2\frac{1}{r}.$$

Constants C_1 and C_2 are found using boundary conditions (6.8.1). As a result the velocity distribution is

$$w = \frac{(\omega_2 R_2^2 - \omega_1 R_1^2)r^2 + (\omega_1 - \omega_2)R_1^2 R_2^2}{(R_2^2 - R_1^2)r}. \qquad (6.8.5)$$

When only internal pipe rotates, then $w_2 = 0$ and

$$w = \frac{\omega_1 R_1^2}{R_2^2 - R_1^2}\left(\frac{R_2^2}{r} - r\right) = \omega_1\frac{\delta^2}{1-\delta^2}\left(\frac{R_2^2}{r} - r\right), \qquad (6.8.6)$$

where $\delta = R_1/R_2$.

Pressure distribution is determined from equation (6.8.3). Stresses caused by friction forces in considered flow follows from (6.8.2) with regard to (6.8.5)

$$\tau = -\frac{2\mu(\omega_1 - \omega_2)R_1^2 R_2^2}{(R_2^2 - R_1^2)r^2}. \qquad (6.8.7)$$

The moment generated at the surface of radius r and height H relative z-axis due to the action of friction force (6.8.7) is

$$M = \pm 2\pi r^2 H\tau = \mp\frac{4\pi\mu H(\omega_1 - \omega_2)R_1^2 R_2^2}{R_2^2 - R_1^2}. \qquad (6.8.8)$$

Upper signs of this formula refer to exterior surface of interior pipe while lower signs to interior surface of exterior pipe. From (6.8.8) it is seen that the moment of friction forces is independent of the current radius r. Moments acting on each pipe are equal in absolute value but opposite in sign.

Relations (6.8.8) will be used in Section 16.1 in determining rheological characteristics of fluids with rotary viscometer. Note that in drilling the drill-stem rotates in motionless well ($\omega_2 = 0$). In this case the resistance moment factor of interior pipe λ_i may be determined if to represent (6.8.8) as

$$M = -\lambda_i \frac{\rho v_1^2}{2} \pi R_1^2 H, \tag{6.8.9}$$

where $v_1 = \omega_1 R_1$.

Comparing (6.8.9) with (6.8.8), we find λ_i for laminar flow caused by interior pipe rotation

$$\lambda_i = \frac{f(\delta)}{Ta}, \tag{6.8.10}$$

where

$$f(\delta) = \frac{8\sqrt{1-\delta}}{\delta^{3/2}(1+\delta)}; \quad Ta = \frac{1}{2}\sqrt{\frac{1-\delta}{\delta}}\frac{v_1 d_h \rho}{\mu}. \tag{6.8.11}$$

For all flow regimes the factor λ_i depending on Taylor number Ta was experimentally investigated and it was shown that the formula (6.8.10) is true up to $Ta = 41.3$ when Taylor eddies begin to generate with rise of pipe angular velocity, that is with increase of Taylor number, then gradually break down and the flow goes into chaotic (turbulent) regime at about $Ta = 400$. In Fig. 6.25, it is plotted dependence λ_i on Ta at $(1-\delta)/\delta = 0.028$ (points refer to Taylor empirical data).

Eddies are produced as a result of instability occurrence at $Ta > 41.3$ caused by rotary motion. The first equation (6.8.3) indicates that at each instant of time in fluid volume unit there is an equality between pressure gradient $\partial p/\partial r$ in the radial direction and centrifugal force $\rho w^2/r$.

At $Ta \geq Ta_{cr} = 400$ owing to flow perturbations caused by irregularity of the annular channel width (ovality, cavityness, occurrence of eccentricity, and so on) and at the action of internal cylinder surface the equality (6.8.3) can periodically be disturbed.

In the case of fluid circulation, that is at axial flow and simultaneous rotation of the interior cylinder, the transition from laminar into turbulent

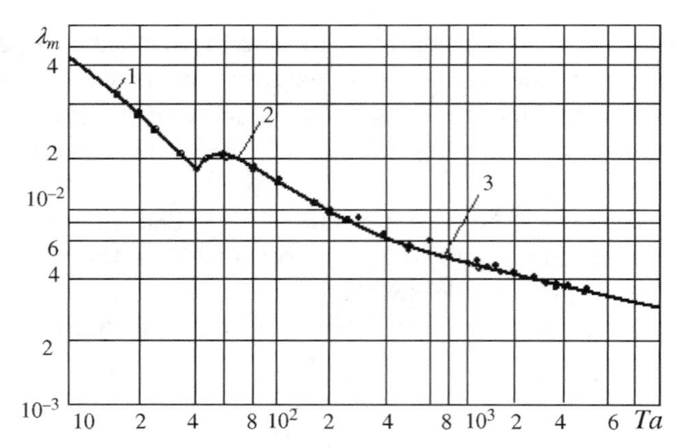

FIGURE 6.25 Dependence of resistance moment factor on Taylor number: 1—region of laminar flow at $Ta < 41.3$; 2—curve $\lambda = 0.27/Ta^{0.635}$ at $Ta > 60$; 3—curve $\lambda = 0.02/Ta^{0.2}$ for turbulent flow at $Ta > 400$.

flow regime is no longer determined by constant value of Reynolds number $Re_{cr} = 2100$ for viscous fluid flow but is a function of Taylor number. This function is shown in Fig. 6.26. The curve $Re_{cr} = f(Ta)$ separates the region of laminar flow from the turbulent one.

In the case of turbulent flow the rotation of internal flow affects the pressure drop in axial direction at fluid circulation. Since the critical regime sets earlier, the hydraulic resistance factor takes greater value than calculated by Blasius formula.

Owing to experimental data for viscous fluids the factor λ can be determined by

$$\lambda = \frac{0.3385}{\sqrt[4]{Re}} \left[1 + \frac{0.5\delta}{1-\delta} \left(\frac{Ta}{Re} \right)^2 \right]^{0.535}. \tag{6.8.12}$$

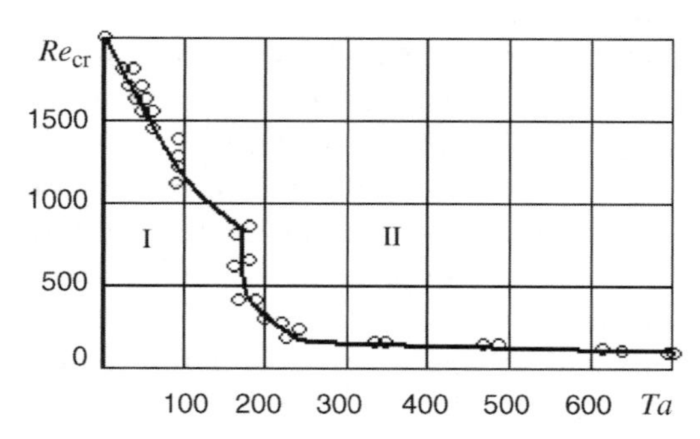

FIGURE 6.26 Empirical dependence of critical Reynolds number on Taylor number: I, II—regions of laminar and turbulent flows, respectively.

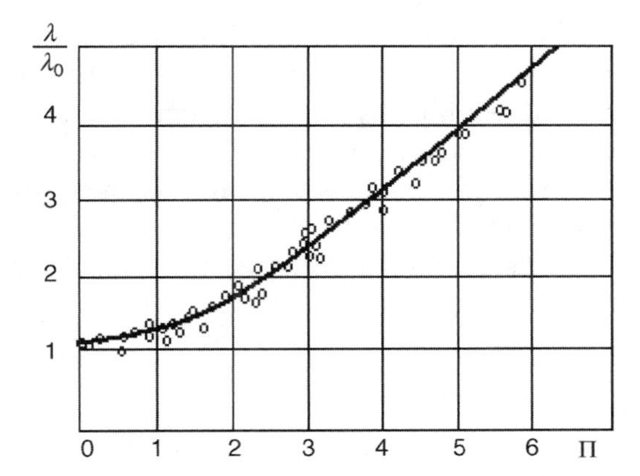

FIGURE 6.27 Experimental dependence of λ/λ_0 on parameter Π.

The formula Darcy–Weisbach then gives the pressure drop

$$\Delta p = \lambda \frac{\rho v^2}{2d_{\text{h}}} L. \tag{6.8.13}$$

Figure 6.27 shows the dependence of ratio λ calculated by (6.8.12) to λ_0 determined by (6.5.58) at $k_r = 0$ on $\Pi = \sqrt{\delta/(1-\delta)}(Ta/Re)$. Thus if $Re > Re_{\text{cr}}$, the pipe rotation increases the pressure drop in axial direction and it is calculated by the formula (6.8.13) with regard to (6.8.12). If $Re < Re_{\text{cr}}$, the rotation does not affect the pressure drop in the axial direction that can be found from corresponding formulas for laminar flow in a channel with motionless walls. The critical Reynolds number Re_{cr} can be obtained from Fig. 6.26.

6.9 PRESSURE DROP IN LOCAL RESISTANCES OF CIRCULATION SYSTEM

Local hydraulic resistances are produced by circulation system elements with variable form and size of channels causing variation of flow velocity, generation of large eddies, and counter-flows. Such elements are: surface equipment (kelly, swivel, boring hose, ascending pipe); lock connections; joints; subs; down-hole motors; bits; centralizers; expanders, and so on.

As shown by experiments, pressure drop Δp in local resistances can be determined independently of element relative length by the formula

$$\Delta p = \alpha \rho Q^2, \tag{6.9.1}$$

where α is hydraulic resistance factor which for each element to a first approximation may be taken as constant. The greatest value of the factor α is achieved in the down-hole hydraulic motors, hydraulic monitor bits, and loch connections of the type 3H. The factor α could be theoretically obtained only for the simplest models of local resistances, for example, for abrupt pipe enlargement. Commonly it is found experimentally.

The factor α of surface equipment is determined by the formula (Sheberstov and Leonov, 1968)

$$\alpha = \alpha_{\text{ap}} + \alpha_{\text{bh}} + \alpha_{\text{sw}} + \alpha_{\text{ks}}, \tag{6.9.2}$$

where $\alpha_{\text{ap}}, \alpha_{\text{bh}}, \alpha_{\text{sw}}, \alpha_{\text{ks}}$ are resistance factors of surface elements given by Table 6.1.

In view of variable channel geometry, the turbo-drills may be considered as local resistances and the pressure drop in them may be calculated from the formula (6.9.1) with

$$\alpha = \frac{\Delta p_{\text{td}}}{\rho_{\text{C}} Q_{\text{td}}^2}, \tag{6.9.3}$$

TABLE 6.1

Surface Element	Relative Size (m)	Diameter of the Flow Area (m)	Notation in (6.9.2)	Values of $\alpha \times 10^{-5}$ (m^{-4})
Ascending pipe	0.114	—	α_{ap}	3.4
	0.140			1.1
	0.168			0.4
Boring hose		0.038	α_{bh}	38
		0.076		1.2
		0.080		0.93
		0.090		0.53
		0.102		0.3
Swivel		0.032	α_{sw}	27
		0.075		0.9
		0.080		0.7
		0.090		0.44
		0.100		0.3
Kelly (square)	0.065	0.032	α_{ks}	11
	0.080	0.040		7.0
	0.112	0.074		1.8
	0.140	0.085		0.9
	0.155	0.100		0.4

where Δp_{td} is pressure drop in turbo-drill at operating condition in accordance with its nominal data at given fluid flow rate Q_{td} and fluid density ρ_f (see Table 6.2).

EXERCISE 6.9.1

It is required to determine the pressure drop in the turbo-drill A6Ш at fluid flow rate $Q_{td} = 0.021\ m^3/s$, fluid density $\rho_5 = 1400\ kg/m^3$. Reference data of the turbo-drill (Table 6.2) are $\Delta p_{td} = 45 \times 10^5$ Pa, $\rho_f = 1000\ kg/m^3$, and $Q_{td} = 0.02\ m^3/s$.

SOLUTION Determine previously

$$\alpha = \frac{\Delta p_{td}}{\rho_f Q_{td}^2} = \frac{45 \times 10^5}{1000 \times 0.02^2} = 1.13 \times 10^7\ m^{-4}.$$

The pressure drop Δp is obtained from the formula (6.9.1)

$$\Delta p = \alpha \rho Q^2 = 1.13 \times 10^7 \times 1400 \times 0.021^2 = 69.8 \times 10^5\ \text{Pa}.$$

The formula (6.9.1) to calculate pressure losses in local resistances including engines is convenient to represent as follows:

$$\Delta p = \xi \frac{\rho v^2}{2}, \tag{6.9.4}$$

where $\xi = 2\alpha S^2$ is resistance factor; S and v are, respectively, characteristic channel cross section area and mean velocity.

For helical motors the factor ξ could be found from the formula

$$\xi = 2(a + b \cdot \bar{M}^c), \tag{6.9.5}$$

where $v = Q/S$ is mean velocity; S is the area of flow section between stator and rotor; $\bar{M} = M \cdot S^2/(\rho Q^2 V)$ is the dimensionless torque; M is the torque at the engine shaft; V is the displacement volume of the engine; a, b, and c are the first part of ratings obtained by the factory of origin or in the factory shop of bit engines on the basis of engine tests.

The flow section area S of the channel between rotor and stator is determined by the dependence (Gukasov and Kochnev, 1991)

$$S = \frac{\pi \cdot D_c^2}{2} \frac{(r_0/e_t)(z_t + 1) + 1}{[(r_0/e_t)(z_t + 1) + 2]^2}, \tag{6.9.6}$$

TABLE 6.2 Main Characteristics of Turbo-Drills in Accordance with Russian Government Standard 26673-85

Type of the Down-Hole Motor (in Russian Designations)	External Diameter of the Motor Frame (mm)	Fluid Flow Rate (Water) $Q_{td} \times 10^3$ (m^3/s)	Pressure Drop Δp_{td} (MPa)	Torque on the Shaft at Maximal Power (kN m)	Rotation Frequency of the Shaft at Maximal Power (rpm)	Length (m)
Single-Section Spindleless Turbo-Drills of the Type T12						
Т12М3Е-172	172	28	3.5	0.687	700	7.9
Т12М3Б-195	195	45	2.9	1.06	580	8.1
Т12РТ-240	240	55	8.1	2.5	690	8.2
Multisection Spindle Turbo-Drills of the Type 3TCШ						
3ТСШ1-172	172	25	9.1	1.17	560	25.4
3ТСШ-195	195	30	5.3	1.06	385	25.7
3ТСШ-195ТЛ	195	40	4.0	1.48	320	25.7
3ТСШ1-240	240	32	5.5	2.65	445	23.2
Multisection Spindle Turbo-Drills of the Type A with Inclined Pressure Line						
А6Ш	164	20	4.2	0.563	400	17.3
А7ГТШ	195	30	10.5	1.80	350	25.0
А9ГТШ	240	45	8.4	1.95	250	23.3
Turbo-Drill with Floating Stator						
ТПС-172	172	25	4.8	1.57	400	26.3
Turbo-Bits of Core Type КТД						
КТД4С-172	172	22	8.3	1.88	490	9.2
КТД4С-195	195	28	5.5	1.21	464	10.1

where D_c is contour diameter, z_t is the number of rotor teeth, r_0 is the radius of initial circle, and e_t is the gearing eccentricity. For engines of the type Д, the ratio $r_0/e_t \approx 1.175$.

As distinct from pressure losses in turbo-drills pressure losses in helical down-hole motors (HDM) at constant fluid flow rate depend significantly on rotation moment at the bit that is difficult to determine in the practice. Therefore, when calculating pressure drop in circulation system, the pressure drop in HDM should be determined at braking operation regime of the motor (Gukasov and Kochnev, 1991). At this design pressure conditions could be somewhat overrated being quite acceptable.

The rotation frequency of the motor rotor may be calculated from the formula

$$n = \frac{Q}{V^{1/3}S}(a_1 + b_1 \cdot \bar{M}^{c_1}), \tag{6.9.7}$$

where factors a_1, b_1, c_1 represent the second part of ratings. From this formula ensues the expression for the moment $\bar{M}$ at braking operation regime ($n = 0$)

$$\bar{M} = \left(\frac{-a_1}{b_1}\right)^{1/c_1}. \tag{6.9.8}$$

Insertion of (6.9.8) in (6.9.5) gives the local resistance factor

$$\xi = 2\left[a + b\left(\frac{-a_1}{b_1}\right)^{c/c_1}\right] \tag{6.9.9}$$

and using it in (6.9.4), the pressure drop at braking operation regime $\Delta p_{\rm br}$ is obtained

$$\Delta p_{\rm br} = \xi \cdot \rho \frac{Q^2}{2S^2}, \tag{6.9.10}$$

being maximal at given values of ρ and Q.

EXERCISE 6.9.2

It is required to determine the pressure drop at braking operation regime of the helical down-hole motor Д2-195 at fluid flow rate $Q = 0.032\,\text{m}^3/\text{s}$ with $\rho = 1000\,\text{kg/m}^3$. Reference data for HDM (see Table 6.3) are $V = 0.01326\,\text{m}^3$;

TABLE 6.3 Characteristics of Positive Displacement Motors Size Standard of the Type Д When Operating with Water (in Russian Designations)

	Engine					
Parameters	Д1-54	Д-85	Д1-127	Д5-172	Д2-195	Д1-240
Fluid flow rate (*l*/s)	1.0–2.5	4.8	15–20	25–35	25–35	30–50
Rotation frequency of the shaft (rpm)	180–366	132	198–258	78–108	90–114	72–132
Pressure drop in HDM (MPa)	4.5–5.5	5.5	5.5–8.5	4.5–7.0	4.3–6.7	6.0–8.0
Rotation moment (kN m)	0.07–0.11	0.5	2.2–3.0	4.5–6.0	5.2–7.0	10.0–14.0
Kinematic relation	5:6	9:10	9:10	9:10	9:10	7:8
Active volume, (10^3 m^3)	0.2	1.08	4.932	13.24	13.26	19.1
Contour diameter D_c (m)	0.036	0.0584	0.94	0.135	0.135	0.168
Diameter of recommended bits (mm)	59.0–76.0	98.4–120.6	139.7–158.7	190.5–215.9	215.9–244.5	269.9–295.3
Length (mm)	1890	3230	5800	6220	6550	7570
Mass (kg)	27	110	420	770	110	1740

$D_c = 0.135$; $z_t = 9$; $r_0/e_t \approx 1.175$; tension in operation pair $\delta = -6 \times 10^{-4}$ m. Factors in formulas (6.9.5)–(6.9.8) are (Mosesyan and Leonov, 2002)

$$\begin{aligned}
a &= 6.163026 + 627.66617\bar{\delta} + 85377.97\bar{\delta}^2;\\
b &= 9.927779 - 454.1393 \times \bar{\delta} - 92.05217 \times \bar{\delta}^2;\\
c &= 1.213759 + 69.59739 \times \bar{\delta} + 24997.64 \times \bar{\delta}^2;\\
a_1 &= 0.031315 + 0.848314 \times \bar{\delta} + 43.7371 \times \bar{\delta}^2;\\
b_1 &= -0.00481635 + 2.6671053 \times \bar{\delta} - 412.39984 \times \bar{\delta}^2;\\
c_1 &= 3.486074 + 805.2927 \times \bar{\delta} + 81326.33 \times \bar{\delta}^2,
\end{aligned}$$

where $\bar{\delta} = \delta/\sqrt[3]{V}$ is dimensionless tension.

SOLUTION Obtain the dimensionless tension $\bar{\delta}$ and factors a, b, c, a_1, b_1, c_1

$\bar{\delta} = \delta/\sqrt[3]{V} = -6 \times 10^{-4}/\sqrt[3]{0.01326} = -2.535 \times 10^{-5}$; $a = 5.121$; $b = 11.078$; $c = 1.198$; $a_1 = 0.029$; $b_1 = -0.014$; $c_1 = 1.967$.

Calculate the area S with the formula

$$S = \frac{\pi \cdot D_c^2}{2} \frac{(r_0/e_t)(z_t+1)+1}{[(r_0/e_t)(z_t+1)+2]^2} = \frac{3.14 \times 0.135^2}{2} \frac{1.175(9+1)+1}{[1.175(9+1)+2]^2} = 0.00193\,\text{m}^2.$$

From the formula (6.9.9) we determine the factor ξ

$$\xi = 2\left[a + b\left(\frac{-a_1}{b_1}\right)^{c/c_1}\right] = 2\left[5.121 + 11.078\left(\frac{-0.029}{-0.014}\right)^{1.198/1.967}\right] = 44.745$$

and from (6.9.10) the pressure drop in HDM at braking operation regime is

$$\Delta p_{br} = \xi \times \rho \frac{Q^2}{2S^2} = 44.745 \times 1000 \frac{0.032^2}{2 \times 0.00193^2} = 6.15 \times 10^6 \text{ Pa} = 6.15 \text{ MPa}.$$

As characteristic quantity in channels of lock connections, the area of pipe cross section is taken

$$S = \pi d_{in}^2/4, \tag{6.9.11}$$

where d_{in} is the internal diameter of drill pipes.

Empirical data (Esman, 1982) show that mean value of ξ for lock channels of the type 3H could be taken equal to 7.66 and for lock channels of the type 3Ш $\xi = 1.52$ (Table 6.4).

TABLE 6.4 Resistance Factors for Joints (in Russian Designations)

Type of Joint	3H	3Ш
Empirical resistance factor ξ	4.1; 7; 7.1; 7.5; 8.3; 8.8; 9.1; 9.4	0.62; 0.63; 1.27; 1.6; 2.0; 2.0; 2.43
Mean resistance factor ξ	7.66	1.52

When calculating pressure in pumps the pressure drop in joints has to be often determined as a part of total pressure drop in the circulation system. In such a case in order to get top estimation one should be oriented to the most probable maximal values of ξ for joints being equal to

$$\xi = 9.1 \quad \text{for joints of the type 3H;} \tag{6.9.12}$$

$$\xi = 1.9 \quad \text{for joints of the type 3Ш.} \tag{6.9.13}$$

Channel diameter of joints 3У and welded joints differ a little from internal diameter of drill pipes. Hence pressure drops in them are insignificant and in calculations are not commonly taken into account.

To determine the pressure drop in local resistances, for example, in couplings, joints, and so on, in circular space, the formula (6.9.4) is also used in which as characteristic the cross section area of the circular channel between pipes and well walls is taken

$$S = \pi(d_{\mathrm{w}}^2 - d_{\mathrm{ex}}^2)/4, \tag{6.9.14}$$

where d_{w} is diameter of the well; d_{ex} is the external diameter of the pipe.

At this the factor ξ is calculated with the formula

$$\xi = 2\left(\frac{d_{\mathrm{w}}^2 - d_{\mathrm{ex}}^2}{d_{\mathrm{w}}^{2e} - d_{\mathrm{m}}^2} - 1\right)^2, \tag{6.9.15}$$

where d_{m} is maximal diameter of the joint or the coupling.

To get the total pressure loss in all joints and couplings of given standard size in the drill-stem the pressure drop obtained from (6.9.4) should be multiplied by the number of joints and couplings

$$n = l/l_{\mathrm{T}}, \tag{6.9.16}$$

where l is the length of the drill-stem of one standard size and l_{T} is the length of the pipe between joints and couplings.

EXERCISE 6.9.3

It is required to determine the pressure drop in joints of the type ЗШ in drill pipe and annular space of the well at $d_m = 0.178$ m, $d_{in} = 0.122$ m, $d_{ex} = 0.14$ m, and $d_w = 0.295$ m. The length of the column is $l = 2400$ m, the distance between joints $l_T = 12$ m, fluid density $\rho = 1500$ kg/m^3, and fluid flow rate $Q = 0.022$ m^3/s.

SOLUTION In channels of joints ЗШ, mean value of the factor $\xi = 1.52$. Then taking into account formulas (6.9.4), (6.9.5), and (6.9.16) following is obtained

$$\Delta p = \xi \frac{\rho v^2}{2} n = \xi \frac{\rho Q^2}{2S^2} \frac{l}{l_T} = 1.52 \frac{1500 \times (0.022)^2}{2(\pi \times 0.122^2/4)^2} \times \frac{2400}{12} = 8.08 \times 10^5 \text{ Pa}.$$

The pressure drop at joints in annular space is determined from (6.9.4) with regard to (6.9.14)–(6.9.16):

$$\Delta p = \xi \frac{\rho v^2}{2} n = 2\left(\frac{d_w^2 - d_{ex}^2}{d_w^2 - d_m^2} - 1\right)^2 \frac{\rho Q^2}{2S^2} \frac{l}{l_T} = \left(\frac{0.295^2 - 0.140^2}{0.295^2 - 0.178^2} - 1\right)^2$$

$$\times \frac{1500 \times (0.022)^2}{(3.14 \times (0.295^2 - 0.140^2)/4)^2} \times \frac{2400}{12} = 0.0247 \times 10^5 \text{ Pa}.$$

Determine now the pressure drop in the following local resistance of the circulation system, namely in the bit. Here, the washing fluid flows in a complex channel ended by slush nozzles. Diameters of outlets can be controlled by nozzles of different form—cylindrical, conical, and so on.

To calculate the pressure drop in a bit, the formula (6.9.4) is also used, in which as characteristic area S is taken as total cross section area of all bit slush nozzles; $\xi = 1/\mu^2$, where μ is empirical flow rate factor dependent of the hole shape, physical properties of the fluid, and the pressure at which performs the outflow. On the basis of large body of experimental data it is established that the flow rate factor for a hole in a thin wall is approximately equal to 0.62; for cylindrical nozzle it is 0.82; for conical converged nozzle with cone angle 13° it is 0.945; cone nozzle is 0.98. In Table 6.5 there are

TABLE 6.5 Flow Rate Factors for Drill Bits

Nozzle shape	Cylindrical Drilling with Acute Angle Edges	Drilling with Conical Inlet	Y-Form Slot	Nozzles with Rounded Inlet and Flare
Flow rate factor	0.64–0.66	0.8–0.9	0.7–0.75	0.9–0.95

listed flow rate factors for typical nozzles of drill bits in accordance with Philatov data.

The total hydraulic resistance of a bit of any size standard could be characterized by a factor similar to μ. To do it is necessary to measure pressure drop not only in nozzles but also in the whole bit, that is in channel consisting of nozzles and internal space of the bit.

EXERCISE 6.9.4

It is required to determine the pressure drop in a bit at pump capacity $Q = 0.018\ \mathrm{m^3/s}$ and fluid density $\rho = 1100\ \mathrm{kg/m^3}$. The area of nozzles is $\Phi = 12 \times 10^{-4}\ \mathrm{m^2}$ and $\mu = 0.7$.

SOLUTION From the formula (6.9.4) it follows

$$\Delta p = \xi \frac{\rho v^2}{2} = \frac{1}{\mu^2} \cdot \frac{\rho Q^2}{2\Phi^2} = \frac{1}{0.7^2} 1100 \frac{0.018^2}{2(12 \times 10^{-4})^2} = 2.53 \times 10^5\ \mathrm{Pa}.$$

CHAPTER 7

EQUILIBRIUM AND MOTION OF RIGID PARTICLES IN FLUID, GAS, AND GAS–LIQUID MIXTURE

Rigid particles of rock solids from the bottom and well walls are carried away by the flow of washing fluid, gas, and gas–liquid mixture. Below are considered regularities of interaction between the flow and rock solid particles at the bottom and in the tubular annulus of the well needed for calculation of pumps in washing, compressors in blasting, or both of them in drilling with the use of aerated fluid. A formula is also obtained to calculate the time of hindered sedimentation of a spherical rigid particle in the downflow in a pipe column based on general mechanism of fluid flow around rigid particles.

7.1 WASHING OF THE WELL BOTTOM

Intensity of the rock solid destruction and removal from the bottom is governed by pressure (tension) field in the critical area formed under mechanical action of the bit construction, fluid flows from the bit, and the bed through well bottom and rock pressure. Consider mechanisms of particle tearing off and their further displacement at the bottom.

Applied Hydro-Aeromechanics in Oil and Gas Drilling. By Leonov and Isaev

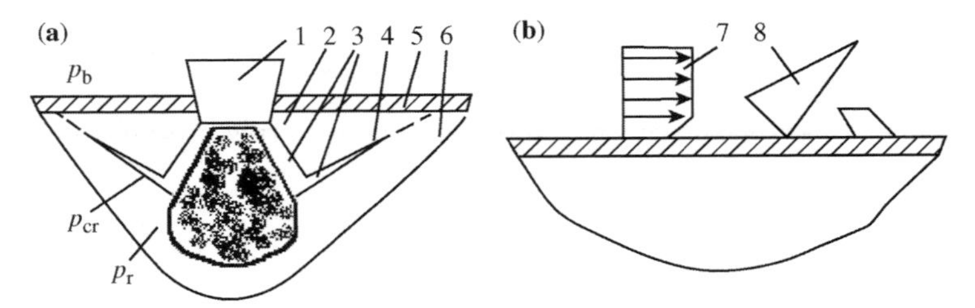

FIGURE 7.1 Schematic pattern of particle location at the bottom at the moment of its detachment. (a) at volume destruction: 1—the tooth of roller bit; 2—particle at tearing off; 3—cracks; 4—rock without cracks; 5—clay coating; 6—matrix. (b) motion along bottom surface: 7—velocity profile of the flow around the particle; 8—particle in the flow.

Figure 7.1 demonstrates two typical positions of a particle at the well bottom in drilling with roller bit at moments of particle tearing off in volume destruction (a) and motion along bottom surface (b).

Separation of a particle from the matrix was preceded by the formation of cracks beneath it in which the pressure p_{cr} is lesser than the bottom-hole p_{bh} and reservoir (pore) p_r pressures. The pressing pressure $(p_{bh} - p_{cr})$ acting on the particle prevents its detachment.

Production of cracks is accompanied by filling them with washing fluid or filtrate, as well as of reservoir fluid. With time, the pressure around the particle levels off. The clay coating formed on the bottom surface retards the fluid inflow to cracks from the well. In drilling out well permeable rock solids, the influence of clay coating on pressure recovery in cracks becomes weaker owing to more intensive inflow of reservoir fluid in it.

Cohesion forces at rock solid regions without cracks, friction forces at regions of contact with clay coating, and the gravity force of a particle along with pressing force retard the detachment of the particle from the matrix.

After particle detachment, the cohesion force between particle and rock solid together with pressing force vanishes, and particle motion along bottom is mostly prevented by gravity force and friction against clay coating, rock solid at the bottom, other particles, and drill bit roller cutters. The motion of the particle from the bottom to the annular space is also hindered by dynamic pressure of fluid flow opposite the main outflow.

Thus, the condition of rock solid particle detachment and its further hindered motion in the critical area result in exceeding the resultant pressure force, further called as withdrawing force, the sum of cohesion, friction, gravity, and pressing forces (holding force) acting on the particle. These forces mostly depend on (1) pressure and velocity of drill roller bit teeth action, (2) differential pressure $(p_{bh} - p_{cr})$, (3) pressing pressure and the rate of its drop, (4) dynamic pressure of washing fluid flow, (5) direction of the flow over the particle, and (6) friction force at the particle surface in its motion to annular space.

The effect of all the above-listed components on the formation of holding and withdrawing forces is different in particle detachment and motion. If forces exert the greatest effect on the particle detachment from the matrix caused by action of drill roller teeth, the motion of suspended particles in the critical area of formation mostly depends on dynamic pressure and direction of the washing fluid flow. The influence of teeth pressure and velocity is studied in another disciplines. Below are considered only hydro-aerodynamic forces on a qualitative level since quantitative consideration is too difficult.

The rise of differential pressure increases the holding force in particle detachment due to growth of the pressing pressure and hardening of the rock solid elevating its compression stress at the bottom; withdrawing force enhancing the rate of pressure recovery in cracks beneath particles, that is, decreasing the time of pressing between particles. These are two opposite factors. The first prevents whereas the second promotes particle detachment from the matrix. A lot of field observations and laboratory experiments showed that the first factor prevails.

The withdrawing force increases with particle detachment and holding forces decrease with enhancement of reduction rate of pressing pressure acting on particles through growth of permeability and fall of clay coating thickness, rheological properties of the solution, and its filtrate. The use of lubricant additives decreases the holding force with decline of friction force between particles, clay coating, bit surface, and bottom. The rise in dynamic pressure leads to an increase in the withdrawing force acting on the particle. It could be enhanced by building up the ratio between hydraulic power N delivered to the bit and area of the bottom.

The power is determined by expression

$$N = Q\Delta p_{\text{bit}}. \tag{7.1.1}$$

By inserting the pressure drop $\Delta p_{\text{bit}} = \rho v^2/(2\mu^2)$ in (7.1.1) and dividing both parts by bottom area S, we get

$$\frac{N}{S} = \frac{Q}{S}\frac{\rho v^2}{2\mu^2}, \tag{7.1.2}$$

where Q and ρ are, respectively, flow rate and fluid density, v is the velocity of fluid outflow from nozzles, and μ is the discharge factor.

Experimental investigations show that the best bottom-hole cleaning is achieved by definite values of Q/S and v in the right part of the formula (7.1.2).

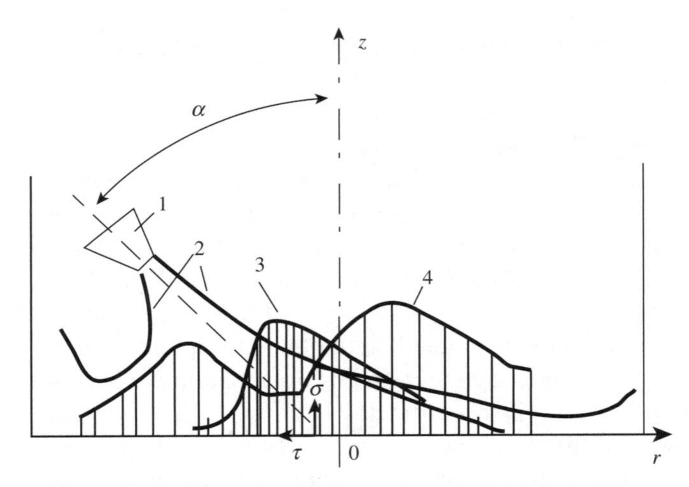

FIGURE 7.2 Scheme of submerged jet propagation in a deadlock: 1—nozzle; 2—jet boundary; 3, 4—plots of normal and tangential stresses at the bottom.

Recommended values of velocity and specific flow rate are $v = 80$–120 m/s and $Q/S = 0.35$–0.7 m/s. At this, the pressure drop in nozzles with serial bits can reach 12–13 MPa conditioned by bit strength. In high-pressure jet bits, the pressure drop can be 30–40 MPa. The range of change Q/S in many cases coincides with its values needed for advantageous transport of cuttings along annular channel to the surface. At given and above specified values of Q/S, the rate of outflow achieves values close to the values at the bottom with the help of special hydro-monitor nozzles with reduced diameters.

The role of fluid jet in producing withdrawing force can be qualitatively estimated with the help of a model of submerged jet flowing into cylindrical deadlock at an angle α to z-axis at a given instance of time (Fig. 7.2). Under action of the jet at the bottom surface of the deadlock occur normal σ and tangential τ (curves 3 and 4) stresses nonuniformly distributed over the bottom radius r. The nonuniformity is compounded with time by rotation of the nozzle around deadlock axis. At different distances and angles of inclination of the nozzle to the bottom α from 0 up to 90°, the stresses σ and τ should not exceed the maximal value of pressure drop in the nozzle 12–13 MPa. This value is lesser than the hardness of even soft rock solids almost by an order of magnitude and comparable to the shear strength of rock solids. Therefore, normal stresses in the rock solid caused by jet deceleration are of secondary importance in forming the withdrawing force compared to tangential stresses generated by jet flow spreading parallel to the bottom. Of particular importance in ablation of particles previously separated from the matrix with bit teeth is the flow along the bottom. Such positioning of the flow at the bottom is desirable so that each particle is removed from the bottom without reuse of bit teeth.

Bit design plays a great role in reducing kinetic energy losses during submerged jet propagation from the nozzle face up to the bottom and along it. The last factor also determines the field of dynamic pressure in the bottom flow. The best arrangement of flows for different types of drill bits is not yet established. Encouraging results for orderly flow patterns were obtained by the use of drill bits with asymmetric arrangement of nozzles. In them, the main flow parallel to the bottom and washing away particles is decelerated to a lesser degree by unfavorable counter flows. In particular, there are very promising three-roller bits in which the flow through cone pin and vertex of one of the rolling cutters with hydro-monitor nozzle is directed at an angle of several degrees to the bottom. This flow pointedly and orderly moves the cutting from the bottom center to the periphery into opposite aperture between bit legs, being set open, and to above bit-space, with the remaining two other apertures being left closed.

7.2 LEVITATION OF RIGID PARTICLES IN FLUID, GAS, AND GAS–LIQUID MIXTURE FLOWS

Consider the motion of a rigid particle with average velocity v_p in ascending vertical fluid stream flowing with velocity v_f. In general case, velocities of the flow and the particle relative to immovable well walls are different, that is, $v_f \neq v_p$. The density of rock solid particles is generally greater than the density of the fluid; therefore, the relative velocity is

$$v_r = v_f - v_p. \tag{7.2.1}$$

Assume that the particle is levitating in the flow. Then, $v_p = 0$ and the flow velocity in (7.2.1) called levitating velocity of the particle v_s is

$$v_s = v_f = v_r. \tag{7.2.2}$$

To get average flow velocity providing transport of particles in annular channel, one should determine velocities v_f and v_p (see Fig. 7.3a).

The equilibrium of a particle in ascending flow is described by equality of forces acting on the particle (Fig. 7.3b).

The resultant buoyancy and gravity forces act on the particle

$$R = V(\rho_p - \rho_f)g, \tag{7.2.3}$$

where V is the particle volume, ρ_p and ρ_f are the densities of particle and fluid, respectively, and g is the gravity acceleration.

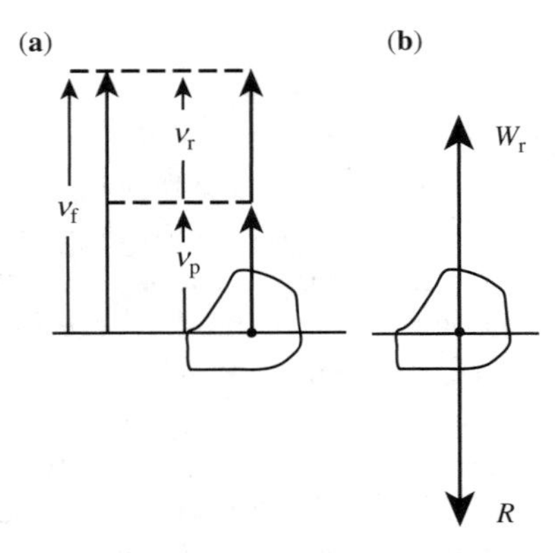

FIGURE 7.3 Definition of particle levitation velocity in fluid (a) and forces acting on the particle (b).

The force R is balanced by resistance force W_r depending rheological properties of fluid, flow regime, concentration, and shape of particles. In what follows, a single spherical particle with equivalent diameter $d_p = (6V/\pi)^{1/3}$ in unbounded media will be considered. It was experimentally established that in this case the particle experienced minimal resistance. The levitation velocity and the flow rate obtained further should be considered as upper estimation. The formula for particle resistance in the relative flow of viscous fluid at small Reynolds numbers $Re = v_s d_p \rho_f / \mu < 1$ is Stokes formula

$$W_r = 3\pi\mu d_p v_s. \tag{7.2.4}$$

For arbitrary Re, the resistance force may be expressed in general form as

$$W_r = C_w \rho_f \frac{v_s^2}{2} S, \tag{7.2.5}$$

where C_w is resistance factor that could be found from the graphics depicted in Fig. 7.4 and S is the area of maximal particle cross section orthogonal to the flow direction far from the particle.

Formula (7.2.5) transforms into (7.2.4) at

$$C_w = 24/Re. \tag{7.2.6}$$

By setting (7.2.3) equal to (7.2.5) and substituting $V = \pi d_p^3/6$ and $S = \pi d_p^2/4$ in resulting relation, one obtains the levitation velocity of a spherical particle

$$v_s = \sqrt{\frac{4}{3}\frac{(\rho_p - \rho_f) d_p g}{\rho_f C_w}}. \tag{7.2.7}$$

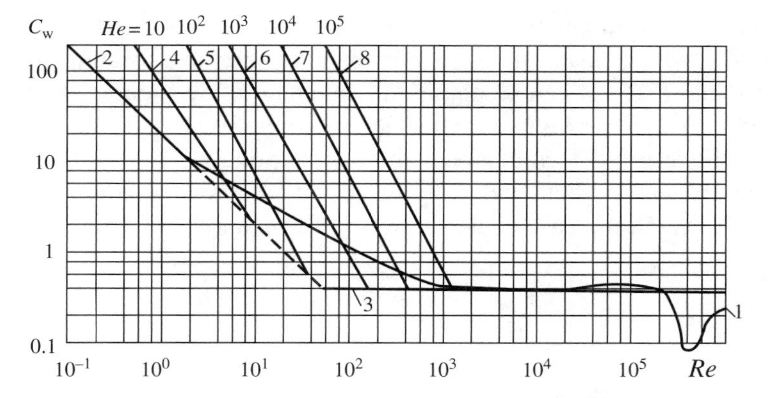

FIGURE 7.4 Dependence of resistance factor in the flow of viscous and viscous-plastic fluid around a particle on Reynolds and Hedström numbers without regard for transient processes: 1—Rayleigh curve; 2—viscous fluid; 3—lower limit of transient regime (approximation); 4–8—viscous-plastic fluid.

Expression (7.2.7) at $C_w = 24/Re$ converts to

$$v_s = \frac{1}{18} d_p^2 \frac{(\rho_p - \rho_f)g}{\mu}. \qquad (7.2.8)$$

This formula is true at $Re < 1$. In the range $1 < Re < 10^3$, another formula should be used

$$C_w = \frac{24}{Re}(1 + 0.17Re^{0.665}). \qquad (7.2.9)$$

In the range $10^3 \leq Re < 10^5$, the resistance factor becomes constant

$$C_w = \text{const} = 0.44. \qquad (7.2.10)$$

Despite the fact that formula (7.2.8) is valid for $Re < 1$, we shall use it for calculating cutting ablation up to $Re = 60$ since it gives at these values of Re only slightly overrated values of the velocity v_s. Formula (7.2.7) can be written as follows:

$$v_s = k\sqrt{d_p\left(\frac{\rho_p}{\rho_f} - 1\right)}, \qquad (7.2.11)$$

where $k = \sqrt{4g/(3C_w)}$ is Rittinger constant. In accordance with experimental data, it could be taken $C_w = 0.4$ at $Re > 60$. Then,

$$k = \sqrt{4g/(3C_w)} = \sqrt{4 \times 9.81/(3 \times 0.4)} = 5.72\ \text{m}^{1/2}/\text{s}. \qquad (7.2.12)$$

To choose a proper formula for levitation velocity v_s, it is required to know *Re*, which in its turn depends on v_s. Therefore, formulas (7.2.9), (7.2.8), and (7.2.11) can be applied to calculate v_s by successive approximations method as follows. After determining the velocity v_s using one of the above-listed formulas and then corresponding Reynolds number, one should check whether *Re* belongs to the range of considered formula applicability. In the case of negative result, use another formula.

EXERCISE 7.2.1

It is required to calculate the levitation velocity of a spherical particle with diameter $d_p = 0.01$ m, density $\rho_p = 2500\,\text{kg/m}^3$ in the flow of water with density $\rho_f = 1000\,\text{kg/m}^3$ and viscosity $\mu = 0.001$ Pa s.

SOLUTION From formula (7.2.11), particle velocity is obtained

$$v_s = k\sqrt{d_p\left(\frac{\rho_p}{\rho_f} - 1\right)} = 5.72\sqrt{0.01\left(\frac{2500}{1000} - 1\right)} = 0.701\ \text{m/s}. \qquad (7.2.13)$$

Corresponding Reynolds number $Re = v_s d_p \rho_f / \mu = 0.701 \times 0.01 \times 10^3 / 0.001 = 7.01 \times 10^3$ lies in the range described by formula (7.2.11).

EXERCISE 7.2.2

It is required to calculate the levitation velocity of the same particle that is flowed about by air with viscosity $\mu = 0.2 \times 10^{-4}$ Pa·s and density $\rho_f = 1.29\,\text{kg/m}^3$.

SOLUTION From formula (7.2.11), the velocity obtained is

$$v_s = k\sqrt{d_p\left(\frac{\rho_p}{\rho_f} - 1\right)} = 5.72\sqrt{0.01\left(\frac{2500}{1.29} - 1\right)} = 25.2\ \text{m/s}. \qquad (7.2.14)$$

Corresponding Reynolds number $Re = v_s d_p \rho_f / \mu = 25.2 \times 0.01 \times 1.29 / 0.00002 = 1.62 \times 10^4$ lies in the range described by formula (7.2.11).

As distinguished from viscous fluid, different spheres in viscous-plastic fluids could also be in equilibrium at $v_s = 0$. The state of a sphere preceding motion is characterized by limiting equilibrium when the gravity force is balanced by the force W_r resulted by the action of stresses over all volume surface. Assume that W_r has form

$$W_r = \pi d_p^2 \tau_0.$$

Then, equating W_r and R (see formula (7.2.3)) one gets that at $v_s = 0$, maximal size of the particle capable to be in equilibrium state in viscous-plastic fluid is (Shischenko et al., 1976)

$$d_p = \frac{6\tau_0}{(\rho_p - \rho_f)g}. \tag{7.2.15}$$

It should be noted that often in practice in calculation using formula (7.2.15) instead of τ_0, static shear stress θ is taken since many solutions do not obey the law of viscous-plastic model in all domains of velocity gradient variability.

Suppose that in motion ($v_s \neq 0$) of a sphere in viscous-plastic fluid the resistances caused by viscous and plastic properties are summarized. Then, total resistance force acting on the particle could be written as

$$W_r = \pi d_p^2 \tau_0 + 3\pi\eta d_p v_s. \tag{7.2.16}$$

Equating (7.2.16) and (7.2.3) levitation velocity of the particle is obtained as

$$v_s = \frac{d_p}{3\eta}\left[\frac{(\rho_p - \rho_f)g d_p}{6} - \tau_0\right]. \tag{7.2.17}$$

From relation (7.2.17) at $\tau_0 = 0$, formula (7.2.8) for the viscous fluid flow around a particle follows. The resistance force at different regimes of viscous-plastic fluid flow around a particle may also be represented in the form (7.2.5) in which the resistance factor C_w is different for each regime.

So for (7.2.16), it is

$$W_r = \pi d_p^2 \tau_0 + 3\pi\eta d_p v_s = C_w \rho_f \frac{v_s^2}{2}\frac{\pi d_p^2}{4}, \tag{7.2.18}$$

from which follows

$$C_w = \frac{\pi d_p^2 \tau_0 + 3\pi\eta d_p v_s}{\rho_f \frac{v_s^2}{2}\frac{\pi d_p^2}{4}} = \frac{8\tau_0}{\rho_f v_s^2} + \frac{24}{Re}$$

or

$$C_w = \frac{24}{Re}\left(1 + \frac{He}{3Re}\right) = \frac{24}{Re}\left(1 + \frac{Se}{3}\right), \tag{7.2.19}$$

where

$$He = \frac{\rho_f \tau_0 d_p^2}{\eta^2}, \quad Se = \frac{\tau_0 d_p}{\eta v_s}.$$

In Fig. 7.4 with solid line 1 is represented dependence of resistance factor C_w on Re at $He = 0$ depicted by experimental data; the line 2 together with its dotted prolongation corresponds to Stokes formula (7.2.4); dotted line 3 corresponds to $C_w = 0.4$. The Stokes law is true at $Re < 1$. At $1 < Re < 10$, inertial forces begin to play a noticeable role and the Stokes law (dotted part of line 2) differs from experimental data. At $Re \approx 10$ separation of laminar boundary layer from the sphere occurs with the production of ring vortex at stern part of the sphere, that is, begins transient regime from laminar to turbulent flow. At $Re \approx 100$, the system of vortexes propagates around the sphere over a distance of an order of sphere diameter. At $Re \approx 150$, called lower critical Reynolds number, the influence of inertial forces increases and the vortex system produces vortex wake behind the sphere. At $Re \approx 10^5$, called upper critical Reynolds number, the laminar boundary layer transits completely into turbulent one.

For practical calculations, approximation of experimental curve 1, that is, straight lines 2–8, is used. This is justified by that taking as a basis for calculations lesser resistance and using formulas (7.2.8), (7.2.11), and (7.2.17) for particle levitation velocity, insignificantly overrated values of levitation velocity are obtained. The field of application formulas (7.2.8), (7.2.11), and (7.2.17) is widened and they could be used, especially owing to uncertainty of information about shape and concentration of particles and rheological properties of the fluid.

Considered approximation requires introduction of one critical Reynolds number Re_{cr} for viscous fluid. Critical Reynolds numbers for viscous-plastic fluid would result from intersection of curves 2, 4–8, and other corresponding to definite Hedström numbers with straight line 3. At these Reynolds numbers with sufficient accuracy, it may be assumed that all non-Newtonian properties of solutions would degenerate.

Determine at which Re_{cr} the formula (7.2.17) is valid. Let us take that at $Re > Re_{cr}$ the resistance factor remains constant equal to $C_w = 0.4$. Substitution of $C_w = 0.4$ in (7.2.19) gives dependence of Re_{cr} on He

$$Re_{cr} = 30(1 + \sqrt{1 + He/45}). \tag{7.2.20}$$

Thus, to calculate particle levitation velocity at $Re \leq Re_{cr}$ and $Re > Re_{cr}$, respectively, formulas (7.2.17) and (7.2.11) are recommended. At $\tau_0 = 0$ from (7.2.20) yields

$$Re_{cr} = 60. \tag{7.2.21}$$

Before going to the next exercise, it should be noticed that formulas (7.2.8), (7.2.11), and (7.2.17) are convenient for calculations when the

Reynolds number Re is known. But for unknown Re, it is impossible to determine the levitation velocity explicitly because in it enters unknown levitation velocity. Therefore, let us transform formula (7.2.7), which is true for any fluid. Substitution in it

$$v_s = \frac{Re\mu}{d_p\rho_f} \tag{7.2.22}$$

gives

$$C_w Re^2 = \frac{4}{3}Ar, \tag{7.2.23}$$

where

$$Ar = \frac{d_p^3 g}{\mu^2}\rho_f(\rho_p - \rho_f) \tag{7.2.24}$$

is Archimedes number.

Insertion of (7.2.6) into (7.2.23) gives for the curve 2 (Fig. 7.4)

$$Re = Ar/18. \tag{7.2.25}$$

After substitution of $C_w = 0.4$ in (7.2.23), the following is obtained for the curve 3 (Fig. 7.4):

$$Re = 1.83\mathrm{Ar}^{0.5}. \tag{7.2.26}$$

Finally, substituting (7.2.19) in (7.2.23) one gets

$$Re = Ar/18 - He/3. \tag{7.2.27}$$

Introduce now critical Archimedes number. To do this, it is needed to insert in (7.2.25) and (7.2.27) successively $Re_{cr} = 60$ and Re_{cr} from (7.2.20). For viscous fluid, it results in

$$Ar_{cr} = 1080 \tag{7.2.28}$$

and for viscous-plastic fluid

$$Ar_{cr} = 18(Re_{cr} + He/3). \tag{7.2.29}$$

In practice, one should calculate the levitation velocity as follows. At first, it is necessary to determine numbers He, Re_{cr}, Ar_{cr}, and Ar and then to compare Ar_{cr} with Ar. If $Ar < Ar_{cr}$, one has to calculate the levitation velocity with formulas (7.2.22) and (7.2.27). If $Ar \geq Ar_{cr}$, one needs to use formula (7.2.26).

EXERCISE 7.2.3

Determine the levitation velocity of a spherical particle with diameter $d_p = 0.01$ m, density $\rho_p = 2300\,\mathrm{kg/m^3}$ in viscous-plastic fluid with density $\rho_f = 1300\,\mathrm{kg/m^3}$, plastic viscosity factor $\eta = 0.015$ Pa s, and dynamic shear stress $\tau_0 = 5$ Pa.

SOLUTION Determine at first Archimedes number with the formula (7.2.24)

$$Ar = \frac{d_p^3 g}{\eta^2}\rho_f(\rho_p - \rho_f) = \frac{0.01^3 \times 9.81}{0.015^2}1300(2300-1300) = 5.67 \times 10^4,$$

He and Re_{cr} with the formula (7.2.20)

$$He = \frac{\rho_f \tau_0 d_p^2}{\eta^2} = \frac{1300 \times 5 \times 0.01^2}{0.015^2} = 2889;$$

$$Re_{cr} = 30(1 + \sqrt{1 + He/45}) = 30(1 + \sqrt{1 + 2889/45}) = 2.72 \times 10^2$$

and then critical Archimedes number from (7.2.29)

$$Ar_{cr} = 18(Re_{cr} + He/3) = 18(272 + 2889/3) = 2.22 \times 10^4.$$

Since $Ar = 5.67 \times 10^4 > Ar_{cr} = 2.22 \times 10^4$, the levitation velocity should be calculated by (7.2.22) with the use of (7.2.26)

$$v_s = \frac{Re\eta}{d_p\rho_f} = \frac{1.83 \times Ar^{0.5}\eta}{d_p\rho_f} = \frac{1.83(5.67 \times 10^4)^{0.5}0.015}{0.01 \times 1300} = 0.503\,\mathrm{m/s}.$$

For power fluid, one can assume that the resistance force also obeys Stokes law with correction to the exponent to be experimentally determined

$$W_r = \pi d_p^2 k \left(\frac{3v_s}{d_p}\right)^n, \tag{7.2.30}$$

where k and n are consistency indices.

Equating (7.2.30) and (7.2.3) for spherical particle, one gets the levitation velocity

$$v_s = \frac{1}{3}\left[\frac{(\rho_p - \rho_f)g d_p}{6k}\right]^{1/n} d_p. \tag{7.2.31}$$

The particle resistance factor for power fluid is found by equating (7.2.30) and (7.2.5)

$$C_w = \frac{\pi d_p^2 k(3v_s/d_p)^n}{\rho_f(v_s^2/2)(\pi d_p^2/4)} = \frac{8 \times (3)^n}{Re}, \tag{7.2.32}$$

where Reynolds number is $Re = \rho_f v_s^{2-n} d_p^n / k$.

At $n = 1$, formula (7.2.32) converts to formula (7.2.6) for viscous fluid (7.2.6). The critical Reynolds number Re_{cr} for power fluid is determined from relation (7.2.32), when in it to accept $C_w = 0.4$

$$Re_{cr} = 20 \cdot (3)^n. \tag{7.2.33}$$

It should be marked that formulas (7.2.8), (7.2.11), (7.2.17), and (7.2.31) are also applicable for floating-up velocity of gas bubble in fluid when the difference between densities of gas and particle $|\rho_f - \rho_p|$ is taken by absolute value. They could also be of use to estimate the sedimentation velocity of mud weighting material particles. Above-listed approximate formulas are true for viscous, viscous-plastic, and power fluids. Application of these formulas to thixotropic solutions is possible with the help of correction factor k' (Shischenko et al., 1976). Then, the formula (7.2.15) should be transformed to

$$d_p = \frac{6 \cdot \theta}{(\rho_p - \rho_f) g k'}, \tag{7.2.34}$$

where parameter θ has to be measured. For spherical particles, $k' \approx 0.3$–0.6. This value is connected with the measurement of the stress θ not in equilibrium but at definite shear rates. Factor k' can be introduced in formula (7.2.17) by replacing τ_0 with θ

$$v_s = \frac{d_p}{3\eta}\left[\frac{(\rho_p - \rho_f) g d_p}{6} - \frac{\theta}{k'}\right]. \tag{7.2.35}$$

As distinct from single-phase medium, the levitation of particles in two-phase (gas–liquid) mixture is governed by two levitation velocities of gas v_g and fluid v_f. The resistance force W_r in ascending two-phase flow around a particle may be expressed by dependence (Leonov, 1973)

$$W_r = f(\rho_g, \rho_f, \rho_p, v_g, v_f, d_p, g). \tag{7.2.36}$$

The resistance factor could be found by equating (7.2.36) and (7.2.3) and using dimension analysis

$$C_w = f(Fr_g, Fr_f, \rho_g/\rho_p, \rho_f/\rho_p), \tag{7.2.37}$$

where $Fr_g = v_g^2/(g d_p)$ and $Fr_f = v_f^2/(g d_p)$ are Froude numbers of gas and fluid, respectively.

Obtained relation (7.2.37) provides a possibility to perform dimensionless treatment of empirical data on particle levitating in turbulent two-phase flow and to get dependence $Fr_g \rho_g/\rho_p$ or $Fr_f \rho_f/\rho_p$ in Fig. 7.5. The curve

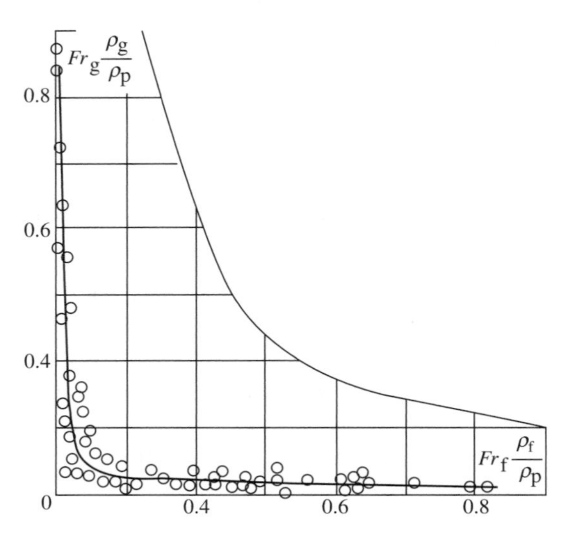

FIGURE 7.5 Dependence between dimensional parameters in particle levitation in vertical ascending flow of aerated fluid (empirical results are denoted by circles).

in this figure is adequately described by

$$(Fr_g\rho_g/\rho_p + 0.008)(Fr_f\rho_f/\rho_p + 0.008) = 0.011. \tag{7.2.38}$$

Resolving (7.2.38) with regard to (4.3.5) and (4.3.6) with respect to one of the velocities, for example, gas velocity, a formula is obtained that is capable to calculate the levitation velocity of one of the phases at another given phase under condition of process isothermality

$$v_g = \sqrt{\frac{gd_p\rho_p p_0}{\rho_0 p}\left(\frac{0.011d_p\rho_p g}{v_f^2\rho_f + 0.008d_p\rho_p g} - 0.008\right)}, \tag{7.2.39}$$

where v_g is the gas velocity at operation pressure p, ρ_0 is the gas density at normal pressure p_0. The dependence (7.2.39) for particles of various diameters is shown in Fig. 7.6.

From (7.2.39), one can find the levitation velocity of a particle v_s in gas flow at normal conditions ($p = p_0$). Really, taking in (7.2.39) $v_f = 0$, the following is obtained:

$$v_g = v_s = \sqrt{1.39\frac{gd_p\rho_p}{\rho_0}}.$$

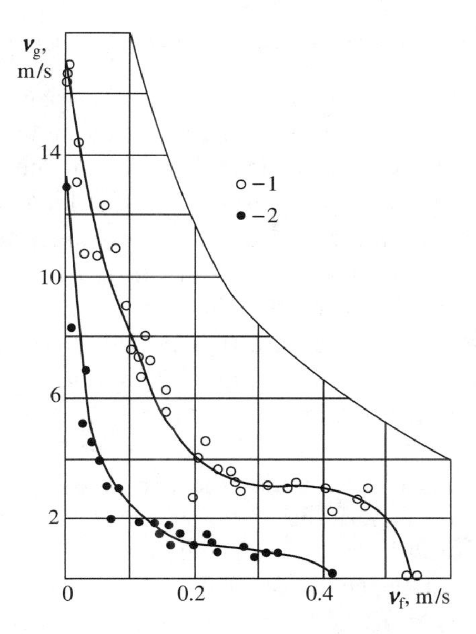

FIGURE 7.6 Dependence of gas velocity on fluid velocity in levitation of particles in the flow of aerated fluid. Empirical data are given for spherical duralumin particles with diameter (mm): 1–10, 2–4.

For conditions of Exercise 7.2.2, the velocity of levitation is

$$v_g = v_s = \sqrt{1.39\frac{gd_p\rho_p}{\rho_0}} = \sqrt{1.39\frac{9.81 \times 0.01 \times 2500}{1.29}} = 16.3\ \mathrm{m/s}.$$

Similarly, the levitation velocity in single-phase fluid flow can be obtained. For conditions of Exercise 7.2.1, the velocity is $v_f = v_s = 0.584\ \mathrm{m/s}$.

When comparing results of calculation with formulas (7.2.13), (7.2.14), and (7.2.39), it is seen that the last formula provides lesser value than both previous ones. This is explained by that formula (7.2.39) was obtained in accordance with experimental data carried out at hindered conditions whereas formula (7.2.14) represents an approximation with velocity overstating as the result of acceptance $C_w = \mathrm{const} = 0.4$ for $Re > 60$.

In hydraulic rotary drilling, the removal of rock solid is performed in rotating pipe column; therefore, particles of cuttings have, besides vertical velocity component of the flow around particles, horizontal component. At this, the removal of cuttings is improved, which is supported by experiments.

7.2.1 Flow Around a Particle of Cutting Near the Wall

The flow of single- and multiphase media around a particle near the wall is distinguished from the flow around a particle in unbounded media. Replacing the flow around a particle near the wall on the flow around it in unbounded media and taking the particle of irregular (nonspherical) shape as spherical one, we in doing so reduce the resistance factor C_w. Furthermore, other suggestions leading to understating the factor C_w, and overstating the starting velocity v_s, that is, beginning of particle motion in required direction assuring particle transport, will be made. As a result, required estimations of quantities would be obtained, since the purpose of calculations is to determine fluid flow rates needed to start particle motion and its transport.

Particles in the fluid flow under the action of gravity force tend to sediment on the lower generating line (generatrix) of inclined, in particular horizontal, part of circular channel. The fluid velocity in the annular space varies from 0 to v_{max} at the center of the gap. Assume that the particle begins to move when the velocity of fluid at the point of particle center location exceeds the velocity of particle levitating. The scheme of forces acting on the particle in this case is shown in Fig. 7.7.

At the instance of spherical particle start, the force R acting on the particle is

$$R = \mu_{fr} \cdot G \cdot \sin\alpha + G \cdot \cos\alpha, \qquad (7.2.40)$$

where μ_{fr} is the factor of friction between particle, wall, and other particles; G is the weight of the particle; and α is the zenith angle.

It should be noted that besides friction force between particle and wall, other interaction forces could be revealed. In general, they may be taken

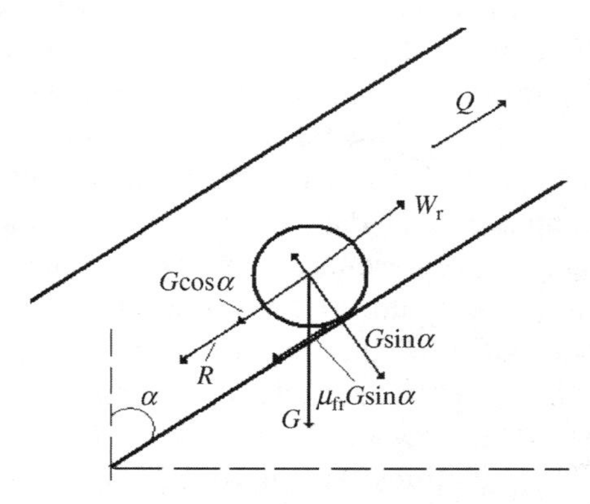

FIGURE 7.7 Scheme of forces applied to the particle in inclined channel.

into account through appropriate correction of factor μ_T, which should be experimentally found.

Suppose that resistance force of spherical particle is determined by formula (7.2.5)

$$W_r = C_w \rho_f v_c^2 \frac{\pi \cdot d_p^2}{8}. \tag{7.2.41}$$

In equilibrium state, equality $R = W_r$ is obeyed. Then, from (7.2.40) and (7.2.41) follows

$$(\mu_T \cdot \sin\alpha + \cos\alpha) \cdot \frac{\pi \cdot d_p^2}{6} \cdot (\rho_p - \rho_f) \cdot g = C_w \rho_f v_c^2 \frac{\pi \cdot d_p^2}{8}, \tag{7.2.42}$$

where ρ_p and d_p are particle density and diameter, respectively; $v_s = v_c$ is the velocity of the fluid at particle center.

From (7.2.42), the velocity of particle start is obtained

$$v_s = \sqrt{\frac{4 \times (\rho_p - \rho_f) \cdot d_p \cdot g \cdot (\mu_{fr} \cdot \sin\alpha + \cos\alpha)}{3 \times C_w \rho_f}}. \tag{7.2.43}$$

Consequently, in finding fluid flow rate ensuring particle transport, the flow velocity at distance $d_p/2$ from the well wall should be equal to

$$v_s = K\sqrt{\frac{4 \cdot (\rho_p - \rho_f) \cdot d_p \cdot g \cdot (\mu_{fr} \cdot \sin\alpha + \cos\alpha)}{3 \times C_w \rho_f}}, \tag{7.2.44}$$

where K is safety factor ($K > 1.0$).

For flow of fluid of arbitrary rheology, dependence (7.2.43) should be held in the form

$$v_s = K\sqrt{\frac{4 \cdot (\rho_p - \rho_f) \cdot d_p \cdot \bar{g}}{3 \cdot C_w \rho_f}}, \tag{7.2.45}$$

where $\bar{g} = g(\mu_{fr} \times \sin\alpha + \cos\alpha)$ is effective gravity acceleration.

For vertical pipes ($\alpha = 0$), valid equality is $\bar{g} = g$ and formula (7.2.45) coincides with (7.2.7).

Hence, the velocity of cutting start in inclined wells may be calculated from formula (7.2.45) similar in structure to the formula for vertical channel but with effective gravity acceleration $\bar{g}$ depending on angle α.

In laminar flow of viscous fluid around a ball, the resistance factor C_w is determined by formula (7.2.6)

$$C_w = 24/Re, \tag{7.2.46}$$

where $Re_s = v_s d_p \rho_f / \mu$ is Reynolds number of the flow around particle and μ is the dynamic viscosity factor.

In this case, formula (7.2.45) takes the form

$$v_s = \frac{1}{18} \frac{(\rho_p - \rho_f) \cdot d_p^2 \cdot \bar{g}}{\mu}. \quad (7.2.47)$$

From this formula, it follows that maximal velocity v_s is achieved at $\alpha = \arctan(\mu_{fr})$. For example, $\alpha = 45°$ at $\mu_{fr} = 1$.

In turbulent flow of viscous fluid around a particle, one should in (7.2.45) assume $C_w = 0.4$, and the formula takes the form

$$v_s = 1.83 \sqrt{\frac{(\rho_p - \rho_f) \cdot d_p \cdot \bar{g}}{\rho_f}}. \quad (7.2.48)$$

In the case of laminar flow, the levitation velocity v_s around a particle is determined by formula (7.2.47). But at this it should be verified whether the flow is laminar. To do so, one has to calculate Re_s with formula (7.2.46). If $Re_s < Re_{cr} = 60$, then calculations made from (7.2.47) remain valid, while at $Re_s \geq Re_{cr}$, calculations with (7.2.48) should be performed.

To find the flow rate Q at particle start, it is necessary to consider velocity profile replacing annular space by a slot. The velocity profile of viscous fluid flow in a slot is represented in Fig. 7.8.

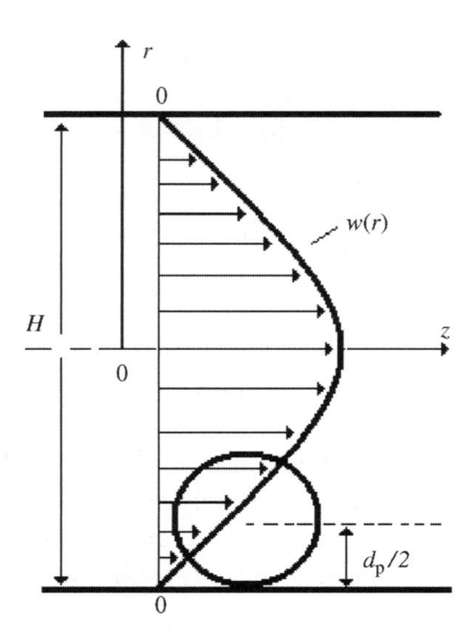

FIGURE 7.8 The flow in plane slot: d_p—equivalent diameter of cutting particle; H–gap width; $W(r)$—fluid flow velocity profile.

Velocity profile may be obtained from momentum equation

$$\frac{\partial p}{\partial z} = \mu \frac{\partial^2 w}{\partial r^2}, \tag{7.2.49}$$

condition of longitudinal pressure gradient constancy

$$\frac{\partial p}{\partial z} = A = \text{const} \quad \text{or} \quad \mu \frac{\partial^2 w}{\partial r^2} = A = \text{const}$$

and boundary conditions (see Fig. 7.8)

$$w = 0 \text{ at } r = \pm H/2.$$

Integration of equation (7.2.49) with boundary conditions gives

$$w(r) = \frac{A}{2\mu}\left[r^2 - \left(\frac{H}{2}\right)^2\right], \tag{7.2.50}$$

where $A = \Delta p/L$ is pressure drop per pipe unit length ($\Delta p < 0$).

Replace in equation (7.2.50) r on coordinate of spherical particle center $r = -H/2 + d_p/2$ and resolve it relative to A changing velocity w on v_s from (7.2.45). As a result, the following is obtained

$$A = \frac{2\mu v_s}{(-H/2 + d_p/2)^2 - (H/2)^2}. \tag{7.2.51}$$

The flow rate in annular space is equal to

$$Q = -\frac{\pi \cdot d_h^3 (d_{ex} + d_{in}) \cdot A}{128\mu \cdot f(\delta)}, \tag{7.2.52}$$

where $f(\delta) = -(1-\delta)^2/(1 + \delta^2 + (1-\delta^2)/\ln\delta)$; $\delta = d_{in}/d_{ex}$; $d_h = d_{ex} - d_{in}$ is the hydraulic diameter.

Calculations with formula (7.2.52) in practice are performed when the flow in annular space is laminar; that is, Reynolds number

$$\text{Re} = \frac{Q \cdot d_h \cdot \rho_f}{S_\kappa \mu} < \text{Re}_{cr} = 2100. \tag{7.2.53}$$

If the flow in annular space is turbulent, that is, $Re \geq Re_{cr}$, it is required to calculate Q in turbulent flow. In turbulent flow, the velocity profile $w(r)$ looks like (6.5.47) or (6.5.48)

$$w_1(r) = \varphi_1 (r - R_1)^{1/N}, \quad R_1 \leq r \leq R_a; \tag{7.2.54}$$

$$w_2(r) = \varphi_2 (R_2 - r)^{1/N}, \quad R_a \leq r \leq R_2, \tag{7.2.55}$$

where

$$\varphi_1 = 0.98C(N)\left[\frac{\Delta p(R_2^2-R_a^2)}{2\rho_f L R_2}\right]^{(N+1)/2N}\left[\left(\frac{R_2-R_a}{R_a-R_1}\right)\frac{\rho_f}{\mu}\right]^{1/N};$$

$$\varphi_2 = 0.98C(N)\left[\frac{\Delta p(R_2^2-R_a^2)}{2\rho_f L R_2}\right]^{(N+1)/2N}\left[\frac{\rho_f}{\mu}\right]^{1/N};$$

$$R_a = R_2(\delta + 0.5(1-\delta)\delta^{0.225}); \quad \delta = R_1/R_2.$$

$C(N)$ is a constant depending on N.

Replacing in (7.2.54) and (7.2.55) w_1 and w_2 with v_s, inserting for φ_1 and φ_2 their expressions, and taking $r = R_2 - d_p/2$, one gets pressure drop per pipe unit length A and fluid flow rate

$$A = \frac{v_c^{2N/(N+1)}\dfrac{2\rho_f R_2}{R_2^2-R_a^2}}{0.98C(N)^{2N/(N+1)}(R_2-R_1-d_p/2)^{2/(N+1)}\left[\dfrac{\rho_f}{\mu}\left(\dfrac{R_2-R_a}{R_a-R_1}\right)\right]^{\frac{2}{N+1}}}$$
$$\text{at}\, d_p/2 < R_2-R_a,$$

$$A = \frac{v_c^{2N/(N+1)}\dfrac{2\rho_f R_2}{R_2^2-R_a^2}}{0.98C(N)^{2N/(N+1)}(d_p/2)^{2/(N+1)}\left[\dfrac{\rho_f}{\mu}\right]^{2/(N+1)}}\ \text{at}\, d_p/2 > R_2-R_a,$$

$$Q = 2\pi\int_{R_1}^{R_2} wr\mathrm{d}r = 2\pi\varphi_2 R_2^{1/(N+2)}\frac{N}{(N+1)(2N+1)}(1-\delta_a)^{1/N}(1-\delta)[\delta_a + N(\delta+1)].$$

For practical calculations, one can take $N = 7$ and $C(N) = 8.74$. Then, the above relations give

$$A = \frac{2\rho_f R_2 v_s^{1.75}}{43.54(R_2^2-R_a^2)(R_2-R_1-d_p/2)^{0.25}\left[\dfrac{\rho_f}{\mu}\left(\dfrac{R_2-R_a}{R_a-R_1}\right)\right]^{0.25}}, \tag{7.2.56}$$

$$A = \frac{2\rho_f R_2 v_s^{1.75}}{43.54(d_p/2)^{0.25}\left[\dfrac{\rho_f}{\mu}\right]^{0.25}}, \tag{7.2.57}$$

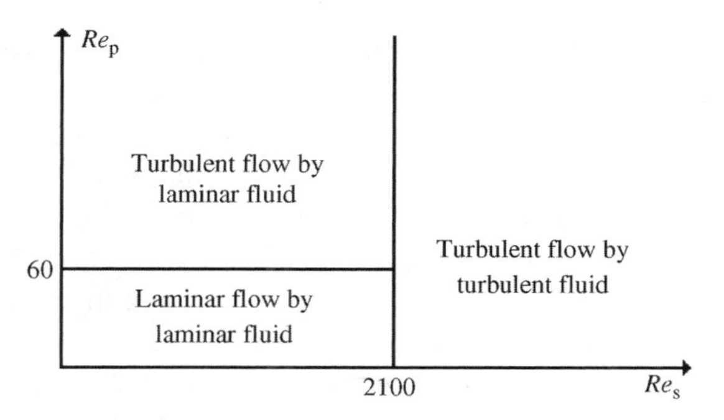

FIGURE 7.9 Regimes of viscous fluid flow around a particle.

$$Q = 2\pi\varphi_2 R_2^{2.143} \times 0.0583(1-\delta_a)^{0.143}(1-\delta)[\delta_a + N(\delta+1)]. \quad (7.2.58)$$

7.2.2 Graphical Illustration of Non-Newtonian Fluid Flow Regimes Around Particles

Flow regimes around particles performed by all fluids can be as follows: laminar flow by laminar fluid, turbulent flow by laminar fluid, and turbulent flow by turbulent fluid.

These regimes exist in certain ranges of Reynolds numbers appropriate to fluid flow around a particle and fluid flow in a pipe given by formulas (7.2.21), (7.2.22), and (7.2.53). The domains of Newtonian fluid flow regimes are depicted in Fig. 7.9.

Calculation of particle start velocity in flows of non-Newtonian fluids is presented by Leonov et al. (2001).

7.3 FLOW RATES OF FLUID, GAS, AND GAS–LIQUID MIXTURE NEEDED FOR REMOVAL OF CUTTING FROM WELL BORE

By choosing flow rates needed for cutting removal from the annular space, it is required to give flow velocity exceeding levitation velocity; that is, particle velocity v_p should be positive. In practice, this velocity is taken equal to 20–30% of the levitation velocity, namely,

$$v_p = (0.2\text{–}0.3)v_s, \quad (7.3.1)$$

where v_s is calculated by one of the formulas (7.2.8), (7.2.11), (7.2.17), (7.2.22), or (7.2.45).

In drilling with roller bits of rock solids capable of brittle fracture, mostly particles with equivalent diameter less than 10 mm are formed. Therefore, in appropriate formulas the diameter of removed particle is taken equal to $d_p = 0.01$ m. Particles of lesser diameter would be easily removed, whereas particles of greater diameter, which are several percentage points larger, would be repeatedly crushed by drill bit. The velocity v_p is very often calculated using formula (7.2.11) taking in advance $Re > 60$.

For example, while drilling with water washing in accordance with (7.3.1) and (7.2.13), it is taken $v_f = 0.3 \times 0.701 = 0.201$ m/s. Then, the flow velocity should be $v_f = 0.701 + 0.211 = 0.912$ m/s.

Depending on the specific character of drilling in each region, the velocity v_f is chosen in the range 0.4–1.4 m/s. Values of v_f closer to the lower boundary of this velocity range may be obtained by the use of weighted drilling solution with elevated rheological properties. When employing service water and other fluids with lowered rheological properties, especially in drilling out argillaceous rock solids, the velocity v_f is raised in order to eliminate balling.

The fluid flow rate in circular space needed for cutting removal is calculated with formula

$$Q = v_f \times S_{as}, \tag{7.3.2}$$

where S_{as} is cross section area of the annular space.

For example, while drilling with gas blasting under conditions of Exercise 7.2.2, the particle velocity is $v_p = 0.3 \times 25.2 = 7.56$ m/s and the flow velocity $v_f = 7.56 + 25.2 = 32.8$ m/s.

Compressor delivery at normal pressure should be no less than $Q = v_f \cdot S_{as} = 32.8 S_{as}$.

In drilling with gas–liquid mixture washing, flow rates of phases providing levitation of particles are chosen in accordance with equation (7.2.39) written in flow rate notations as (Leonov, 1973)

$$Q_0 = S_{as}\sqrt{\frac{g d_p \rho_p p_0}{\rho_0 p}\left(\frac{0.108 S_{as}^2 d_p \rho_p}{Q_f^2 \rho_f + 0.0785 S_{as}^2 d_p \rho_p} - 0.008\right)}. \tag{7.3.3}$$

EXERCISE 7.3.1

In washing with aerated fluid, it is required to find minimal needed compressor delivery at normal pressure $p = p_0$ to carry over cutting particles with diameter $d_p = 0.01$ m and density $\rho_p = 2500\ \text{kg/m}^3$ from the annular space with area $S_{as} = 0.053\ \text{m}^2$. Compressor delivery is $Q_f = 0.01\ \text{m}^3/\text{s}$ and solution density $\rho_f = 1200\ \text{kg/m}^3$.

SOLUTION In accordance with the formula (7.3.3) for cutting levitation, air flow rate is required

$$Q_0 = 0.053 \times \sqrt{\frac{9.81 \times 0.01 \times 2500}{1.29} \times \left(\frac{0.108 \times 0.053^2 \times 0.01 \times 2500}{0.01^2 \times 1200 + 0.0785 \times 0.053^2 \times 0.01 \times 2500} - 0.008 \right)} = 0.167\,\mathrm{m^3/s}.$$

Based on technical possibilities of equipment to carry out cutting from circular space in accordance with (7.3.1), the needed total flow rate of gas and fluid should provide the flow rate of one of the phases 20–30% higher than the flow rate presented in above-considered exercises.

It should be noted in conclusion that in designing hydraulic program of drilling flow rates, providing cleaning of the downhole from cutting and its transport along circular space should not contradict demands 1–12 in Chapter 2. Especially, often at given flow rates one has to comply with conditions for the flow in circular space to produce minimal possible differential pressure in order for the pressure in the flow to be less than the pressure of reservoir hydraulic fracturing and for downhole hydraulic motor to develop torque needed for drill bit to crush the rock solid.

7.4 CALCULATION OF BALL DROP TIME IN DESCENDING FLOW OF WASHING FLUID IN A COLUMN OF PIPES

With formulas of Section 7.2, one can determine the velocity of ball drop in vertical pipe under condition that the ball sinks with velocity v_s in quiescent fluid (Leonov et al., 1972). But the motion of a particle with a diameter exceeding $0.1d_p$, where d_p is pipe diameter, is hindered by pipe walls and is therefore called hindered sedimentation. Hindered sedimentation may be taken into account by introduction of experimentally obtained hinderness factor k_h (Fig. 7.10). Hence, the velocity of spherical particle sedimentation v_{sed} is equal to

$$v_{sed} = v_s / k_h. \tag{7.4.1}$$

Taking the ball velocity relative to pipe wall as a sum of descending fluid flow velocity v_f and ball drop velocity in quiescent fluid v_{sed}, the formula for ball drop time is obtained as

$$T = L/(v_f + v_{sed}), \tag{7.4.2}$$

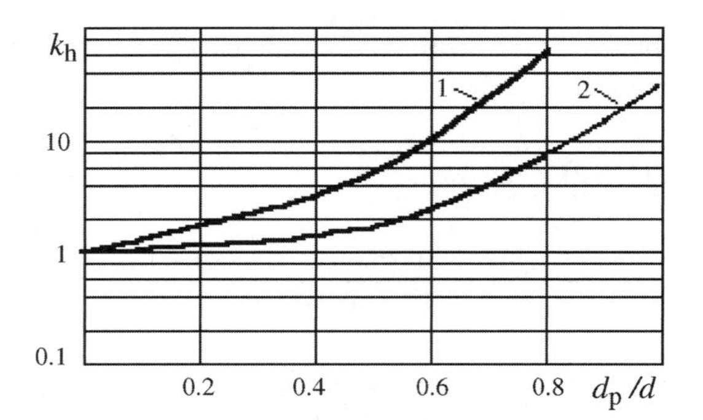

FIGURE 7.10 Dependence of hinderness factor k_h on d_p/d at laminar (1) and turbulent (2) flow around a particle.

where L is the length of the pipe column and

$$v_f = 4Q/(\pi d^2), \tag{7.4.3}$$

where d is the internal diameter of the pipe.

EXERCISE 7.4.1

It is required to calculate the time of ball drop in descending flow of washing fluid in pipes under the following conditions: $L = 4800$ m, $d = 0.094$ m, $d_p = 0.05$ m, $\rho_p = 7800$ kg/m^3, $\rho_f = 1260$ kg/m^3, $\tau_0 = 6.66$ Pa, $\eta = 0.015$ Pa s, and $Q = 0.00625$ m^3/s.

SOLUTION At first, we calculate numbers He, Re_{cr}, Ar_{cr}, Ar:

$$Ar = \frac{d_p^3 g}{\eta^2}\rho_f(\rho_p - \rho_f) = \frac{0.05^3 \times 9.81}{0.015^2} 1260(7800 - 1260) = 4.49 \times 10^7.$$

$$He = \frac{\rho_f \tau_0 d_p^2}{\eta^2} = \frac{1260 \times 6.66 \times 0.05^2}{0.015^2} = 9.33 \times 10^4.$$

$$Re_{cr} = 30(1 + \sqrt{1 + He/45}) = 30(1 + \sqrt{1 + 9.33 \times 10^4/45}) = 1.4 \times 10^3.$$

$$Ar_{cr} = 18(Re_{cr} + He/3) = 18(0.14 \times 10^4 + 9.33 \times 10^4/3) = 5.85 \times 10^5.$$

Since $Ar = 4.49 \times 10^7 > Ar_{cr} = 5.85 \times 10^5$, the levitation velocity is calculated by formula (7.2.22) with the use of (7.2.26):

$$v_s = \frac{Re\eta}{d_p\rho_f} = \frac{1.83 \times Ar^{0.5}\eta}{d_p\rho_f} = \frac{1.83 \times (4.49 \times 10^7)^{0.5} 0.015}{0.05 \times 1260} = 2.92 \text{ m/s}.$$

Determine ratio

$$\frac{d_p}{d} = \frac{0.05}{0.096} = 0.521.$$

From Fig. 7.8 (curve 2) at $d_p/d = 0.521$, $k_h = 1.6$ is obtained. Formula (7.4.1) gives $v_{sed} = v_s/k_h = 2.92/1.6 = 1.83$ m/s. Calculate v_f with (7.4.3): $v_f = 4Q/(\pi d^2) = 4 \times 0.00625/(3.14 \cdot (0.094)^2) = 0.901$ m/s. The time of particle drop in accordance with (7.4.2) is $T = L/(v_f + v_{sed}) = 4800/(0.901 + 1.83) = 1756$ s.

7.5 HYDRAULIC CALCULATION OF CIRCULAR SYSTEM IN DRILLING WITH INCOMPRESSIBLE FLUID WASHING

Trustworthiness of source information has a great effect on the accuracy of hydraulic calculation of well washing. But some information needed for calculation by virtue of some reasons could be determined only approximately. Among these are diameter of noncased wellbore, rheological properties of washing fluid, roughness of the pipe and well walls, and so on. Therefore, in calculations it is needed to use estimations that satisfy all technological and geological conditions of drilling. So as lower estimation for hydrodynamic pressure in well circular space with the aim to produce counterpressure on clean sands is hydrostatic pressure of washing fluid column. For upper estimation of pressure distribution for the purpose of nonadmission of bed hydraulic fracturing (absorption) and in determining pump pressure, it is worthwhile to use calculation relations and source data giving slightly overrated pressure drops in various elements of circulation system (Leonov and Isaev, 1978; Leonov et al., 1984).

7.5.1 Determination of Fluid Flow Rate Providing Washing of the Well Bottom and Cutting Transport in Annulus

When solving this problem, it is required to know mean velocity of fluid in hole annulus v_{as} providing carryout of drill solids from the well. In washing of the first wells on the area, the velocity v_{as} is chosen according to the recommendations presented in Section 7.3. With drilling the area and accumulation of experience, the value of v_{as} would be defined more exactly with regard to other factors, for example, the type of drilled rock solid, method of drilling, construction of bits, and so on.

For given v_{as}, flow rate of washing fluid Q needed for cutting removal is determined

$$Q = \frac{\pi}{4}(d_w^2 - d_{in}^2)v_{as}, \tag{7.5.1}$$

where d_w is well diameter and d_{in} minimal external diameter of drill column pipes, m.

Obtained values of Q are determined more exactly by checking up condition providing bottom cleaning from cutting

$$Q \geq a\frac{\pi}{4}d_w^2, \tag{7.5.2}$$

where $a = 0.35$–0.5 m/s in rotor- and electrodrilling and $a = 0.5$–0.7 m/s in drilling with hydraulic downhole motor.

Further calculations are performed at flow rate determined by formulas (7.5.1) and (7.5.2).

7.5.2 Selection of Pump Cylindrical Bushing Diameter

Refraining from definite value of Q satisfying condition of bottom cleaning and cutting carryout, it is needed to choose diameters of cylindrical bushings for drill pump with the help of Table 7.1. At this, the delivery of pumps will be

$$Q = nmQ_{pnd}, \tag{7.5.3}$$

where m is the operating efficiency of the pump, Q_{pnd} is the nominal pump delivery at given diameter of bushings ($m = 1$) (m^3/s), and n is the number of pumps.

The factor m is chosen depending on conditions of the pump suction. With overpressure in suction in hand, it is $m = 1.0$. If the suction occurs from ground capacity, then $m = 0.9$ in washing with water and $m = 0.8$ in washing with mud solution.

7.5.3 Selection of Washing Fluid Density

The density of washing fluid used in drill out of given interval should be determined on the basis of two conditions: creating a counterpressure preventing inflow of reservoir fluids and gases in the well and keeping the weakest bed from hydraulic fracturing.

The first condition is

$$\rho_{wf} = \min\left\{\frac{k_r p_r}{gL_t}, \frac{p_r + \Delta p_r}{gL_t}\right\}, \tag{7.5.4}$$

where ρ_{wf} is the washing fluid density (kg/m^3), p_r is the reservoir pressure (Pa), g is the gravity acceleration (m/s^2), L_t is the depth of reservoir top bedding with maximal reservoir pressure gradient (m), and k_r is the reserve factor.

TABLE 7.1 Nominal Data of Drill Pumps at $m = 1$ (in Russian Designations)

	У8-7М		У8-6М		БРН-1		НТБ-600		УНБТ-950		УНБ-1250	
	p_p	Q_{pnd}	p_p	Q_{pnd}	p_p	Q_{pnd}	p_p	Q_{pnd}	p_p	Q_{pnd}	p_p	Q_{pnd}
200	14.2	50.9	10.0	50.9	—	—	—	—	—	—	21.0	51.4
190	15.9	45.5	11.1	45.5	—	—	—	—	—	—	23.6	45.4
180	18.0	40.4	12.5	40.4	9.8	31.0	11.3	42.9	19.0	46.0	26.5	40.7
170	20.4	35.5	14.0	35.5	11.0	27.2	12.6	38.3	20.8	41.0	30.5	35.7
160	23.4	31.0	16.3	31.0	12.5	24.0	14.3	33.9	23.0	37.0	35.0	31.1
150	27.2	26.8	19.0	26.7	14.0	20.8	16.2	29.8	26.0	33.0	40.0	26.7
140	32.0	22.7	22.3	22.7	16.9	17.8	18.6	26.0	32.0	27.6	—	—
130	—	—	25.0	18.9	20.0	15.0	21.6	22.4	—	—	—	—
120	—	—	—	—	—	—	25.0	19.0	—	—	—	—

p_p—pressure (MPa); Q_{pnd}—pump delivery (l/s); D—diameter of cylindrical bushing (mm).

In accordance with the existing safety codes in oil and gas industry (in Russia), the following values of k_r and Δp_r are recommended:

$$\begin{aligned} &k_r \geq 1.1, \quad \Delta p_r \leq 1.5\,\text{MPa} \quad \text{at} \quad L_t < 1200\,\text{m}; \\ &k_t \geq 1.05, \quad \Delta p_r \leq 2.5{-}3.0\,\text{MPa} \quad \text{at} \quad L_t \geq 1200\,\text{m}. \end{aligned} \tag{7.5.5}$$

The density ρ_{wf} calculated with (7.5.4) should be checked for consistency with the second condition from which follows that the pressure of washing fluid in hole annulus located opposite each reservoir has to be less than the pressure needed for hydraulic fracturing of given reservoir. This condition may be written as follows:

$$\rho_{wf} < \frac{p_{hf} - \Sigma(\Delta p_{as}) - (1-\varphi)\rho_{cut} g L_t}{\varphi g L_t}, \tag{7.5.6}$$

where $\varphi = Q/((\pi/4)v_d d_{bb}^2 + Q)$ is the fluid content in cutting-liquid flow without regard for relative velocities of phases, p_{hf} is the reservoir hydraulic fracture pressure (Pa), $\Sigma(\Delta p_{as})$ is the pressure loss in fluid flow in circular space from the bottom-hole of the considered reservoir up to the well mouth (Pa), ρ_{cut} is the cutting density (kg/m^3), L_t is the depth of the considered bottom from the mouth (m), v_d is the mechanical rate of drilling (m/s), and d_{bb} is the diameter of borehole bottom.

Since values of $\Sigma(\Delta p_{as})$ and φ depend on washing fluid flow rate, it is possible to check the second condition only after getting pump delivery.

7.5.4 Selection of the Downhole Motor

In drilling with the use of downhole motors, the chosen flow rate of washing fluid Q besides downhole cleaning and cutting carryout should provide motor operation with given moment M_p sufficient to crush the rock solid. Therefore, it is necessary to select a motor obeying the following conditions: diameter of the body to be 10 mm more than the diameter of the bit and the torque of the motor M to be more than the given moment M_p needed for rock solid fracture.

The moment of the turbo-drill shaft should obey the condition

$$M = M_{tr} \frac{\rho Q^2}{\rho_f Q_f^2} \geq M_p, \tag{7.5.7}$$

where M_{tr}, ρ_f, and Q_f are reference data of the turbo-drill shaft moment in the regime of maximal power at given density ρ_f and flow rate Q_f of the fluid.

For the positive displacement motor the condition for needed moment with regard to the formula (6.9.7) may be written as follows:

$$M=\frac{\rho VQ^2}{S^2}\left[\frac{1}{b_1}\left(\frac{nS\times V^{1/3}}{Q}-a_1\right)\right]^{1/c_1}\geq M_{\mathrm{p}}. \qquad (7.5.8)$$

7.5.5 Calculation of Pressure Losses in Elements of Circulation System

The total pressure loss Δp in washing fluid flow in elements of circulation system is determined by expression

$$\Delta p=\Sigma(\Delta p_i)=\Sigma(\Delta p_{\mathrm{p}})+\Sigma(\Delta p_{\mathrm{as}})+\Delta p_{\mathrm{lp}}+\Delta p_{\mathrm{las}}+\Delta p_{\mathrm{sb}}+\Delta p_{\mathrm{dhl}}+\Delta p_{\mathrm{bit}}+\Delta p_{\mathrm{hp}}, \qquad (7.5.9)$$

where $\Sigma(\Delta p_{\mathrm{p}})$ and $\Sigma(\Delta p_{\mathrm{as}})$ are the friction loss along the pipe and annular space length (Pa), Δp_{lp} and Δp_{las} are the pressure loss in local resistances along the pipe and annular space length (Pa), Δp_{sb} is the pressure loss in surface binding (Pa), Δp_{dhm} is the pressure loss in downhole motor (Pa), Δp_{bit} is the pressure loss in water courses of the bit (Pa), and Δp_{hp} is the difference between hydrostatic pressures in annular space and pipes (Pa).

To calculate pressure loss owing to friction of washing fluid flow without cutting in pipes and circular space, it is required to determine the flow regime on which depends the choice of formulas. In order to do this, one should at first calculate critical Reynolds number Re_{cr} for the washing non-Newtonian fluid flow using Fig. 6.20 or the formula

$$Re_{\mathrm{cr}}=2100+7.3He^{0.58}, \qquad (7.5.10)$$

where $He=\rho d_{\mathrm{h}}^2\tau_0/\eta^2$ is Hedsröm number, η is the plastic (dynamic) viscous factor of the washing fluid (Pa s), and τ_0 is the dynamic shear stress (Pa).

In the fluid flow inside the drill pipe string, the value of d_{h} is taken equal to the internal diameter of the drill pipe d_{d}. In the hole annulus, d_{h} is taken as the difference between well diameter d_{w} and external diameter of the pipe d_{ex}. When the Reynolds number of the fluid flow in pipes Re_{p} or in annular space Re_{as} is more than calculated value Re_{cr}, the flow regime is turbulent. In the opposite case, the flow regime is laminar.

Values of Re_{p} and Re_{as} are determined by formulas

$$Re_{\mathrm{p}}=\frac{\rho v_{\mathrm{p}}d_{\mathrm{p}}}{\eta}=\frac{4\rho Q}{\pi d_{\mathrm{p}}\eta}, \qquad (7.5.11)$$

$$Re_{as} = \frac{\rho v_{as}(d_w - d_{ex})}{\eta} = \frac{4\rho Q}{\pi(d_w + d_{ex})\eta}, \qquad (7.5.12)$$

where $v_p = 4Q/\pi d_p^2$ and $v_{as} = 4Q/\pi(d_w^2 - d_{ex}^2)$ are the mean velocities in pipes and annular channel, d_p and d_{ex} are the internal and external diameters of the drill-stem section consisting of pipes of identical sizes m.

Pressure losses along pipe length in turbulent flow are determined by Darcy–Weisbach formula

inside pipes

$$\Delta p_p = \lambda_p \frac{\rho v_p^2}{2d_p} l = \lambda_p \frac{8\rho Q^2 l}{\pi^2 d_p^5}, \qquad (7.5.13)$$

inside annular space

$$\Delta p_{as} = \lambda_{as} \frac{\rho v_{as}^2}{2(d_w - d_{ex})} l, \qquad (7.5.14)$$

where l is section length of drill pipes having identical diameters d_p and d_{ex} (m); λ_p and λ_{as} are the factors of hydraulic friction resistance in pipes and annular spaces, respectively, equal to

$$\lambda_p = 0.1\left(\frac{1.46k}{d_p} + \frac{100}{Re_p}\right)^{0.25}; \qquad (7.5.15)$$

$$\lambda_{as} = 0.107\left(\frac{1.46k}{d_w - d_{ex}} + \frac{100}{Re_{as}}\right)^{0.25}. \qquad (7.5.16)$$

The roughness k (see Section 6.5) for walls of pipe and cased sections of the annular space is taken as equal to 3×10^{-4} m, whereas for uncased sections of the annular space equal to 3×10^{-3} m. Formulas (7.5.15) and (7.5.16) are obtained for turbulent flows of viscous fluid in pipes and annular channels. We shall use these formulas also for turbulent flows of non-Newtonian fluids, by virtue of the fact that analogous formulas being experimentally justified are absent. Pressure losses along pipe length in laminar flow of viscous-plastic fluid are determined by the formulas

$$\Delta p_p = \frac{4\tau_0 l}{\beta_p d_p}; \qquad (7.5.17)$$

$$\Delta p_{as} = \frac{4\tau_0 l}{\beta_{as}(d_w - d_{ex})}, \qquad (7.5.18)$$

where β_p and β_{as} are factors determined from the plot in Fig. 6.7 with previously calculated Saint Venant numbers for pipe Se_p and annular space Se_{as} by formulas

$$Se_p = \frac{\tau_0 d_p}{\eta v_p} = \frac{\pi \tau_0 d_p^3}{4\eta Q}, \tag{7.5.19}$$

$$Se_{as} = \frac{\tau_0 (d_w - d_{ex})}{\eta v_{as}} = \frac{\pi \tau_0 (d_w - d_{ex})^2 (d_w + d_{ex})}{4\eta Q}. \tag{7.5.20}$$

Pressure losses in annular channel between well walls and downhole motor are determined by formulas (7.5.14) and (7.5.18). Values of d_{ex} and l in these formulas are equal to external diameter of the motor body d_m and motor length l_m, respectively. Local pressure losses from joints in the annular space are determined by expression

$$\Delta p_{lj} = \frac{l}{l_p} \left(\frac{d_w^2 - d_{ex}^2}{d_w^2 - d_{exj}^2} - 1 \right)^2 \rho v_{as}^2, \tag{7.5.21}$$

where l_p is the mean pipe length of given section of the drill-stem (m), d_{exj} is the external diameter of the joint (m), and $l$is the length of identical pipes in drill-stem section (m).

For section of a drill-stem consisting of pipes with internal heading, pressure losses in local resistances inside pipes are determined with the formula

$$\Delta p_{lp} = \xi \frac{\rho v_p^2}{2} \frac{l}{l_p}. \tag{7.5.22}$$

Pressure losses in the surface bending are found from the formula

$$\Delta p_{sb} = (\alpha_{ap} + \alpha_{bh} + \alpha_{sw} + \alpha_{ks}) \rho Q^2, \tag{7.5.23}$$

where α_{ap}, α_{bh}, α_{sw}, and α_{ks} are resistance factors listed in Table 6.1.

Pressure losses in the downhole motor (turbo-drill) are calculated on the basis of kinematic similarity with the formula

$$\Delta p_{dhm} = \Delta p_{td} \frac{\rho Q^2}{\rho_f Q_f^2}, \tag{7.5.24}$$

where Δp_{td} and Q_f are reference data of the turbo-drill when operating with fluids of given density ρ_f.

Pressure drop in downhole motor is calculated by formula (6.9.4) with regard to (6.9.5) and (6.9.6).

Pressure drop Δp_h is determined by the formula $\Delta p_h = (1-\varphi)\times(\rho_{cut}-\rho)gL$. In washing without deepening, when the density of the solution at the entrance and the exit of the well would be equal, Δp_h vanishes.

7.5.6 Determination of Pressure Loss in a Drill Bit: Selection of Hydro-Monitor Head

Pressure reserve Δp_r that can be realized in the drill bit is defined as difference between the pressure bp_p developed by a pump (or pumps) at chosen diameter of bushes and the sum of above-listed pressure losses $\Delta p = \sum(\Delta p_i)$ in elements of the circulation system

$$\Delta p_r = bp_p - \sum(\Delta p_i), \tag{7.5.25}$$

where $b = 0.75-0.80$. Factor b takes into account the desirability of operation of discharge pressure of pumps to be less than the reference pressure on 20–25%.

Based on Δp_r one should gauge the ability to use hydro-monitor effect in drilling the given well interval. To do this, it is necessary to calculate the velocity of fluid flow in washing holes of the bit with the formula

$$v_b = \mu\sqrt{\frac{2\Delta p_r}{\rho}}, \tag{7.5.26}$$

where μ is the flow rate factor that in accordance with data presented in Section 6.9 should be taken as equal to 0.95. If the obtained velocity is $v_b \geq 80\ \text{m/s}$, it means that the interval under consideration could be drilled with use of hydro-monitor drilling bit.

It should be kept in mind that the pressure drop developed in hydro-monitor bit heads should not surpass a certain limiting value Δp_{cr} bounded by the ability of the downhole motor to be started as well as by the strength of bit structural elements. At present, this limit can be taken as $\Delta p_{cr} = 7$ MPa. Hence, using formula (7.5.26) one should select values of v_b and Δp_b, the following conditions to be obeyed

$$v_b \geq 80\ \text{m/s}, \quad \Delta p_b < \Delta p_{cr}. \tag{7.5.27}$$

After conditions (7.5.27) have been satisfied, one calculates the total area of hydro-monitor bit heads Φ with the formula

$$\Phi = \frac{Q-Q_l}{v_b}, \tag{7.5.28}$$

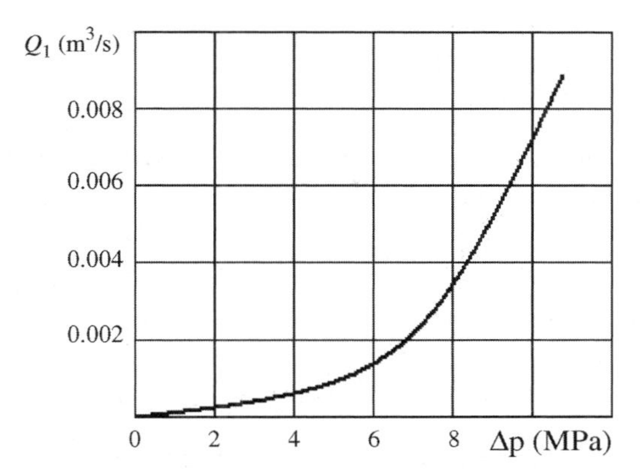

FIGURE 7.11 Dependence of leakage through an abut of the stuffing box of the turbo-drill 3TCШ-195 TЛ on pressure drop in drilling bit.

where $Q_l = \sqrt[n]{\Delta p_r / k\rho}$ is flow rate (leakage) of washing fluid through the seal of turbo-drill shaft (m^3/s); k, n denote empirical factors characterizing leakage of concrete turbo-drill. After finding Q_l, it is necessary to verify conditions on cutting removal and downhole cleaning. If the difference $Q-Q_l$ exceeds flow rates calculated by formulas (7.5.1) and (7.5.2), the above-mentioned conditions would be obeyed.

It is easy to get experimentally the dependence of Q_l on Δp_r for each concrete motor. For example, the approximate value of Q_l could be obtained for the turbo-drill 3TCШ-195TЛ from Fig. 7.11.

In accordance with Φ, diameters of hydro-monitor bit heads are chosen. If for a given bit there is $v_b < 80\,\text{m}/s$, one should conclude that the drilling of given interval with the use of hydro-monitor effect would be impossible. In this case, it is needed to calculate the pressure drop in the bit with formula (6.9.4)

$$\Delta p_b = \frac{\rho v_b^2}{2\mu^2}, \qquad (7.5.29)$$

taking flow rate factor μ in accordance with data given in Section 6.9.

7.5.7 Main Formulas to Calculate Pressure at the Mouth of Drilling Pipe, at the Juncture of Circulation System Elements, and Bottom Hole

Let us give a summary of main formulas to calculate pressures in a well.

Hydrostatic pressure of a solution p_{hyd} without regard for cutting is determined by the formula in which vertical height H independent of well profile is used

$$p_{\rm hyd} = \rho g H. \tag{7.5.30}$$

Hydrostatic pressure of the solution $p'_{\rm hyd}$ with regard to cutting is calculated by the formula

$$p'_{\rm hyd} = [\varphi\rho + (1-\varphi)\rho_{\rm cut}]gH. \tag{7.5.31}$$

The pressure in an ascending pipe $p_{\rm ap}$ can be obtained from the formula

$$p_{\rm ap} = \sum \Delta p_{\rm dp} + \sum \Delta p_{\rm as} + \Delta p_{\rm h}. \tag{7.5.32}$$

The pressure in the annular space at junction of drilling pipe and drilling collar annular space $p_{\rm dpdcas}$ is given by the formula

$$p_{\rm dpdcas} = \Delta p_{\rm dpas} + \Delta p_{\rm tjas} + \rho_{\rm as} \cdot g \cdot H_{\rm dp}. \tag{7.5.33}$$

Calculation of pressure in the circular space at the junction of DC and turbo-drill should be performed with the formula

$$p_{\rm dcdas} = \Delta p_{\rm dpas} + \Delta p_{\rm tjas} + \Delta p_{\rm dcas} + \rho_{\rm as} \cdot g \cdot (H_{\rm dp} + H_{\rm dc}). \tag{7.5.34}$$

Bottom hole pressure $p_{\rm bot}$ is calculated by the formula

$$p_{\rm bot} = p'_{\rm hyd} + \sum \Delta p_{\rm as}. \tag{7.5.35}$$

The pressure in the drill-stem before the bit $p_{\rm bds}$ is found from the formula

$$p_{\rm bds} = p_{\rm bh} + \Delta p_{\rm bit}. \tag{7.5.36}$$

To calculate the pressure in the drill-stem before the turbo-drill (or other motor) $p_{\rm dstb}$ the following formula is used:

$$p_{\rm dstb} = p_{\rm bh} + \Delta p_{\rm bit} + \Delta p_{\rm td} - \rho \cdot g \cdot (H_{\rm td} + H_{\rm bit}). \tag{7.5.37}$$

Pressure in the drill-stem before DC $p_{\rm dcds}$ is determined with the formula

$$p_{\rm dcds} = p_{\rm bot} + \Delta p_{\rm b} + \Delta p_{\rm td} + \Delta p_{\rm dc} - \rho \cdot g \cdot (H_{\rm td} + H_{\rm b} + H_{\rm dc}). \tag{7.5.38}$$

Pressure at the entrance into DP is obtained with the formula

$$p_{\rm dp} = p_{\rm dcds} + \Delta p_{\rm dp} - \rho \cdot g \cdot H_{\rm dp}. \tag{7.5.39}$$

If the construction of the well or drill-stem has sections with another diameters, one should continue calculations with formulas analogous to (7.5.38) adding appropriate pressure drops and taking into account the enhancement of hydrostatic component.

For the sake of convenience in the construction of curves of pressure distribution in circulation system, the obtained data should be summarized in a table.

7.5.8 Construction of Pressure Chart

Construction of pressure distribution curve in drilling circulation system represents an illustration of hydraulic calculation. The calculation should be performed in the following order.

1. On the basis of calculated pressures accumulated in the table with the help of one of the graphical programs, pressure distribution profile in the well and hydrostatic pressure distribution over vertical depth of the well $p = p(H)$ are plotted.
2. To the left of the plot is pictured the scheme of the head part of the well circulation system.

Below, an example of the calculation and the pressure distribution profile in a well circulation system is given.

7.5.9 An Example of Hydraulic Calculation of Well Washing

Initial data for calculation	
Depth of well drilling H (m)	2700
Depth of occurrence of bed roof with maximal bed pressure gradient H_k (m)	2670
Pressure in bed with maximal bed pressure gradient p_r (MPa)	30.43
Depth of occurrence of bed foot with minimal gradient of hydraulic fracturing L_t (m)	2450
Pressure of hydraulic fracturing (absorption) p_{hf} (MPa)	39
Density of drilling-out rock ρ_{cut} (kg/m^3)	2400
Mechanical drilling rate v_M (m/s)	0.015
Turbo-drill moment needed for rock fracture M_P (N m)	800
Minimal velocity of fluid lift in hole clearance provided removal of cuttingv_{cut} (m/s)	0.85
Rheological properties of fluid	
Dynamic shear stress τ_0 (Pa)	15
Plastic viscosity η (Pa s)	0.01
Type of drilling pump	У8–7М

Number of drilling pumps	1
Well diameter d_w (m)	0.225
Elements of drill pipe string (DS)	
Drill collar (DC)	
Length l (m)	70
External diameter d_{ex} (m)	0.146
Internal diameter d_{in} (m)	0.075
Drill pipe (DP)	
Length l (m)	2600
External diameter d_{ex} (m)	0.127
Internal diameter d_{in} (m)	0.107
External diameter of lock joint d_{joint} (m)	0.170
Elements of surface binding	
Conventional size of the ascending pipe (mm)	140
Flow area diameter (mm) of	
Drill hose	102
Swivel	75
Leading pipe	40

1. As it is used turbine method in drilling, factor a in (7.5.2) is taken as equal to 0.65 m/s. Based upon the suction condition, the filling factor m is taken as equal to 0.9.
2. Determine with formula (7.5.1) the flow rate of the flushing fluid from the condition of slurry removal at minimal value of external diameter of drill pipe string $d_{ex} = 0.127$ m and given velocity of fluid lift in the hole clearance $v_{cut} = 0.85$ m/s:

$$Q = \frac{3.14}{4}(0.225^2 - 0.127^2)0.85 = 0.0230\ \text{m}^3/\text{s}.$$

3. Determine with formula (7.5.2) the flow rate of the flushing fluid from the condition of bottom clearance:

$$Q = a\frac{\pi}{4}d_w^2 = 0.65\frac{3.14}{4}0.225^2 = 0.0258\ \text{m}^3/\text{s}.$$

4. From Table 6.2, we select diameter of drilling pump bushes based on maximal value $Q = 0.0258\ \text{m}^3/\text{s}$. In practice, one of the two pumps is used, while the other one is kept in reserve. Nevertheless, if the hydraulic part of pumps is reliable, it is profitable to exploit both pumps to input greater hydraulic power to the bit providing condition

$Q \geq 0.0258\,\text{m}^3/\text{s}$. In the example under consideration, calculations are performed when operating with one pump. The diameter of bushes is taken 160 mm. The capacity of one pump ($n = 1$) with filling factor $m = 0.9$ is determined by formula (7.5.3)

$$Q = nmQ_\text{p} = 1.0 \times 0.9 \times 0.031 = 0.0279\,\text{m}^3/\text{s}.$$

The capacity obtained is acceptable because it is no lesser than capacities given by formulas (7.5.1) and (7.5.2). Then, minimal velocity of fluid in annular channel after DP is

$$v_\text{as} = \frac{4Q}{\pi(d_\text{w}^2 - d_\text{ex}^2)} = \frac{4 \times 0.0279}{3.14(0.225^2 - 0.127^2)} = 1.03\,\text{m/s}.$$

5. Determine with formula (7.5.4) the density of the flushing fluid from the condition of backpressure preventing inflow of the bed fluid to the well:

$$\rho = k_\text{r}\frac{p_\text{r}}{gL_\text{t}} = 1.05\frac{30.43 \times 10^6}{9.81 \times 2670} = 1220\,\text{kg/m}^3;$$

$$\rho = \frac{p_\text{r} + \Delta p_\text{P}}{gL_\text{K}} = \frac{30.43 \times 10^6 + 2.5 \times 10^6}{9.81 \times 2670} = 1257\,\text{kg/m}^3.$$

In further calculations, $\rho = 1220\,\text{kg/m}^3$ is accepted.

6. In accordance with data given by Isaev and Markov (2006) and Table 6.3, we select the turbo-drill 3ТСШ-195ТЛ, which when operating on water with density $\rho_\text{w} = 1000\,\text{kg/m}^3$ in the regime of maximal power has moment $M_\text{w} = 1480\,\text{N·m}$ at flow rate $Q_\text{w} = 0.040\,\text{m}^3/\text{s}$ and pressure drop $\Delta p_\text{w} = 4.0\,\text{MPa}$. The length of the turbo-drill is $l_\text{T} = 25.7\,\text{m}$ and its external diameter $d_\text{T} = 0.195\,\text{m}$.

   At given Q determine with formula (7.5.6) the torque of the selected turbo-drill needed to crush the rock:

$$M_\text{tr} = M_\text{w}\frac{\rho Q^2}{\rho_\text{w} Q_\text{w}^2} = 1480\frac{1220 \times 0.0279^2}{1000 \times 0.040^2} = 878\,\text{N} \cdot \text{m}.$$

   Since the moment M_tr exceeds given moment $M_\text{P} = 800\,\text{N m}$ needed to destroy the rock, we can use this turbo-drill and bushes with diameter 160 mm of the pump У8-7М.

7. Determine with formula (7.5.5) the critical density of the flushing fluid at which can occur the fracturing of the weakest formation of the interval constituent formations to be drilled. In order to do this,

it is necessary to calculate previous parameters φ and $\Sigma(\Delta p_{as})$. The value of φ is determined from the given rate of mechanical drilling $v_M = 0.015$ m/s and accepted flow rate $Q = 0.0279$ m^3/s:

$$\varphi = \frac{Q}{\frac{\pi}{4} v_M d_w^2 + Q} = \frac{0.0279}{\frac{3.14}{4} 0.015 \times 0.225^2 + 0.0279} = 0.979.$$

From here it is easy to get the content of slurry in the flushing fluid $1 - \varphi = 1 - 0.0979 = 0.021$. Such content of the slurry is the characteristic of drilling wells with high mechanical rate. At low rates of mechanical drilling, the concentration of slurry in fluid is small and its influence on pressure of flushing fluid flow in annular space decreases.

To determine $\Sigma(\Delta p_{as})$, linear and local pressure losses in the hole clearance up to the bottom of the weak bed are calculated. Determine at first with formula (7.5.10) critical values of flushing fluid Reynolds numbers Re_{cr}, at which transition from laminar to turbulent flow happens in annular channel:

after DP

$$Re_{cr} = 2100 + 7.3 \left[\frac{1220(0.225 - 0.127)^2 15}{0.01^2} \right]^{0.58} = 32675;$$

after DC

$$Re_{cr} = 2100 + 7.3 \left[\frac{1220(0.225 - 0.146)^2 15}{0.01^2} \right]^{0.58} = 25912;$$

after turbo-drill (TD)

$$Re_{cr} = 2100 + 7.3 \left[\frac{1220(0.225 - 0.195)^2 15}{0.01^2} \right]^{0.58} = 9845.$$

Calculate with formula (7.5.12) actual Reynolds number of the fluid flow in annular space:

after DP

$$Re_{as} = \frac{4 \times 0.0279 \times 1220}{3.14(0.225 + 0.127)0.01} = 12318;$$

after DC

$$Re_{as} = \frac{4 \times 0.0279 \times 1220}{3.14(0.225 + 0.146)0.01} = 11687;$$

after TD

$$Re_{as} = \frac{4 \times 0.0279 \times 1220}{3.14(0.225 + 0.195)0.01} = 10324.$$

Since after DP and DC there is $Re_{as} < Re_{cr}$ and after turbo-drill $Re_{as} > Re_{cr}$, the flow in annular channel is laminar whereas the flow after turbo-drill is turbulent.

Determine with formula (7.5.20) Saint Venant numbers in annular space after DP and DCafter DP

$$Se_{as} = \frac{3.14 \times 15(0.225 - 0.127)^2(0.225 + 0.127)}{4 \times 0.0279 \times 0.01} = 143;$$

after DC

$$Se_{as} = \frac{3.14 \times 15(0.225 - 0.146)^2(0.225 + 0.146)}{4 \times 0.0279 \times 0.01} = 98.$$

Parameter β is determined from Fig. 6.7 curve 2:

fluid flow in annular channel after DP $\beta_{as} = 0.8$; after DC $\beta = 0.75$.

Calculate with (7.5.18) the pressure drop along the length of annular channel after DP in the case when the equipment is located above the roof of weak bed

$$\Delta p_{as} = \frac{4 \times 15 \times (2450 - 95.7)}{0.8(0.225 - 0.127)} = 1.802 \text{ MPa}.$$

Local pressure loss from locks in the annular space (due to initial data there are $l_T = 12$ m and $d_{joint} = 0.170$ m) is determined from formula (7.5.21)

$$\Delta p_{lock} = \frac{2350}{12}\left(\frac{0.225^2 - 0.127^2}{0.225^2 - 0.170^2} - 1\right)^2 1220 \times 1.03^2 = 0.0878 \text{ MPa}.$$

Pressure loss at the section after DC is

$$\Delta p_{as} = \frac{4 \times 15 \times 70}{0.75(0.225 - 0.146)} = 0.0709 \text{ MPa}.$$

Δp_{as} after the turbo-drill is determined by Darcy–Weisbach formula

$$\Delta p_{as}=\lambda\frac{\rho v^2}{2d_h}l_p;\ v_{as}=\frac{Q}{S_{as}}=\frac{4Q}{\pi(d_w^2-d_{ex}^2)}=\frac{4\times 0.0279}{3.14(0.225^2-0.195^2)}=2.82\,\text{m/s};$$

$$\lambda=\frac{0.316}{\sqrt[4]{Re}}+\frac{10He}{Re^2}=\frac{0.316}{\sqrt[4]{10324}}+\frac{10\times 164700}{10324^2}=0.0468;$$

$$\Delta p_{as}=\lambda\frac{\rho v_{as}^2}{2d_h}l_T=0.0468\frac{1220\times 2.82^2}{2\times 0.03}25.76=0.1945\,\text{MPa}.$$

Summing values of Δp_{as}, we get $\Sigma(\Delta p_{as})$ up to the depth of the weak formation needed to calculate ρ_{cr} from condition (7.5.6):

$$\Sigma(\Delta p_{as})=(1.802+0.0878+0.0709+0.1945)\times 10^6=2.15\,\text{MPa}.$$

Formula (7.5.6) yields ρ_{cr}

$$\rho_{cr}=\frac{39\times 10^6-2.15\times 10^6-(1-0.979)2400\times 9.81\times 2450}{0.979\times 9.81\times 2450}=1515\,\text{kg/m}^3.$$

Since obtained value of ρ_{cr} is more than the given density $\rho=1220\,\text{kg/m}^3$, the condition of formation fracturing absence is obeyed.

8. Calculate pressure losses inside drill pipes. To do this, determine with (7.5.10) critical Reynolds numbers of the fluid flow in the drill pipe string:

   in DP

$$Re_{cr}=2100+7.3\left[\frac{1220\times 0.107^2\times 15}{0.01^2}\right]^{0.58}=35955;$$

   in DC

$$Re_{cr}=2100+7.3\left[\frac{1220\times 0.075^2\times 15}{0.01^2}\right]^{0.58}=24520.$$

   Calculate now with (7.5.11) actual Reynolds numbers of fluid flow in pipes and lock joints constituting the drill pipe:

   in DP

$$Re_{as}=\frac{4\times 0.0279\times 1220}{3.14\times 0.107\times 0.01}=40520;$$

in DC

$$Re_{as} = \frac{4 \times 0.0279 \times 1220}{3.14 \times 0.075 \times 0.01} = 57810.$$

Since in the drill pipe there is $Re_T > Re_{cr}$, pressure losses are obtained by Darcy–Weisbach formulas in accordance with which the hydraulic resistance factor λ is calculated from (7.5.15):

in DP

$$\lambda_T = 0.1\left(\frac{1.46 \times 3 \times 10^{-4}}{0.107} + \frac{100}{40520}\right)^{0.25} = 0.0285;$$

in DC

$$\lambda_T = 0.1\left(\frac{1.46 \times 3 \times 10^{-4}}{0.075} + \frac{100}{57810}\right)^{0.25} = 0.0295.$$

From (7.5.13), one can find pressure losses inside DP and DC of the drill pipe string:
in DP

$$\Delta p_T = 0.0285\frac{8 \times 1220 \times 0.0279^2 \times 2600}{3.14^2 \times 0.107^5} = 4.071 \text{ MPa};$$

in DC

$$\Delta p_T = 0.0295\frac{8 \times 1220 \times 0.0279^2 \times 70}{3.14^2 \times 0.075^5} = 0.671 \text{ MPa}.$$

Local pressure losses in welded ends of DP are neglected.

9. Calculate with formula (7.5.23) pressure losses in the surface binding, previously determined from Table 6.1 factors:

$$\alpha_{ap} = 1.1 \times 10^5 \text{ m}^{-4}, \quad \alpha_{bh} = 0.3 \times 10^5 \text{ m}^{-4},$$
$$\alpha_{sw} = 0.9 \times 10^5 \text{ m}^{-4}, \quad \alpha_{ks} = 7 \times 10^5 \text{ m}^{-4};$$

$$\Delta p_0 = (1.1 + 0.3 + 0.9 + 7.0)10^5 \times 1220 \times 0.0279^2 = 0.883 \text{ MPa}.$$

10. With formula (7.5.24) determine pressure drop in the turbo-drill:

$$\Delta p_{\rm dhm} = 4.0 \times 10^6 \frac{1220 \times 0.0279^2}{1000 \times 0.040^2} = 2.374\,\text{MPa}.$$

11. Pressure losses in the annular space after DP were before obtained for the section with length 2450 – 70 – 25.7 = 2354 m. Recalculate pressure losses to total length $l = 2600$ m of DP:

$$\Delta p_{\rm as} = \frac{\Delta p_{\rm as} \times 2600}{2354} = \frac{17.98 \times 10^5 \times 2600}{2354} = 1.99\,\text{MPa};$$

$$\Delta p_{\rm lj} = \frac{0.876 \times 10^5 \times 2600}{2354} = 0.0968\,\text{MPa}.$$

12. Pressure drop $\Delta p_{\rm A}$ on account of difference of solution and cutting densities is

$$\begin{aligned}\Delta p_{\rm A} &= (1-\varphi)(\rho_{\rm cut}-\rho)gL \\ &= (1-0.979)(2400-1220)9.81 \times 2700 = 0.656\,\text{MPa}.\end{aligned}$$

13. Calculate with formula (7.5.9) the sum of pressure losses in all elements of the circulation system except pressure losses in the drill bit:

$$\begin{aligned}\left(\sum \Delta p - \Delta p_{\rm bit}\right) &= (4.071 + 1.99 + 0.671 + 0.0709 + 2.374 + 0.194 \\ &\quad + 0.833 + 0.0968 + 0.656)10^6 = 11\,\text{MPa}.\end{aligned}$$

14. Calculate with formula (7.5.25) at $b = 0.8$ the pressure reserve on the drill bit:

$$\Delta p_{\rm r} = bp_{\rm p} - ((\Sigma\Delta p) - \Delta p_{\rm bit}) = 0.8 \times 23.4 \times 10^6 - 11 \times 10^6 = 7.72\,\text{MPa}.$$

15. Determine the possibility to use the hydro-monitor effect through calculation with formula (7.5.26) at $\mu = 0.95$ of the fluid flow velocity in bit nozzles:

$$v_{\rm b} = 0.95\sqrt{\frac{2 \times 7.72 \times 10^6}{1220}} = 107\,\text{m/s}.$$

Since $v_{\rm b} > 80$ m/s, the drilling of given interval is possible with the use of hydro-monitor bits.

16. Calculate pressure drop in the bit with formula (7.5.29) accepting $v_b = 80$ m/s:

$$\Delta p_{bit} = \frac{1220 \times 80^2}{2 \times 0.95^2} = 4.326\,\text{MPa} < \Delta p_{cr} = 7\,\text{MPa}.$$

Thus, the design operation pressure in the pump is $p_p = 11 + 4.33 = 15.33$ MPa.

17. From a plot of Fig. 7.11, we determine the leakage Q_l depending on the obtained values of $\Delta p_{bit} = 4.33$ MPa: $Q_l = 0.0008\,\text{m}^3/\text{s}$.

Assure us that the difference $Q - Q_l = 0.0279 - 0.0008 = 0.0271\,\text{m}^3/\text{s}$ obeys condition (7.5.1) of the slurry removal and (7.5.2) of the bottom cleaning because $0.0271 > 0.0258$.

18. Determine with formula (7.5.28) the area of flushing holes:

$$\Phi = \frac{Q - Q_l}{v_b} = \frac{0.0279 - 0.0008}{80} = 0.000339\,\text{m}^2.$$

Select three nozzles with internal diameter 12 mm.

19. Determine additional data needed to construct pressure plot.

Calculate with formula (7.5.30) hydrostatic pressure of the solution without cutting:

$$p_{hyd} = 1220 \times 9.81 \times 2700 = 32.3\,\text{MPa}.$$

Calculate with formula (7.5.31) hydrostatic pressure with regard to cutting:

$$p'_{hyd} = [0.979 \times 1220 + (1 - 0.979)2400] \cdot 9.81 \times 2700 = 32.97\,\text{MPa}.$$

Calculate with formula (7.5.32) the pressure in the ascending pipe p_{ap}:

$$\begin{aligned} p_{ap} &= \sum \Delta p_{dp} + \sum \Delta p_{as} + \Delta p_h \\ &= [(4.071 + 0.671 + 2.374 + 4.326 + 0.883) \\ &+ (1.99 + 0.0709 + 0.195 + 0.0968) + 0.656] \times 10^6 \\ &= 15.33\,\text{MPa}. \end{aligned}$$

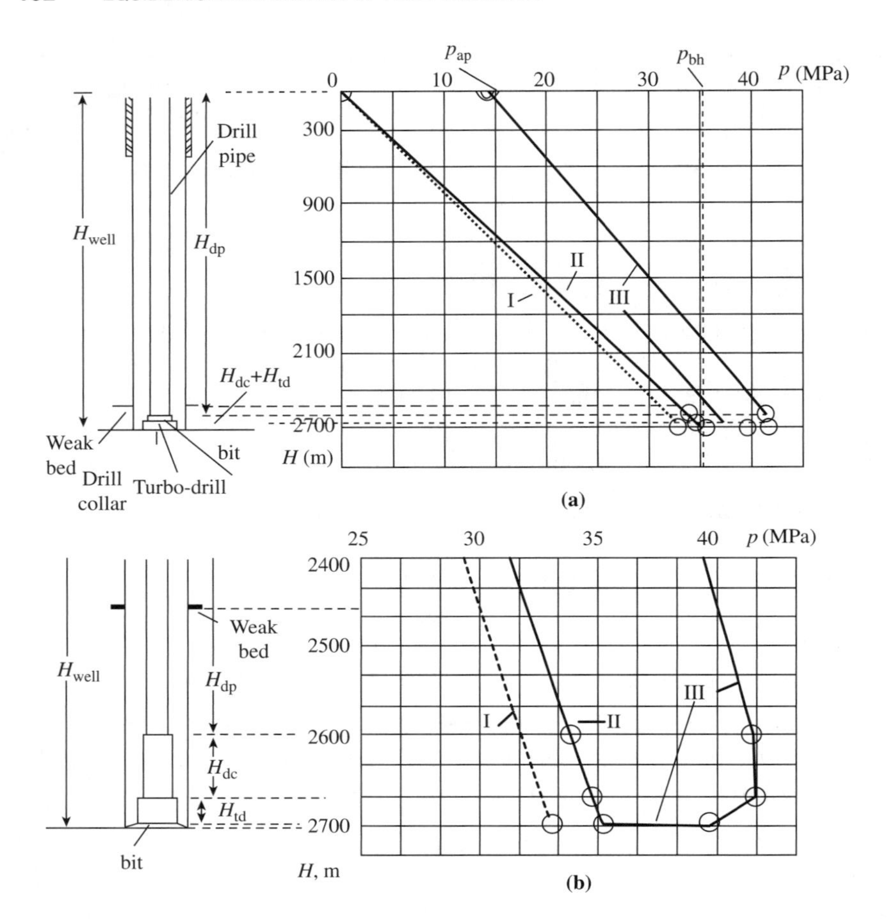

FIGURE 7.12 (a) Plots of pressure distribution in circulation system: I—hydrostatic pressure in annular space; II—pressure distribution in annular space; III—pressure distribution in drill-stem. By circles are shown pressures from the table. The pressure below the depth 2400 m is shown for clearness on an enlarged scale in (b). (b) Plots of pressure distribution in the well below the depth 2400 m. Designations are the same as in Figure 7.12a.

Calculate with formula (7.5.33) the pressure p_{dpdcas} in annular space at the place of DP and DC connection:

$$p_{dpas} = \Delta p_{dpas} + \Delta p_{lj} + \rho_{as} \cdot g \cdot H_{dp} = 1.99 + 0.0968 + 1.245 \times 0.00981 \times 2600 = 33.84\,\text{MPa}.$$

Calculate with formula (7.5.34) the pressure in annular space at the place of DC and TD connection:

$$p_{\text{dctdas}} = \Delta p_{\text{dpas}} + \Delta p_{\text{tjas}} + \Delta p_{\text{dcas}} + \rho_{\text{as}} \cdot g \cdot (H_{\text{dp}} + H_{\text{dc}})$$
$$= 1.99 + 0.0968 + 0.709 + 1.245 \times 0.00981 \times (2600 + 70)$$
$$= 34.76\,\text{MPa}.$$

Calculate with formula (7.5.35) the bottom-hole pressure p_{bot} in the annular space:

$$p_{\text{bot}} = p'_{\text{c}} + \sum \Delta p_{\text{as}} = 32.97 + (1.99 + 0.0709 + 0.194 + 0.0968)$$
$$= 35.32\,\text{MPa}.$$

Calculate with formula (7.5.36) the pressure in drill-stem before the bit p_{bitds}:

$$p_{\text{bitds}} = p_{\text{bot}} - \Delta p_{\text{bit}} = 35.32 + 4.326 = 39.65\,\text{MPa}.$$

Calculate with formula (7.5.37) the pressure before the motor (turbo-drill) p_{tdbs}:

$$p_{\text{tdbs}} = p_{\text{bot}} + \Delta p_{\text{bit}} + \Delta p_{\text{td}} - \rho \cdot g \cdot (H_{\text{td}} + H_{\text{bit}})$$
$$= 35.32 + 4.326 + 2.374 - 1.22 \times 0.00981 \times (25.7 + 0.5)$$
$$= 41.8\,\text{MPa}.$$

Calculate with formula (7.5.38) the pressure in DS before DC p_{dcds} at the depth 2600 m:

$$p_{\text{dcds}} = p_{\text{bot}} + \Delta p_{\text{bit}} + \Delta p_{\text{td}} + \Delta p_{\text{dc}} - \rho \cdot g \cdot (H_{\text{td}} + H_{\text{bit}} + H_{\text{dc}})$$
$$= 35.32 + 4.326 + 2.374 + 0.671 - 1.22 \times 0.00981$$
$$\times (25.7 + 0.5 + 70) = 41.64\,\text{MPa}.$$

Calculate with formula (7.5.39) the pressure at the entrance of DP in drill-stem p_{DP} at the depth 0 m:

$$p_{\text{dp}} = p_{\text{dcbs}} + \Delta p_{\text{dp}} - \rho \cdot g \cdot H_{\text{dp}}$$
$$= 41.64 + 4.071 - 1.22 \times 0.00981 \times 2600$$
$$= 14.5\,\text{MPa}.$$

Obtained data of pressure losses in the circulation system are tabulated as follows:

	Pressure in annular space (MPa)		Pressure in drill-stem (MPa)	
Vertical depth H (m)	Hydrostatic	Drilling	Hydrostatic	Drilling
0	0	0	0	14.5
2600	31.7	33.84	31.12	41.64
2670	32.6	34.76	31.96	41.8
2696.5	32.97	35.32	32.31	39.65
2700	32.97	35.32	32.31	35.32

20. Construct the plot in accordance with table data connecting the neighboring points by straight lines (Fig. 7.12a, b).

CHAPTER 8

STATIONARY FLOW OF GAS AND GAS-CUTTING MIXTURE IN ELEMENTS OF WELL CIRCULATION SYSTEM[1]

8.1 PRESSURE DISTRIBUTION IN ASCENDING FLOW OF GAS AND GAS-CUTTING MIXTURE IN ANNULAR CHANNEL OF A WELL

Flow of gas-cutting mixture in annular channel of the well happens in course of blasting drilling. Such flow represents two-phase flow in which the first phase is cutting and the second one is gas. In the case of one-dimensional two-phase stationary flow the system of equations (4.6.33)–(4.6.39) with regard to the well axis slope takes the following form:

momentum equation

$$\frac{\mathrm{d}p}{\mathrm{d}z} = g\cos\alpha(\rho_{\mathrm{cut}}\varphi + \rho(1-\varphi)) + \frac{\lambda_{\mathrm{c}}}{2d_{\mathrm{h}}}\left[\rho_{\mathrm{cut}}v_{\mathrm{cut}}^{2}\varphi + \rho v^{2}(1-\varphi)\right] - \rho_{\mathrm{cut}}v_{\mathrm{cut}}\varphi\frac{\mathrm{d}v_{\mathrm{cut}}}{\mathrm{d}z} - \rho(1-\varphi)v\frac{\mathrm{d}v}{\mathrm{d}z}; \tag{8.1.1}$$

[1]Chapter 8 is written in collaboration with S.Yu. Samochvalow.

Applied Hydro-Aeromechanics in Oil and Gas Drilling. By Leonov and Isaev

mass conservation equation

$$\varphi S\rho_{\mathrm{cut}}v_{\mathrm{cut}} = Q_{\mathrm{cut}}\rho_{\mathrm{cut}} = m_{\mathrm{cut}} = \mathrm{const},$$

$$(1-\varphi)S\rho v = Q\rho = m = \mathrm{const}; \tag{8.1.2}$$

thermodynamic equations of state

$$\rho_{\mathrm{cut}} = \mathrm{const},$$

$$p = \rho RT\bar{z}g; \tag{8.1.3}$$

concentration equation

$$\varphi = \varphi(\rho_{\mathrm{cut}}, \rho, Q_{\mathrm{cut}}, Q, \lambda_{\mathrm{c}}, p); \tag{8.1.4}$$

hydraulic resistance factor equation

$$\lambda_{\mathrm{c}} = \lambda_{\mathrm{c}}(\rho_{\mathrm{cut}}, \rho, Q_{\mathrm{cut}}, Q, \varphi, p). \tag{8.1.5}$$

Here z is the current well depth with reference point at the well mouth; $p(z)$ is the current pressure; g is the gravity acceleration; α is the angle between z-axis and the direction of gravity force; Q_{cut}, Q, ρ_{cut}, ρ, v_{cut}, v are the volumetric flow rate, density, velocity of cutting, and gas, respectively; φ is the volume concentration of cutting; $d_{\mathrm{h}} = d_{\mathrm{ex}} - d_{\mathrm{in}}$, where d_{ex} and d_{in} are the external and internal diameters of annular space; S is the area of annular space cross section; R is the gas constant; T is the absolute temperature; λ_{c} is the hydraulic resistance factor; $\bar{z}$ is the overcompressibility factor.

In equations (8.1.1)–(8.1.5), $\bar{z} = \mathrm{const}$ and $T = \mathrm{const}$ are assumed. To determine these parameters, a method having been used before for gas wells is recommended. To calculate the hydraulic resistance factor λ_{c} for a flow in annular channel, we use the formula (6.10.14) in which the Reynolds number is taken for gas flow without regard for cutting.

Reynolds number for annular space is taken equal to

$$\mathrm{Re} = \frac{v\rho d_{\mathrm{h}}}{\mu} = \frac{Q\rho d_{\mathrm{h}}}{S(1-\varphi)\mu}.$$

In the approximation $\varphi \ll 1$, the Reynolds number may be calculated by formula

$$\mathrm{Re} = \frac{Q\rho d_{\mathrm{h}}}{S\mu}.$$

At constant values of S, d_{h}, μ with regard to (8.1.2), $Re = \mathrm{const}$ and consequently $\lambda_{\mathrm{c}} = \mathrm{const}$ follow.

When considering flow of gas-cutting mixture in annular space, we neglect inertial terms in equation (8.1.1) and suggest that the slip between cutting and gas is absent, that is, $v_{cut} = v$ and the flow regime is turbulent. Then, (8.1.2) gives $\varphi = Q_{cut}/(Q + Q_{cut})$, that is, the true concentration φ is equal to the flow rate concentration β.

Then equations (8.1.1), (8.1.2), (8.1.4), and (8.1.5) are transformed into

$$\frac{dp}{dz} = g\cos\alpha(\rho_{cut}\beta + \rho(1-\beta)) + \frac{\lambda_c}{2d_h}\left[\rho_{cut}v_{cut}^2\beta + \rho v^2(1-\beta)\right]; \quad (8.1.6)$$

$$\beta S\rho_{cut}v_{cut} = Q_{cut}\rho_{cut} = m_{cut} = \text{const}, \quad (1-\beta)S\rho v = Q\rho = m = \text{const}; \quad (8.1.7)$$

$$\beta = Q_{cut}/(Q_{cut} + Q); \quad (8.1.8)$$

$$\lambda_c = \text{const}. \quad (8.1.9)$$

From (8.1.7) and (8.1.8), it follows that $v = v_{cut} = (Q_{cut} + Q)/S$. Inserting expressions for β, v, v_{cut} in (8.1.6) and bringing the result to dimensionless form we get the following with regard to (8.1.3)

$$\frac{d\bar{p}}{d\xi} = \frac{(1+\eta)\bar{p}}{1+\eta\bar{p}}\cos\alpha + \frac{k_1^2(1+\eta)(1+\eta\bar{p})}{\bar{p}}, \quad (8.1.10)$$

where $\bar{p} = p/(\bar{z}\cdot g\cdot R\cdot T\cdot\rho_{cut})$ is dimensionless pressure; $\xi = z/(RT\bar{z})$ is the dimensionless coordinate; $\eta = Q_{cut}\rho_{cut}/Q\rho = m_{cut}/m$ is the mass factor reflecting the cutting existence and $k_1^2 = \frac{(Q\rho)^2\lambda_c}{(S\rho_{cut})^2 2d_h g}$.

Integration of equation (8.1.10) with initial condition $\bar{p}(0) = p_0$ yields

$$\frac{1-k_2^2\eta^2}{2(1+k_2^2\eta^2)^2}\ln\frac{\bar{p}^2 + k_2^2(1+\eta\bar{p})^2}{\bar{p}_0 + k_2^2(1+\eta\bar{p}_0)^2} - \frac{2k_2\eta}{(1+k_2^2\eta^2)^2}$$
$$\times\left(\arctan\frac{(1+k_2^2\eta^2)\bar{p} + \eta k_2^2}{k_2} - \arctan\frac{(1+k_2^2\eta^2)\bar{p}_0 + \eta k_2^2}{k_2}\right) \quad (8.1.11)$$
$$+ \frac{\eta}{1+k_2^2\eta^2}(\bar{p} - \bar{p}_0) = (1+\eta)\xi\cos\alpha,$$

where $k_2^2 = k_1^2/\cos\alpha$.

From (8.1.11) at given $\bar{p}$, one can determine ξ. The transcendental equation (8.1.11) would be approximately solved. Note that $\eta\cdot\bar{p} = Q_{cut}\rho_{cut}\rho/(Q\rho\rho_{cut}) = Q_{cut}/Q \ll 1$. Then, neglecting in (8.1.10) $\eta\cdot\bar{p}$ as compared to unity, we get

$$\frac{d\bar{p}}{d\xi} = \cos\alpha(1+\eta)\cdot\bar{p} + \frac{k_1^2(1+\eta)}{\bar{p}}. \tag{8.1.12}$$

The solution of (8.1.12) at $\bar{p}(0) = \bar{p}_0$ gives

$$\bar{p} = \sqrt{(\bar{p}_0^2 + \frac{k_1^2}{\cos\alpha})\, e^{2\xi(1+\eta)\cos\alpha} - \frac{k_1^2}{\cos\alpha}}. \tag{8.1.13}$$

The pressure distribution in flow of pure gas is obtained from (8.1.13) taking $\eta = 0$

$$\bar{p} = \sqrt{\left(\bar{p}_0^2 + \frac{k_1^2}{\cos\alpha}\right) e^{2\xi\cos\alpha} - \frac{k_1^2}{\cos\alpha}}. \tag{8.1.14}$$

When considering the flow of gas without cutting in the whole circulation system, it is more reasonable to calculate dimensionless parameters k_1 and $\bar{p}$ substituting in them the gas density at normal conditions instead of cutting density ρ_{cut}.

8.2 PRESSURE DISTRIBUTION IN DESCENDING FLOW OF GAS IN PIPES

The equation for the downflow of gas can be obtained from (8.1.10) at $\eta = 0$ taking into account that the term of friction force has opposite sign in the downflow. Hence, we have

$$\frac{d\bar{p}}{d\xi} = \bar{p}\cos\alpha - \frac{k_1^2}{\bar{p}}. \tag{8.2.1}$$

The parameter k_1 in this equation would be the same as in equation (8.1.10) if to replace d_h with d and to calculate λ_c by the formula (6.10.13).

If the pressure $\bar{p}_0$ in the downflow is given at certain depth ξ_0, that is, $\bar{p}(\xi_0) = \bar{p}_0$, the solution of the equation (8.2.1) takes form

$$\bar{p} = \sqrt{\left(\bar{p}_0^2 - \frac{k_1^2}{\cos\alpha}\right) e^{-2(\xi_0-\xi)\cos\alpha} + \frac{k_1^2}{\cos\alpha}}. \tag{8.2.2}$$

From this relation, it follows that at the wellhead $\xi = 0$, there is

$$\bar{p} = \sqrt{\left(\bar{p}_0^2 - \frac{k_1^2}{\cos\alpha}\right) e^{-2\xi_0\cos\alpha} + \frac{k_1^2}{\cos\alpha}}. \tag{8.2.3}$$

$$\bar{p} = \bar{p}_0 \, e^{\xi \cos \alpha} \tag{8.2.4}$$

or in dimensional form

$$p = p_0 \, e^{\frac{z}{\bar{z}RT} \cos \alpha}.$$

In the flow of pure gas ($\eta = 0$) in horizontal pipe ($\alpha = \pi/2$) from (8.1.10), it follows

$$\frac{d\bar{p}}{d\xi} = \frac{k_1^2}{p}$$

The solution of this equation at $\bar{p}(0) = p_0$ is

$$\bar{p} = \sqrt{2k_1^2 \xi + \bar{p}_0^2}. \tag{8.2.5}$$

The formula (8.2.5) is needed to calculate losses in the binding.

8.3 PRESSURE LOSSES IN BIT HEADS AND PIPE JOINTS

Let us derive a formula to calculate pressure p_0 above a bit in a column for given bottom pressure p_{bot}. To do this, we consider the gas flow in bit heads. Let v_0, ρ_0, T_0 and v, ρ, T, p be the values of parameters at the bit entrance before the head and at the head exit. Assume that in gas motion in bits, inertial forces play the main role. Neglecting gravity and friction forces in equation (8.1.1) and taking into account $\varphi = 0$, we get

$$\frac{1}{\rho}\frac{dp}{dz} + v\frac{dv}{dz} = 0. \tag{8.3.1}$$

The process of gas outflow from heads is taken to be adiabatic. Then,

$$\frac{p}{p_0} = \left(\frac{\rho}{\rho_0}\right)^k, \tag{8.3.2}$$

where k is adiabatic index (for air $k = 1.4$).

Substitution of (8.3.2) in (8.3.1) and further integration gives

$$v^2 = \frac{2k}{k-1}\frac{p_0}{\rho_0}\left[1 - \left(\frac{p}{p_0}\right)^{\frac{k-1}{k}}\right] + v_0^2. \tag{8.3.3}$$

At $v_0 \ll v$ from (8.3.3) follows approximate expression

$$v = \sqrt{\frac{2k}{k-1}\frac{p_0}{\rho_0}\left[1-\left(\frac{p}{p_0}\right)^{\frac{k-1}{k}}\right]}. \tag{8.3.4}$$

From (8.1.2), it ensues that the mass rate of gas is $m = Q\rho$. Then, the mass rate through a single head is equal to $m_n = m/n$, where n is number of heads. Multiplying both parts of the equation by $\Phi\rho$, where Φ is the area of the head cross section, we get

$$m_n = v\Phi\rho = \Phi\sqrt{\frac{2k\rho^2}{k-1}\frac{p_0}{\rho_0}\left[1-\left(\frac{p}{p_0}\right)^{\frac{k-1}{k}}\right]}.$$

In the same manner as when deriving formulas for incompressible fluid, let us introduce correction factor μ (flow rate factor). As a result, the following equation is obtained

$$m_n = \Phi\mu\sqrt{\frac{2kp_0\rho^2}{(k-1)\rho_0}\left[1-\left(\frac{p}{p_0}\right)^{\frac{(k-1)}{k}}\right]}. \tag{8.3.5}$$

The velocity of sound in gas is determined by the formula

$$a_s = \sqrt{\frac{\partial p}{\partial \rho}}. \tag{8.3.6}$$

Sonic flow is a flow whose velocity v at given cross section is equal to sound velocity, that is, $v = a_s$.

Equating (8.3.4) and (8.3.6) and using (8.3.2), we get

$$\frac{p}{p_0} = \left(\frac{2}{k+1}\right)^{\frac{k}{k-1}}. \tag{8.3.7}$$

It is known that at subsonic gas outflow $p = p_s$, whereas at sonic outflow $p > p_s$.

Thus, in accordance with (8.3.7), the sonic regime is defined by the inequality

$$\frac{p_s}{p_0} \leq \left(\frac{2}{k+1}\right)^{\frac{k}{k-1}}, \tag{8.3.8}$$

whereas the subsonic regime by

$$\frac{p_s}{p_0} > \left(\frac{2}{k+1}\right)^{\frac{k}{k-1}}. \tag{8.3.9}$$

Insertion in (8.3.5) values of ρ from (8.3.2), p from (8.3.7), and taking into account $\rho_0 = \frac{p_0}{RT_0 \bar{z} g}$, we get

$$p_0 = \frac{m_n \sqrt{\bar{z} R T_0 g}}{\mu \Phi \sqrt{k \left(\frac{2}{k+1}\right)^{\frac{k+1}{k-1}}}}. \tag{8.3.10}$$

For subsonic outflow regime, one should substitute the value of ρ in (8.3.5) from (8.1.2), p_s instead of p, and take into account $\rho_0 = \frac{p_0}{RT_0 \bar{z} g}$. Then,

$$p_0 = p_s \left(\frac{1 + \sqrt{1 + 4a}}{2}\right)^{\frac{k}{k-1}}, \tag{8.3.11}$$

where

$$a = \left(\frac{m_n}{\Phi p_s \mu}\right)^2 \frac{\bar{z} R T_0 g (k-1)}{2k}.$$

The formula (8.3.10) is true at condition (8.3.8), whereas the formula (8.3.11) at condition (8.3.9). Hence, one can calculate the pressure p_0 before the bit entrance with formulas (8.3.10) and (8.3.11) under condition that the pressure p_s is given.

To determine pressure losses in joints Δp_{joint} inside drill pipes, one has to use formulas (6.9.4), (6.9.5), and (6.9.10) being valid for incompressible fluid

$$\Delta p_{\text{joint}} = \xi \frac{8 \rho_G Q_G^2}{\pi^2 d_{\text{in}}^4} n_{\text{joint}}, \tag{8.3.12}$$

where d_{in} is the internal diameter of drill pipes, n_{joint} is the number of joints; ρ_G and Q_G are density and flow rate of gas in drill pipes averaged over well depth.

Since $\rho_G Q_G^2 = m Q_G$, the relation (8.3.12) may be written as

$$\Delta p_{\text{joint}} = \xi \frac{8 m Q_G}{\pi^2 d_{\text{in}}^4} n_{\text{joint}}. \tag{8.3.13}$$

Q_G in the first approximation could be calculated by mean pressure $p_G = (p_l + p_u)/2$, where p_l and p_u are pressures in lower and upper parts of the drill-stem. Thus,

$$Q_{\mathrm{G}}=\frac{m}{\rho_{\mathrm{G}}}=\frac{m\bar{z}RTg}{p_{\mathrm{G}}}=\frac{2m\bar{z}RTg}{p_{\mathrm{l}}+p_{\mathrm{u}}}. \tag{8.3.14}$$

As a result, the following equation is obtained for $\Delta p_{\mathrm{joint}}$

$$\Delta p_{\mathrm{joint}}=\xi\frac{16m^2 n_{\mathrm{joint}}}{\pi^2 d_{\mathrm{in}}^4}\frac{\bar{z}RTg}{(p_{\mathrm{l}}+p_{\mathrm{u}})}. \tag{8.3.15}$$

Pressure losses from joints in the annular space owing to their negligibility can be neglected.

8.4 CALCULATION PROCEDURE OF PUMP CAPACITY AND COMPRESSOR PRESSURE IN DRILLING WITH BLASTING

In order to choose compressor characteristic properly, it is required to know what gas flow rate is needed to clean the annular channel from slag and what pressure would be at this in compressor. A question is raised on the choice of gas mass flow rate m. Find the ratio of particle soaring velocity v_{s} and gas velocity v. From (7.2.7) and (8.1.2) we have

$$\frac{v_{\mathrm{s}}}{v}=\sqrt{\frac{4gd_{\mathrm{cut}}}{3C_{\mathrm{cut}}}\left(\frac{\rho_{\mathrm{cut}}}{\rho}-1\right)}\frac{S\rho(1-\varphi)}{m}. \tag{8.4.1}$$

Neglecting in (8.4.1) ρ_{cut}/ρ and φ as compared to unity, we get

$$\bar{v}=\frac{v_{\mathrm{s}}}{v}=C\sqrt{\bar{p}}, \tag{8.4.2}$$

where

$$C=\sqrt{\frac{4g}{3C_{\mathrm{w}}}d_{\mathrm{cut}}}\frac{S\rho_{\mathrm{cut}}}{m};\quad \bar{p}=\frac{\rho}{\rho_{\mathrm{cut}}}=\frac{p}{\bar{z}\rho_{\mathrm{cut}}gRT}.$$

From (8.4.2), it is seen that $v_{\mathrm{s}}/v=\bar{v}$ grows with pressure. Therefore, the worst condition of slag removal from annular channel with invariable cross section would be at bottom. To clean well, the annular channel from the slag the condition (5.3.1). $v_{\mathrm{p}}\geq 0.2v_{\mathrm{s}}$ should be obeyed. Since $v-v_{\mathrm{p}}=v_{\mathrm{s}}$, (5.3.1) is equivalent to

$$\frac{v_{\mathrm{s}}}{v}=C\sqrt{\bar{p}}\leq\frac{1}{1.2}=0.83. \tag{8.4.3}$$

If at the bottom $C\sqrt{\bar{p}} = 0.83$, then $C\sqrt{\bar{p}} < 0.83$ at all other cross sections of the channel and the condition (8.4.3) is satisfied.

In practice, the area of annular channel cross section S frequently changes with depth owing to use of bits with different diameters, drill pipes, drill collars, and so on. With change of S varies $\bar{v}$, therefore, the inequality (8.4.3) should be checked in the lower sections of the annular space with constant S. If the inequality (8.4.3) is everywhere satisfied, it means that the given mass flow rate of gas m is sufficient to remove particles with diameter d_{cut}. Otherwise, one has to find a section in which $\bar{v}$ is maximal and to increase the flow rate until the condition (8.4.3) would be obeyed at this section. At this in calculation of $\bar{v}$, one should use formulas (8.4.2) and (8.1.13).

Hence, to get the needed mass flow rate, it is required to solve the following equation

$$f(m) = \bar{v}(m) - 0.83 = 0. \tag{8.4.4}$$

The equation (8.4.4) is solved by the chord method. We choose such flow rates m_1 and m_2 that $f(m_1) < 0, f(m_2) > 0$ and calculate m from the formula

$$m = m_1 - \frac{(m_2 - m_1)f(m_1)}{f(m_2) - f(m_1)}. \tag{8.4.5}$$

If $f(m) = 0$ is accurate for a given small quantity, the value m is the sought flow rate. Otherwise, the calculation should be repeated by the formula (8.4.5) with new values of m_1 and m_2 one of which is m and another one is taken from m_i $(i = 1, 2)$ obeying the condition $f(m) \cdot f(m_i) < 0$.

After obtaining m, we successively determine pressures at the bottom from the formula (8.1.13), above the bit from formulas (8.3.10) and (8.3.11), at the wellhead from formulas (8.2.2) and (8.2.3), pressure losses in joints of drill pipes from the formula (8.3.13). Summing pressure losses in joints and wellhead pressure, we determine from (8.2.5) the pressure at the end of the binding system equal to the sought pressure at the compressor.

EXAMPLE 8.4.1

Initial data are (Mezshlumov and Makurin, 1967) well diameter $d_w = 0.22$ m; well depth $L = 2000$ m; temperature $T = 310$ K; wellhead pressure $p_0 = 9.8 \times 10^4$ Pa; gas constant $R = 28.7$ m/K; adiabatic index $k = 1.4$; overcompressibility factor $\bar{z} = 1$; on-bottom drilling rate $v_{dr} = 18.3$ m/h $= 0.00508$ m/s; cutting density $\rho_{cut} = 2700$ kg/m^3; diameter of drill pipes: $d_h = 0.141$ m, $d_{ex} = 0.123$ m; length of drill collar $l = 50$ m, diameter of drill collar $d_{dc} = 0.197$ m, $d_{ex} = 0.09$ m; joints ЗУ-185; number of joints $n_{joint} = 156$; equivalent roughness of drill pipes, drill collars, and annular space up to the depth 1500 m is $k_{eq} = 3 \times 10^{-4}$ m; equivalent roughness of the uncased annular space in the interval 1500–2000 m is $k_{eq} = 3 \times 10^{-3}$ m; total area of bit heads $\Phi = 0.0004\ \text{m}^2$; dynamic viscosity factor of air $\mu = 1.8 \times 10^{-5}$ Pa s; length and diameter of binding system pipes $L_{bp} = 250$ m; $d_{ex} = 0.067$ m. It is required to determine the flow rate of air and the compressor pressure needed to clean the bottom from the cutting with effective diameter $d_{cut} = 0.0014$ m.

SOLUTION Determine the mass flow rate m. For annular space of the well (interval 0–1500 m), we have

$$S = \frac{\pi}{4}(d_{ex}^2 - d_{in}^2) = \frac{\pi}{4}(0.22^2 - 0.141^2) = 0.0224\ \text{m}^2;$$

$$\rho_0 = \frac{p_0}{\bar{z}RTg} = \frac{9.81 \times 10^4}{1 \times 28.7 \times 310 \times 9.81} = 1.1\ \text{kg/m}^3;$$

$$Q = \frac{\pi d_w^2 v_{dr}}{4} = \frac{\pi(0.22)^2 \times 0.00508}{4} = 1.93 \times 10^{-4}\ \text{m}^3/\text{s}.$$

We take $Q_1^0 = 0.5\ \text{m}^3/\text{s}$. Then,

$$m_1 = Q_1^0 \rho_0 = 0.55\ \text{kg/s};$$

$$\eta = Q_{cut}\rho_{cut}/m_1 = 1.93 \times 10^{-4} \times 2.7 \times 10^3/0.55 = 0.948;$$

$$\bar{\rho}_0 = \frac{\rho_0}{\rho_{cut}} = \frac{1.1}{2700} = 4.07 \times 10^{-4};$$

$$d_h = d_{ex} - d_{in} = 0.22 - 0.141 = 0.079\ \text{m};$$

$$\text{Re} = \frac{m_1 d_h}{S\mu} = \frac{0.55 \times 0.079}{0.0224 \times 1.8 \times 10^{-5}} = 1.08 \times 10^5;$$

$$\lambda = 0.107\left(\frac{1.46k_{eq}}{d_h} + \frac{100}{\text{Re}}\right)^{0.25} = 0.0303;$$

$$k_1^2 = \left(\frac{m_1}{F\rho_{cut}}\right)^2 \frac{\lambda}{2gd_h} = \left(\frac{0.55}{0.0224 \times 2700}\right)^2 \frac{0.0303}{2 \times 9.81 \times 0.079} = 1.62 \times 10^{-6};$$

$$\xi = \frac{L}{RT\bar{z}} = \frac{1500}{28.7 \times 310 \times 1.0} = 0.169;$$

in accordance with the formula (8.1.13)

$$\bar{p}_1 = \sqrt{\left(\bar{p}_0^2 + k_1^2\right) e^{2\xi(1+\eta_1)} - k_1^2} = 10^{-4}\sqrt{(4.07^2 + 162)\, e^{2\times 0.169\times 1.948} - 162}$$
$$= 1.35 \times 10^{-3}.$$

The pressure $\bar{p}_1$ is needed to determine $\bar{p}_2$. In the range of depth 1500–1950 m in annular space, we have $\xi = 0.051$; $\lambda = 0.052$; $k_1^2 = 2.78 \times 10^{-6}$; $\bar{p}_2 = 1.68 \times 10^{-3}$. Pressure $\bar{p}_2$ is calculated by the same formula as was calculated $\bar{p}_1$ with the exception that instead of $\bar{p}_0$ it is taken $\bar{p}_1$ and values of ξ and k_1 are taken correspondent to given interval

$$C = \sqrt{\frac{4g}{3c_w} d_{cut} \frac{S\rho_{cut}}{m}} = \sqrt{\frac{4 \times 9.81}{3 \times 0.4} 0.0014 \frac{0.0224 \times 2700}{0.55}} = 23.5.$$

Here, $C_w = 0.4$ since the regime is turbulent;

$$C\sqrt{\bar{p}_2} = 23.5\sqrt{1.68 \times 10^{-3}} = 0.96$$

consequently, the removal of the cutting above the collar pipe is absent.

At the bottom

$$d_{dp} = 0.023\ \text{m}; \qquad S = 0.00753\ \text{m}^2; \qquad \text{Re} = 0.93 \times 10^5;$$

$$\lambda = 0.071; \qquad \xi = 0.0056; \qquad k_1^2 = 1.16 \times 10^{-4};$$

$$\bar{p}_3 = 2.36 \times 10^{-3}; \qquad C = 7.9; \qquad C\sqrt{\bar{p}_3} = 0.38 < 0.83;$$

and the removal of the slag from the bottom takes place.

Hence, for the cutting removal, the section above the collar pipe is the worst.

Now, calculate the pressure $\bar{p}_2$ at the depth $L = 1950$ m with flow rate $Q_2^0 = 1\ \text{m}^3/\text{s}$.

For interval 0–1500 m, we have

$$m_2 = 1.1\ \text{kg/s}; \qquad \eta = 0.474; \qquad \text{Re} = 2.16 \times 10^5;$$

$$\lambda = 0.03; \qquad k_1^2 = 6.48 \times 10^{-6}; \qquad \bar{p}_1 = 2.1 \times 10^{-3};$$

for the interval 1500–1950 m, we have

$$\lambda = 0.052; \qquad \xi = 0.051; \qquad k_1^2 = 11.1 \times 10^{-6}; \qquad \bar{p}_2 = 2.62 \times 10^{-3};$$

$$C = 11.8; \quad C\sqrt{\bar{p}_2} = 0.604 < 0.83,$$

and at the flow rate $m_2 = 1.1$ kg/s the cutting removal above the collar pipe takes place. Specify the flow rate with the formula (8.4.5)

$$m = 0.55 - \frac{(1.1-0.55)(0.96-0.83)}{0.604-0.96} = 0.75\ \text{kg/s}; \qquad Q_0 = 0.68\ \text{m}^3/\text{s}.$$

At this flow rate, it is necessary to calculate $\bar{p}_1$ and $\bar{p}_2$ anew, after which one can calculate the pressure $\bar{p}_s$. For interval 0–1500 m, we have

$$\eta = 0.7; \quad \text{Re} = 1.47 \times 10^5; \quad \lambda = 0.03; \quad k_1^2 = 3 \times 10^{-6}; \quad \bar{p}_1 = 1.6 \times 10^{-3};$$

for interval 1500–1950 m, there are

$$\lambda = 0.052; \quad k_1^2 = 5.13 \times 10^{-6}; \quad \bar{p}_2 = 2 \times 10^{-3}; \quad C = 17.4;$$

$$C\sqrt{\bar{p}_2} = 0.78 < 0.83.$$

Thus, at the flow rate $m = 0.75$ kg/s, the cutting is removed and all further calculations of pressure are performed with this flow rate.

Bottom pressure

$$\lambda = 0.07; \quad k_1^2 = 2.15 \times 10^{-4}; \quad \bar{p}_3 = 2.87 \times 10^{-3};$$

$$p_3 = \bar{p}_3 \bar{z} RT \rho_{\text{cut}} g = 2.87 \times 10^{-3} \times 28.7 \times 310 \times 2700 \times 9.81 = 0.69\ \text{MPa}.$$

Pressure at the bit entrance. Turn to one of the formulas (8.3.10) or (8.3.11). The formula (8.3.11) gives

$$a = \left(\frac{m_n}{\Phi p_3 \mu}\right)^2 \frac{\bar{z} RTg(k-1)}{2k} = \left(\frac{0.75}{0.0004 \times 6.9 \times 10^5}\right)^2 \times \frac{28.7 \times 310 \times 0.4 \times 9.81}{2.8}$$

$$= 0.092;$$

$$p_4 = p_3 \left(\frac{1 + \sqrt{1 + 4a}}{2}\right)^{\frac{k}{k-1}} = 0.69 \left(\frac{1 + \sqrt{1 + 4 \times 0.092}}{2}\right)^{3,5}$$

$$= 0.92\ \text{MPa}; \quad \left(\frac{2}{k+1}\right)^{\frac{k}{k-1}} = 0.528;$$

since $p_3/p_4 > 0.528$, the calculation is made with suitable formula.

Inside the pipe space above the collar pipe, we have

$$S = 0.00636\ \text{m}^2; \quad \text{Re} = 5.76 \times 10^5; \quad \lambda = 0.027;$$

$$\xi_0 - \xi = 0.0056; \quad k_1^2 = 2.9 \times 10^{-5};$$

$$\bar{p}_5 = \sqrt{(\bar{p}_4^2 - k_1^2)\, e^{-2(\xi - \xi_0)} + k_1^2} = 3.92 \times 10^{-3}; \quad p_5 = 0.93\ \text{MPa};$$

at the wellhead of drill pipes, there are

$$S = 0.0119\ \text{m}^2; \quad \text{Re} = 4.31 \times 105; \quad \lambda = 0.025; \quad k_1^2 = 5.6 \times 10^{-6};$$

$$\xi_0 - \xi = 0.22; \quad \bar{p}_6 = 3.45 \times 10^{-3}; \quad p_6 = 0.81\ \text{MPa}.$$

Since the joints of the type ЗУ are equal passable, the losses in them may be neglected.

TABLE 8.1 Characteristics of Compressors (in Russian Designations)

Compressor	Feed Q (m^3/min)	Pressure (MPa)
Movable with diesel drive		
УКП-80	8	8
КПУ-16/100	16	10
КПУ-16/250	16	25
ДКС-7/200А	7	20
4НО/2а[a]	40	1.2
КС-20/45	20	4.5
VBC-3438W3[a]	40	4.0
СД9/101	9	10.1
Stationary with electric drive[b]		
7ВП-20/220	20	22
305ВП-12/220	12	22

[a]Austrian and Italian compressor plants used in domestic drilling.
[b]Mobile bases to plants used in cluster drilling.

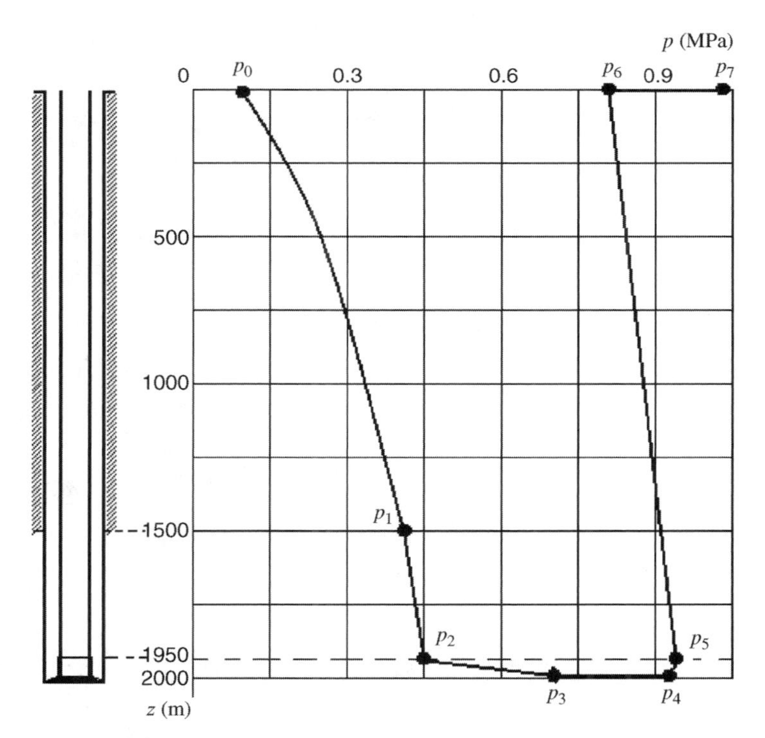

FIGURE 8.1 Scheme of arrangement and pressure distribution profile in the well with blasting: (1 ÷ 7) pressure indices at characteristic points of the circulation system.

Losses in the binding:

$$S = 0.00353\ \mathrm{m}^2; \qquad \mathrm{Re} = 7.9 \times 10^5; \qquad \lambda = 0.029; \qquad k_1^2 = 1.37 \times 10^{-4};$$

$$\xi = 0.028; \qquad \bar{p}_7 = \sqrt{\bar{p}_6^2 + 2k_1^2\xi} = 4.4 \times 10^{-3}; \qquad p_7 = 1.04\ \mathrm{MPa}.$$

Hence, in drilling with gas blow, compressor with mass flow rate is needed, which is no less than 0.75 kg/s (or feed $0.68\ \mathrm{m}^3/\mathrm{c} = 40.8\ \mathrm{m}^3/\mathrm{min}$) and has pressure 1.04 MPa. Table 8.1 contains various types of compressors.

Figure 8.1 shows the pressure distribution in the well circulation system resulted from the above considered example.

CHAPTER 9

STATIONARY FLOWS OF GAS–LIQUID MIXTURES IN A WELL

Washing wells is frequently performed by drilling fluid containing gas. Cementation can be also carried out with gas cut flushing fluids.

Gas–liquid media have great compressibility due to the presence of gas. It can come into the mud solution by natural way, for example, when drilling out gas-bearing stratum or may be introducing into the fluid by artificial means and when washing wells with aerated fluid. The knowledge of pressure distribution with well depth in circulation of gas–liquid mixtures permits to determine the pressure drop between well and surrounding beds. Timely controlling the pressure drop gives possibility to prevent complications such as gas absorptions, showings, and other.

Joint flow of gas and fluid can happen with different flow structures visually determined mainly by the form and mutual arrangement of phases in the flow.

Four structures of gas–liquid flow in vertical pipes and annular channels are being recognized (Fig. 9.1): bubble flow characterized by almost uniform distribution of gas bubbles in the fluid; plug flow characterized by alternation of gas or fluid plugs in the flow, at which the size of gas plugs can be commensurable to the channel diameter; annular or film flow when the flow happens to be near-stratified flow with pure gas in the center of the

---

Applied Hydro-Aeromechanics in Oil and Gas Drilling. By Leonov and Isaev

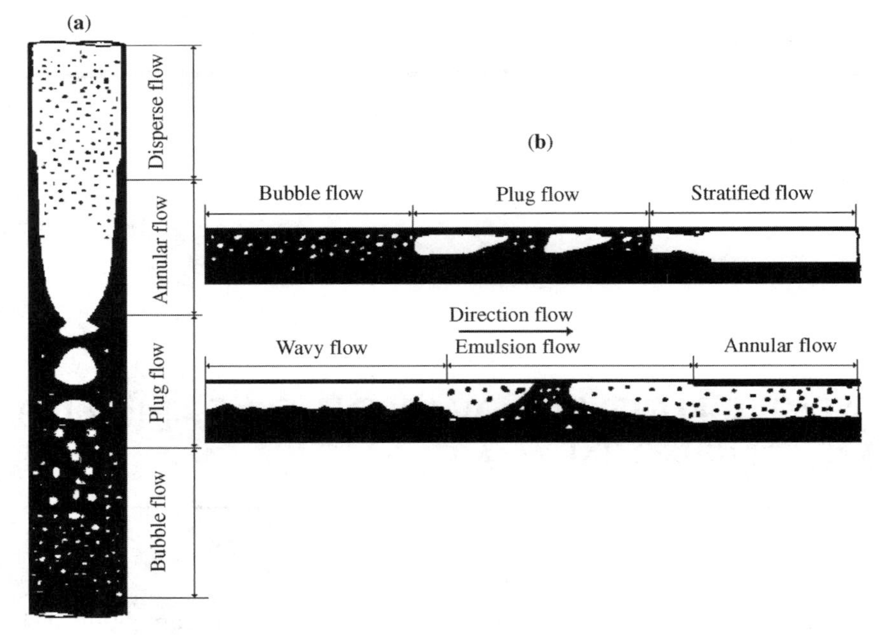

FIGURE 9.1 Principal structure forms of gas–liquid flows: (a) vertical rising flow and (b) horizontal flow.

channel and the bulk fluid in the form of a film at the channel wall; disperse flow characterized by uniform distribution of fluid drops in gas.

The structure of the gas–liquid flow depends primarily on physical properties of gas and fluid, volume content of gas, and velocities of both phases. In rise of mixture in the well as the pressure falls, different structures of the flow at which the flow can move in laminar or turbulent regimes may exist. The latter is most commonly encountered, since relative displacement (slippage) of phases owing to the difference of their densities, for example, in air–water mixtures, favors mixing and as a result flow turbulization.

Relative velocity between gas and fluid phases decreases in mixtures having non-Newtonian properties.

In air–water flows, the slippage between phases can be reduced to the point of disappearance by adding structure building additives such as mud, surfactants, and so on. In doing so, the flow of air–water mixture calms and in the mixture the laminar regime can be settled, which would be reached faster to give air–water mixture non-Newtonian properties. In particular, in foam flow, the displacement of gas bubbles is restricted by fluid films of enhanced strength between bubbles.

9.1 EQUATIONS OF GAS–LIQUID MIXTURE FLOW

Stationary laminar or turbulent flow in the absence of phase transitions is described by equations (4.6.33)–(4.6.39) under condition that the first phase is gas and the second one incompressible fluid (Isaev and Leonov, 1976; Sheberstov and Leonov, 1968):

momentum (motion) equation

$$\frac{\mathrm{d}p}{\mathrm{d}z} = g(\varphi\rho_1 + (1-\varphi)\rho_2) \pm \frac{\lambda_c}{2d}(\varphi\rho_1 v_1^2 + (1-\varphi)\rho_2 v_2^2) - \left(\varphi\rho_1 v_1 \frac{\mathrm{d}v_1}{\mathrm{d}z} + (1-\varphi)\rho_2 v_2 \frac{\mathrm{d}v_2}{\mathrm{d}z}\right), \quad (9.1.1)$$

the plus sign is taken for ascending flow while minus sign for descending flow, z-axis coincides with the direction of gravity force;

equations of mass conservation

$$S\varphi\rho_1 v_1 = Q_1\rho_1 = m_1 = \text{const}, \quad (9.1.2)$$

$$S(1-\varphi)\rho_2 v_2 = Q_2\rho_2 = m_2 = \text{const}; \quad (9.1.3)$$

thermodynamic equations of state

$$p = \bar{z}R\bar{T}\rho_1 g; \quad (9.1.4)$$

$$\rho_2 = \text{const}, \quad (9.1.5)$$

where $\bar{z}$ and $\bar{T}$ are values of overcompressibility factor and temperature averaged over well depth;

concentration equation

$$\varphi = \varphi(v_1, v_2, \rho_1, \rho_2, \mu_1, \mu_2, g, d, \sigma); \quad (9.1.6)$$

equation for hydraulic resistance factor of the mixture

$$\lambda_c = \lambda_c(v_1, v_2, \rho_1, \rho_2, \mu_1, \mu_2, g, d, \sigma, \varphi). \quad (9.1.7)$$

Introduction of dimensionless variables $\bar{p} = p/(\bar{z}R\bar{T}\rho_2 g)$ and $\xi = z/(\bar{z}R\bar{T})$, transforms the equation (9.1.1) to the form

$$\frac{\mathrm{d}\bar{p}}{\mathrm{d}\xi} = 1-\varphi(1-\bar{p}) \pm \frac{\lambda_c k^2}{2}\left(\frac{\eta^2}{\varphi\bar{p}} + \frac{1}{(1-\varphi)}\right) - \frac{d}{\bar{z}R\bar{T}}k^2\left[\eta^2\frac{\mathrm{d}(1/(\varphi\bar{p}))}{\mathrm{d}\xi} + \frac{\mathrm{d}(1/(1-\varphi))}{\mathrm{d}\xi}\right], \quad (9.1.8)$$

where $k^2 = Q_2^2/(g\,\mathrm{d}S^2)$, $\eta = Q_0\rho_0/(Q_2\rho_2) = a\rho_0/\rho_2$ is the mass factor of aeration; Q_0, ρ_0 are the volumetric flow rate and density of gas at normal conditions (T_0, p_0); $a = Q_0/Q_2$ is the flow rate factor of aeration.

Differentiate terms in brackets of the last term in the right part (9.1.8) with respect to the coordinate and resolve anew the result relative derivative of pressure. We get

$$\frac{\mathrm{d}\bar{p}}{\mathrm{d}\xi} = \frac{1-\varphi(1-\bar{p}) \pm \frac{\lambda_c k^2}{2}\left(\frac{\eta^2}{\varphi\bar{p}} + \frac{1}{(1-\varphi)}\right) - \frac{d}{\bar{z}R\bar{T}}k^2\left(\frac{1}{1-\varphi} - \frac{\eta^2}{\varphi^2\bar{p}}\right)\frac{\mathrm{d}\varphi}{\mathrm{d}\xi}}{1 - \frac{d}{\bar{z}R\bar{T}}k^2\frac{\eta^2}{\varphi\bar{p}^2}}. \tag{9.1.9}$$

This equation can be resolved also for relative derivative of true gas content

$$\frac{\mathrm{d}\varphi}{\mathrm{d}\xi} = \frac{1-\varphi(1-\bar{p}) \pm \frac{\lambda_c k^2}{2}\left(\frac{\eta^2}{\varphi\bar{p}} + \frac{1}{(1-\varphi)}\right) - \left(1 - \frac{d}{\bar{z}R\bar{T}}k^2\frac{\eta^2}{\varphi\bar{p}^2}\right)\frac{\mathrm{d}\bar{p}}{\mathrm{d}\xi}}{\frac{d}{\bar{z}R\bar{T}}k^2\left(\frac{1}{1-\varphi} - \frac{\eta^2}{\varphi^2\bar{p}}\right)}. \tag{9.1.10}$$

With differential equations (9.1.9) and (9.1.10), one can perform calculations of pressure and true gas content along a channel if distribution of true gas content or pressure is known.

9.2 LAMINAR ASCENDING FLOW OF GAS–LIQUID MIXTURES IN PIPES AND ANNULAR CHANNELS

Consider laminar flow of the gas–liquid viscous-plastic mixture, for example, foam, with equal velocities of phases. To get formulas to calculate pressures in pipes and annular channels, it is needed to solve the system of equations (9.1.1)–(9.1.7).

Determine before functions (9.1.6) and (9.1.7). Since phases move with equal velocities, we have

$$\varphi = \beta = \frac{Q_1}{Q_1 + Q_2} = \frac{\eta}{\eta + \bar{p}}. \tag{9.2.1}$$

As far as the mixture as a whole is viscous-plastic, the factor λ_c can be approximately described by formula (6.3.50)

$$\lambda_c = 64/\mathrm{Re}^*, \tag{9.2.2}$$

where

$$\mathrm{Re}^* = \frac{v d_h \rho}{\mu\left(1 + \frac{\tau_0 d_h}{6\mu v}\right)}; \tag{9.2.3}$$

$v = v_1 = v_2$; d_h is the diameter of the pipe or hydraulic diameter of the annular space; τ_0 is the dynamic shear stress; μ is the plastic viscosity factor of the mixture; ρ is the density of the mixture (see (3.6))

$$\rho = \beta\rho_1 + (1-\beta)\,\rho_2. \tag{9.2.4}$$

In accordance with relations (3.6) and (9.2.1), the number Re^* can be strongly dependent on the height of the well because of the gas phase compressibility.

Let us leave some assumptions made when deriving the system of equations (9.1.1)–(9.1.7), and suppose temperature, overcompressibility factor, dynamic shear stress, and plastic viscosity factor to be constant along the well height L and equal to correspondent average values $\bar{T}$, $\bar{z}$, τ_0, and μ.

Transform (9.2.3) to

$$\mathrm{Re}^* = \frac{v d_h \rho}{\mu\left(1 + \frac{\tau_0 d_h}{6\mu v}\right)} = \frac{v^2 d_h \rho}{\frac{\tau_0 d_h}{6} + \mu v}. \tag{9.2.5}$$

Substitution of (9.2.5) in (9.1.8) yields

$$\frac{d\bar{p}}{d\xi} = 1 - \frac{\eta(1-\bar{p})}{\eta+\bar{p}} + n\left(\frac{\eta}{\bar{p}} + 1\right) + m, \tag{9.2.6}$$

where $n = 32\mu Q_2/(S d_h^2 \rho_2 g)$; $m = 16\tau_0/(3 d_h \rho_2 g)$.

Separating variables and integrating both parts of equation (9.2.6) from $\xi = 0$ to ξ and from $\bar{p} = \bar{p}_1$ (pressure at the wellhead) to $\bar{p}$, respectively, we obtain the following solution

$$\xi = \frac{(\bar{p}-\bar{p}_1)}{A} + \frac{A\eta - B}{2A^2}\ln\frac{D}{E} + \frac{B^2 - AB\eta - 2AC}{A^2\sqrt{\Delta}} \times \arctan\left|\frac{2A(\bar{p}-\bar{p}_1)\sqrt{\Delta}}{\Delta + (2A\bar{p}+B)(2A\bar{p}_1+B)}\right|, \quad \Delta > 0; \tag{9.2.7}$$

$$\xi = \frac{(\bar{p}-\bar{p}_1)}{A} + \frac{A\eta - B}{2A^2}\ln\frac{D}{E} - \frac{B^2 - AB\eta - 2AC}{2A^2\sqrt{-\Delta}} \times \ln\left|\frac{(B + 2A\bar{p} + \sqrt{-\Delta})(B + 2A\bar{p}_1 - \sqrt{-\Delta})}{(B + 2A\bar{p} - \sqrt{-\Delta})(B + 2A\bar{p}_1 + \sqrt{-\Delta})}\right|, \quad \Delta < 0; \tag{9.2.8}$$

$$\xi = \frac{(\bar{p}-\bar{p}_1)}{A} + \frac{A\eta - B}{2A^2}\ln\frac{D}{E} + \frac{B^2 - AB\eta - 2AC}{A^2}\left(\frac{1}{B+2A\bar{p}_1} - \frac{1}{B+2A\bar{p}}\right), \quad \Delta = 0. \tag{9.2.9}$$

Here we have

$$A = 1+\eta+n+m; \qquad B = \eta(2n+m); \qquad C = n\eta^2;$$
$$D = A\bar{p}^2 + B\bar{p} + C; \qquad E = A\bar{p}_1^2 + B\bar{p}_1 + C; \qquad \Delta = 4AC - B^2. \tag{9.2.10}$$

Open the expression for Δ

$$\Delta = 4AC - B^2 = 4(1+\eta+n+m)n\eta^2 - (2n+m)^2\eta^2 = \eta^2[4n(1+\eta) - m^2]$$

Since the sign of Δ coincides with the sign of the expression in square brackets, the formula (9.2.7) is valid at $4n(1+\eta) - m^2 > 0$. Insertion in these inequality values of n and m from (9.2.6) gives that the formula (9.2.7) is obeyed when

$$\frac{2d_\mathrm{h}\tau_0^2}{g\mu^2} < 9\mathrm{Re}_2(1+\eta), \tag{9.2.11}$$

where $\mathrm{Re}_2 = Q_2\rho_2 d_\mathrm{h}/(S\mu)$ is Reynolds number of the fluid phase.

Respectively, the formula (9.2.9) is true when

$$\frac{2d_\mathrm{h}\tau_0^2}{g\mu^2} = 9\mathrm{Re}_2(1+\eta). \tag{9.2.12}$$

and the formula (9.2.8), when

$$\frac{2d_\mathrm{h}\tau_0^2}{g\mu^2} > 9\mathrm{Re}_2(1+\eta). \tag{9.2.13}$$

Since at $\tau_0 = 0$, the relation (9.2.11) is always true, only the formula (9.2.7) gives pressure distribution in laminar flow of viscous gas–liquid mixture at $\varphi = \beta$. The inequality (9.2.11) is typical in the practice of drilling. The formula (9.2.7) permits to calculate pressure distribution over the well height. Numerical analysis of this formula has shown that at $\eta \le 0.2$, $m < 0.2$, $n < 0.003$, and $\xi \ge 0.01$, encountered in practice the last term in the right part can be neglected owing to its negligibility.

Generally the incompressible gas–liquid viscous-plastic mixture can flow in laminar regime in the lower section and in turbulent regime in the upper section of the annular space. Therefore, it is important to clarify whether at one or at both regimes, the mixture in the well flows and at what depth, the change of regime happens. Since values of τ_0 and μ are given for the mixture as a whole, the flow occurs with equal velocities of phases $v = v_1 = v_2$; to determine flow regime, one can use results obtained in Section 6.6, where it was shown that laminar regime exists at

$$\mathrm{Re} < \mathrm{Re}_{\mathrm{cr}} = 2100 + 7.3He^{0.58}. \tag{9.2.14}$$

The inequality (9.2.14) added by appropriate expression for Reynolds number Re

$$\mathrm{Re} = \frac{vd_h\rho}{\mu} = \left(1 + a\frac{\bar{p}_0}{\bar{p}}\right)(1+\eta)\frac{\bar{p}}{\eta+\bar{p}}\frac{Q_2 d_h \rho}{S\mu} \tag{9.2.15}$$

can be used to determine the flow regime.

The velocity v and density ρ

$$v = \left(1 + a\frac{\bar{p}_0}{\bar{p}}\right)\frac{Q_2}{S}, \quad \rho = (1+\eta)\frac{\bar{p}}{\eta+\bar{p}}\rho_2$$

are found with the help of mass conservation equations (9.1.2) and (9.1.3) at $\varphi = \beta$ and equations of state (9.1.4) and (9.1.5).

Substitution of (9.2.15) and (9.2.14) gives the value of p^* at which the regime change occurs. Thus, to calculate the pressure one can use the formula (9.2.7) at $p > p^*$, where p^* is the pressure at $Re = Re_{\mathrm{cr}}$. At $p < p^*$ turbulent flow regime takes place.

9.3 CALCULATION OF PRESSURE IN PIPES AND ANNULAR SPACE IN ASCENDING VERTICAL TURBULENT FLOWS OF GAS–LIQUID MIXTURES

Let us derive a formula for pressure distribution over the channel length at given pressure at the upper end and given flow rates of gas and fluid.

In Sheberstov and Leonov (1968), it was shown that when solving the equation (9.1.8) for ascending flow in pipes and hole annulus the factor, λ_c can be taken as a constant equal to 0.05, whereas the function of true gas content for turbulent flow can be approximated as follows

$$\varphi = 0.81\beta \quad \text{at} \quad \frac{k}{1-\beta} \geq 1.93; \tag{9.3.1}$$

$$\varphi = \frac{2.2k\eta}{\bar{p}(1+2.2k)+2.2k\eta} \quad \text{at} \quad \frac{k}{1-\beta} < 1.93, \tag{9.3.2}$$

where $\beta = \frac{Q_1}{Q_1+Q_2} = \frac{\eta}{\eta+\bar{p}}$ is the flow rate gas content.

Then ignoring inertial term in the equation (9.1.8), using (9.3.1), (9.3.2), and integrating the resulting equation in limits from p' to p, the following equation is obtained

$$z = \frac{1}{\rho_2 g A_1}\left[p - p' + p_0 \frac{\bar{z}\bar{T}}{z_0 T_0} a \frac{A_1 - B_1}{A_1} \ln \frac{p + p_0 a \frac{B_1}{A_1}}{p' + p_0 a \frac{B_1}{A_1}}\right]; \tag{9.3.3}$$

$$z = \frac{n}{\rho_2 g A_1}\left[p - p' + p_0 \frac{\bar{z}\bar{T}}{z_0 T_0} a \frac{A_2 - nB_2}{A_2 n} \ln \frac{p}{p'}\right], \tag{9.3.4}$$

where $A_1 = 1 - 0.81\eta + \frac{\lambda_c k^2}{2}\frac{\eta + 0.81}{0.81}$;

$$B_1 = 0.19 + \frac{\lambda_c k^2}{2}\left[1 + \frac{\eta}{\eta + 0.81}\right]; \tag{9.3.5}$$

$$A_2 = n + \eta + n^2 m; \qquad B_2 = 2nm; \quad n = 1 + \frac{1}{2.2k}; \quad m = \frac{\lambda_c k^2}{2}\left[\frac{\eta n + 1}{n}\right].$$

The formula (9.3.3) is true for $\frac{k}{1-\beta} \geq 1.93$ or

$$p \leq \frac{k a p_0 \frac{\bar{z}\bar{T}}{z_0 T_0}}{1.93 - k} = p^* \tag{9.3.6}$$

and formula (9.3.4) for $p > p^*$.

To simplify calculations, the solution of equations (9.3.3) and (9.3.4) is reduced to

$$N = M + \log M, \tag{9.3.7}$$

where M is a certain linear function of the sought pressure p and N can be calculated by given data.

At $N \geq 5$, $N \leq -2$ can be found as approximate solutions of the equation (9.3.7)

$$\begin{aligned} M &= N - \log N \text{ at } N \geq 5; \\ M &= 10^N \qquad \text{at } N \leq -2. \end{aligned} \tag{9.3.8}$$

At $-2 < N < 5$, the solution for M may be obtained from curves in Fig. 9.2 corresponding to $\alpha = 0$. Once M has been found, it is easy to determine p.

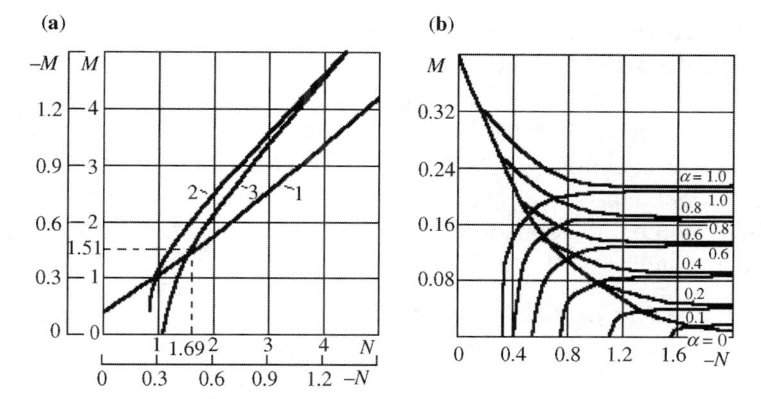

FIGURE 9.2 Graphics to determine M:(a) 1—$0 < N < 5$ (after drill pipe, drill collar, and turbo-drill, in a bit); 2—$N < 5$ (in turbo-drill), 3—$\alpha > 1$, $M_{00} < 0$, $-1.5 < N < 0$ (inside drill pipe, drill collar), curves 1 and 2 are built in positive coordinates, curve 3 in negative coordinates; (b) upper curves $\alpha = 0$ (after drill pipe, drill collar, and turbo-drill), $\alpha \leq 1$, $M_{00} > 0$ (in drill pipe and drill collar), lower curves $\alpha \leq 1$, $M_{00} < 0$ (in drill pipe and drill collar.

Let us cite formulas to calculate N and p. They are for (9.3.3)

$$N = \frac{\rho_2 g z A_1}{S} + \frac{p' + p_0 a(B_1/A_1)}{S} + \log \frac{p' + p_0 a(B_1/A_1)}{S},$$

$$S = 2.3 a p_0 \frac{\bar{z}\bar{T}}{z_0 T_0} \frac{A_1 - B_1}{A_1}, \qquad p = MS - p_0 \frac{aB_1}{A_1}; \tag{9.3.9}$$

for (9.3.4)

$$N = \frac{\rho_2 g z A_2}{n S_1} + \frac{p'}{S_1} + \log \frac{p'}{S_1};$$

$$S_1 = 2.3 a p_0 \frac{\bar{z}\bar{T}}{z_0 T_0} \frac{A_2 - nB_2}{A_2 n}, \qquad p = MS_1. \tag{9.3.10}$$

When going with well depth to pipes of another diameters, for example, from drill pipes to drill collars, calculations should be repeated taking pressure obtained from the previous calculation as initial one.

Hence, one can find the bottom pressure p_{bot} in the direct circulation or before the bit in the inverse circulation.

9.4 PRESSURE DROP IN BIT HEADS IN FLOW OF GAS–LIQUID MIXTURE

The flow in heads of aerated fluid is determined for the most by inertial terms of the equation (9.1.1) on the assumption that true and flow rate contents are equal.

Then momentum equation (9.1.1) at $v = v_1 = v_2$ for the case under consideration has the form

$$\frac{dp}{dz} = -\left(\beta\rho_1 v_1 \frac{dv_1}{dz} + (1-\beta)\rho_2 v_2 \frac{dv_2}{dz}\right). \tag{9.4.1}$$

The equation (9.4.1) can be rewritten as follows

$$\frac{dp}{dz} = -\frac{\rho}{2}\frac{dv^2}{dz}, \tag{9.4.2}$$

where $\rho = \rho_1\beta + \rho_2(1-\beta)$ is the mixture density; $v = (Q_1 + Q_2)/\Phi_c$ is the mixture velocity; Φ_c is the total area of all bit head cross sections.

Supposing that the gas–liquid flow from heads is completely braked, we integrate equation (9.4.2) in limits from calculated pressure p_{bot} to the pressure sought p_{bit}.

At the entrance of the bit we have

$$p_{bit} + ap_0 \ln p_{bit} = \frac{\left(\frac{p_0}{p_{bot}} Q_0 + Q_2\right)^2 (\rho_2 + ap_0)}{2\Phi_c^2\mu^2} + p_{bot} + ap_0 \ln p_{bot}. \tag{9.4.3}$$

Here, the correction factor μ having the sense of flow rate factor determined experimentally is introduced. In calculations one can accept μ equal to its value for single-phase fluid.

The solution of the equation (9.4.3) can be reduced to the form of (9.3.7) with

$$N = \frac{\left(\frac{p_0}{p_{bot}} Q_0 + Q_2\right)^2 (\rho_2 + ap_0)}{2\Phi_c^2\mu^2 2.3ap_0} + \frac{p_{bot}}{2.3ap_0} + \log\frac{p_{bot}}{2.3ap_0}. \tag{9.4.4}$$

Thus, the number M for flows in heads is found by the same rule as it was used before for ascending flow in annular space. The pressure sought p_{bit} before the bit is calculated by the formula

$$p_{bit} = 2.3M \cdot ap_0. \tag{9.4.5}$$

9.5 PRESSURE DROP IN TURBO-DRILLS

Let the flow in turbo-drills (Isaev and Leonov, 1976) is replaced by the flow in a pipe. It is assumed that the true and flow rate gas content are equal, and the effect of mixture column weight and inertia forces is insignificant. Then the momentum equation takes the form

$$\frac{\mathrm{d}p}{\mathrm{d}z} = -\frac{\lambda_{\mathrm{c}}}{2d}\rho v^2, \tag{9.5.1}$$

where

$$\lambda_{\mathrm{c}} = -\frac{2d \cdot \Delta p_{\mathrm{td}} S^2}{l \rho_2 Q_{\mathrm{td}}^2}. \tag{9.5.2}$$

In (9.5.1), the minus sign is taken because the flow in the turbo-drill is descending. In (9.5.2), d is fictive diameter of the turbo-drill; Δp_{td} and Q_{td} are pressure drop and flow rate in the turbo-drill at optimal operation regime on pure liquid with density ρ_2; S is the fictive area of the turbo-drill cross section; l is the length of the turbo-drill.

Integration of the equation (9.5.1) gives

$$p_{\mathrm{td}} = p_{\mathrm{bit}} + \Lambda g(m_1 + m_2)Q_2 + ap_0 \ln \frac{p_{\mathrm{td}} + ap_0}{p_{\mathrm{bit}} + ap_0}, \tag{9.5.3}$$

where $\Lambda = \Delta p_{\mathrm{td}}/(g\rho_2 Q_{\mathrm{td}}^2)$; $m_1 + m_2 = Q_0\rho_0 + Q_2\rho_2$; p_{td} is the pressure at the turbo-drill entrance.

If Δp_{td}, Λ, Q_{td} are known, then with (9.5.3), one can calculate pressure p_{T} and pressure drop $\Delta p = p_{\mathrm{td}} - p_{\mathrm{bit}}$ in the turbo-drill. Note, that the solution of the equation (9.5.3) can be reduced to equation

$$N = M - \log M, \tag{9.5.4}$$

where

$$M = \frac{p_{\mathrm{td}} + ap_0}{2.3ap_0}; \qquad N = \frac{\Lambda g(m_1 + m_2)Q_2}{2.3ap_0} + \frac{p_{\mathrm{bit}} + ap_0}{2.3ap_0} - \ln \frac{p_{\mathrm{bit}} + ap_0}{2.3ap_0}. \tag{9.5.5}$$

From (9.5.5), it follows that inequalities $M \geq 1/2.\ 3$; $N > M$ are always obeyed. The solution at $N \leq 5$ may be found from Fig. 9.2a , curve 2. At $N > 5$, it is approximately

$$M = N + \log N. \tag{9.5.6}$$

After determining M, the pressure sought is easy to obtain from (9.5.5)

$$p_{\mathrm{td}} = ap_0(2.3M-1), \tag{9.5.7}$$

which is further taken as initial condition for the flow in the next element of the circulation system.

9.6 CALCULATION OF PRESSURE IN PIPES IN DESCENDING VERTICAL TURBULENT FLOW OF GAS–LIQUID MIXTURE

The momentum equation (9.1.8) for descending flow has the same form as for the ascending flow except the sign before the term characterizing friction losses. Neglecting inertial terms, we write the equation (9.1.8) as follows

$$\frac{\mathrm{d}\bar{p}}{\mathrm{d}\xi} = 1-\varphi(1-\bar{p})-\frac{\lambda_c k^2}{2}\left(\frac{\eta^2}{\varphi\bar{p}}+\frac{1}{(1-\varphi)}\right). \tag{9.6.1}$$

On the basis of experimental data for descending flow (Isaev and Leonov, 1976), there is

$$\varphi = \frac{\sqrt{Fr}}{\sqrt{Fr}-0.45}\beta, \tag{9.6.2}$$

where $Fr = k^2/(1-\beta)^2$ is Froude number.

Substitution of numbers Fr and β in (9.6.2) yields

$$\varphi = \frac{\eta'}{\eta'+\bar{p}}, \tag{9.6.3}$$

where $\eta' = \eta k/(k-0.45)$.

The hydraulic resistance factor λ_c for descending flow is somewhat higher than that for ascending flow (Sheberstov and Leonov, 1968). Let it be constant equal to 0.06 in the mean.

Using the value obtained for φ and introducing designations

$$i^2 = k_2^2\frac{\eta^2/\eta'+1}{\eta'+1}, \quad k_2^2 = \frac{\lambda_c k^2}{2}, \tag{9.6.4}$$

we rewrite the momentum equation (9.6.1) in the form

$$\frac{\mathrm{d}\bar{p}}{\mathrm{d}\xi} = (\eta'+1)\left[\frac{\bar{p}}{\eta'+\bar{p}}-i^2\frac{\eta'+\bar{p}}{\bar{p}}\right]. \tag{9.6.5}$$

From here it follows

$$(1+\eta')\xi = \frac{(\bar{p}_{td}-\bar{p})}{1-i^2} + \frac{\eta'}{2}\left[\frac{1}{(1-i)^2}\ln\frac{(1-i)\bar{p}_{td}-i\eta'}{(1-i)\bar{p}-i\eta'} + \frac{1}{(1+i)^2}\ln\left|\frac{(1+i)\bar{p}_{td}+i\eta'}{(1+i)\bar{p}+i\eta'}\right|\right]. \quad (9.6.6)$$

Thus, the formula (9.6.6) permits to get the pressure in the ascending pipe $\bar{p}$ at given pressure $\bar{p}_{td}$ at the turbo-drill entrance.

As distinct from the ascending flow when the right part of the momentum equation is always positive and thus the pressure gradient is also positive, then in descending flow it may be the case of negative or zero pressure gradient. Therefore, the pressure in the descending flow can fall as well as rise with well depth increase. This fact was experimentally pointed out in Mezshlumov (1976). It is connected with redistribution of forces defined by terms in the right part of the momentum equation.

For the sake of convenience, let us introduce

$$\alpha = 2i/(1+i) \quad (9.6.7)$$

and reduce (9.6.6) to the equation with respect to M

$$N = M + 0.5\log|M-0.217\alpha| + 0.5(1-\alpha)^2\log|M+0.217\alpha(1-\alpha)|, \quad (9.6.8)$$

where

$$N = M_0 + 0.5\log|M_0-0.217\alpha| + 0.5(1-\alpha)^2\log|M_0+0.217\alpha(1-\alpha)| - \frac{1+\eta'}{2.3\eta'}(1-i)^2\xi;$$

$$M_0 = \frac{\bar{p}_{td}}{2.3\eta'}(1-\alpha); \qquad M = \frac{\bar{p}}{2.3\eta'}(1-\alpha).$$

At $\alpha = 0$, this relation turns into (9.3.7).

The sequence of pressure calculation with the formula (9.6.8) is as follows. For given data (L, d_1, d_2, p_{td}, and other), we determine numbers η', M_0, α, and the sign of $M_{00} = M_0 - 0.217\alpha$ coinciding with the sign of pressure gradient.

At $\alpha < 1$, $M_{00} > 0$, and $0 < N < 5$, the influence of the parameter α is insignificant and one can use the relation (9.3.7) and curve 1 (Fig. 9.2a) instead of the equation (9.6.8). If $N > 5$, the formula (9.3.8) for M is valid.

When $\alpha < 1$, $M_{00} > 0$ and $N < 0$, or $M_{00} < 0$ and $N < 0$, the value of M is found from Fig. 9.2b.

At $\alpha > 1$, $M_{00} < 0$, and $-2 < N < 0$, the value of M can be received from Fig. 9.2a, and at $N < -2$ from approximate formula

$$M = N - 0.5 \log|N - 0.217\alpha| - 0.5(1-\alpha)^2 \log|N + 0.217\alpha(1-\alpha)|. \tag{9.6.9}$$

After this, one can get the pressure sought

$$\bar{p} = 2.3M\eta'/(1-\alpha) \tag{9.6.10}$$

or in dimensional form

$$p = \bar{p}(\bar{z}R\bar{T}\rho_2 g). \tag{9.6.11}$$

The case $M_0 - 0.217 \cdot \alpha = 0$ corresponds to zero pressure gradient. At this, the flow is unstable and the momentum equation takes form $\mathrm{d}\bar{p}/\mathrm{d}\xi = 0$, from which it follows $p = \text{const}$. It means that the pressure in descending flow does not vary with the pipe section length at given pipe diameter.

For pipes with another diameter located below, the calculation should be repeated with initial pressure corresponding to the end pressure obtained from the calculation of the previous pipe section.

9.7 METHOD OF CALCULATION OF DELIVERY AND PRESSURE OF PUMPS AND COMPRESSORS IN DRILLING WITH AERATED FLUID WASHING

In Section 7.3, relation (7.3.3) was given between gas flow rate Q_0 at normal conditions and delivery of pump Q_2 for certain sizes of annular space, densities of gas ρ_0, fluid ρ_2, cutting ρ_p, and maximal diameter d_p of cutting particles to be removed

$$Q_0 = S_\mathrm{as}\sqrt{\frac{gd_\mathrm{p}\rho_\mathrm{p}p}{\rho_0 p_0}\left(\frac{0.108S_\mathrm{as}^2 d_\mathrm{p}\rho_\mathrm{p}}{Q_2^2\rho_2 + 0.0785S_\mathrm{as}^2 d_\mathrm{p}\rho_\mathrm{p}} - 0.008\right)}. \tag{9.7.1}$$

With the relation (9.7.1), one can calculate flow rates of phases providing levitation of cutting particles in the flow of aerated fluid.

In Fig. 9.3 (curves 2–6), a family of curves $Q_0 = \varphi(Q_2, p)$ obtained through substitution of fixed values of pressure $p \geq p_p$ in (9.7.1) are plotted,

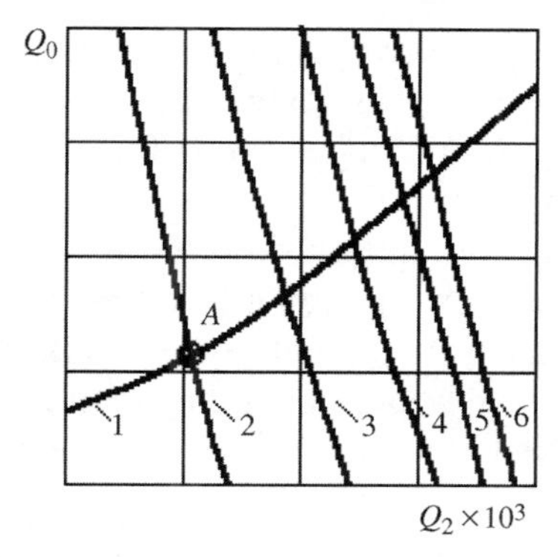

FIGURE 9.3 Graphics of gas and fluid flow rates needed to support given pressure in well against lost circulation horizon and cutting removal at different bottom pressures: (1) flow rates of phases providing constant pressure against lost circulation horizon; (2–6) flow rates of phases providing cutting removal at pressure p_i.

where p_p is given pressure in the well against lost circulation horizon. Later, p_p will be taken equal to reservoir pressure p_r in the lost circulation horizon. It is also taken that the curve 2 in Fig. 9.3 is plotted at pressure p_p.

A scheme of the well–intake reservoir system is shown in Fig. 9.4. Budgeted depth of the well and intake reservoir is designated at the scheme

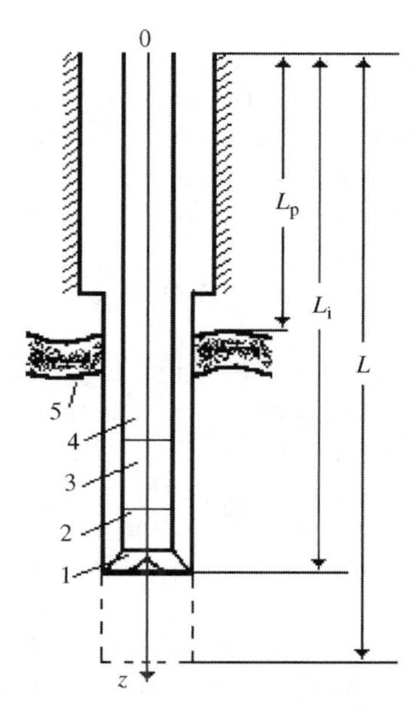

FIGURE 9.4 A scheme of the system well–intake reservoir:(1) bit; (2) turbo-drill; (3) drill collar; (4) drill pipe; (5) intake reservoir.

by L and L_p, L_i is the depth corresponding to intermediate bottom position. However, only equation (9.7.1) is not enough to select flow rates of phases.

It is seen from Fig. 9.3 that if the pressure in the well is equal to the reservoir pressure, the removal of cutting particles from the well bore during intake reservoir drilling can be performed at any relations between gas and fluid flow rates corresponding to different points at the curve 2. To choose concrete values of Q_0 and Q_2 for drilling the intake reservoir, it is also necessary to take into account the pressure produced by the flow of aerated fluid in the hole annulus with the help of equations (9.3.3) and (9.3.4).

Combined solution of equations (9.7.1) and (9.3.3) or (9.7.1), (9.3.3), and (9.3.4) gives unique combination of flow rates of phases providing concurrent levitating of particles in the bottom zone and given pressure at the intake reservoir. The system of equations (9.7.1), (9.3.3), and (9.3.4) could be solved with the help of computer (Leonov and Isaev, 1980). Below, semigraphical method permitting to calculate flow rates of phases by hand is described. The sequence of operations is as follows.

For each of the given values of pressures p_i $(i = 1, 2, 3, \ldots)$ from the range of its possible variation from p_1 to $\rho_2 g L$ in well drilling at corresponding depth L_i, we get a plethora of solutions of the equation (9.7.1) in the form of a function $Q_0 = f(Q_2, p_i)$ and display them in Fig. 9.3 with curves 2–6.

From initial data $(L_1, d_1, d_2, \rho_0, \rho_2, Q_2, p_0, p_m, \bar{T}, T_0, \bar{z}, z_0)$ at $p = p_1 = p_p$, we obtain solutions of equations (9.3.3) and (9.3.4) in form $Q_0 = f_1(Q_2, L_1, p_1)$. In Fig. 9.3, the curve 1 corresponds to these solutions, characterizing the relation between flow rates of gas and fluid at which on the intake reservoir from the flow acts at a given pressure equal to the reservoir one.

Calculations from equations (9.3.3) and (9.3.4) are performed as follows.

We get p^* with the formula (9.3.6). At $p_p < p^*$, the sought flow rate Q_0 is determined from the equation (9.3.3) in which $z = L_p, p' = p_m$ is taken (pressure at the wellhead) and $p = p_r = p_p$. Otherwise, (at $p_p > p^*$) after substitution $p = p^*, p' = p_m$ in the equation (9.3.3), the depth z_1 at which achievement the dependence (9.3.3) is changed to (9.3.4) is found. In this case Q_0 is determined from the equation (9.3.4) in which $z = L_p - z_1$, $p' = p^*$, $p = p_p$ should be taken. In Fig. 9.3, the intersection point A of curves 1 and 2 corresponds to the result of combined solution of equations (9.7.1) and (9.3.3) or (9.7.1), (9.3.3), and (9.3.4).

When drilling below the bottom of the intake reservoir in the zone of its bedding, it is required to hold the pressure $p_p = p_r$ and at the same time to change the relation between flow rates of gas and fluid so that with pressure rise at the bottom conserves the lifting force of the flow of aerated fluid needed to carry away the cutting.

Curves 2–6 of Fig. 9.3 characterizing conditions (9.7.1) of cutting removal are plotted at $p_i \geq p_{\mathrm{p}}$. Therefore, the quantities Q_{0i}, Q_{2i} corresponding to intersection points of curves 2–6 with the curve 1 ensure the removal of the cutting from the well depth L_i. At this on cutting particles carried away act along the whole well bore including the interval of intake reservoir bedding the lifting force exceeding the velocity needed for particle soaring.

Depths L_i are determined from equation (9.3.3) or (9.3.4). At $p = p_i$, $Q_0 = Q_{0i}$, $Q_2 = Q_{2i}$ the solution has the form

$$L_i = f_2(p_i, Q_{0i}, Q_{2i}). \tag{9.7.2}$$

Curves 1–3 of Fig. 9.5 show qualitative features of the dependence (9.7.2).

Once the flow rates of gas and fluid providing concurrent particle soaring in the bottom zone and given pressure on the intake reservoir have been got, it is possible to determine pressure losses in bit, turbo-drill (for turbine drilling) and column of drill pipes with step-by-step calculations.

In accordance with formulas (9.3.9), (9.3.10), (9.4.5), (9.5.7), (9.6.10), the calculation of pressure beginning from the wellhead pressure p_{m} in annulus hole up to the pressure at the end of each element is determined

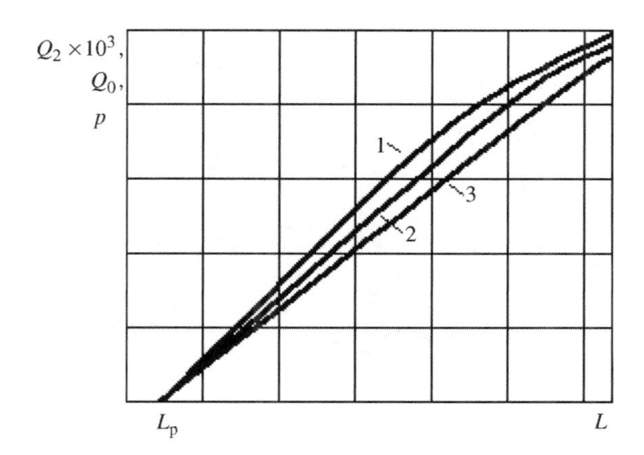

FIGURE 9.5 Dependence of phase flow rates and bottom pressure on the well depth needed to support given pressure against intake reservoir and cutting removal: (1, 2, 3) curves of fluid and gas flow rates and bottom pressure variation, respectively.

by the formula

$$p = AM + B, \tag{9.7.3}$$

where M is obtained from Fig. 9.2a and b for each corresponding element.

Accept $z\bar{T} = z_0 T_0$. Determine at first initial data: aeration factor $a = Q_0/Q_2$, mass aeration factor $\eta = a\rho_0/\rho_2$, factor $S_2 = 2.3ap_0$, pressure $p^* = kap_0/(1.93-k)$, number $k = Q_2/(F\sqrt{gd_h})$. Then determine

I. Coefficients A and B in the formula (9.7.3) for elements of annular space

(a) $A = (S_2 - 2.3C_1), \qquad B = -C_1 \quad \text{for } z_1 > L,$ (9.7.4)

where z_1 is calculated with (9.3.3) at $p' = p_m$, $p = p^*$

$$z_1 = \frac{1}{\rho_2 g A_1}\left[p^* - p_m + A\log\frac{p^* + C_1}{p_{bot} + C_2}\right],$$

$$A_1 = 1 + 0.81\eta + \frac{\lambda_c}{2}k^2\frac{\eta + 0.81}{0.81};$$

$$B_1 = 0.19 + \frac{\lambda_c}{2}k^2\left(1 + \frac{\eta}{\eta + 0.81}\right); \qquad C_1 = ap_0\frac{B_1}{A_1}.$$

To get the number M we determine

$$N = \frac{\rho_2 g L A_1}{A} + \frac{p_m + C_1}{A} + \log\frac{p_m + C_1}{A}; \tag{9.7.5}$$

(b) $A = S_2\left(\frac{1}{n} - \frac{B_2}{A_2}\right), \qquad B = 0 \quad \text{for } z_1 < L,$ (9.7.6)

where $A_2 = n + \eta + n^2 m$; $B_2 = 2mn$; $n = 1 + \frac{1}{2.2k}$; $m = \frac{\lambda_c}{2}k^2\frac{\eta n + 1}{n}$. To get the number M, we determine

$$N = \frac{\rho_2 z g A_1}{nA} + \frac{p^*}{A} + \log\frac{p^*}{A}, \tag{9.7.7}$$

where $z = L - z_1$.

(c) $A = 1, \qquad M = p^*, \qquad B = 0 \quad \text{for } z_1 = L,$ (9.7.8)

p^* is the pressure at the depth $z_1 = L$. If the length of the element coincides with z_1, then $p = p^*$.

II. To calculate the pressure at the bit entrance with given pressure above the bit, numbers A and B in (9.7.3) should be taken as $A = S_2, B = 0$. To get the number M we determine

$$N = \frac{\left(\frac{p_0}{p_{\text{bot}}}Q_0 + Q_2\right)^2 (\rho_2 + a\rho_0)}{2\mu^2\Phi^2 S_2} + \frac{p_{\text{bot}}}{S_2} + \log\frac{p_{\text{bot}}}{S_2}, \tag{9.7.9}$$

where Φ is total area of cross sections of heads, μ is the flow rate factor.

III. To determine the pressure before the turbo-drill entrance p_{td} with given pressure at the exit of the turbo-drill p_{bit}, the numbers A and B in (9.7.3) should be taken as $A = S_2, B = -S_2/2.3$. To determine the number M we find

$$N = \frac{\Lambda g(Q_0\rho_0 + Q_2\rho_2)Q_2}{S_2} + \frac{p_{\text{td}} + a\rho_0}{S_2} - \log\frac{p_{\text{td}} + a\rho_0}{S_2}, \tag{9.7.10}$$

where $\Lambda = \Delta p_{\text{td}}/(g\rho_2 Q_{\text{td}}^2)$.

IV. To determine the pressure at upper cross sections of pipe elements with given pressure p_{lp} at lower cross sections in descending flow, the numbers A and B in (9.7.3) should be taken equal to

$$A = \frac{\bar{z}R\bar{T}\rho_2 g 2.3\eta'}{(1-\alpha)}, \qquad B = 0, \tag{9.7.11}$$

where $\eta' = \frac{\eta k}{k-0.45}$; $\alpha = \frac{2i}{1+i}$; $i = k_2\sqrt{\frac{\eta^2/\eta' + 1}{\eta' + 1}}$; $k_2 = \sqrt{\lambda_c/2k}$.

To determine the number M we calculate

$$\begin{aligned} N = M_0 + 0.5\log|M_{00}| + 0.5(1-\alpha)^2 \log|M_0 \\ + 0.217\alpha(1-\alpha)| - \frac{\eta' + 1}{2.3\eta'}(1-i)^2 \frac{l}{\bar{z}R\bar{T}}, \end{aligned} \tag{9.7.12}$$

where $M_0 = \frac{\bar{p}_{\text{lp}}}{2.3\eta'}(1-\alpha)$; $M_{00} = M_0 - 0.217\alpha$.

Thus, through step-by-step calculations we get the pressure at the upper end of the last element of the drill-stem accepted equal to the pressure in the ascending pipe p_{ap}.

For turbine method of drilling, supplementary checking of the obtained mixture flow rate and pressure drop in the turbo-drill is required to elucidate whether the turbo-drill is able to operate at given flow rates of phases and to develop sufficient moment M. The latter should be equal to or greater than the moment M_p needed to breakdown the rock solid.

Checking conditions are as follows

$$Q_2 \geq 0.5Q_{td}, \quad M_p \leq M = M_{td2}\frac{\Delta p}{\Delta p_T}, \tag{9.7.13}$$

where Q_2, M_{td2}, Δp_{td} are reference data of the turbo-drill operation regime (flow rate, breaking torque, pressure drop); $\Delta p = p_{td} - p_{bit}$ is pressure drop in turbo-drill operation at gas and liquid flow rates calculated with the formula (9.5.3). If conditions (9.7.13) are not obeyed, another turbo-drill should be taken or the flow rate of gas and fluid should be increased. Flow rates ought to be so chosen that they would provide constant pressure p_r in the well against intake reservoir. In such case the removal of the cutting would be wittingly provided, since the values of gas and liquid flow rates should lie on the curve 1 of Fig. 9.3 and satisfy conditions (9.7.13).

Thus, to conduct concrete calculations one should plot a curve similar to the curve 1 of Fig. 9.3a and to retrieve from it values Q_0 and Q_2 that lie above the intersection point of the curve 1 with the curve 2 mapping dependence (9.7.1). Moving up along the curve one should dwell on those values of Q_0 and Q_2 that obey conditions (9.7.13).

All the above reasoning are true when the well and drill-stem have constant diameters with the depth. For annular channel with different geometrical sizes, calculation is complicated and it should be performed with the help of computer. In such case as a basis of calculation, the above given algorithm with the following additions is taken.

1. The cutting removal from any element of the annular space would be wittingly realized if in the formula (9.7.1) when calculating Q_0 and Q_2 to take maximal pressure p, that is bottom pressure p_b, and maximal area of the annular channel cross section. At this Q_0 and Q_2 would be overstated as compared to needed flow rates for cutting removal. These values of flow rates may be obtained more accurately with increase of calculation time and number of iterations by substitution in the formula (9.7.1), the calculated pressure at the end of each annular element, and its area at fixed flow rates Q_2. As a result we would get flow rates Q_{0i} maximum of which gives desirable result.
2. The equation (9.7.1) with previous comments should be solved together with equations (9.3.3) and (9.3.4) because they are true for each element of the annular space of the well circulation system.

EXAMPLE 9.7.1

It is required to calculate pressure losses in the circulation system of the well at given flow rates of gas and fluid. Initial data are taken from experimental data for turbine drilling (Mezshlumov, 1976). The scheme of the circulation system corresponds to the scheme depicted in Fig. 9.4.

Data common for all elements are $\rho_2 = 1000\,\text{kg/m}^3$, $\rho_0 = 1.29\,\text{kg/m}^3$, $Q_0 = 0.267\,\text{m}^3/\text{s}$, $Q_2 = 0.0287\,\text{m}^3/\text{c}$, $p_m = p_0 = 10^5\,\text{Pa}$, $T = T_0 = 300\,\text{K}$, $\bar{z} = z_0 = 1$, $R = 29.27\,\text{m/K}$. Data for ascending flow in the annulus: $d_1 = 0.305\,\text{m}$, $d_2 = 0.141\,\text{m}$, $L = 250\,\text{m}$, $\lambda_c = 0.05$.

SOLUTION

1. Determine at first governing parameters: $a = Q_0/Q_2 = 0.267/0.0287 = 9.3$; $n = a \cdot \rho_0/\rho_2 = 9.3 \times 1.29/1000 = 0.012$; $S = 2.3a \cdot p_0 = 2.3 \times 9.3 \times 10^5 = 21.39 \times 10^5$;

$$k = \frac{Q_2}{S_{as}\sqrt{g(d_1 - d_2)}} = \frac{4 \times 0.0287}{3.14(0.305^2 - 0.141^2)\sqrt{9.81 \times (0.305 - 0.141)}} = 0.394.$$

2. With formula (9.3.6) we get

$$p^* = \frac{kap_0}{1.93 - k} = \frac{0.394 \times 9.3 \times 10^5}{1.93 - 0.394} = 2.386 \times 10^5\ \text{Pa}.$$

3. With formula (9.7.4) we obtain

$$A_1 = 1 + 0.81\eta + \frac{\lambda_c k^2}{2}\frac{\eta + 0.81}{0.81} = 1 + 0.81 \times 0.012$$

$$+ \frac{0.05 \times 0.0394^2}{2}\frac{0.012 + 0.81}{0.81} = 1.014;$$

$$B_1 = 0.19 + \frac{\lambda_c k^2}{2}\left[1 + \frac{\eta}{\eta + 0.81}\right]$$

$$= 0.19 + \frac{0.05 \times 0.394^2}{2}\left[1 + \frac{0.012}{0.012 + 0.81}\right] = 0.194;$$

$$z_1 = \frac{1}{\rho_2 g A_1}\left[p^* - p_y + p_0 a \frac{A_1 - B_1}{A_1}\ln\frac{p^* + p_0 a \frac{B_1}{A_1}}{p_y + p_0 a \frac{B_1}{A_1}}\right]$$

$$= \frac{1}{10^3 \times 9.81 \times 1.014}\left[2.386 \times 10^5 - 10^5 + 10^5 \times 9.3\frac{1.014-0.194}{1.014}\right.$$

$$\left. \times \ln \frac{2.386 \times 10^5 + 10^5 \times 9.3\frac{0.194}{1.014}}{10^5 + 10^5 \times 9.3\frac{0.194}{1.014}}\right] = 44.5\ \text{m}.$$

4. Since $L = 250\,\text{m} > z_1 = 44.5\,\text{m}$, to get the pressure we use formulas (9.7.6)–(9.7.7) in which $z = z_2 = L - z_1 = 250 - 44.5 = 205.5\,\text{m}$:

$$n = 1 + \frac{1}{2.2k} = 1 + \frac{1}{2.2 \times 0.394} = 2.154;$$

$$m = \frac{\lambda_c k^2}{2}\left[\frac{\eta n + 1}{n}\right] = \frac{0.05 \times 0.394^2}{2}\left[\frac{0.012 \times 2.154 + 1}{2.154}\right] = 0.00185;$$

$$A_2 = n + \eta + n^2 m = 2.154 + 0.012 + 2.154^2 \times 0.00185 = 2.17;$$

$$B_2 = 2nm = 2 \times 0.00185 \times 2.154 = 0.00797;$$

$$A = 2.3ap_0\frac{A_2 - nB_2}{A_2 n} = 2.3 \times 9.3 \times 10^5\frac{2.17 - 2.154 \times 0.00797}{2.17 \times 2.154}$$

$$= 9.85 \times 10^5;$$

$$N = \frac{\rho_2 g z_2 A_2}{nA} + \frac{p^*}{A} + \log\frac{p^*}{A} = \frac{1000 \times 9.81 \times 205.5 \times 2.17}{2.154 \times 9.85 \times 10^5}$$

$$+ \frac{2.386 \times 10^5}{9.85 \times 10^5} + \log\frac{2.386 \times 10^5}{9.85 \times 10^5} = 1.69.$$

5. From Fig. 9.2a (curve 1) of equation (9.3.7), we receive $M = 1.51$ and with formula (9.7.3) at $B = 0$

$$p_1 = M \times A = 1.51 \times 9.85 \times 10^5 = 1.487\ \text{MPa}.$$

Initial data for ascending flow behind the drill collar and turbo-drill are $d_1 = 0.305\,\text{m}$, $d_2 = 0.178\,\text{m}$, $L = 45\,\text{m}$, $\lambda_c = 0.05$. Calculations are repeated taking for initial pressure $p_1 = 14.87 \times 10^5\,\text{Pa}$. As a result we get $p_{bot} = 18.35 \times 10^5\,\text{Pa}$.

Initial data for the flow in the bit are $\mu = 0.8$, $\Phi_c = 26.74 \times 10^{-4}\,\text{m}^2$, $p_{bot} = 18.35 \times 10^5\,\text{Pa}$.

1. From the formula (9.7.9) it follows

$$N = \frac{\left(\frac{p_0}{p_{\text{bot}}} Q_0 + Q_2\right)^2 (\rho_2 + a\rho_0)}{2\mu^2 \Phi_c^2 2.3 a p_0} + \frac{p_{\text{bot}}}{2.3 a p_0} + \log \frac{p_{\text{bot}}}{2.3 a p_0}$$

$$= \frac{\left(\frac{10^5}{18.35 \times 10^5} 0.267 + 0.0287\right)^2 (10^3 + 9.3 \times 1.29)}{2 \times 0.8^2 \times 0.002674^2 \times 21.39 \times 10^5}$$

$$+ \frac{18.35 \times 10^5}{21.39 \times 10^5} + \log \frac{18.35 \times 10^5}{21.39 \times 10^5} = 0.888$$

2. From Fig. 9.2a (curve 1) or equation (9.3.7), we get $M = 0.923$.
3. From the formula (9.7.3) at $B = 0, A = 2.3 a p_0$, the pressure at the bit entrance or the turbo-drill exit $p_{\text{td}} = 0.923 \times 21.39 \times 10^5 = 1.97$ MPa is obtained.

Initial data for the flow in the turbo-drill are $\Delta p_{\text{td}} = 4$ MPa, $Q_{\text{T}} = 0.045\ \text{m}^3/\text{s}$, $\rho_2 = 1000\ \text{kg/m}^3$.

1. With the formula (9.7.10) we receive

$$\Lambda = \Delta p_{\text{td}}/(g\rho_2 Q_{\text{td}}^2) = 4 \times 10^6/[9.81 \times 10^3 \times 0.045^2] = 2.01 \times 10^5;$$

$$N = \frac{\Lambda g(m_1 + m_2)Q_2}{2.3 a p_0} + \frac{p_{\text{bit}} + a p_0}{2.3 a p_0} - \log \frac{p_{\text{bit}} + a p_0}{2.3 a p_0}$$

$$= \frac{2.01 \times 10^5 \times 9.81(0.267 \times 1.29 + 0.0287 \times 10^3)0.0287}{21.39 \times 10^5}$$

$$+ \frac{19.7 \times 10^5 + 9.3 \times 10^5}{21.39 \times 10^5} - \log \frac{19.7 \times 10^5 + 9.3 \times 10^5}{21.39 \times 10^5} = 1.99.$$

2. From Fig. 9.2a (curve 2) or equation (9.5.4), we obtain $M = 2.37$.
3. From the formula (9.7.3) at $B = -a p_0$ and $A = 2.3\, a p_0$, we get the pressure at the entrance of the turbo-drill

$$p_{\text{td}} = M \cdot A - a p_0 = 2.37 \times 21.39 \times 10^5 - 9.3 \times 10^5 = 4.14\ \text{MPa}.$$

Initial data for descending flow in the drill collar are

$$d_1 = 0.09\ \text{m}, \qquad d_2 = 0\ \text{m}, \qquad L = 20\ \text{m}, \qquad \lambda_c = 0.06.$$

Calculate dimensionless variables

$$\xi = L/(\bar{z}R\bar{T}) = 20/(1 \times 29.27 \times 300) = 0.00228;$$

$$\bar{p}_{\text{td}} = p_{\text{td}}/(\bar{z}R\bar{T}\rho_2 g) = 41.4 \times 10^5/(1 \times 29.27 \times 300 \times 10^3 \times 9.81) = 0.0481.$$

1. With formulas (9.7.11) and (9.7.12) we get

$$k = \frac{Q_2}{S\sqrt{g(d_1 - d_2)}} = \frac{0.0287}{3.14 \times \frac{0.09^2}{4}\sqrt{9.81(0.09-0)}} = 4.8;$$

$$\eta' = \frac{\eta k}{k - 0.45} = \frac{0.012 \times 4.8}{4.8 - 0.45} = 0.01324;$$

$$i = \sqrt{k_2^2 \frac{\eta^2/\eta' + 1}{\eta' + 1}} = \sqrt{0.69^2 \frac{0.012^2/0.01324 + 1}{0.01324 + 1}} = 0.83;$$

$$k_2^2 = \frac{\lambda_c k^2}{2} = \frac{0.06}{2} 4.8^2 = 0.69; \qquad \alpha = \frac{2i}{1+i} = \frac{2 \times 0.83}{1 + 0.83} = 0.91 < 1;$$

$$M_0 = \frac{\bar{p}_{td}}{2.3\eta'}(1-\alpha) = \frac{0.0481}{2.3 \times 0.01324}(1 - 0.91) = 0.1422;$$

$$M_{00} = M_0 - 0.217\alpha = 0.1422 - 0.217 \times 0.91 = -0.0553 < 0;$$

$$\begin{aligned} N &= M_0 + 0.5 \log|M_0 - 0.217\alpha| + 0.5(1-\alpha)^2 \log|M_0 + 0.217\alpha(1-\alpha)| \\ &\quad - \frac{1+\eta'}{2.3\eta'}(1-i)^2 \xi \\ &= 0.1422 + 0.5 \log|0.0553| + 0.5(1-0.91)^2 \log|0.1422 \\ &\quad + 0.217 \times 0.91(1-0.91)| \\ &\quad - \frac{1 + 0.01324}{2.3 \times 0.01324}(1-0.83)^2 0.00228 = -0.492. \end{aligned}$$

2. Since $\alpha < 1$, $M_{00} < 0$, $N = -0.492$, the number M is found from Fig. 9.2b (lower family of curves) or from equation (9.6.8): $M = 0.145$.
3. Desired pressure at the entrance of the drill collar is received from the formula (9.7.11):

$$\begin{aligned} p &= \frac{\bar{z}R\bar{T}\rho_2 g 2.3 \times \eta'}{1-\alpha} M \\ &= \frac{1 \times 29.27 \times 300 \times 10^3 \times 9.81 \times 2.3 \times 0.01324}{1 - 0.91} 0.145 = 4.23\,\text{MPa}. \end{aligned}$$

Initial data for descending flow in pipes are

$$d_1 = 0.117\,\text{m}, \qquad d_2 = 0\,m, \qquad L = 250\,m, \qquad \lambda_c = 0.06.$$

Calculations are repeated using the same formulas

1. $\xi = L/(\bar{z}R\bar{T}) = 250/(1 \times 29.27 \times 300) = 0.0285;$

   $\bar{p} = p/(\bar{z}R\bar{T}\rho_2 g) = 42.3 \times 10^5/(1 \times 29.27 \times 300 \times 10^3 \times 9.81) = 0.049.$

2. With formulas (9.7.11) and (9.7.12) it is obtained

$$k = \frac{Q_2}{S\sqrt{g(d_1 - d_2)}} = \frac{0.0287}{3.14 \times \frac{0.117^2}{4}\sqrt{9.81(0.117 - 0)}} = 2.49;$$

$$\eta' = \frac{\eta k}{k - 0.45} = \frac{0.012 \times 2.49}{2.49 - 0.45} = 0.0146;$$

$$i = \sqrt{k_2^2 \frac{\eta^2/\eta' + 1}{\eta' + 1}} = \sqrt{0.186^2 \frac{0.012^2/0.0146 + 1}{0.0146 + 1}} = 0.43;$$

$$k_2^2 = \frac{\lambda_c k^2}{2} = \frac{0.06}{2} 2.49^2 = 0.186; \qquad \alpha = \frac{2i}{1+i} = \frac{2 \times 0.43}{1 + 0 \times 43} = 0.6 < 1;$$

$$M_0 = \frac{\bar{p}_{td}}{2.3\eta'}(1 - \alpha) = \frac{0.049}{2.3 \times 0.146}(1 - 0.6) = 0.584;$$

$$M_{00} = M_0 - 0.217\alpha = 0.584 - 0.217 \times 0.6 = 0.454 > 0;$$

$$N = M_0 + 0.5\log|M_0 - 0.217\alpha| + 0.5(1-\alpha)^2 \log|M_0 + 0.217\alpha(1-\alpha)| - \frac{1+\eta'}{2.3\eta'}(1-i)^2\xi$$

$$= 0.584 + 0.5\log|0.454| + 0.5(1-0.6)^2 \log|0.584 + 0.217 \times 0.6(1-0.6)| - \frac{1+0.0146}{2.3 \times 0.0146}(1-0.43)^2 0.0285 = 0.117.$$

3. Since $\alpha < 1, M_{00} > 0, N > 0$, the number M is found from Fig. 9.2a (curve 1) or equation (9.3.7): $M = 0.457$.

The pressure in the ascending pipe is found from the formula (9.7.11):

$$p_c = \frac{\bar{z}R\bar{T}\rho_2 g 2.3 \cdot \eta'}{1 - \alpha} M = \frac{1 \times 29.27 \times 300 \times 10^3 \times 9.81 \times 2.3 \times 0.0146}{1 - 0.6} 0.457$$

$$= 3.3 \text{ MPa}.$$

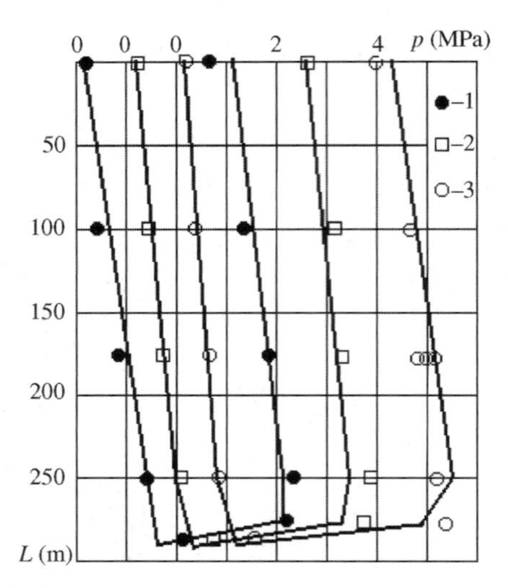

FIGURE 9.6 Experimental (Mezshlumov, 1976) and calculated pressure for different flow regimes. Diameter of drill pipes 0.141 m, air flow rate Q_0, m^3/s: (1) 0.267; (2) 0.533; (3) 0.8; water flow rate расхо™ $Q_2 = 0.0287$ m^3/s.

Figure 9.6 represents results of the above example (black points) as well as results of another calculations of pressure drop in elements of the circulation system performed at conditions of experiments carried out in Mezshlumov (1976).

9.8 EFFECT OF GAS SOLUBILITY IN FLUID ON PRESSURE OF MIXTURE IN WELL

In flow of gas–liquid mixtures, for example, gas–oil mixtures, significant role can play solubility of gas in fluid. In this case right parts of equations (9.1.2), (9.1.3), and (9.1.5) are functions of the pressure p.

Suppose that at each cross section of the channel a part of gas phase is dissolved in the gas phase in accordance with Henry law, then

$$Q_{10} = k_{sol} Q_{20}(p - p_0), \tag{9.8.1}$$

where Q_{10} is a part of gas phase flow rate dissolved in the fluid at pressure p at normal conditions; k_{sol} is the solubility factor; Q_{20} is the fluid flow rate at normal conditions (flow rate of degassed fluid); p_0 is the atmospheric pressure.

Assume that gas is dissolved in fluid up to molecular level. At this, as shown by experiments, the fluid becomes more compressible and its equation of state may be written as

$$\rho_2 = a - b \cdot p, \tag{9.8.2}$$

where a and b are constant factors.

Determining flow rate of gas Q_1 and of fluid Q_2 with density ρ_2 with regard to formulas (9.8.1) and (9.8.2) at average values of pressure, one can get pressure drop using formulas (9.3.3) and (9.3.4) obtained regardless solubility. Thus, dividing the depth of the well by small sections and specifying at each of them its own average pressure, it is able to find total pressure drop with regard to solubility.

EXAMPLE 9.8.2

With regard to the solubility of gas in oil it is required to calculate pressure in a column of tubings at the depth $L = 1600$ m and to compare the result obtained with pressure 108.9×10^5 Pa measured by bottom-hole pressure gauge.

Initial data for this example are taken from the Reference Book on Designing, Development and Exploitation of Oil Fields (Gimatudinov, 1983): at normal condition are oil discharge $Q_{20} = 0.001673\ \mathrm{m^3/s}$, gas flow rate $Q_0 = 0.08472\ \mathrm{m^3/s}$, oil density $\rho_{20} = 844\ \mathrm{kg/m^3}$, gas density $\rho_0 = 1.3\ \mathrm{kg/m^3}$; pressure at the wellhead $p_m = 10 \times 10^5$ Pa, overcompressibility factor averaged over the depth $\bar{z} = z_0 = 1$, gas constant $R = 26.76$ m/K. At saturation pressure $p_{sat} = 97 \times 10^5$ Pa, the whole gas is dissolved and density of oil is $\rho_2 = 790\ \mathrm{kg/m^3}$. Temperature at the mouth is T_m 288 K and at the depth 1600 m $T_{bot} = 302$ K, diameter of tubing is $d_t = 0.062$ m.

Oil density is described by dependence (9.8.2) as

$$\rho_2 = 845 - 5.62 \times 10^{-6} p.$$

SOLUTION In accordance with (9.8.1)

$$k_{sol} = \frac{Q_0}{Q_{20}(p_{sol} - p_0)} = \frac{0.08472}{0.001673(9.7 \times 10^6 - 0.1 \times 10^6)} = 5.28 \times 10^{-6}\ \mathrm{Pa^{-1}}.$$

Thus, at given pressure p, the mass of dissolved gas entering together with oil in a unit time is equal to

$$\begin{aligned} Q_0 \cdot \rho_0 &= k_{sol} Q_{20}(p - p_0)\rho_0 = 5.28 \times 10^{-6} \times 0.001673(p - 0.1 \times 10^6)1.3 \\ &= 0.0115(p - 0.1 \times 10^6) \times 10^{-6}\ \mathrm{kg/s}. \end{aligned}$$

The temperature averaged over the depth is

$$\bar{T} = \frac{T_m + T_{bot}}{2} = \frac{288 + 302}{2} = 295\ \mathrm{K}.$$

Volumetric flow rate of free gas is

$$\begin{aligned}Q_1 &= \frac{[Q_0 - k_p Q_{20}(p - p_0)]\rho_0 \bar{z} R \bar{T} g}{p}\\ &= \frac{[0.08472 - 5.28 \times 10^{-6} \times 0.001673(p - 0.1 \times 10^6)] \times 1.3 \times 1.0 \times 26.76 \times 295 \times 9.81}{p}\\ &= \frac{8529 - 889 \times 10^{-6}(p - 0.1 \times 10^6)}{p}\,\mathrm{m^3/s}.\end{aligned}$$

Total volumetric flow rate of oil together with dissolved gas is

$$\begin{aligned}Q_2 &= \frac{Q_{20}\rho_{20} + k_{sol}(p - p_0)Q_{20}\rho_0}{\rho_0}\\ &= \frac{0.001673 \times 844 + 5.28 \times 10^{-6}(p - 0.1 \times 10^6) \times 0.001673 \times 1.3}{845 - 5.62 \times 10^{-6} p}\\ &= \frac{1.41 + 1.148 \times 10^{-8}(p - 0.1 \times 10^6)}{845 - 5.62 \times 10^{-6} p}\,\mathrm{m^3/s}.\end{aligned}$$

Calculate flow rates Q_1 and Q_2 for given p in the range from p_m to p_{sat}, plot the graph of function $Q_1 + Q_2 = f(p)$ (Fig. 9.7), determine from this graphic pressure p^* at which the dependence (9.3.3) changes on (9.3.4).

At $p = p^*$, the flow rate of the mixture is determined from the formula (9.3.1)

$$Q_1 + Q_2 = 1.93 \frac{\pi}{4} d^2 \sqrt{gd} = 1.93 \frac{3.14}{4} 0.062^2 \sqrt{9.81 \times 0.062} = 4.54 \times 10^{-3}\,\mathrm{m^3/s}.$$

In Fig. 9.7, to this flow rate corresponds pressure $p^* = 24 \times 10^5$ Pa.

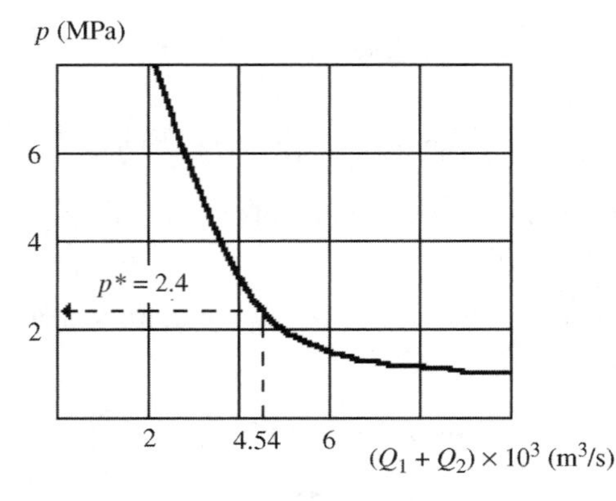

FIGURE 9.7 Dependence of pressure on total flow rate of mixture.

In lifting of fluid in oil well tubings, the dissolved gas liberated from the oil. Therefore, the flow rate Q_1 increases while Q_2 decreases. Further calculations will be performed at averaged values of Q_1, Q_2, and ρ_2.

At the upper interval z from the mouth to the cross section of the tubing where the pressure is equal to p^*, average pressure is

$$p_{av} = \frac{p_m + p^*}{2} = \frac{10 \times 10^5 + 24 \times 10^5}{2} = 17 \times 10^5 \text{ Pa}.$$

The pressure in fluid is

$$\rho_2 = 845 - 5.62 \times 10^{-6} \times 1.7 \times 10^6 = 835 \text{ kg/m}^3.$$

Flow rates of gas and fluid are

$$Q_1 = \frac{8529 - 889 \times 10^{-6}(1.7 \times 10^6 - 0.1 \times 10^6)}{1.7 \times 10^6} = 0.00418 \text{ m}^3/\text{s};$$

$$Q_2 = \frac{1.41 + 1.148 \times 10^{-8}(1.7 \times 10^6 - 0.1 \times 10^6)}{845 - 5.62 \times 10^{-6} \times 1.7 \times 10^6} = 0.00171 \text{ m}^3/\text{s}.$$

Volumetric flow rate of free gas at $p_{av} = 1.7$ MPa at normal conditions is

$$Q_0 = \frac{Q_1 p_{av}}{p_0} = \frac{0.00418 \times 1.7 \times 10^6}{0.1 \times 10^6} = 0.07106 \text{ m}^3/\text{s}.$$

With the formula (9.3.3), calculate the length z_1 of the upper interval. Determine previously factors

$$k^2 = \frac{Q_2^2}{gdS_t^2} = \frac{0.0171^2}{9.81 \times 0.062(0.785 \times 0.062^2)^2} = 0.5285;$$

$$k = 0.727; \qquad a = Q_0/Q_2 = 0.07106/0.00171 = 41.6;$$

$$\eta = a \cdot \rho_0/\rho_2 = 41.6 \times 1.3/835 = 0.0648;$$

$$A = 1 + 0.81\eta + \frac{\lambda_c k^2}{2}\frac{\eta + 0.81}{0.81} = 1 + 0.81 \times 0.0648$$

$$+ \frac{0.05 \times 0.5285}{2}\frac{0.0648 + 0.81}{0.81} = 1.067;$$

$$B = 0.19 + \frac{\lambda_c k^2}{2}\left[1 + \frac{\eta}{\eta + 0.81}\right] = 0.19 + \frac{0.05 \times 0.5285}{2}\left[1 + \frac{0.0648}{0.0648 + 0.81}\right] = 0.204.$$

Then,

$$z_1 = \frac{1}{\rho_2 g A}\left[p^* - p_{\mathrm{m}} + p_0 \frac{\bar{z}\bar{T}}{z_0 T_0} a \frac{A-B}{A} \ln \frac{p^* + p_0 a \dfrac{B}{A}}{p_y + p_0 a \dfrac{B}{A}}\right]$$

$$= \frac{1}{835 \times 9.81 \times 1.067}\left[2.4 \times 10^6 - 10^6 + 10^5 \times \frac{1 \times 295}{1 \times 293} 41.6\right.$$

$$\left. \times \frac{1.067 - 0.204}{1.067} \ln \frac{2.4 \times 10^6 + 10^5 \times 41.6 \dfrac{0.204}{1.067}}{10^6 + 10^5 \times 41.6 \dfrac{0.204}{1.067}}\right] = 383.4\ \mathrm{m}.$$

We calculate the length of the lower interval z_2, with the formula (9.3.4). Previously let us obtain average values of p_{av}, ρ_2, Q_1, Q_2, Q_0, and needed factors

$$p_{\mathrm{av}} = \frac{p^* + p_{\mathrm{bot}}}{2} = \frac{24 \times 10^5 + 97 \times 10^5}{2} = 60.5 \times 10^5\ \mathrm{Pa};$$

$$\rho_2 = 845 - 5.62 \times 10^{-6} \times 6.05 \times 10^6 = 811\ \mathrm{kg/m^3};$$

$$Q_1 = \frac{8529 - 889 \times 10^{-6}(6.05 \times 10^6 - 0.1 \times 10^6)}{6.05 \times 10^6} = 0.000535\ \mathrm{m^3/s};$$

$$Q_2 = \frac{1.41 + 1.148 \times 10^{-8}(6.05 \times 10^6 - 0.1 \times 10^6)}{811} = 0.00182\ \mathrm{m^3/s};$$

$$k^2 = \frac{Q_2^2}{g d S_{\mathrm{t}}^2} = \frac{0.00182^2}{9.81 \times 0.062(0.785 \times 0.062^2)^2} = 0.5981;$$

$$k = 0.7734; \qquad a = Q_0/Q_2 = 0.032/0.00182 = 17.6;$$

$$\eta = a \cdot \rho_0/\rho_2 = 17.6 \times 1.3/811 = 0.0282;$$

$$n = 1 + \frac{1}{2.2k} = 1 + \frac{1}{2.2 \times 0.7734} = 1.588;$$

$$m = \frac{\lambda_{\mathrm{c}}}{2} k^2 \frac{\eta n + 1}{n} = \frac{0.05}{2} 0.7734^2 \frac{0.0282 \times 1.588 + 1}{1.588} = 0.00984;$$

$$B_1 = 2mn = 2 \times 0.00984 \times 1.588 = 0.0313;$$

$$A_1 = n + \eta + n^2 m = 1.588 + 0.0282 + 1.588^2 \times 0.00984 = 1.64;$$

$$
\begin{aligned}
z_2 &= \frac{n}{\rho_2 g A_1}\left[p_{sat} - p^* + p_0 \frac{\bar{z}\bar{T}}{z_0 T_0} a \frac{A_1 - nB_1}{A_1 n} \ln \frac{p_{sat}}{p^*}\right] \\
&= \frac{1.588}{811 \times 9.81 \times 1.64}\left[9.7 \times 10^6 - 2.4 \times 10^6 + 10^5 \frac{1 \times 295}{1 \times 293} 17.6\right. \\
&\quad \left. \times \frac{1.64 - 1.588 \times 0.0313}{1.64 \times 1.588} \ln \frac{9.7 \times 10^6}{2.4 \times 10^6}\right] = 1072\,\text{m}.
\end{aligned}
$$

The length of the section along which flows oil with fully dissolved gas is

$$L - z_1 - z_2 = 1600 - 383.4 - 1072 = 144.6\,\text{m}.$$

Hence, the calculated pressure of fluid is

$$p = p_{sat} + (L - z_1 - z_2)\rho_2 g = 97 \times 10^5 + 144.6 \times 790 \times 9.81 = 108.2 \times 10^5\ \text{Pa}.$$

The measured pressure at the depth 1600 m is 10.89 MPa (Reference Book on Designing, Development and Exploitation of Oil Fields, Gimatudinov, 1983), which is greater only on 0.7×10^5 Pa or on 0.64% than that of the calculated one.

CHAPTER 10

NONSTATIONARY FLOWS OF SINGLE-PHASE FLUIDS IN A WELL

Under nonstationary flow of a fluid in the well, we shall understand such flows the characteristics of which depend on time. In this chapter, we will consider nonstationary flows of single-phase fluids in pipes and annular channels occurring in main technological operations, in particular in carrying down- and up-pipes, in restoring fluid circulation in a well, and in seating a ball (ball cage) on the saddle (baffle collar) in the drill-stem.

10.1 EQUATIONS OF NONSTATIONARY SINGLE-PHASE FLOWS

Equations for nonstationary flows are determined from equations (4.6.1)–(4.6.4) by substitution of $N = 1$ corresponding to a single-phase flow:

momentum equation without regard for gravity force

$$\frac{\partial p}{\partial z} = -\rho \frac{\mathrm{d}w}{\mathrm{d}t} + \frac{1}{r}\frac{\partial r\tau}{\partial r}; \qquad (10.1.1)$$

equation of mass conservation

$$\frac{\partial \rho}{\partial t} + \frac{\partial \rho w}{\partial z} = 0; \qquad (10.1.2)$$

equation of state

$$p = p(\rho, \bar{T}); \tag{10.1.3}$$

rheological equation

$$\tau = \tau\left(\frac{\partial w}{\partial r}\right). \tag{10.1.4}$$

System of averaged equations can be obtained from (4.6.33)–(4.6.39) by taking $\varphi = 1$ and omitting gravity force:

momentum equation

$$\frac{\partial p}{\partial z} = -\rho \frac{\mathrm{d}v}{\mathrm{d}t} + \frac{\lambda}{2d}\rho v^2; \tag{10.1.5}$$

equation of mass conservation

$$\frac{\partial \rho}{\partial t} + \frac{\partial \rho v}{\partial z} = 0; \tag{10.1.6}$$

equation of state

$$p = p(\rho, \bar{T}); \tag{10.1.7}$$

equation for hydraulic resistance factor

$$\lambda = \lambda(v, d, \rho, \mu). \tag{10.1.8}$$

In the case of incompressible fluid, equations (10.1.3) and (10.1.7) reduce to $\rho = \text{const}$ and equations (10.1.1)–(10.1.8) are simplified. In Sections 10.2–10.5, some solutions of the system (10.1.1)–(10.1.4) and their application to the flow of incompressible fluid in round trip operations will be considered. To account for compressibility ($\rho \neq \text{const}$), let us transform equations (10.1.5)–(10.1.8).

Differentiating pressure p with respect to time

$$\frac{\partial p}{\partial t} = \frac{\partial p}{\partial \rho}\frac{\partial \rho}{\partial t} \tag{10.1.9}$$

and denoting $\partial p/\partial \rho = c^2$, we get from equation (10.1.7)

$$\frac{\partial \rho}{\partial t} = \frac{1}{c^2}\frac{\partial p}{\partial t}. \tag{10.1.10}$$

Then, the system of equations (10.1.5)–(10.1.8) with regard to (10.1.10) takes the form

$$\frac{\partial p}{\partial z} = -\rho \frac{\mathrm{d}v}{\mathrm{d}t} + \frac{\lambda}{2d}\rho v^2; \tag{10.1.11}$$

$$\frac{\partial p}{\partial t} = -c^2 \frac{\partial \rho v}{\partial z}; \qquad (10.1.12)$$

$$p = p(\rho, \bar{T}); \qquad (10.1.13)$$

$$\lambda = \lambda(v, d, \rho, \mu). \qquad (10.1.14)$$

For given equation (10.1.13), one can find from (10.1.9) the velocity of wave propagation. The density ρ may be expressed through specific volume V

$$\rho = 1/V. \qquad (10.1.15)$$

Then, the density increment is

$$\delta\rho = -\delta V/V^2. \qquad (10.1.16)$$

Since the relative increment of specific volume $\delta V/V$ in one-dimensional flow coincides with relative deformation $\delta\varepsilon$

$$\delta\varepsilon = -\delta V/V, \qquad (10.1.17)$$

the only normal stress σ coincides with p

$$\sigma = p. \qquad (10.1.18)$$

Using (10.1.16)–(10.1.18), we find

$$c^2 = \frac{\partial p}{\partial \rho} = \frac{\delta\sigma}{-\delta V/V^2} = \frac{\delta\sigma}{\delta\varepsilon/V} = \frac{\delta\sigma}{\delta\varepsilon}\frac{1}{\rho}. \qquad (10.1.19)$$

If the flow obeys Hook law, then linear relation $\sigma = E\varepsilon$ is satisfied, where E is elastic modulus of fluid ($E = 2.1 \times 10^3$ MPa for water and can be increased up to 2.62×10^3 MPa for mud solutions). Then, from (10.1.19) it follows

$$c = \sqrt{E/\rho}. \qquad (10.1.20)$$

In getting c in the form (10.1.20), the deformability of walls was not taken into account. The effect of walls on the velocity c is taken into account by the formula

$$c = \sqrt{E''/\rho}, \qquad (10.1.21)$$

where $E'' = \frac{1}{\frac{1}{E} + \frac{d_h}{\Delta}\frac{1}{E'}}$, E' is elastic modulus of the pipeline material (for steel $E' = 2.1 \times 10^5$ MPa); d_h and Δ are hydraulic diameter and wall depth of the pipeline.

Consider the inertial component of equation (10.1.11). In accordance with expression (4.2.17) for total derivative, it can be written as

$$\rho\frac{\mathrm{d}v}{\mathrm{d}t} = \rho\left(\frac{\partial v}{\partial t} + v\frac{\partial v}{\partial z}\right) = \rho\frac{\partial v}{\partial t} + \rho v\frac{\partial v}{\partial z} + v\frac{\partial \rho}{\partial t} - v\frac{\partial \rho}{\partial t} = \frac{\partial \rho v}{\partial t} + v\left(\rho\frac{\partial v}{\partial z} - \frac{\partial \rho}{\partial t}\right). \tag{10.1.22}$$

If in (10.1.11) we neglect the convective component $v(\partial v/\partial z)$, in (10.1.12) ignore $v(\partial \rho/\partial z)$, and assume $\lambda = 0$, we get Zhukowski equations (Zhukowski, 1948)

$$\frac{\partial p}{\partial z} = -\rho\frac{\partial v}{\partial t}; \tag{10.1.23}$$

$$\frac{\partial p}{\partial t} = -\rho c^2\frac{\partial v}{\partial z}. \tag{10.1.24}$$

When the second term in (10.1.22) is negligibly small, then substitution of (10.1.22) in (10.1.11) gives Charniy equations (Charniy, 1975)

$$\frac{\partial p}{\partial z} = -\frac{\partial \rho v}{\partial t} + \frac{\lambda}{2d}\rho v^2; \tag{10.1.25}$$

$$\frac{\partial p}{\partial t} = -c^2\frac{\partial \rho v}{\partial z}. \tag{10.1.26}$$

In Sections 10.6–10.9, some problems on the flow of compressible fluid at $\lambda = 0$ will be considered and approximate method taking into account the influence of friction term ($\lambda \neq 0$) on the basis of equations (10.1.25)–(10.1.26) will be shown.

10.2 NONSTATIONARY FLOWS OF INCOMPRESSIBLE FLUID IN ROUND TRIP OPERATIONS

Round trip operations of drill-stems in different technological processes of well building frequently cause complications such as hydraulic fracturing of well walls, absorption of washing fluid, cavings, shows, and others.

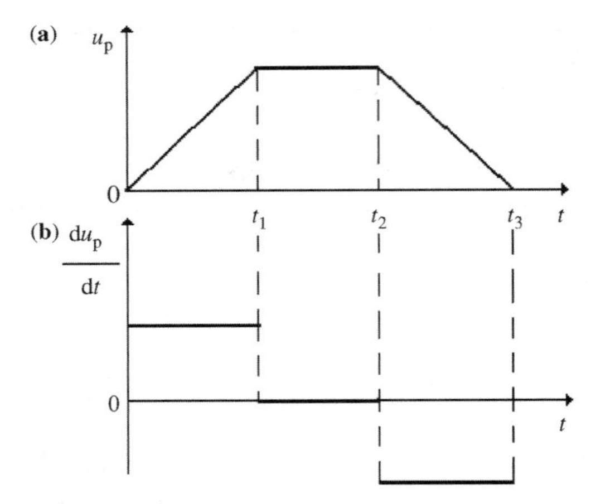

FIGURE 10.1 Schematic graphics of velocity (a) and acceleration (b) of drill-stem.

Sometimes, these complications occur owing to impermissible hydrodynamic pressures caused by motion of drill-stems in fluid filling the well (Gukasov, 1976; Sereda and Solov'ev, 1974; Shischenko et al., 1976).

During round trip operation, the drill-stem moves commonly with variable velocity u_{ds}. Graphics of the velocity and the acceleration of the drill-stem with time are schematically shown in Fig. 10.1. The range 0–t_1 corresponds to drill-stem acceleration, t_1–t_2 to drill-stem motion with constant velocity, t_2–t_3 to drill-stem braking.

Round trip operations are conducted with open or closed lower drill-stem end. At this, in calculating hydrodynamic pressure distribution in pipe and annulus, the washing fluid is frequently assumed to be incompressible. For incompressible fluid, equation (10.1.3) takes the form

$$\rho = \text{const.} \tag{10.2.1}$$

Then, from (10.1.6) it follows

$$\partial v/\partial z = 0 \tag{10.2.2}$$

and in accordance with (10.1.22), the total derivative $\mathrm{d}v/\mathrm{d}t$ in equation (10.1.5) could be replaced with partial one $\partial v/\partial t$.

Expressions (10.1.4) for stresses τ depend on rheological model of the fluid.

The system of equations (10.1.1)–(10.1.4) with regard to conditions (10.2.2) and (10.2.1), added by relation for τ and appropriate initial and boundary conditions, describes nonstationary flow of incompressible fluid in a well (Leonov and Isaev, 1980a).

To analyze the pressure distribution in the well and its influence on the formation of hydrodynamic parameters of fluid flow, we rewrite the momentum equation (10.1.1) with regard to (10.2.2) as follows

$$-\frac{\partial p}{\partial z} = \rho\frac{\partial w}{\partial t} - \frac{1}{r}\frac{\partial r\tau}{\partial r}. \qquad (10.2.3)$$

The first term in the right part of equation (10.2.3) expresses the rate of momentum change in a unit volume, and for incompressible fluid, it is equal to the product of density and acceleration. It is known as inertial component of pressure gradient as distinct from the second term that in stationary flows determines the pressure distribution and reflects interaction of external forces and friction force between fluid layers.

Initial and boundary conditions for velocities in pipes and annular space are

$$\begin{aligned} &w = w(r) && \text{at} \quad t = 0; \\ &w = \pm u_{\text{p}}(t) && \text{at} \quad r = R_1,\ t > 0; \\ &w = 0 && \text{at} \quad r = R_2,\ t \geq 0; \\ &w = \pm u_{\text{p}}(t) && \text{at} \quad r = R_0,\ t > 0, \end{aligned} \qquad (10.2.4)$$

where $w(r)$ is the velocity of stationary flow caused by pump delivery before beginning of the round trip operation; $u_{\text{p}}(t)$ the velocity of pipe motion.

Consider the problem on displacement of the drill-stem in the well. Fig. 10.2a and b shows typical velocity profiles of laminar and turbulent flows in annular space and Fig. 10.3 shows the possible velocity profiles inside the moving pipe.

In accordance with (10.2.2), the velocity of fluid flow $w = w\ (r,\ t)$ is a function only of radius and time. Therefore, the flow rate

$$q = q(t) = 2\pi\int_{R_1}^{R_2} w(r,t)r\,\mathrm{d}r \qquad (10.2.5)$$

should be equal to

$$q = q_{\text{pd}} \pm q_{\text{fd}}, \qquad (10.2.6)$$

where q_{pd} is pump delivery and q_{fd} the flow rate of the fluid displaced by the lower end of the drill-stem in its descent or filled the space below the drill-stem at its ascent.

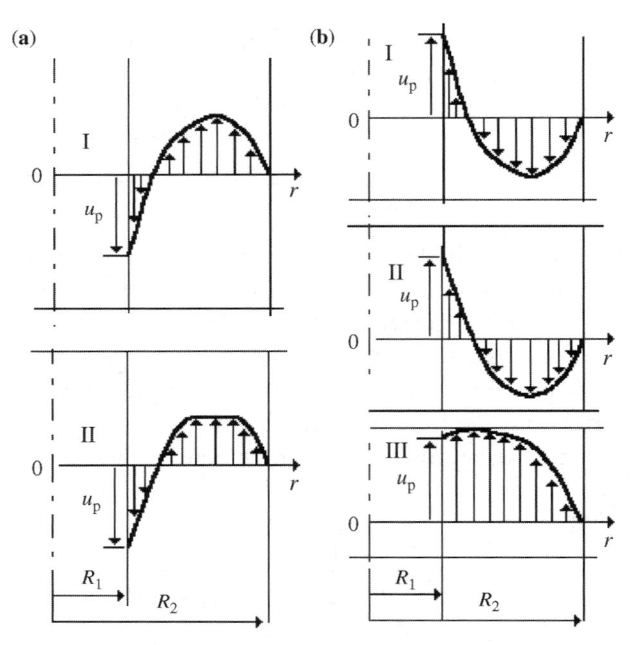

FIGURE 10.2 Typical velocity profiles in annular space in carrying down (a) and up (b) pipes: I—turbulent flow; II—laminar flow; III—possible velocity profile at $q_{as} > q_1$.

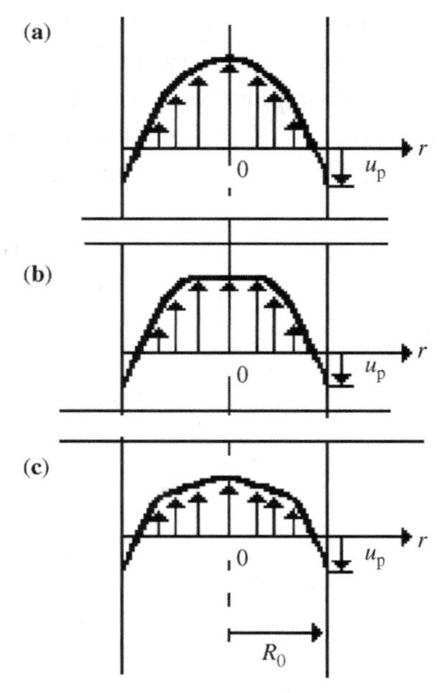

FIGURE 10.3 Velocity profiles inside moving pipes in carrying down: (a) for viscous fluid; (b) for viscous-plastic fluid; and (c) for turbulent flow.

10.3 HYDRODYNAMIC PRESSURE IN ROUND TRIP OPERATION IN A WELL FILLED BY VISCOUS FLUID

Consider a flow of viscous fluid in annular space in descending drill-stem with closed lower end at $t_1 \leq t \leq t_2$ (see Fig. 10.1). Formulas obtained will also be true for the calculation of pressure distribution in drill-stem ascent. The motion of the drill-stem during time $t_2 - t_1$ is stationary. Assuming geometrical sizes of the drill-stem to be invariable, one should take in (10.2.3) $\mathrm{d}w/\mathrm{d}t = 0$. Then,

$$\frac{\partial p}{\partial z} = \frac{1}{r}\frac{\partial r\tau}{\partial r}.$$

Solution of this equation with regard to dependence of τ for viscous fluid yields

$$w = \frac{A}{4\mu}r^2 + b\ln r + c, \qquad (10.3.1)$$

where

$$A = \frac{\partial p}{\partial z} = -\frac{\Delta p}{L}, \qquad (10.3.2)$$

L is the distance considered well cross section measured from the fluid surface at the well mouth, and z-axis is directed upward.

Boundary conditions (10.2.4) for the flow in this section are as follows:

$$w = 0 \quad \text{at} \quad r = R_2; \qquad (10.3.3)$$

$$w = -u_\mathrm{p} \quad \text{at} \quad r = R_1. \qquad (10.3.4)$$

Determining coefficients b and c in (10.3.1) and (10.3.2) with the help of boundary conditions (10.3.3) and (10.3.4), we get velocity distribution of the flow of viscous fluid in annular channel (Targ, 1951)

$$w = \frac{A}{4\mu}(r^2 - R_2^2) - \frac{u_\mathrm{p} + \frac{A}{4\mu}(R_1^2 - R_2^2)}{\ln(R_1/R_2)}\ln\frac{r}{R_2}. \qquad (10.3.5)$$

From (10.3.5) at $u_\mathrm{p} = 0$ ensues Boussinesq formula for viscous fluid flow in the annular space under the action of pressure drop Δp. At $\Delta p = 0$ (i.e., at $A = 0$), it gives the velocity profile of the flow in the infinite pipe.

Integration of (10.3.5) from R_1 to R_2 yields fluid flow rate in the annular channel

$$q = 2\pi\int_{R_1}^{R_2} wr\,\mathrm{d}r = \frac{\pi}{4\mu}\left\{A\left[\frac{R_2^4-R_1^4}{2}+R_2^2(R_1^2-R_2^2)\right]\right.$$
$$\left.-\frac{4\mu u_{\mathrm{p}}+A(R_1^2-R_2^2)}{\ln(R_1/R_2)}\left[\frac{R_1^2}{2}-\frac{R_2^2}{2}-R_1^2\ln\frac{R_1}{R_2}\right]\right\}. \quad (10.3.6)$$

The flow rate represented by formula (10.3.6) corresponds to the flow with given pressure gradient and velocity of drill-stem descent. In this case from (10.2.6) at $q_{\mathrm{H}}=0$, one gets

$$q = q_{\mathrm{fd}}. \quad (10.3.7)$$

The flow q_{fd} is determined by the formula

$$q_{\mathrm{fd}} = \pi u_{\mathrm{p}} R_1^2. \quad (10.3.8)$$

Substitution of (10.3.8) and (10.3.6) in (10.3.7) gives

$$\pi u_{\mathrm{p}} R_1^2 = \pi\frac{A}{4\mu}\left[\frac{R_2^4-R_1^4}{2}+R_2^2(R_1^2-R_2^2)\right]$$
$$-\pi\frac{4\mu u_{\mathrm{p}}+A(R_1^2-R_2^2)}{4\mu\ln(R_1/R_2)}\left[\frac{R_1^2-R_2^2}{2}-R_1^2\ln\frac{R_1}{R_2}\right].$$

Resolving this relation with respect to Δp and taking into account (10.3.2), we obtain

$$\Delta p = \frac{4\mu}{R_2^2}L\frac{u_{\mathrm{p}}}{-\left[1+\left(\frac{R_1}{R_2}\right)^2\right]\ln\frac{R_1}{R_2}-1+\left(\frac{R_1}{R_2}\right)^2}. \quad (10.3.9)$$

Now, introduce mean velocity

$$v_{\mathrm{av}} = \frac{q}{S_{\mathrm{as}}} = \frac{q_{\mathrm{fd}}}{\pi(R_2^2-R_1^2)} = u_{\mathrm{p}}\frac{(R_1/R_2)^2}{1-(R_1/R_2)^2} = u_{\mathrm{p}}\frac{\delta^2}{1-\delta^2} \quad (10.3.10)$$

and rewrite (10.3.9) in the form

$$\Delta p = \frac{4\mu}{R_2^2}\frac{1-\delta^2}{\delta^2\left[-(1+\delta^2)\ln\delta-1+\delta^2\right]}v_{\mathrm{av}}L \quad (10.3.11)$$

or

$$\Delta p = \lambda \frac{\rho v_{\mathrm{av}}^2}{2d_{\mathrm{h}}} L, \tag{10.3.12}$$

where

$$\lambda = \frac{64}{\mathrm{Re}} f(\delta); \quad \mathrm{Re} = \frac{\rho v_{\mathrm{av}} d_{\mathrm{h}}}{\mu};$$

$$f(\delta) = \frac{(1-\delta^2)(1-\delta)^2}{2\delta^2[-(1+\delta^2)\ln\delta - 1 + \delta^2]}; \tag{10.3.13}$$

$$v_{\mathrm{av}} = \psi(\delta) u_{\mathrm{p}}; \qquad \psi(\delta) = \delta^2/(1-\delta^2); \quad \delta = R_1/R_2.$$

Data for hydraulic resistance factor λ are given in Grachev et al. (1980). Formula (10.3.12) in designations of Grachev et al. (1980) is

$$\Delta p = \lambda \frac{\rho u_{\mathrm{p}}^2}{2d_{\mathrm{h}}} L, \tag{10.3.14}$$

where

$$\lambda = \frac{64}{\mathrm{Re}_{\mathrm{p}}} f(\delta); \quad \mathrm{Re}_{\mathrm{p}} = \frac{\rho u_{\mathrm{ds}} d_{\mathrm{h}}}{\mu};$$

$$f(\delta) = \frac{(1-\delta)^2}{2[-(1+\delta^2)\ln\delta - 1 + \delta^2]} = \frac{k(\delta)}{1-\delta}.$$

The graphic of the function $k(\delta)$ is shown in Fig. 10.4. For values $0.4 \leq \delta \leq 1$ occurring in drilling practice, the function $k(\delta)$ may be represented as

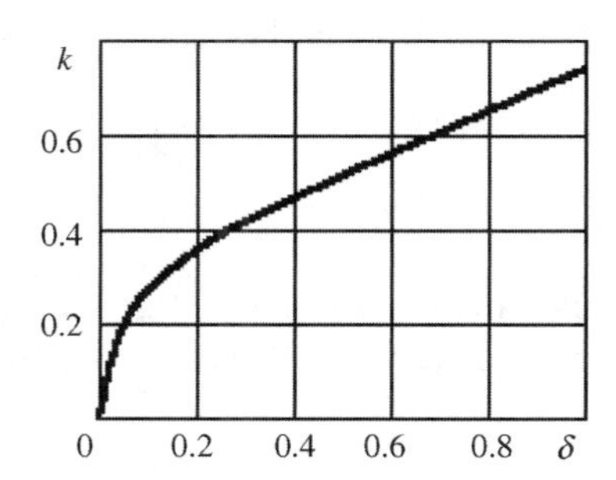

FIGURE 10.4 Graphic of function $k(\delta)$ for laminar flow.

$k(\delta) = 0.45\delta + 0.3$. Then, λ in (10.3.14) can be calculated by simplified formula

$$\lambda = \frac{64}{\text{Re}_\text{p}} \frac{0.45\delta + 0.3}{1-\delta}.$$

Results of comparison of calculations of the quantity λRe_p obtained from the dependence (10.3.14) and experimental formula from Grachev et al. (1980)

$$\lambda = \frac{58}{\text{Re}_\text{p}} \frac{\delta}{1-\delta} \tag{10.3.15}$$

are tabulated below

δ	0.552	0.684	0.738	0.79	0.84
λRe_p from					
(10.3.14)	78	123	154	200	271
(10.3.15)	71	125	164	218	304
Relative discrepancy (%)	9.4	1.5	6.1	8.25	10

As seen from the data, the convergence of calculation is satisfactory and formula (10.3.14) may be used to calculate hydraulic losses due to friction in laminar flow in round trip operations of drill-stems with closed end.

In round trip operations of drill-stems with lower and upper open ends, one can write the formula similar to (10.3.9) accounting for fluid flow in pipes (Leonov and Isaev, 1982). If the pipe moves with the velocity u_T, the velocity distribution inside the pipe is also expressed by formula (10.3.1).

Boundary conditions for the flow in the pipe in the drill-stem descent are

$$w \neq \infty \quad \text{at} \quad r = 0;$$

$$w = -u_\text{p} \quad \text{at} \quad r = R_0.$$

Using these conditions instead of (10.3.1) gives the following velocity distribution in pipes in their movement with velocity u_p:

$$w = \frac{A}{4\mu}(r^2 - R_0^2) - u_\text{p}.$$

Integration of this expression over radius in limits from 0 to R_0 yields the flow rate in pipes

$$q_\mathrm{p} = -\pi\left(\frac{AR_0^4}{8\mu} + u_\mathrm{p}R_0^2\right). \tag{10.3.16}$$

The flow rate in the annulus q_as in drill-stem motion with the same velocity is determined by formula (10.3.6).

The mass conservation law analogous to (10.2.6) for flows in pipes and annular space at $q_\mathrm{pd}=0$ takes the form

$$q = q_\mathrm{fd} = q_\mathrm{p} + q_\mathrm{as}, \tag{10.3.17}$$

where $q_\mathrm{fd} = \pi u_\mathrm{p}(R_1^2 - R_0^2)$.

Inserting relations (10.3.16) and (10.3.6) in (10.3.17) and resolving them with respect to Δp being identical in pipes and annulus, we obtain

$$\Delta p = \frac{4\mu}{R_2^2} L \frac{u_\mathrm{p}}{\left[-(1+\delta^2)\ln\delta - 1 + \delta^2 - \frac{\delta_0^4}{1-\delta^2}\ln\delta\right]}, \tag{10.3.18}$$

where $\delta_0 = R_0/R_2$.

Setting $R_0=0$ in (10.3.18), we get (10.3.9). After comparing (10.3.18) and (10.3.9), one can conclude that the quantity Δp at invariable R_1 and R_2 has lesser values in the descent of the drill-stem with open lower and upper ends.

In descent of the drill-stem with operating pumps, equation (10.2.6) has the form

$$q = q_\mathrm{pd} + \pi u_\mathrm{p} R_1^2. \tag{10.3.19}$$

Equating (10.3.19) to (10.3.6), we receive a formula for pressure drop in the annular space

$$\Delta p = \frac{4\mu}{R_2^2} L \frac{u_\mathrm{p} - \frac{2q_\mathrm{pd}}{\pi(R_2^2 - R_1^2)}\ln\delta}{[-(1+\delta^2)\ln\delta - 1 + \delta^2]}. \tag{10.3.20}$$

Above, laminar flows of viscous fluid in pipes and annular channels in motion of the drill-stem with constant velocity were considered. There are possible cases of not only turbulent flow in pipes but also laminar flow in annular space and the inverse.

For the purpose of calculations, let us consider the change of flow characterized by critical Re_{cr}. Turbulent flow regime of viscous fluid begins at $Re_{cr} = 2100$ calculated at average velocity $v_{av} = q/S$

$$\mathrm{Re}_{cr} = \frac{\rho v_{av} d_h}{\mu}. \tag{10.3.21}$$

Pass now from Re_{cr} to Re'_{cr} calculated with the velocity $u_p \neq 0$ and $q_{pd} = 0$

$$Re'_{cr} = \frac{\rho u_p d_h}{\mu} = \frac{\rho v_{av} d_h}{\mu} \frac{u_p}{v_{av}} = \mathrm{Re}_{cr} \frac{u_p}{v_{av}}. \tag{10.3.22}$$

One should insert in (10.3.22) v_{av} with (10.3.10) and $Re_{av} = 2100$

$$Re'_{cr} = 2100 \frac{1 - \delta^2}{\delta^2}. \tag{10.3.23}$$

In the range $0.55 \leq \delta \leq 0.84$, approximate relation (Grachev et al., 1980) corresponds to this formula

$$Re'_{cr} \approx 5500 \frac{R_2 - R_1}{R_1}. \tag{10.3.24}$$

Thus, the demarcation line of flow regimes is determined by formula (10.3.23) or (10.3.24).

Consider a problem on turbulent fluid flow in an annular channel descending with constant velocity of a drill-stem with closed lower end. Similar to (10.3.5), the velocity of the turbulent flow in the annular channel is assumed approximately equal to

$$\begin{aligned} w &= f(A, r, R_1, R_2) + \varphi(u_p, r, R_1, R_2) \\ &= f(A, r, R_1, R_2) - u_p \left(\frac{R_2 - r}{R_2 - R_1} \right)^n, \end{aligned} \tag{10.3.25}$$

where the first term is the velocity in the annular channel with immovable walls ($u_p = 0$) and the second one is the velocity of motion of infinitely long drill-stem ($\Delta p = 0$). The function φ in (10.3.25) is represented in the form of power dependence. Since expressions (10.2.6) and (10.3.8) are conserved for turbulent flow, we get

$$q = q_{fd} = \pi u_p R_1^2 = q_2 + q_1, \tag{10.3.26}$$

where

$$q_2 = 2\pi \int_{R_1}^{R} f(A, r, R_1, R_2) r \, dr;$$
$$q_1 = -2\pi \int_{R_1}^{R_2} u_p \left(\frac{R_2 - r}{R_2 - R_1} \right)^n r \, dr. \quad (10.3.27)$$

Then, from (10.3.26) it is obtained

$$q_2 = q_{fd} - q_1 = \pi(R_2^2 - R_1^2) u_p \left[\frac{\delta^2}{1-\delta^2} + \frac{2}{1+\delta} \left(\frac{1}{n+1} - \frac{1-\delta}{n+2} \right) \right]. \quad (10.3.28)$$

For practical calculations in (10.3.28) it is taken $n = 1$.

Equation (10.3.26) cannot be resolved with respect to Δp without knowing the form of the function $f(A, r, R_1, R_2)$ in (10.3.25), that is, the velocity profile in the annular channel at $u_p = 0$ and its connection with the pressure drop in turbulent fluid. In accordance with results obtained in Section 6.5 the function $f(A, r, R_1, R_2)$ in (10.3.27) should be taken in the form of (6.5.41) and (6.5.42), whereas q_2 is equal to the flow rate (10.3.28). Equating expressions for q_2 written in the form of (10.3.28) and (6.5.52), we get

$$\Delta p = f_1(\delta, N, n) \frac{u_p^{2N/(N+1)} \nu^{2/(N+1)}}{2 d_h^{(N+3)/(N+1)}} \rho L \quad (10.3.29)$$

or

$$\Delta p = \lambda \frac{\rho u_p^2}{2 d_h} L, \quad (10.3.30)$$

where

$$\lambda = f_1(\delta, N, n) / \mathrm{Re}_p^{2/(N+1)}; \quad (10.3.31)$$

$$f_1(\delta, N, n) = f(\delta, N) \left[\frac{\delta^2}{1-\delta^2} + \frac{2}{1+\delta} \left(\frac{1}{n+1} - \frac{1-\delta}{n+2} \right) \right]^{2N/(N+1)} \quad (10.3.32)$$

and the function $f(\delta, N)$ is given by formula (6.5.54).

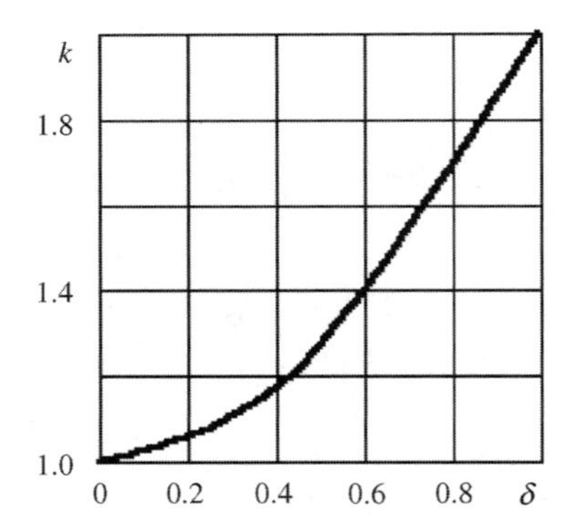

FIGURE 10.5 Graphic of the function $k(\delta)$ for turbulent flow.

At $n = 1$ and $N = 7$, we have

$$\lambda = \frac{f_1(\delta, 7, 1)}{\mathrm{Re}_{\mathrm{p}}^{0.25}}, \tag{10.3.33}$$

where

$$f_1(\delta, 7, 1) = 0.0488 \frac{k(\delta)}{(1-\delta)^{1.75}},$$
$$k(\delta) = \left(\frac{1+\delta+\delta^2}{1+\delta}\right)^{1.75}. \tag{10.3.34}$$

The graphic of $k(\delta)$ is shown in Fig. 10.5.

Thus, formula (10.3.29) makes it possible to calculate losses due to friction. Formula (10.3.29) may be rewritten in the form of (10.3.12) in which v_{av} will be the same whereas the factor λ is changed. From (10.3.33) at $N = 7$ and δ taken from $0.55 \leq \delta \leq 0.85$ corresponding experimental data of Grachev et al. (1980), we get approximate formula

$$\lambda = \frac{0.175}{\mathrm{Re}_{\mathrm{p}}^{0.25}} \left(\frac{d}{d_{\mathrm{h}}}\right)^{1.5}. \tag{10.3.35}$$

The expression for λ in (10.3.35) coincides with λ obtained by Grachev et al. (1980). Results of calculations with the formula (10.3.33) at $N = 7$ for hydraulic smooth pipes and with the formula (10.3.35) are exhibited below. It is seen that the discrepancy of results is insignificant.

δ	0.552	0.684	0.790	0.840
$\lambda \mathrm{Re}_{\mathrm{p}}^{0,25}$ calculated with formula				
(10.3.33)	0.259	0.555	1.260	2.130
(10.3.35)	0.239	0.555	1.280	2.100
Relative discrepancy (%)	2.3	0.0	1.6	1.4

If a drill-stem with open end is descended, one should also take into account fluid motion inside pipes while calculating pressure drop. The preceding calculation method of pressure in round trip operations of drill-stem with columns of identical sizes can be extended to round trip operations of drill-stem with compound columns. In the case of different flow regimes in different sections of the compound pipe, the pressure drop Δp should be calculated successively.

Determine now the inertial component of the pressure in drill-stem descent described by the first term in the right part of equation (10.2.3).

Equation (10.2.3) of nonstationary single-phase flow averaged over the cross section of the annular channel is determined from equation (4.6.33) at $\varphi = 1$ with regard to

$$\frac{\partial p}{\partial z} = -\rho \frac{\partial v_{\mathrm{av}}}{\partial t} + \frac{\lambda \rho}{2 d_{\mathrm{h}}} v_{\mathrm{av}}^2. \tag{10.3.36}$$

Since the right part of the last equation is independent of z, the left part can be written as

$$\frac{\partial p}{\partial z} = \mathrm{const}(t). \tag{10.3.37}$$

Integration of (10.3.36) gives pressure regardless of the hydrostatics at given depth of the annular channel

$$p = \pm \Delta p_{\mathrm{in}} + \Delta p_{\mathrm{fr}} + p_{\mathrm{m}}, \tag{10.3.38}$$

where
$\Delta p_{\mathrm{in}} = \rho \left| \frac{\partial v_{\mathrm{av}}}{\partial t} \right| L$, $\Delta p_{\mathrm{mp}} = \frac{\lambda \rho}{2 d_{\mathrm{h}}} v_{\mathrm{av}}^2 L$, and p_{m} is the pressure at the well mouth. In formula (10.3.38), the sign plus at Δp_{in} is taken in the drill-stem acceleration and the sign minus in its deceleration. In the calculation of $p(t)$ with formula (10.3.38), formulas (10.3.12), (10.3.18), and (10.3.29) can be used for Δp_{fr} depending on flow conditions and (10.3.10) for v_{av} for all kinds of flow. In these formulas, it is assumed that $u_{\mathrm{p}} = u_{\mathrm{p}}(t)$, that is, the velocity of the drill-stem, depends on time. The same approach will be used further in considering flows with other rheological properties.

Find now $\partial v_{av}/\partial t$ for all cases. Differentiation of (10.3.10) gives

$$\frac{\partial v_{av}}{\partial t} = \frac{\delta^2}{1-\delta^2}\frac{\mathrm{d}u_p}{\mathrm{d}t} = \psi(\delta)\frac{\mathrm{d}u_p}{\mathrm{d}t}. \tag{10.3.39}$$

Relations (10.3.8) and (10.3.39) do not depend on the flow regime and rheologic properties of the washing fluid.

Consider the term Δp_{in} in (10.3.38)

$$\Delta p_{in} = \rho\left|\frac{\partial v_{av}}{\partial t}\right|L. \tag{10.3.40}$$

With (10.3.39) it can be rewritten as

$$\frac{\Delta p_{in}}{\rho L\left|\frac{\mathrm{d}u_p}{\mathrm{d}t}\right|} = \psi(\delta) = \frac{\delta^2}{1-\delta^2}. \tag{10.3.41}$$

In a similar manner, the empirical formula (8) from Grachev et al. (1980) may be represented

$$\frac{\Delta p_{in}}{\rho L\left|\frac{\mathrm{d}u_p}{\mathrm{d}t}\right|} = 0.25\frac{\delta}{1-\delta}. \tag{10.3.42}$$

Calculation made on the basis of formula (10.3.41) gives overrated results on average about 60% compared to experimental ones (Grachev et al., 1980). This can be explained by that in formula (10.3.38), the first and second terms are calculated independently regardless of their interactions. The effect of compressibility can also play a certain role.

Consider now inertial pressure p_2 generated inside the drill-stem by closed lower end in the process of round trip operation at depth L. Pressure p_2 can be calculated as the product of liquid column mass and acceleration of the column divided by pipe cross section

$$p_2 - p_{cm} = \pm\Delta p_{in} = \frac{\rho S_p L}{S_p}\frac{\mathrm{d}u_p}{\mathrm{d}t} = \rho L\frac{\mathrm{d}u_p}{\mathrm{d}t}, \tag{10.3.43}$$

where p_{cm} is the pressure at the column mouth.

Thus, pressure p_2 results in the column at a depth L determined by formula (10.3.43) and outside the column in the annular space pressure p_2' calculated by formula (10.3.38)

$$p_2' - p_{am} = \pm\Delta p_{in} + \Delta p_{fr}, \tag{10.3.44}$$

where p_{am} is pressure at the well mouth in the annulus.

The sum of pressure drops $(p_2-p_2')+(p_{am}-p_{cm})$ characterizes the action on the drill-stem walls (bearing stress or break), which should be taken into account in strength design calculation of the drill-stem

$$(p_2-p_2')+(p_{am}-p_{cm})=\rho L\frac{du_p}{dt}\Delta p_{in}\mp\Delta p_{fr}. \tag{10.3.45}$$

Pressure drop in joints is taken as

$$\Delta p_{joint}=\xi\rho\frac{v_{av}^2}{2}n_{joint}, \tag{10.3.46}$$

where ξ and n_{joint} are calculated with formulas (6.9.9) and (6.9.10); v_{av} is the velocity determined by (10.3.10).

10.4 HYDRODYNAMIC PRESSURE GENERATING IN DRILL-STEM DESCENT IN A WELL FILLED BY VISCOUS-PLASTIC FLUID

Reasoning used to determine hydrodynamic pressure is similar to that outlined in Section 10.3. First, we determine pressure losses due to friction in the annular space and then calculate average velocity v_{av} and inertial component of the pressure drop. The total pressure drop is obtained by summing components of pressure drops.

To get the pressure drop in nonstationary flow in the annular space during drill-stem descent, one should solve the system of equations (6.1.1)–(6.1.4) with regard to (6.3.31) and (6.3.32) under the following boundary conditions:

$$\begin{aligned} w&=-u_p \quad \text{at} \quad r=R_1;\\ w&=0 \quad \text{at} \quad r=R_2. \end{aligned} \tag{10.4.1}$$

The corresponding velocity profile is shown in Fig. 10.2,II. The flow rate q is (Golubev, 1979)

$$q=\frac{\tau_0 d_h S_{as}}{8\eta\beta}\left[\frac{1+\delta^2}{(1-\delta)^2}-\frac{2\xi}{1-\delta}\left(\frac{\xi}{1-\delta}-\beta\right)\right.$$
$$\left.-\frac{4}{3}\frac{1-\delta+\delta^2}{(1-\delta)^2}\beta+\frac{1}{3}\left(\frac{2\xi}{1-\delta}-\beta\right)^3\beta\frac{1-\delta}{1+\delta}\right]+\pi u_T R_1^2; \tag{10.4.2}$$

$$\xi\beta(1-\delta)+\xi(\xi-\beta(1-\delta))\ln\frac{\xi\delta}{\xi-\beta(1-\delta)}+\frac{1}{2} \times(1-2\beta(1-\delta)-(\beta(1-\delta)+\delta)^2)+\beta(1-\delta)\frac{\eta u_{\mathrm{p}}}{R_2\tau_0}=0, \tag{10.4.3}$$

where ξ, δ, and β are the same quantities as in (6.3.45).

Equations (10.4.2) and (10.4.3) may be transformed to dimensionless form by insertion of dimensionless parameter

$$\mathrm{Se}=\frac{8\beta}{\frac{1+\delta^2}{(1-\delta)^2}-\frac{2\xi}{1-\delta}\left(\frac{\xi}{1-\delta}-\beta\right)-\frac{4}{3}\frac{1-\delta+\delta^2}{(1-\delta)^2}\beta+\frac{1}{3}\left(\frac{2\xi}{1-\delta}-\beta\right)^3\beta\frac{1-\delta}{1+\delta}+8\beta\frac{\delta^2}{1-\delta^2}\bar{u}_{\mathrm{p}}}; \tag{10.4.4}$$

$$\xi\beta(1-\delta)+\xi(\xi-\beta(1-\delta))\ln\frac{\xi\delta}{\xi-\beta(1-\delta)} +\frac{1}{2}(1-2\beta(1-\delta)-(\beta(1-\delta)+\delta)^2)+2\beta(1-\delta)^2\bar{u}_{\mathrm{p}}=0, \tag{10.4.5}$$

where $\mathrm{Se}=\tau_0 d_{\mathrm{h}}S_{\mathrm{as}}/(\eta q)$ and $\bar{u}_{\mathrm{p}}=u_{\mathrm{p}}\eta/(\tau_0 d_{\mathrm{h}})$

At $u_{\mathrm{p}}=0$ expressions (10.4.4) and (10.4.5) turn into Frederickson–Bird formulas (6.3.44) and (6.3.45) whereas at $\tau_0\to 0$ and $u_{\mathrm{p}}\neq 0$ (10.3.6) turns into Targ solution for viscous fluid.

To calculate the parameter *Se*, one should first get q with (10.2.6)

$$q=q_{\mathrm{pd}}+q_{\mathrm{fd}}=q_{\mathrm{pd}}+\pi u_{\mathrm{p}}R_1^2. \tag{10.4.6}$$

Then, with the help of relations (10.4.4) and (10.4.5) to build the nomographic chart $\beta=\beta(\delta,\ \bar{u}_{\mathrm{p}},\ Se)$.

For the descent of the drill-stem with pumps switched off ($q_{\mathrm{pd}}=0$), equation (10.4.6) is simplified

$$q=q_{\mathrm{fd}}=\pi u_{\mathrm{p}}R_1^2. \tag{10.4.7}$$

Equating (10.4.2) and (10.4.7), we get

$$\frac{1+\delta^2}{(1-\delta)^2}-\frac{2\xi}{1-\delta}\left(\frac{\xi}{1-\delta}-\beta\right)-\frac{4}{3}\frac{1-\delta+\delta^2}{(1-\delta)^2}\beta+\frac{1}{3}\left(\frac{2\xi}{1-\delta}-\beta\right)^3\beta\frac{1-\delta}{1+\delta}=0; \tag{10.4.8}$$

$$\mathrm{Se}=\frac{2\beta(1-\delta)^2(1-\delta^2)}{\delta^2\left[\xi(\xi-\beta(1-\delta))\ln\frac{\xi-\beta(1-\delta)}{\xi\delta}-\xi\beta(1-\delta)-\frac{1}{2}(1-2\beta(1-\delta)-(\beta(1-\delta)+\delta)^2)\right]}, \tag{10.4.9}$$

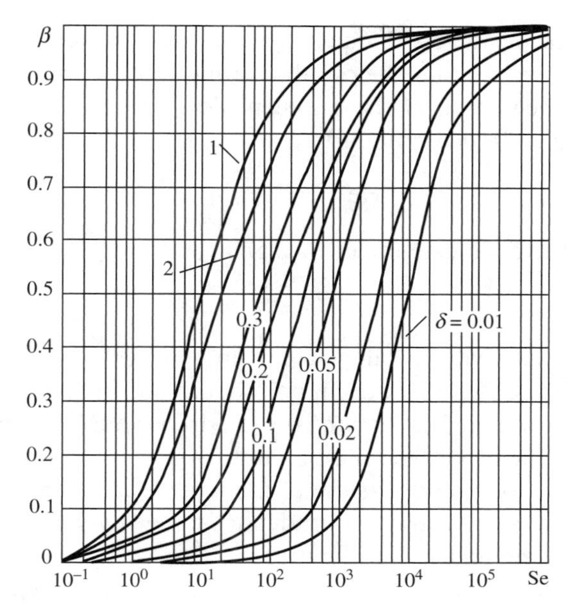

FIGURE 10.6 Graphic of the function $\beta = \beta(Se,\ \delta)$ at $q_{\text{pd}} = 0$. (Graphic is re-presented by V.Z. Digalev through dimensionless form of equations (10.4.2) and (10.4.3).)

where $\text{Se} = \tau_0 d_{\text{h}}/(\eta v_{\text{av}})$; $v_{\text{av}} = u_{\text{p}} \frac{\delta^2}{1-\delta^2}$; $\delta = R_1/R_2$;

$$\beta = 4\tau_0 L/(d_{\text{h}}\, \Delta p). \tag{10.4.10}$$

Formulas (10.4.2) and (10.4.3) at $q_{\text{pd}} = 0$ in Golubev (1979) were brought in a dimensionless form on the basis of which Digalev et al. (1980) plotted a graphic of the function $\beta = \beta(\delta, Se)$ shown in Fig. 10.6. In the same figure are displayed curves 1 and 2 from Fig. 6.7 at $v_{\text{av}} = Q/S$ in channels with immovable walls.

Pressure losses due to friction are calculated by the following method. First, we determine δ, v_{av}, and Se with formulas (10.4.10), then β from Fig. 10.6 and finally the pressure drop from the formula

$$\Delta p_{\text{p}} = \frac{4\tau_0 L}{\beta d_h}. \tag{10.4.11}$$

The problem is solved with respect to pressures coming into being in drill-stem descent. Nevertheless, formula (10.4.11) can be used to calculate pressure losses in drill-stem ascent.

To determine the inertial component of the pressure, relation (10.3.40) is used. To get the pressure below the string shoe, formula (10.3.38) is suitable. The beginning of the turbulent flow can be obtained from the graphic in Fig. 6.20 determining previously Hedström number

$$\mathrm{He} = \tau_0 d_\mathrm{h}^2 \rho / \eta^2 \tag{10.4.12}$$

The critical Reynolds number Re_cr gives v_cr

$$v_\mathrm{cr} = \mathrm{Re}_\mathrm{cr} \eta / (\rho d_\mathrm{h}). \tag{10.4.13}$$

If it turns out that v_cr calculated with (10.3.10) is greater than v_cr calculated with formula (10.4.13), one should use formulas for turbulent flow. Formulas obtained in Section 10.3 for viscous fluid are applicable also for turbulent flow of viscous-plastic fluid.

In the descent of the drill-stem at moderate depth, one should in the formulas for pressure drop insert the depth attained by the drill-stem end to the instant of time under consideration.

Let the descent of the drill-stem is occurring from the depth L_1 to L_2 in accordance with Fig. 10.1. Then, the current depth L and the time of its achievement are connected by the following relations:

for the section of uniformly accelerated motion

$$L = L_1 + \frac{\mathrm{d}u_\mathrm{p}}{\mathrm{d}t}\frac{t^2}{2};$$

for the section of descent with constant velocity

$$L = L_\mathrm{p} + u_\mathrm{p}(t - t_1);$$

for the section of uniformly decelerated motion

$$L = L_\mathrm{n} + a(t - t_2) + \frac{\mathrm{d}u_\mathrm{p}}{\mathrm{d}t}\frac{t^2 - t_2^2}{2},$$

where L_p and L_n are depths of the drill-stem section end at the beginning of its motion, t_1 and t_2 are instants of time (see Fig. 10.1), a is the factor in the expression for the velocity $u_\mathrm{p} = a + (\mathrm{d}u_\mathrm{p}/\mathrm{d}t)t$.

10.5 EXAMPLES OF PRESSURE CALCULATION IN ROUND TRIP OPERATIONS

EXAMPLE 10.5.1

Determine the pressure distribution in the descent of the drill-stem with closed lower end on one drill stand length in a well filled by viscous-plastic fluid.

Initial data are	
Diameter of the well d_w (m)	0.224
External diameter of drill pipes d_{ex} (m)	0.141
Length of pipes lowered down L (m)	1192
Length of a pipe between joints l_p (m)	12.5
Length of one drill pipe stand l_{dp} (m)	25
External diameter of the joint ЗШ-178 d_{joint} (m)	0.178
dynamic shear stress τ_0 (Pa)	6.3
Plastic viscosity factor η (Pa·s)	0.037
Fluid density ρ (kg/m^3)	1710

Velocity of the drill-stem (m/s) at different instants of time (see Fig. 10.1)	
At the section of acceleration	$u_p = a_p t = 1.2t \quad 0 \leq t \leq 4;$
At the section of lowering down with constant velocity	$u_p = 4.8 \quad 4 \leq t \leq 5.2;$
At the section of deceleration	$u_p = 11.04 - 1.2t \quad 5.2 \leq t \leq 9.2.$

SOLUTION To determine the pressure drop due to friction for viscous-plastic washing fluid, we use formula (10.3.29) for turbulent flow and formula (10.4.11) for laminar flow. Get Hedström number

$$\mathrm{He} = \tau_0 d_p^2 \rho / \eta^2 = 6.3(0.224\text{–}0.141)^2 1710/0.037^2 = 5.42 \times 10^4,$$

critical Reynolds number from Fig. 6.20

$$\mathrm{Re}_{cr} = 6 \times 10^3,$$

critical velocity with formula (10.4.13)

$$v_{cr} = \mathrm{Re}_{cr}\eta/(\rho d_h) = 6 \times 10^3 \times 0.037/(1710(0.224\text{–}0.141)) = 1.56\,\mathrm{m/s}.$$

Calculate $\delta = d_{ex}/d_w = 0.141/0.224 = 0.6295$ and with formula (10.3.10)

$$v_{av} = u_p \frac{\delta^2}{1-\delta^2} = u_p \frac{(0.6295)^2}{1-(0.6295)^2} = 0.656 u_p.$$

Then, $v_{cr} = 1.56$ m/s will correspond to the following velocity of the drill-stem descent:

$$(u_p)_{cr} = v_{cr}/0.656 = 1.56/0.656 = 2.38\,\mathrm{m/s}.$$

So, at $u_p \geq 2.38$ m/s, the flow regime is turbulent whereas at $u_p < 2.38$ m/s, it is laminar. Since in accordance with initial data u_T varies in limits $0 \leq u_T \leq 4.8$ m/s, in the range $0 \leq u_p \leq 2.38$ m/s there is laminar flow and at $u_p > 2.38$ m/s turbulent flow throughout the whole descent of the drill-stem.

Determine now the time of laminar flow existence. Since $u_p = 1.2t$ in the acceleration section, the laminar regime takes place up to the instant of time $t = (u_p)_{cr}/1.2 = 2.38/1.2 = 1.98$ s.

Transition to laminar flow at the deceleration section begins in time

$$t = (11.04 - (u_p)_{cr}/1.2 = (11.04 - 2.38)/1.2 = 7.22 \text{ s}.$$

Determine the pressure drop caused by friction force in turbulent fluid flow in sections of acceleration, descent with constant velocity and deceleration. Using formula (10.3.29) with regard to (10.3.34), we get

$$f_1(\delta, 7, 1) = 0.0488\left(\frac{1+\delta+\delta^2}{1-\delta^2}\right)^{1.75} = 0.0488\left(\frac{1+0.6295+(0.6295)^2}{1-(0.6295)^2}\right)^{1.75} = 0.406;$$

$$\Delta p = 0.406\frac{u_p^{1.75}(0.037/1710)^{0.25}}{2(0.224-0.141)^{1.25}}1710L = 531u_p^{1.75}L.$$

Results for different times of drill-stem descent, that is, for different values of descent velocity, are given in Table 10.1.

For example, at $t = 1.98$ s

$$\Delta p_{fr} = 531(1.2t)^{1.75}L = 531(1.2 \times 1.98)^{1.75}1192 = 2.88 \text{ MPa}.$$

TABLE 10.1

Regime	t (s)	Δp_{in} (MPa)	Δp_{fr} (MPa)	$\Delta p_{fr} + \Delta p_{in}$ (MPa)	Δp_{joint} (MPa)	p (MPa)
Laminar	0	1.61	0.362	1.972	0	1.97
	1.98	1.61	1.57	3.18	0.161	3.34
Turbulent	1.98	1.61	2.88	4.49	0.161	4.65
	3	1.61	5.96	7.57	0.37	7.94
	4	1.61	9.86	11.47	0.657	12.13
	4	0	9.86	9.86	0.657	10.52
	5.2	0	9.86	9.86	0.657	10.52
	t (s)	Δp_{in} (MPa)	Δp_{fr} (MPa)	$\Delta p_{fr} - \Delta p_{in}$ (MPa)	Δp_{joint} (MPa)	p (MPa)
Turbulent	5.2	1.61	9.86	8.25	0.657	8.91
	6	1.61	6.67	5.06	0.422	5.48
	7.22	1.61	2.88	1.27	0.161	1.43
Laminar	7.22	1.61	1.57	−0.04	0.161	0.121
	9.2	1.61	0.362	−1.248	0	−1.248

The same value of Δp_{fr} will be at $t = 7.22$ s, since velocities at these instants of time are equal.

At $t = 6$ s

$$\Delta p_{fr} = 531(11.04 - 1.2 \times 6)^{1.75} 1192 = 6.67 \text{ MPa}.$$

Let us determine pressure drops in sections of acceleration ($0 \le t \le 1.98$) and deceleration ($7.22 \le t \le 9.2$) in laminar regimes of flow using formulas (10.4.7), (10.4.11) and the graphic in Fig. 10.6. At $t = 0$, there is $u_p = 0$, so that $Se \to \infty$ at $u_p \to 0$, and from Fig. 10.6 it follows $\beta \to 1$. Then, in (10.4.11) one should take $\beta = 1$. At this

$$\Delta p_{fr} = \frac{4\tau_0}{d_h} L = \frac{4 \times 6.3}{0.224 - 0.141} 1192 = 0.362 \text{ MPa}.$$

The pressure drop $\Delta p_{fr} = 0.362$ MPa found at $t = 0$ means that this pressure drop is required to begin the flow of viscous-plastic fluid.

The velocity at $t = 1.98$ s is obtained from (10.3.10):

$$v_{av} = u_p \frac{\delta^2}{1 - \delta^2} = 1.2t \frac{(0.6295)^2}{1 - (0.6295)^2} = 1.2 \times 1.98 \frac{(0.6295)^2}{1 - (0.6295)^2} = 1.56 \text{ m/s};$$

$$Se = \frac{\tau_0 d_h}{\eta v_{av}} = \frac{6.3(0.224 - 0.141)}{0.037 \times 1.56} = 9.1.$$

From Fig. 10.6 at $\delta = 0.6295 \approx 0.63$ it ensues $\beta = 0.23$ and from (10.4.11) we get

$$\Delta p_{fr} = \frac{4\tau_0}{\beta d_h} L = \frac{4 \times 6.3}{0.23(0.224 - 0.141)} 1192 = 1.57 \text{ MPa}.$$

The same value $\Delta p_{fr} = 1.57$ MPa will be at $t = 7.22$ s because average velocities at $t = 1.98$ s and $t = 7.22$ s are equal. In a similar manner Δp_{fr} may be found for another times of laminar regime existence.

The inertial component of the pressure drop at drill-stem acceleration can be found from formulas (10.3.40) and (10.3.39)

$$\Delta p_{in} = \rho \left| \frac{\partial v_{av}}{\partial t} \right| L = \rho \frac{\delta^2}{1 - \delta^2} \left| \frac{du_p}{dt} \right| L = 1710 \frac{(0.6295)^2}{1 - (0.6295)^2} 1.2 \times 1192 = 1.61 \text{ MPa}.$$

The overpressure ($p_{ov} = 0$) in laminar flow without losses in joints and hydrostatic component can be found from (10.3.38)

at $t = 1.98$ s

$$p = \Delta p_{in} + \Delta p_{fr} = 1.61 + 1.57 = 3.18 \text{ MPa};$$

at $t = 7.22$ s

$$p = -\Delta p_{\text{in}} + \Delta p_{\text{fr}} = -1.61 + 1.57 = -0.04 \text{ MPa};$$

at $t = 9.2$ s (after completion of drill-stem descent)

$$p = -\Delta p_{\text{in}} + \Delta p_{\text{fr}} = -1.61 + 0.362 = -1.248 \text{ MPa}.$$

The rest of the calculations are tabulated in Table. 10.1.

Determine pressure losses in joints Δp_{joint} having previously calculated the factor ξ by formula (10.3.46)

$$\xi = 2\left[\frac{1-(d_{\text{ex}}/d_{\text{w}})^2}{1-(d_{\text{joint}}/d_{\text{w}})^2} - 1\right]^2 = 2\left[\frac{1-(0.141/0.224)^2}{1-(0.178/0.224)^2} - 1\right]^2 = 0.814;$$

$$\Delta p_{\text{joint}} = \xi\rho\frac{v_{\text{av}}^2}{2}\frac{L(t)}{l_{\text{P}}} = 0.814 \times 1710\frac{v_{\text{av}}^2}{2}\frac{L(t)}{12.5} = 55.68 v_{\text{av}}^2 L(t).$$

By substituting here the velocity from (10.3.10), we get

$$\Delta p_{\text{joint}} = 55.68\left(\frac{0.6295^2}{1-0.6295^2}\right)^2 u_{\text{p}}^2 1192 = 2.86 \times 10^4 u_{\text{p}}^2.$$

Results of calculation are displayed in Table 10.1 and in Fig. 10.7. From Fig. 10.7, it is seen that the pressure at lower end of the drill-stem in the course of its descent on one drill stand length may become more but less than the hydrostatic pressure. Such variations in pressure can lead to absorption of the washing fluid and show of reservoir fluids.

EXAMPLE 10.5.2

It is required to determine pressure distribution with time during drill-stem descent in a well filled by viscous fluid. Initial data are the same as in Example 10.5.1 except that instead of viscous-plastic fluid viscous washing fluid is used, that is, $\tau_0 = 0$. Since the fluid is viscous, the pressure drop due to friction in laminar flow is calculated by formula (10.3.9) and for inertial pressure formula (10.3.40) is used. For turbulent flow, the same formulas are applied as in Example 10.5.1.

SOLUTION Critical Reynolds number for viscous fluids is equal to 2100. Therefore, the critical velocity in accordance with (10.4.13) is

$$v_{\text{cr}} = \text{Re}_{\text{cr}}\mu/(\rho d_{\text{h}}) = 2100 \times 0.037/(1710(0.224-0.141)) = 0.547 \text{ m/s}.$$

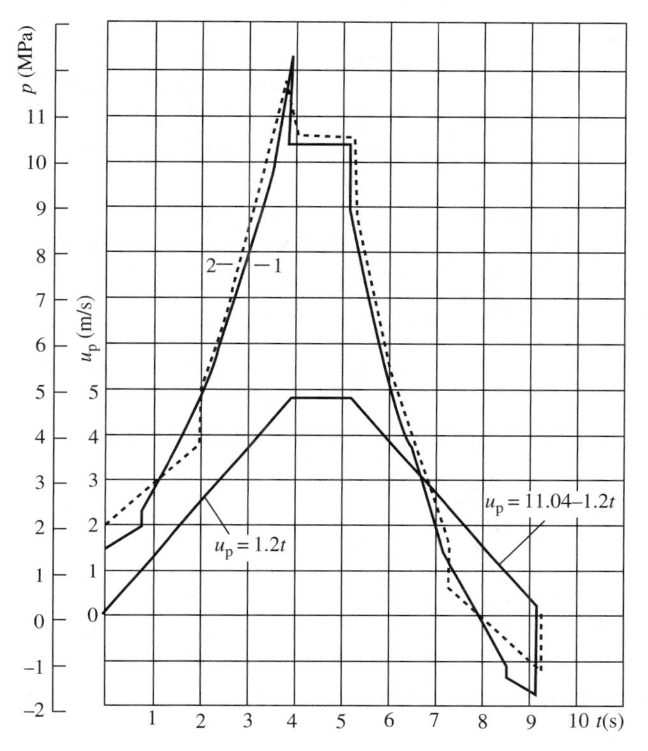

FIGURE 10.7 Pressure change in time in drill-stem descent in viscous (curve 1) and viscous-plastic (curve 2) fluids.

If v_{cr} calculated by (10.3.10) is greater than $v_{cr} = 0.547$ m/s, the regime is turbulent. As applied to our case, find the velocity of the drill-stem $(u_p)_{cr}$ corresponding to v_{cr}. The average velocity v_{av} is the same as in the previous example, since geometric sizes are not changed, that is, $v_{av} = 0.656u_p$. Comparison with $v_{cr} = 0.547$ m/s gives that laminar regime is observed at $u_p < v_{cr}/0.656 = 0.547/0.656 = 0.834$ m/s.

Determine the time of laminar flow existence. Since $u_p = 1.2t$, the laminar flow at the section of acceleration takes place at $t \leq 0.834/1.2 = 0.695$ s. Comparison of times of turbulent regime beginning for both examples yields that for viscous fluid it begins earlier. At the section of deceleration, the transition to laminar regime begins at $t = (11.04 - 0.834)/1.2 = 8.5$ s.

For pressure drop due to friction force in turbulent flow in sections of acceleration, descending with constant velocity and deceleration, one and the same formula (10.3.29) as in the previous example is valid. Since geometric sizes are not changed, there is $\Delta p = 531u_p^{1,75}L$. At $t = 0.695$ s we have $\Delta p = 0.461$ MPa. The same value of Δp will be at $t = 8.5$ s. Inertial component of pressure drop is the same as in Example 10.5.1, namely, $\Delta p_{in} = 1.61$ MPa.

TABLE 10.2

Regime	t (s)	Δp_{in} (MPa)	Δp_{fr} (MPa)	$\Delta p_{fr}+\Delta p_{in}$ (MPa)	Δp_{joint} (MPa)	p (MPa)
Laminar	0	1.61	0	1.61	0	1.61
	0.695	1.61	0.276	1.89	0.02	1.91
Turbulent	0.695	1.61	0.461	2.07	0.02	2.09
	1.98	1.61	2.88	4.49	0.161	4.65
	3	1.61	5.96	7.57	0.37	7.94
	4	1.61	9.86	11.47	0.657	12.13
	4	0	9.86	9.86	0.657	10.52
	5.2	0	9.86	9.86	0.657	10.52
	t (s)	Δp_{in} (MPa)	Δp_{fr} (MPa)	$\Delta p_{fr}-\Delta p_{in}$ (MPa)	Δp_{joint} (MPa)	p (MPa)
Turbulent	5.2	1.61	9.86	8.25	0.657	8.91
	6	1.61	6.67	5.06	0.422	5.48
	7.22	1.61	2.88	1.27	0.161	1.43
	8.5	1.61	0.461	−1.15	0.02	−1.13
Laminar	8.5	1.61	0.276	−1.33	0.02	−1.31
	9.2	1.61	0	−1.61	0	−1.61

Getting the pressure drop due to friction force in laminar flow from (10.3.9)

$$\Delta p_{fr} = \frac{4\mu}{R_2^2}L\frac{u_m}{-\left[1+\left(\frac{R_1}{R_2}\right)^2\right]\ln\left(\frac{R_1}{R_2}\right)-1+\left(\frac{R_1}{R_2}\right)^2}$$

$$= \frac{4\times 0.037}{(0.224/2)^2}L\frac{u_p}{-[1+0.6295^2]\ln 0.6295-1+(0.6295)^2} = 277u_pL$$

At $t=0.695$ s it is $\Delta p = 277\times 1.20\times 0.695\times 1192 = 0.276$ MPa.

The pressure drop in joints is the same as in Example 10.5.1.

Results of calculations are listed in Table 10.2. In Fig. 10.7, variation of the pressure with time due to friction forces and inertia in drill-stem descent in viscous-plastic fluid is shown through dotted lines. It should be noted that in descent as well as in ascent of the drill-stem this pressure can become equal even to hydrostatic pressure of the fluid column with formation of vacuum zones.

10.6 NONSTATIONARY FLUID FLOW IN A WELL AS WAVE PROCESS

In this chapter, the solution of equations (10.1.25)–(10.1.26) at $\lambda=0$ is considered and graphical method is outlined to get pressure distribution in nonstationary flow of slightly compressible fluid in a well.

Let us introduce dimensionless variables

$$\bar{p} = pS/(\rho_0 c q_0), \qquad \bar{v} = \bar{q} = \rho v/(\rho_0 v_0) = \rho q/(\rho_0 q_0),$$
$$\bar{z} = z/L, \qquad \bar{t} = ct/L, \tag{10.6.1}$$

where ρ_0, q_0, S, and L are characteristic fluid density, fluid flow rate, area of pipe cross section, and pipe length, respectively. These parameters will be specified further by concrete examples.

Using variables (10.6.1), the system of equations (10.1.25)–(10.1.26) is reduced to the following equations:

$$\frac{\partial \bar{p}}{\partial \bar{z}} = -\frac{\partial \bar{q}}{\partial \bar{t}}; \tag{10.6.2}$$

$$\frac{\partial \bar{p}}{\partial \bar{t}} = -\frac{\partial \bar{q}}{\partial \bar{z}}. \tag{10.6.3}$$

Differentiating equation (10.6.2) with respect to $\bar{z}$ and equation (10.6.3) to $\bar{t}$ and equating identical parts, we receive a wave equation for dimensionless pressure $\bar{p}$

$$\frac{\partial^2 \bar{p}}{\partial \bar{t}^2} = \frac{\partial^2 \bar{p}}{\partial \bar{z}^2}. \tag{10.6.4}$$

In the same manner, differentiation of (10.6.2) with respect to $\bar{t}$ and (10.6.3) to $\bar{z}$ yields wave equation for the mass flow rate $\bar{q}$

$$\frac{\partial^2 \bar{q}}{\partial \bar{t}^2} = \frac{\partial^2 \bar{q}}{\partial \bar{z}^2}. \tag{10.6.5}$$

In order to get solution of this equation, introduce new dimensionless variables

$$\xi = \bar{z} - \bar{t}; \qquad \varsigma = \bar{z} + \bar{t}. \tag{10.6.6}$$

Then,

$$\frac{\partial \bar{p}}{\partial \bar{z}} = \frac{\partial \bar{p}}{\partial \xi} + \frac{\partial \bar{p}}{\partial \varsigma}; \qquad \frac{\partial \bar{p}}{\partial \bar{t}} = -\frac{\partial \bar{p}}{\partial \xi} + \frac{\partial \bar{p}}{\partial \varsigma}; \tag{10.6.7}$$

$$\frac{\partial^2 \bar{p}}{\partial \bar{z}^2} = \frac{\partial^2 \bar{p}}{\partial \xi^2} + 2\frac{\partial^2 \bar{p}}{\partial \xi \partial \varsigma} + \frac{\partial^2 \bar{p}}{\partial \varsigma^2}; \qquad \frac{\partial^2 \bar{p}}{\partial \bar{t}^2} = \frac{\partial^2 \bar{p}}{\partial \xi^2} - 2\frac{\partial^2 \bar{p}}{\partial \xi \partial \varsigma} + \frac{\partial^2 \bar{p}}{\partial \varsigma^2}.$$

Relations similar to (10.6.7) could be written for $\bar{q}$.

Replacing in (10.6.4) second derivatives of $\bar{p}$ with (10.6.7) and second derivatives of $\bar{q}$ in (10.6.5) with the same relations for $\bar{q}$, we get

$$\frac{\partial^2 \bar{p}}{\partial \xi \partial \varsigma} = 0; \tag{10.6.8}$$

$$\frac{\partial^2 \bar{q}}{\partial \xi \partial \varsigma} = 0. \tag{10.6.9}$$

Integration of (10.6.8) gives

$$\bar{p} = f_1(\bar{z} - \bar{t}) + f_2(\bar{z} + \bar{t}), \tag{10.6.10}$$

where f_1 and f_2 are arbitrary twice differentiated functions; $\bar{p}$ is the pressure at time $\bar{t}$ in cross section $\bar{z}$.

Since the form of equation (10.6.9) is the same as (10.6.8), we obtain for the mass flow rate $\bar{q}$

$$\bar{q} = f_3(\bar{z} - \bar{t}) + f_4(\bar{z} + \bar{t}). \tag{10.6.11}$$

Substitution of (10.6.10) and (10.6.11) in (10.6.2) and (10.6.3) yields

$$f_1(\bar{z} - \bar{t}) = f_3(\bar{z} - \bar{t}), \tag{10.6.12}$$

$$f_2(\bar{z} + \bar{t}) = -f_4(\bar{z} + \bar{t}). \tag{10.6.13}$$

Then, solutions (10.6.10) and (10.6.11) take forms

$$\bar{p} = f_1(\bar{z} - \bar{t}) + f_2(\bar{z} + \bar{t}), \tag{10.6.14}$$

$$\bar{q} = f_1(\bar{z} - \bar{t}) - f_2(\bar{z} + \bar{t}), \tag{10.6.15}$$

From (10.6.14) and (10.6.15) it follows

$$f_1(\bar{z} - \bar{t}) = (\bar{p} + \bar{q})/2 \tag{10.6.16}$$

and

$$f_2(\bar{z} + \bar{t}) = (\bar{p} - \bar{q})/2. \tag{10.6.17}$$

Let in the fluid flow in cross section z at time t happen the change of parameters p and q. Perturbations from these changes propagate along z-axis with sound velocity c in the form of a direct wave. Then,

$$c = \frac{\mathrm{d}z}{\mathrm{d}t} \tag{10.6.18}$$

or in dimensionless variables of (10.6.1)

$$\frac{d\bar{z}}{d\bar{t}} = 1. \tag{10.6.19}$$

It means that the velocity of wave propagation in dimensionless variables is equal to 1. Integration of (10.6.19) gives

$$\bar{z} - \bar{t} = \text{const.} \tag{10.6.20}$$

Analogous to waves propagating in the reverse direction there is

$$\bar{z} + \bar{t} = \text{const.} \tag{10.6.21}$$

Thus, formula (10.6.16) connects parameters $\bar{p}$ and $\bar{q}$ for direct wave, whereas (10.6.17) for reverse wave. Since waves obey conditions (10.6.19), (10.6.20) and (10.6.21), it should be valid relations

$$\bar{p} + \bar{q} = 2f_1(\bar{z} - \bar{t}) = \text{const}; \tag{10.6.22}$$

$$\bar{p} - \bar{q} = 2f_2(\bar{z} + \bar{t}) = \text{const.} \tag{10.6.23}$$

In coordinates $\bar{p}$ and $\bar{q}$, equations (10.6.22) and (10.6.23) give two crossing families of parallel straight lines. Family (10.6.22) is perpendicular to family (10.6.23), the slope of straight lines (10.6.23) is 45°. Formulas (10.6.22) and (10.6.23) lie in the basis of graphic method being more visual and free from bulky calculations.

Let us consider flows in pipes and annular channels that are also accepted as pipes with respective hydraulic radius.

The circulation system of a concrete problem will be represented as a system of several pipes connected in a certain succession. At the beginning and end of such complex pipeline, as well as at junctions of pipes and other cross sections on which act flow perturbations or flow restrictions, there should be given dependences $q = q(t)$ and $p = p(t)$ or $p = p(q)$ added by given initial conditions.

The graphical picture of the pressure distribution is found in accordance with the following rules (Bergeron, 1950; Digalev et al., 1987; Shischenko et al., 1976):

1. build the pipeline scheme appropriate for the given circulation system;
2. at the pipeline scheme, choose positive direction of the z-axis coinciding with the direction of the initial fluid flow rate;
3. at the pipeline scheme, mark cross sections (A, B, C, and so on) at which it is desirable to know pressure variation in time including

cross sections at which initial conditions are given (as a rule among these are beginning, end, and junctions of the pipeline);

4. perform graphical construction sketch with coordinates $\bar{p}$, $\bar{q}$, at the upper part and $\bar{q}$, $\bar{t}$ at the lower part;
5. build given dependences of $\bar{q} = \bar{q}(\bar{t})$ and $\bar{p} = \bar{p}(\bar{q})$ at the lower and upper parts of the sketch, respectively;
6. build straight lines (10.6.22) and (10.6.23). These straight lines are mutually perpendicular and have slope 45° to the $\bar{q}$-axis. The first straight lines express connection between initial pressure and flow rate in the system up to the time chosen as calculation step. Having determined $\bar{q}$ at this instant of time and marked the point at the straight line, we raise a perpendicular to the $\bar{q}$-axis. Its intersection with the straight line found gives $\bar{p}$, which is taken as initial for the following wave. The next straight line is drawn perpendicular to the first one and so on;
7. local resistances could also be taken into account by graphic method. To do it, let us build in the upper part of the graphic the curve $\bar{p} = \Delta\bar{p}(\bar{q})$ corresponding to local resistance law. At the cross section where the local resistance has to be found, the condition $\bar{p}' = \Delta\bar{p} + \bar{p}''$, where $\bar{p}'$and $\bar{p}''$ are pressures in waves coming from both sides to the resistance under consideration, should be obeyed. It means that on the graphic at each step one has to build summary curve of $\bar{p} = \Delta\bar{p}(\bar{q})$ and the straight line appropriate to parameters of respective wave. The intersection point of this curve and the state straight line of another wave is taken as initial one for further calculations.
8. dispose next to the first draft the graphic of pressure variation $\bar{p}(\bar{t})$ in cross sections to be interested;
9. time step and number of steps are chosen by desirable accuracy for concrete problem. As first approximation for each pipeline section, time step $\Delta t = l/c$ is assumed equal to the run time along a section with length l.

10.7 PRESSURE CALCULATION IN DETERIORATION OF THE SAFETY BYPASS

Safety bypass in the circulation system deteriorates at the pressure above which the weakest of the circulation system devices (diaphragm, pump, swivel, drill branch, etc.) fails. The clogging of slush nozzles may also be one of the reasons of safety bypass deterioration.

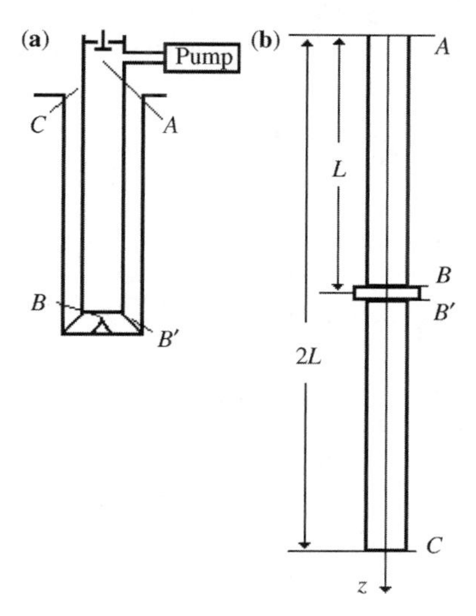

FIGURE 10.8 Well circulation system (a) and corresponding calculation scheme (b).

EXAMPLE 10.7.1

Consider pressure variation in the characteristic cross sections of the well circulation system (Fig. 10.8) in the case of instantaneous clogging of the slush nozzles.

Initial data	
Well length L (m)	4000
External diameter of the drill-stem d_{ex} (m)	0.141
Internal diameter of the drill-stem d_{in} (m)	0.119
Well diameter d_w (m)	0.220
Pump delivery Q_{pd} (m^3/s)	0.025
Fluid density ρ (kg/m^3)	1260
Diaphragm ultimate strength σ (MPa)	20.48
Modulus of elasticity E (MPa)	2.43×10^3
Modulus of elasticity E' (MPa)	2.10×10^5
Pressure in washing (MPa)	
in ascending pipe p_{ap}	16.73
above the bit p_{bit0}	$14.59 + p_{hyd}$*
at the bottom $p_{bot'0}$	$10.94 + p_{hyd}$
in the annular space at the wellhead p_{asm}	0.1

Here, p_{hyd} is hydrostatic pressure.

Pressure data may be obtained by calculation of stationary flow with formulas given in Section 6.

SOLUTION Calculate sound velocity in pipes and annular space with formulas (10.1.21)

in pipes

$$E'' = \frac{1}{\frac{1}{E} + \frac{d_{\rm h}}{\Delta}\frac{1}{E'}} = \frac{1}{\frac{1}{2.43\times10^3\times10^6} + \frac{0.119}{0.011}\times\frac{1}{2.1\times10^5\times10^6}} = 2.16\times10^9\ \text{Pa};$$

$$C = \sqrt{\frac{E''}{\rho}} = \sqrt{\frac{2.16\times10^9}{1260}} = 1309\ \text{m/s}. \tag{10.7.1}$$

in annular space

$$E'' = \frac{1}{\frac{1}{E} + \frac{d_{\rm h}}{\Delta}\frac{1}{E'}} = \frac{1}{\frac{1}{2.43\times10^3\times10^6} + \frac{0.220-0.141}{0.011}\times\frac{1}{2.1\times10^5\times10^6}} = 2.244\times10^9\ \text{Pa};$$

$$\tilde{\eta} = \sqrt{\frac{E''}{\rho}} = \sqrt{\frac{2.244\times10^9}{1260}} = 1335\ \text{m/s}. \tag{10.7.2}$$

Time Δt of wave run of pipe length L is approximately equal to time $\Delta t_{\rm as}$ of wave run in annular space of the same length L. Therefore, for simplicity we take

$$\Delta t \approx \Delta t_{\rm as} = 3\ \text{s}. \tag{10.7.3}$$

This value Δt is taken as dimensionless characteristic time in (10.6.1)

$$\bar{t} = t/\Delta t = tc/L. \tag{10.7.4}$$

The circulation pipeline scheme is represented in Fig. 10.8b. Local resistances are not taken into account. Let $\rho \approx \rho_0$. Stationary pressure distribution is taken as zero-order approximation for the sake of convenience. Then, in accordance with (10.6.1), we have $\bar{p} = 0$, $\bar{q} = \rho Q_{\rm pd}/(\rho_0 Q_{\rm pd0}) = 1$. The diaphragm ultimate strength is

$$\bar{\sigma} = \frac{20.48\times10^6\times\frac{3.14\times0.0119^2}{4}}{1260\times1309\times0.025} = 2.$$

Initial and boundary conditions are taken as follows:

$$\bar{q} = 0 \quad \text{at} \quad z = L, \quad \bar{t} \geq 0; \tag{10.7.5}$$

$$\bar{p} = 0, \quad \bar{q} = 1 \quad \text{at} \quad z = L, \quad \bar{t} = 0; \tag{10.7.6}$$

$$\bar{p} = 0, \quad \bar{q} = 0 \quad \text{at} \quad z = L, \quad \bar{t} = 0; \tag{10.7.7}$$

$$\bar{q} = 1 \quad \text{at} \quad z = 0, \quad \bar{t} < \bar{t}_p; \tag{10.7.8}$$

$$\bar{p} = 0 \quad \text{at} \quad z = 2L, \quad \bar{t} \geq 0; \tag{10.7.9}$$

$$\bar{p} = 0 \quad \text{at} \quad z = 0, \quad \bar{t} \geq \bar{t}_p, \tag{10.7.10}$$

where $\bar{t} = \bar{t}_p$ is instant of time of diagram break obtained in the course of the problem solution. This time comes when the pressure exceeds the diaphragm ultimate strength.

Condition (10.7.5) means that at $\bar{t} = 0$ happens sudden clogging of the bit and the flow rate at cross sections B and B' vanishes ($\bar{q} = 0$). Conditions (10.7.6) and (10.7.7) show that parameters of stationary flow at $\bar{t} = 0$ are conserved in all circulation system except the flow rate at cross sections B and B'. Condition (10.7.8) reflects the fact that the pump delivers the fluid into the circulation system up to the moment of diaphragm break. Conditions (10.7.9) and (10.7.10) mean that the pressure at corresponding cross sections at certain time intervals is equal to the atmospheric pressure.

The instant of time $\bar{t} = 0$ is the beginning of nonstationary wave process. From this time in all system begins generation of waves to which parameters are connected by relations (10.6.22) and (10.6.23). Determine variation in time of parameters $\bar{p}$ and $\bar{q}$ in four characteristic cross sections A, B, B', and C. In order to do it, let us trace waves leaving these sections at $\bar{t} = 0$.

Direct wave leaves cross section A at $\bar{t} = 0$. At $\bar{t} = 1$, this wave reaches cross section B and is reflected from it in the form of back wave. After reaching once again the section A, the wave turns into direct wave and so on. Waves emerging from other cross sections B, B', and C at $\bar{t} = 0$ behave in a similar manner.

In accordance with initial conditions (10.7.6), parameters of the direct wave in cross section A at $\bar{t} = 0$ will be coordinates of point A_0 ($\bar{p} = 0$, $\bar{q} = 1$). Here and later on, the index of cross section notation corresponds to the instant of time at which wave parameters in given cross section are considered. Since the wave is a direct wave, it should obey the relation (10.6.22). Determine in the latter the constant at $\bar{p} = 0$, $\bar{q} = 1$: const $= 0 + 1$. Then, equation (10.6.22) for the direct wave in the time segment $0 \leq \bar{t} \leq 1$ takes the form

$$\bar{p} + \bar{q} = 1. \tag{10.7.11}$$

To equation (10.7.11) corresponds the straight line A_0B_1 in Fig. 10.9a. At $\bar{t} = 1$, the direct wave reaches cross section B, in which $\bar{q} = 0$ in accord with (10.7.5). Therefore, from (10.7.11) it follows that $\bar{p} = 1 - \bar{q} = 1 - 0 = 1$ at $\bar{t} = 1$. Thus, to parameters of the direct wave in cross section B at $\bar{t} = 1$ correspond coordinates of the point B_1 ($\bar{p} = 1$, $\bar{q} = 0$). These parameters are initial ones for the wave reflected from cross section B in which at time segment $1 \leq \bar{t} \leq 2$ equation (10.6.23) is satisfied. Substituting in (10.6.23) values $\bar{p} = 1$, $\bar{q} = 0$ at $\bar{t} = 1$, we get const $= 1 - 0 = 1$. Then, equation (10.6.23) takes the form

$$\bar{p} - \bar{q} = 1. \tag{10.7.12}$$

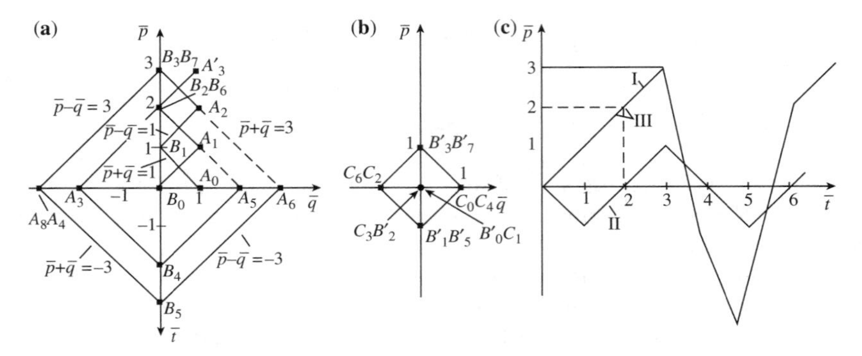

FIGURE 10.9 Schemes of graphic calculations (a) and (b) and pressure variations with time (c) in characteristic cross sections: curves I, II, and III are pressure variation in cross sections B, B', and A, respectively.

To equation (10.7.12) corresponds the straight line B_1A_2 in Fig. 10.9a. After the direct wave reaches at $\bar{t} = 2$, cross section A should have $\bar{q} = 1$ in accordance with condition (10.7.8). Consequently with (10.7.12), one can get pressure $\bar{p}$ at $\bar{t} = 2$: $\bar{p} = 1 + \bar{q} = 1 + 1 = 2$. Parameter of the wave in cross section A at $\bar{t} = 2$ in Fig. 10.9a corresponds to coordinates of the point A_2 ($\bar{p} = 2$, $\bar{q} = 1$). The pressure $\bar{p} = 2$ in dimensional form is $p = 3.75 \times 10^6$ Pa. In accordance with (10.7.10), the pressure in cross section A at $\bar{t} = 2$ instantly decays to $\bar{p} = 0$.

The direct wave having been reflected from cross section A reaches cross section B at $\bar{t} = 3$ with $\bar{q} = 0$ (condition (10.7.5)) and $\bar{p} = 3$ calculated from (10.6.23) at const $= 3$ (the point B_3 of Fig. 10.9a). For the backwave leaving cross section B, values of $\bar{p}$ and $\bar{p}$ lie on the segment B_3A_4 since initial values to determine the constant (const $= 3$) in equation (10.6.23) are parameters of point B_3 ($\bar{p} = 3$, $\bar{q} = 0$), and to determine coordinates of point A_4 from condition (10.7.10) one of the parameters $\bar{p} = 0$ (the diaphragm is broken) is known. Furthermore, the direct wave has values of $\bar{p}$ and $\bar{q}$ lying on the segment A_4B_5. Values of $\bar{p}$ and $\bar{q}$ for the backwave lie on the segment B_5A_6 corresponding to equation (10.6.23) at const $= -3$. The following direct wave has values of $\bar{p}$ and $\bar{q}$ located on the segment A_6B_7. Thus, the cycle is closed, and since the friction is not taken into account, the calculation can be continued to infinity. In practice, the process will decay owing to friction losses.

We have considered propagation of the wave leaving cross section A at $\bar{t} = 0$ and obtained values of $\bar{p}$ and $\bar{q}$ in Fig. 10.9a in this cross section for $\bar{t} = 0, 2, 4, 6$, and so on, then in cross section B at $\bar{t} = 1, 3, 5, 7$, and so on. Now, trace the wave leaving cross section B at $\bar{t} = 0$.

In accordance with boundary conditions (10.7.5) and (10.7.7) values of $\bar{p}$ and $\bar{q}$ of the wave at $\bar{t} = 0$ are characterized by point B_0 ($\bar{p} = 0$, $\bar{q} = 0$)—the beginning of nonstationary flow. The wave leaving B_0 is propagated opposite the direction of initial flow and is a backwave. Insertion of $\bar{p} = 0$ and $\bar{q} = 0$ in (10.6.23) gives const = 0 and equation (10.6.23) reduces to

$$\bar{p} - \bar{q} = 0. \tag{10.7.13}$$

Values of $\bar{p}$ and $\bar{q}$ at $\bar{t} = 1$ for the backwave lie on the segment B_0A_1 (point A_1). One coordinate $\bar{q} = 1$ of point A_1 is known from the boundary condition (10.7.8), another one $\bar{p} = 1$ is determined from (10.7.13). Reasoning similar to the foregoing one allows to build segments A_1B_2, B_2A_3, and others. Thus, we get values of $\bar{p}$ and $\bar{q}$ in cross section A for $\bar{t} = 1, 3, 5, 7$, and so on and in cross section B for $\bar{t} = 2, 4, 6$, and so on. One can also find $\bar{p}$ and $\bar{q}$ for any intermediate values of $\bar{t}$ if needed.

The variation of pressure in cross section B' of the annular space may be obtained from Fig. 10.9b performing constructions for direct and backwaves in annular space with the help of equations (10.6.22) and (10.6.23). Since due to (10.6.22) at $\bar{t} = 0$ in cross section B' there are $\bar{p} = 0$, $\bar{q} = 0$ and const = 0, the direct wave propagating from cross section B' to C is described by equation

$$\bar{p} + \bar{q} = 0. \tag{10.7.14}$$

In cross section C at $\bar{t} = 1$, $\bar{p} = 0$ because the wellhead is open. Then, from (10.7.14) it follows $\bar{q} = 0$. The same is valid for the backwave reflected from cross section C at $\bar{t} = 2$ when it reaches cross section B'. Thus, values of $\bar{p}$ and $\bar{q}$ in cross section B' at $\bar{t} = 0$, 2, 4, and so on as well as in cross section C at $\bar{t} = 1$, 3, 5, 7, and so on always vanish.

Since owing to (10.6.23) and conditions $\bar{p} = 0$ and $\bar{q} = 1$ there is const = −1 in cross section C, the backwave propagating from cross section C at $\bar{t} = 0$ is described by the equation

$$\bar{p} - \bar{q} = -1. \tag{10.7.15}$$

So, values of $\bar{p}$ and $\bar{q}$ for the backwave at $\bar{t} = 1$ lie on the segment $C_0B'_1$. Furthermore, for the direct wave at $\bar{t} = 2$ we get values of $\bar{p}$ and $\bar{q}$ lying on the segment B'_1C_2, and so on. Obtained values of $\bar{p}(t)$ are plotted in Fig. 10.9c, from which one can determine pressure drop acting on the drill-stem as well as increase and decrease of pressure action on cased and open well walls. These pressures should not exceed strength characteristics of the drill-stem walls as well as pressures of absorption, break, and show.

Calculate maximal and minimal pressures in cross sections B and B'. From Fig. 10.9c, we get maximal pressure $\bar{p}_B = 3$ in cross section B or in dimensional form (see (10.6.1)) $p = \bar{p}_B\rho_0 c q_0/S$. For the example under consideration, we have $\rho_0 = \rho = 1260\,\text{kg/m}^3$; $c = 1309$ m/s; $q_0 = Q_{\text{pd}} = 0.025\,\text{m}^3\text{/s}$; $S = \pi d^2/4 = 3.14(0.119)^2/4 = 1.112 \times 10^{-2}\,\text{m}^2$; and $p_B = 3 \times 1260 \times 1309 \times 0.025/(1.112 \times 10^{-2}) = 11.2 \times 10^5$ Pa. In circulation, the initial pressure in cross section B, exclusive hydrostatic pressure, is $p_B = 145.9 \times 10^6$ Pa.

Hence, maximal increase in pressure is comparable to initial one and comes at $\bar{t} = 3$. Dimensional time in accordance with (10.6.1) is $t = \bar{t}L/c = 3\,\Delta t = 3 \times 3 = 9$ s.

In cross section B', maximal pressure is $\bar{p}_{B'} = 1$ or in accordance with (10.6.1)

$$p_{B'} = \bar{p}_{B'}\rho_0 c q_0/S = \bar{p}_{B'}\rho_0 c v_0 = \rho_0 c v_0. \tag{10.7.16}$$

Expression (10.7.16) represents Zhukowski formula, derived for sudden pipeline closing. Insertion of initial data in (10.7.16) yields

$$p_{B'} = 1260 \times 1335 \times 0.025 \times 4/(3.14 \times (0.22^2 - 0.141^2)) = 18.78 \times 10^5 \text{ Pa}.$$

This pressure exceeds twice the initial bottom pressure minus hydrostatic one $p_{B'0} = 10.94 \times 10^5$ Pa.

10.8 CALCULATION OF PRESSURE IN RECOVERY OF CIRCULATION IN A WELL

In restoring circulation in a short span of time t_p (time of pump start), the flow rate of fluid increases up to operating flow rate Q_{pd} (Fig. 10.10). Considering nonstationary flow in well circulation system as a wave process, one can graphically get pressure distributions in time in different well cross sections using relations (10.6.22) and (10.6.23).

EXAMPLE 10.8.1

It is required to determine pressure variation in a pump during closing of the starting valve gate.

Initial data	
Well length L (m)	4000
External diameter of the drill-stem d_{ex} (m)	0.141
Internal diameter of the drill-stem d (m)	0.119
Well diameter d_w (m)	0.220
Pump nominal delivery Q_{pnd} (m^3/s)	0.025
Pump start characteristic $Q(t)$ (m^3/s)	$Q = at = 14.17 \times 10^{-3} t$
Time of pump start (time of gate valve closure t_p (s)	6
Cross section area of slush nozzles Φ (m^2)	1.8×10^{-4}

SOLUTION Schemes of the circulation system and associated pipeline are shown in Fig. 10.11.

With regard to initial data, we accept the following initial and boundary conditions in dimensionless form:

$$\bar{p} = 0, \quad \bar{q} = 0 \quad \text{at} \quad 0 \le z \le 2L, \quad \bar{t} = 0; \tag{10.8.1}$$

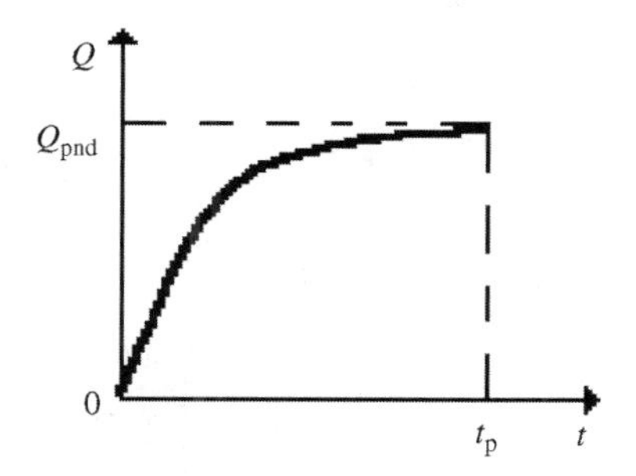

FIGURE 10.10 Graphic of pump delivery variation during pump descending.

$$\bar{p} = 0 \quad \text{at} \quad z = 2L, \quad \bar{t} \geq 0, \tag{10.8.2}$$

$$\bar{q} = \bar{a}\,\bar{t} \quad \text{at} \quad z = 0, \quad 0 \leq \bar{t} \leq \bar{t}_p; \tag{10.8.3}$$

$$\bar{q} = 1 \quad \text{at} \quad z = 0, \quad \bar{t}_p \geq 0, \tag{10.8.4}$$

$$\Delta\bar{p} = \bar{\alpha}\bar{q}^2 \quad \text{at} \quad z = L, \quad \bar{t} \geq 0, \tag{10.8.5}$$

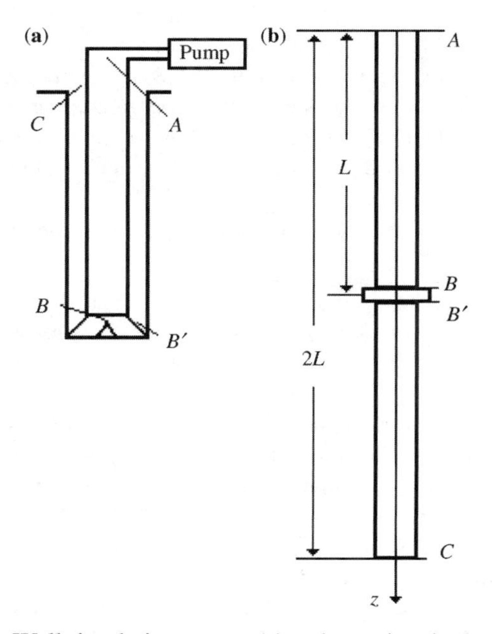

FIGURE 10.11 Well circulation system (a) and associated calculation scheme (b).

where with (10.6.1)

$$\bar{p} = \frac{pS}{\rho c Q_{\text{pd}}} = \frac{p3,14 \times 0.119^2/4}{1260 \times 1309 \times 0.025} = 0.27 \times 10^{-6} p;$$

$$\bar{q} = Q/Q_{\text{pd}} = Q/0.025 = 40Q;$$

$$\bar{t} = \frac{ct}{L} = 0,327t; \qquad \bar{t}_{\text{p}} = \frac{ct_{\text{p}}}{L} = \frac{1309 \times 6}{4000} = 2;$$

$$\bar{a} = \frac{a \times L}{cQ_{\text{pd}}} = \frac{4.17 \times 10^{-3} \times 4000}{1309 \times 0.025} = 0.5;$$

$$\Delta\bar{p} = \frac{\rho Q^2}{2\mu^2\Phi^2} \frac{S}{\rho c Q_{\text{pnd}}} = \bar{\alpha}\bar{q}^2; \quad \bar{\alpha} = \frac{Q_{\text{pnd}} S}{2\mu^2\Phi^2 c} = \frac{0.025 \times 3.14 \times 0.119^2/4}{2 \times 0.95^2 (1.8 \times 10^{-4})^2 1309} = 3.63. \tag{10.8.6}$$

Here, $c = 1309$ m/s is taken from Example 10.7.1 since geometrical sizes of the circulation system in both examples are identical.

Thus, in the given example, the pressure drop in local resistance (bit, cross sections B and B') is

$$\Delta\bar{p} = \bar{\alpha}\bar{q}^2 = 3.63\bar{q}^2. \tag{10.8.7}$$

Turn to the explanation of graphical constructions in Fig. 10.12a and b. In Fig. 10.12a with formula (10.8.7), the curve 0α in coordinates $\bar{p}$ and $\bar{q}$ is built. In the lower part of the same figure in coordinates $\bar{q}$ and $\bar{t}$, dependence is shown (10.8.3)–(10.8.4), characterizing operation of the gate valve. Initial values of $\bar{p}$ and $\bar{q}$, owing to (10.8.1)–(10.8.5) at $\bar{t} = 0$ in cross sections A, B, B', C, are zeroth located at the origin of coordinates. These states are denoted in Fig. 10.12 through A_0, B_0, B'_0, and C_0.

Equation (10.6.23) for the backwave leaving at $\bar{t} = 0$ the cross section is

$$\bar{p} - \bar{q} = 0 \tag{10.8.8}$$

since the substitution of $\bar{p} = 0$ and $\bar{q} = 0$ in (10.6.23) gives const $= 0$. The equation for the backwave leaving the cross section C in annular space at $\bar{t} = 1$ has also the form of (10.8.8).

Thus, the straight line OA_1 represents values of $\bar{p}$ and $\bar{q}$ for backwaves leaving cross sections B and C at $\bar{t} = 0$ and $\bar{t} = 1$, respectively. These waves existed up to instants of time $\bar{t} = 1$ and $\bar{t} = 2$, that is, up to their reflections from cross sections A and B. Point A_1 (0.5; 0.5) for cross section A at $\bar{t} = 1$ can be obtained if from the curve $\bar{q} = 0.5\bar{t}$ to take $\bar{q} = 0.5$ at $\bar{t} = 1$ and after inserting it in (10.8.8) to get $\bar{p} = 0.5$. Graphically, it means to draw through $\bar{t} = 1$ a straight line parallel to axis $\bar{q}$ and then from the intersection point of this line with curve $\bar{q} = 0.5\bar{t}$ to erect a perpendicular up to its intersection with straight line OA_1.

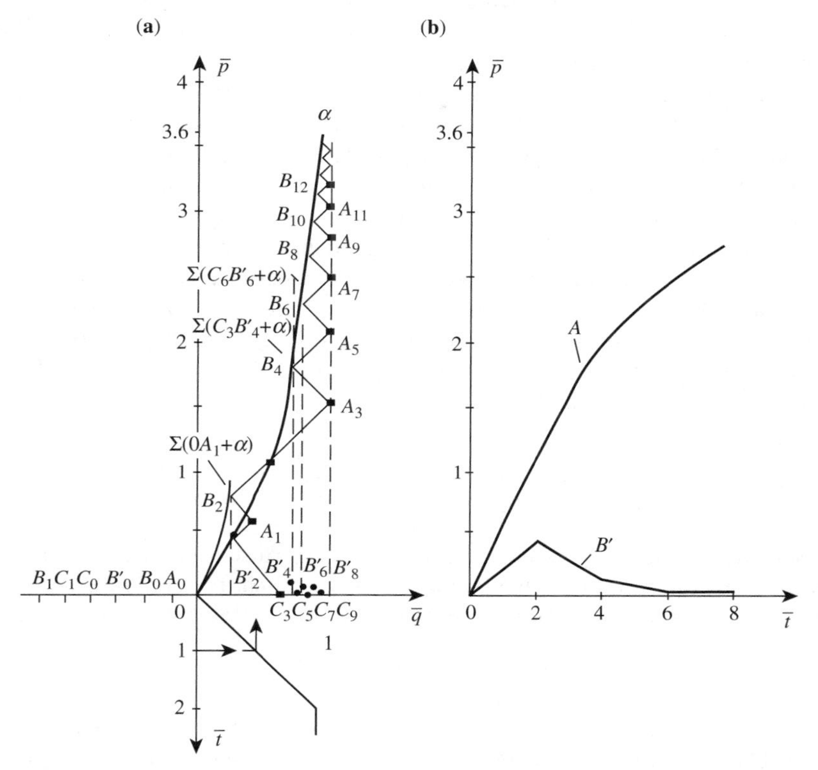

FIGURE 10.12 Scheme of graphical calculations (a) and dependences of pressure variation with time (b) in cross sections A and B'.

The wave reflected at $\bar{t} = 1$ from cross section A will be a direct wave. Its equation can be obtained from (10.6.22) as equation of straight line passing through point A_1 perpendicular to OA_1. The wave having come to cross section B at $\bar{t} = 2$ (point B_2) will have the value $\bar{p}$ equal to the sum of $\bar{p}$ at point B_2' at $\bar{t} = 2$ and $\Delta\bar{p}$ obtained by (10.8.7). Build a curve being the sum of the straight line OA_1 and the curve $O\alpha$. Intersection of the curve $\Sigma(OA_1 + O\alpha)$ with straight line perpendicular to OA_1 at point A_1 gives point B_2. Dropping perpendicular from B_2 to axis $\bar{q}$ up to the intersection with straight line OA_1, we get point B_2'.

The wave reflected from cross section B at $\bar{t} = 2$ is a backwave. It comes at $\bar{t} = 3$ to cross section A where the flow rate in accordance with boundary condition (10.8.4) will be equal to $\bar{q} = 1$. The wave state at this time is graphically expressed by point A_3 being intersection of straight line B_2A_3 perpendicular to A_1B_2 and straight line $\bar{q} = 1$. In annular space, the wave reflected from cross section B' at $\bar{t} = 2$ is a direct wave and (10.6.22) represents equation of straight line $B_2'C_3$ perpendicular to OA_1. The wave comes to cross section C at which in accordance with (10.8.2) the pressure is equal to $\bar{p} = 0$. The wave state is characterized by point C_3. The direct wave having reflected from cross section C transforms into a backwave. Its values $\bar{p}$ and $\bar{q}$ before $\bar{t} = 4$ are located on straight line $C_3B'_4$. Point B_4 describing parameters of the direct wave in pipes at $\bar{t} = 4$ is obtained as intersection of the curve $\Sigma(C_3B'_4 + O\alpha)$ being the sum of the straight line $C_3B'_4$ and the curve $O\alpha$, with the straight line B_4A_3. Further constructions are repeated.

In such a way, other values of pressure and flow rate could be obtained in selected cross sections A, B, B', and C. Values of the flow rate in the whole system tend to $\bar{q} = 1$, that is, to nominal delivery of the pump. At this, wave pressures in the pump, that is, in cross section A, and above the bit, that is, in cross section B, tend to $\bar{p} = 3.63$ in stationary flow, whereas the pressure in the cross section B' tends to be zero. These pressures are supplementary to friction losses and to hydrostatic pressure.

Maximal rise in the pressure in pipes is equal to the pressure drop in the bit Δp_{bit} at nominal flow rate Q_{npd} of the pipe

$$p_{\max} = \Delta p = \frac{\rho Q_{\text{npd}}^2}{2\mu^2\Phi^2} = \frac{1260 \times 0.025^2}{2 \times 0.95^2(1.8 \times 10^{-4})^2} = 13.5 \times 10^6 \text{ Pa}. \qquad (10.8.9)$$

Such increase in the pressure is not dangerous because in stationary regime of the flushing, the pressure in the pump exceeds maximal pressure in accordance with (10.8.9). Really, in stationary flow the pressure in the pump is equal to

$$p_{\text{pump}} = \Delta p_{\text{bit}} + \Delta p_{\text{p}} + \Delta p_{\text{as}}, \qquad (10.8.10)$$

where $\Delta p_{\text{p}} + \Delta p_{\text{as}}$ denotes the pressure loss due to friction in pipes and annular space. No great rise in pressure in pipes as a result of wave process in our example is connected to favorable characteristic of the pump starting gate valve closing. With increase in the rate of the gate valve closing, the wave pressure may become greater than the pressure calculated with formula (10.8.9).

At another value of α and parameter of the gate valve $\bar{q} = \bar{q}(\bar{t})$, the pressure in the pump can grow nonmonotonic as in Fig. 10.12b (cross section A), that is, at first to increase and then to decrease, until it reaches stationary monotonic one calculated by formula (10.8.10).

As it is seen from Fig. 10.12b, the pressure variation in cross section B' can be a reason for the reservoir hydraulic fracturing and absorption of the fluid.

10.9 CALCULATION OF PRESSURE IN A WELL IN SETTING OF A BALL CAGE ON A SEAT (THRUST RING) IN DRILL-STEM

Consider pressure variation with time in characteristic cross section of the well in setting the ball cage on the thrust ring in the drill-stem. The scheme of the circulation system with its attendant pipeline is shown in Fig. 10.13. The ball cage shuts off the orifice in the thrust ring for some time t_s reducing

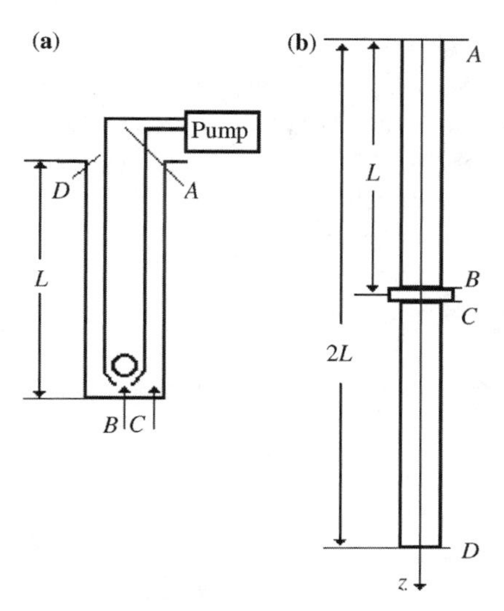

FIGURE 10.13 Well circulation system (a) and attendant to it calculation scheme (b). Characteristic cross sections are denoted by letters.

the operating flow rate Q_{pd} of pumps through the thrust ring to zero. Rise in pressure due to this fact is registered by pumps and at t_c they are cut out from the circuit. So, in time $\Delta t_{pd} = t_p - t_c$, the delivery of pumps decreases from Q_{pd} to zero. This flow rate variation should be known as function of time.

EXAMPLE 10.9.1

Determine pressure variation with time in the pump at the bottom and before the ball cage when setting it on the thrust ring. The hole annulus at the wellhead is assumed to be open.

Initial data	
Drill-stem length L (m)	1000
External diameter of the drill-stem d_{ex} (m)	0.178
Internal diameter of the drill-stem d (m)	0.158
Well diameter d_w (m)	0.250
Density of the cement slurry ρ_{cs} (kg/m^3)	1580
Density of the chaser ρ (kg/m^3)	1450
Delivery (m^3/s)	
of the cementing trailer in driving Q_{ct}	0.005
of the pump at its disconnection $Q(t)$	$Q = [5 - 3.3(t - t_c)] \times 10^{-3}$
Fluid flow rate during blocking an orifice with ball cage $Q(t)$ (m^3/s)	$Q = [5 - 59.5(t)] \times 10^{-3}$

SOLUTION On the scheme of Fig. 10.13 are marked off cross sections A, B, C, and D located before the pump and saddle, underneath the drill-stem, and at the mouth of the hole annulus, respectively. It is required to get pressure variations with time in these cross sections.

Determine sound velocity with formula (10.1.21)
in the hole annulus

$$E'' = \frac{1}{\frac{1}{E} + \frac{d_h}{\Delta}\frac{1}{E'}} = \frac{1}{\frac{1}{2.43\times10^3\times10^6} + \frac{0.250-0.178}{0.01\times2.1\times10^5\times10^6}} = 2.243 \times 10^9 \text{ Pa};$$

$$c = \sqrt{E''/\rho_{cs}} = \sqrt{2.243 \times 10^9/1580} = 1191 \text{ m/s};$$

in pipes

$$E'' = \frac{1}{\frac{1}{E} + \frac{d_h}{\Delta}\frac{1}{E'}} = \frac{1}{\frac{1}{2.43\times10^3\times10^6} + \frac{0.158}{0.01\times2.1\times10^5\times10^6}} = 2.054 \times 10^9 \text{ Pa};$$
$$c = \sqrt{2.054 \times 10^9/1450} = 1190 \text{ m/s}. \tag{10.9.1}$$

Since times of wave travel along pipe and hole annulus with length 1000 m are approximately identical, then

$$\Delta t = \Delta t_1 \approx \Delta t_2 = L/c = 1000/1190 = 0.84 \text{ s}. \tag{10.9.2}$$

It is assumed that pumps are cut out at $t = t_c$. Then,

$$t_c = \Delta t + \Delta t', \tag{10.9.3}$$

where $\Delta t'$ is time expended on decision acceptance to cut off the pump when the pressure rise at cross section A begins in time Δt.

In accordance with characteristic of orifice closing there is $\Delta t' = 0.084$ s. Then, with formula (10.9.3) we have $t_c = 0.84 + 0.084 = 0.924$ s. The time of pump disconnecting is $\Delta t_{pd} = t_p - t_c$. It is determined by pump disconnection characteristic at $Q = 0$ and $t = \Delta t_{pd}$, namely, $\Delta t_{pd} = 1.512$ s. Then, $t_p = \Delta t_{pd} + t_c = 1.512 + 0.924 = 2.436$ s.

Pressure distribution in stationary flow is taken as zero-order approximation. Then, due to initial data and formulas (10.6.1), we accept the following initial and boundary conditions:

$$\bar{p} = 0, \quad \text{at} \quad 0 \le z \le 2L, \quad \bar{t} \le 0; \tag{10.9.4}$$

$$\bar{q} = 0, \quad \text{at} \quad 0 \le z \le 2L, \quad \bar{t} \le 0; \tag{10.9.5}$$

$$\bar{q} = 1 - 10\bar{t} \quad \text{at} \quad z = L, \quad 0 \le \bar{t} \le \bar{t}_p = 0.1; \tag{10.9.6}$$

$$\bar{q} = 0 \quad \text{at} \quad z = L, \quad \bar{t} > 0.1; \tag{10.9.7}$$

$$\bar{q} = 1, \quad \text{at} \quad z = 0, \quad 0 \leq \bar{t} \leq \bar{t}_c = 1.1; \tag{10.9.8}$$

$$\bar{q} = 1.61 - 0.56\bar{t}, \quad \text{at} \quad z = 0, \quad 1.1 \leq \bar{t} \leq \bar{t}_n = 2.9; \tag{10.9.9}$$

$$\bar{q} = 0, \quad \text{at} \quad z = 0, \quad \bar{t} > \bar{t}_p = 2.9; \tag{10.9.10}$$

$$\bar{p} = 0 \quad \text{at} \quad z = 2L, \quad \bar{t} \geq 0. \tag{10.9.11}$$

In the lower part of Fig. 10.14a, we build in dimensionless coordinates curves 1 and 2 corresponding to dependences (10.9.6)–(10.9.7), (10.9.9)–(10.9.11) and expressing variations in flow rates through the thrust ring and pumps with time.

Let us explain constructions in Fig. 10.14. Consider the direct wave leaving the cross section A at $\bar{t} = 0$. Parameters $\bar{p}$ and $\bar{q}$ in accordance with (10.9.4) and (10.9.5) at this instant of time are known. In Fig. 10.14a, point A_0 corresponds to these parameters. Equation (10.6.22) determines the straight line passing through point A_0. At $\bar{t} = 1$, the wave reaches the cross section B. The flow rate in this cross section

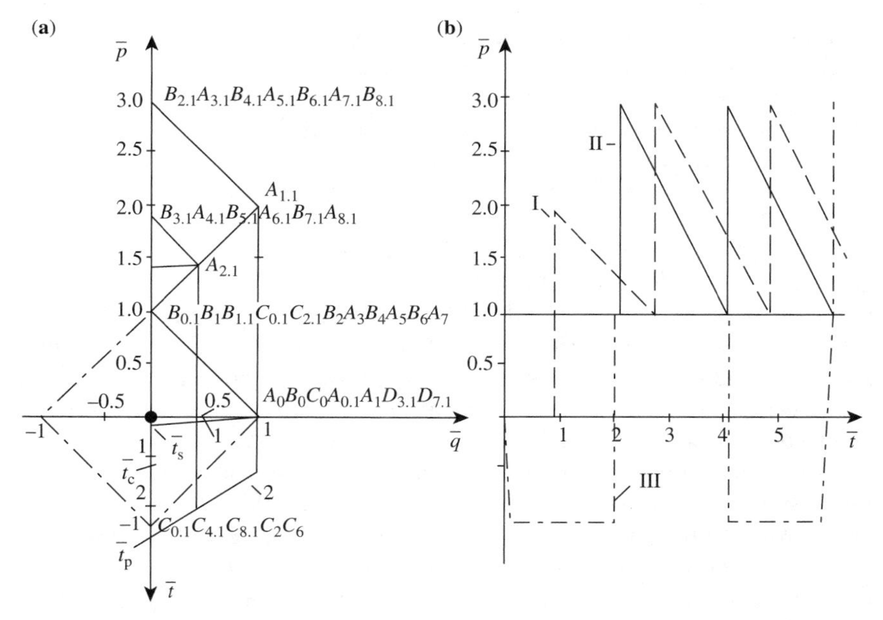

FIGURE 10.14 Scheme of graphical calculations (a) and dependences of pressure variation with time (b) in characteristic cross sections: I, II, and III—curves of pressure variation in cross sections A, B, and C.

is equal to $\bar{q} = 0$ since $\bar{t} = 1 > 0.1$ due to (10.9.7). Consequently, the pressure $\bar{p}$ in cross section B at $\bar{t} = 1$ corresponds to intersection point B_1 of straight lines A_0B_1 and $\bar{q} = 0$.

Consider a wave leaving cross section A at $\bar{t} = \bar{t}_s - 1$. It comes into cross section B at $\bar{t} = \bar{t}_s = 0.1$ associated with complete closing of the thrust ring orifice by the ball cage. This wave is a direct wave and the line determined by equation (10.6.22) should pass through point A_0. Thus, the pressure in cross section B at $\bar{t} = 0.1$ would be determined by the ordinate of the point $B_{0.1}$. So, point $B_{0.1}$ coincides with point B_1.

The backwave leaving cross section B at $\bar{t}_s = 0.1$ comes into cross section A at $\bar{t} = 1.1$. The pressure in the latter cross section is determined as the ordinate of the intersection point $A_{1.1}$ of straight lines $B_{0.1}A_{1.1}$ and $\bar{q} = 1$. At this instant of time, the pump is cut out. In cross section A at $\bar{t} = 1$ conditions (10.9.4)–(10.9.5) will still be obeyed. Really, if to consider the backwave leaving cross section B at $\bar{t} = 0$ and arriving at cross section A at $\bar{t} = 1$, the intersection of lines given by equation (10.6.22) and condition (10.9.5) gives point A_1. Considering the wave leaving cross section A at $\bar{t} = 1$, we find that it comes to cross section B at $\bar{t} = 2$ and wave parameters would correspond to the intersection point B_2 of straight lines A_1B_2 and $\bar{q} = 0$. In a similar manner, points B_{11}, B_2, A_3, B_4, and so on could be found.

The wave leaving cross section B at $\bar{t} = 1.1$ reaches cross section A at $\bar{t} = 2.1$.

In the lower part of the scheme, draw through point $\bar{t} = 2.1$ a straight line parallel to $\bar{q}$-axis up to its intersection with curve 2 and then erect a perpendicular up to intersection with straight line $B_{1.1}A_{1.1}$. As a result, we get point $A_{2.1}$ characterizing pressure and flow rate in cross section A at $\bar{t} = 2.1$. Waves reflected from cross section A at $\bar{t} = 1.1$ and $\bar{t} = 2.1$ arrived at cross section B at $\bar{t} = 2.1$ and $\bar{t} = 3.1$. Values of $\bar{p}$ and $\bar{q}$ in this cross section will be associated with coordinates of points $B_{2.1}$ and $B_{3.1}$.

Waves reflected from cross section B are backwaves. After these waves arriving at cross section A, the values of $\bar{p}$ and $\bar{q}$ would not change because $\bar{t} > 3$ and due to (10.9.10) the flow rate is $\bar{q} = 0$ and in accordance with (10.6.23) $\bar{p} = \text{const}$.

Thus, pressure and flow rate at $\bar{t} = 3, 5, 7$ and so on in cross section A and at $\bar{t} = 2, 4, 6$, and so on in cross section B are displayed by a point with coordinates $\bar{p} = 1, \bar{q} = 0$. At $\bar{t} = 4.1, 6.1$, and so on in cross section A and at $\bar{t} = 3.1, 5.1, 7.1$, and so on in cross section B, there are $\bar{p} = 1.92$ and $\bar{q} = 0$.

The obtained pressures are plotted in Fig. 10.14b. For annular space, straight lines associated with direct and backwaves (see Fig. 10.14a) give intersection points with coordinate axes, with the help of which in Fig. 10.14b pressure variation with time is built. Since in the motion of waves pressure losses due to friction are not taken into account, the graphic of pressure variation in Fig. 10.14b represents an undamped process.

The friction can be taken into account at each step by subtraction from obtained pressure friction losses calculated by Darcy–Weisbach formula for stationary flow at associated instant of time.

10.10 CALCULATION OF PRESSURE IN ROUND TRIP OF DRILL-STEM AS WAVE PROCESS

In Sections 10.3–10.5, calculations of pressure in round trip operations in a well filled with incompressible fluid were considered. Consider now the calculation of pressure on the basis of wave representations applicable to analyze slightly compressible fluid flows.

The circulation system in round trip operation can be represented as a pipeline with given variation in the fluid flow rate in one of the cross sections. Such scheme as applied to the round trip operation of a drill-stem with closed end is shown in Fig. 10.15.

EXAMPLE 10.10.1

Calculate the pressure distribution with time in descending the drill-stem with closed lower end in the well assuming the process to be wave. Initial data are the same as in Example 10.5.1. Schemes of the circulation system and the pipeline associated with it are shown in Fig. 10.15.

SOLUTION The flow rate in the pipeline is determined as flow rate averaged with time using formula (10.3.10).

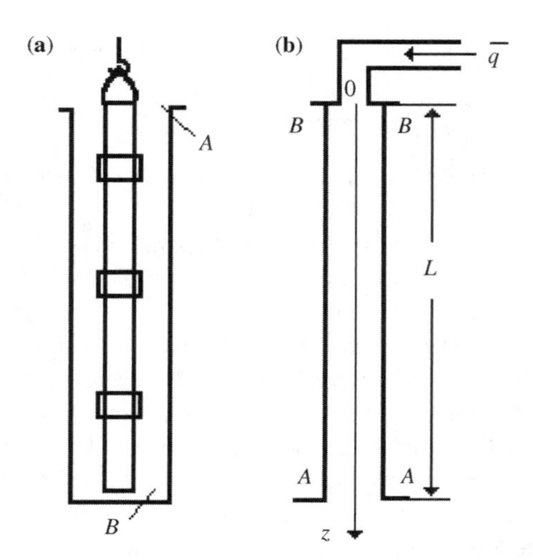

FIGURE 10.15 Well circulation system in drill-stem descent (a) and associated calculation scheme (b).

In accordance with formula (10.1.21), the sound velocity is

$$E'' = \frac{1}{\frac{1}{E} + \frac{d_h}{\Delta}\frac{1}{E'}} = \frac{1}{\frac{1}{2.43\times10^3\times10^6} + \frac{0.224-0.141}{0.011\times2.1\times10^5\times10^6}} = 2.24\times10^9\ \text{Pa};$$

$$c = \sqrt{E''/\rho} = \sqrt{2.24\times10^9/1710} = 1145\ \text{m/s}.$$

Dimensionless parameters are

$$\bar{t} = \frac{ct}{L} = \frac{1145t}{1192} = 0.96t;\ \ \bar{p} = \frac{pS}{\rho c q} = \frac{p}{\rho c v} = \frac{p}{1710\times1145\times3.15} = 1.62\times10^{-7}p;$$

$$\bar{q} = \frac{q}{\pi u_p R_1^2} = \frac{q}{3.14\times4.8(0.141/2)^2} = 13.35q,$$

where $v = 3.15$ m/s is calculated with formula (10.3.10) at $u_{ds} = 4.8$ m/s.

Turn now to get pressure variation in cross section B under the drill-stem. In accordance with initial data of Example 10.4.1 and introduced dimensionless parameters $\bar{p}$, $\bar{q}$, $\bar{t}$, initial and boundary conditions are

$$\bar{p} = 0,\quad \text{at}\quad 0 \le z \le L,\quad \bar{t} = 0; \tag{10.10.1}$$

$$\bar{q} = 0,\quad \text{at}\quad 0 \le z \le L,\quad \bar{t} = 0; \tag{10.10.2}$$

$$\bar{q} = 1 - 0.26\bar{t}\quad \text{at}\quad z = 0;\quad 0 \le \bar{t} \le 3.84; \tag{10.10.3}$$

$$\bar{q} = 1\quad \text{at}\quad z = 0;\quad 3.84 \le \bar{t} \le 5; \tag{10.10.4}$$

$$\bar{q} = 0.26(8.83 - \bar{t})\quad \text{at}\quad z = 0;\quad 5 \le \bar{t} \le 8.83; \tag{10.10.5}$$

$$\bar{p} = 0\quad \text{at}\quad z = L,\quad \bar{t} > 0. \tag{10.10.6}$$

In the lower part of the graphic in Fig. 10.16a, we build curve 1 of the flow rate in accordance with conditions (10.10.3)–(10.10.5).

According to conditions (10.10.1)–(10.10.2), parameters $\bar{p}$ and $\bar{q}$ in cross sections A and B are equal to zero and correspond to the origin of coordinates. These initial states of waves are denoted by points A_0 and B_0. Point A_1 characterizing the wave state in cross section A at $\bar{t} = 1$ also coincides with the origin of coordinates.

The wave leaving at $\bar{t} = 0$ cross section A is a backwave. It arrives at cross section B at $\bar{t} = 1$. Parameters of the wave at this instant of time are determined by coordinates of point B_1, which is an intersection point of straight line A_0B_1 with a straight line passing at $\bar{t} = 1$ through a point of curve 1 perpendicular to $\bar{q}$-axis. In a similar manner, we get point B_2 for $\bar{t} = 2$. Direct waves reflected from cross section B reach the cross section A at $\bar{t} = 2$ and $\bar{t} = 3$ with values of $\bar{p}$ and $\bar{q}$ corresponding to coordinates of points A_2 and A_3 being intersection points of straight lines B_1A_2 and B_2A_3 with axis $\bar{p} = 0$ (see condition (10.10.6)).

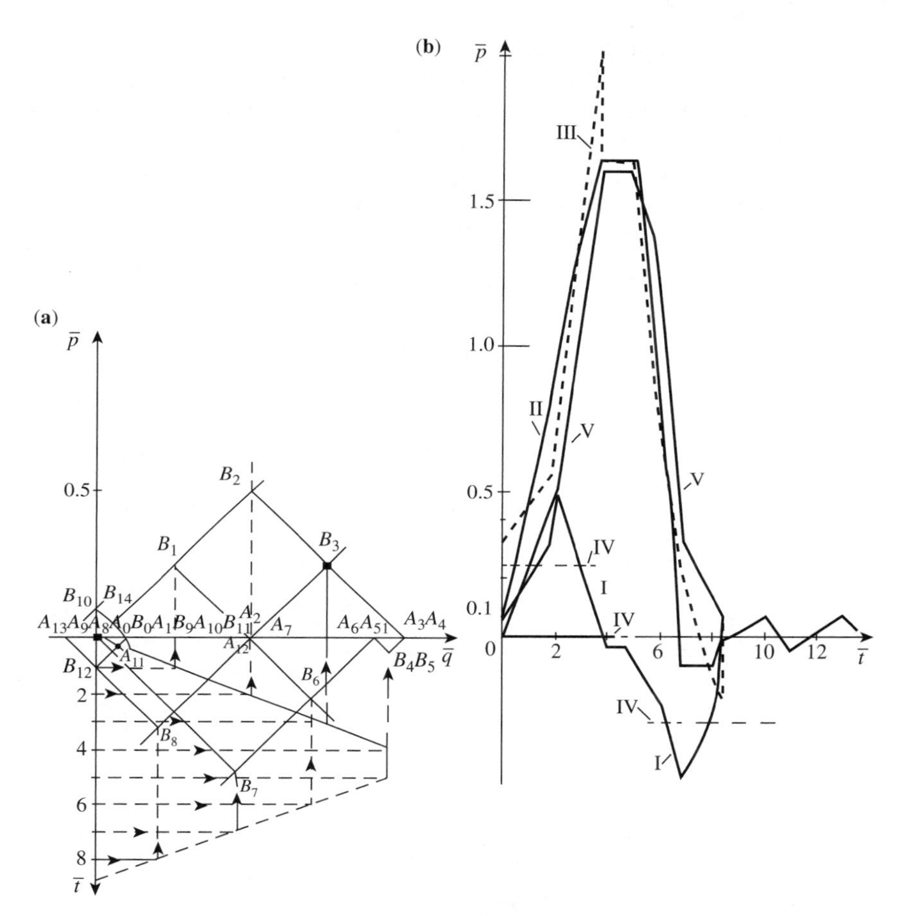

FIGURE 10.16 Scheme of graphical calculations (a) and pressure variations with time (b) in cross section B: I—compressible ideal fluid; II—compressible viscous-plastic fluid; III—incompressible viscous-plastic fluid; IV—incompressible ideal fluid; V—friction losses for compressible and incompressible viscous-plastic fluids.

Waves having been reflected from cross section A reach cross section B at $\bar{t} = 3$ and $\bar{t} = 4$. Flow rates of curve I correspond to them at these instants of time. The following calculations are made in a similar manner.

Values of $\bar{p}$ equal to ordinates of points B_i are carried from Fig. 10.16a over to Fig. 10.16b. As a result, we get pressure distribution with time (curve I) in the cross section B under the drill-stem. In order to compare results of pressure calculation (curve III) for incompressible fluid with results obtained in Example 10.4.1, we take into account the pressure change $\Delta p_{\mathrm{fr}} + \Delta p_{\mathrm{lr}}$ due to friction losses and local resistances assuming them to be independent of inertia forces. Construct in Fig. 10.16b curve V ($\bar{p} = \Delta\bar{p}_{\mathrm{T}}(\bar{t})$) answering data of Example 10.5.1 presented in Table 10.2. Then, we add up ordinates of this curve with ordinates of curves I or IV. As a result, we get desired curves II and III.

As seen from the comparison of curves II and III, there is a certain domain of variability of parameters at which one can use formulas for incompressible fluid to calculate pressures in round trip operations.

CHAPTER 11

FLOWS OF FORMATION FLUIDS AND ROCK SOLIDS

11.1 BASIC EQUATIONS OF FORMATION FLUID AND ROCK SOLID FLOWS

For drilling and exploitation of wells, it is necessary to study mass-exchange processes not only in a well but also between well and drilled formation (Businov and Umrichin, 1973; Pihachev and Isaev, 1973; Shischenko et al., 1976). Interaction of media in well-formation system is chiefly determined by the flow of washing fluids and grouting mortars from the well into the formation (absorption) or of formation fluid into the well (inflow). In some cases, the flow can enclose both fluid and rock skeleton. Then, when the rock skeleton is permeable, its flow and the flow of each phase of the fluid take place with different velocities. If the rock skeleton is practically impermeable, that is, the fluid contained in it exists in the state bounded by adsorptive forces in closed pores (voids), the flow is observed as flow of the whole rock medium without relative velocity of its components, so flow under certain conditions, for example, argillaceous solids and salts narrows down the well bore.

Assume that the formation has immovable horizontal cover and bottom, gravity force and the formation skeleton do not make great impact on the flow, phase transitions are absent, flow happens at constant average temperature of the formation. In general, the pressure in formation p_1 differs from the pressure in fluid p_2. Pressure p in the rock medium can be represented as $p = \varphi_1 p_1 + \varphi_2 p_2$.

Consider a question about relations between real flows in the formations and their descriptions with the system of equations (4.6.12)–(4.6.16) or (4.6.46)–(4.6.52).

A peculiarity of application of these equations to flows in the formation is as follows. At $\varphi_1 = 1$, $\varphi_2 = 0$, they describe a flow in round slot of impermeable formation and $p = p_1$ is pressure of the solid in the skeleton. At $\varphi_1 = 0$, $\varphi_2 = 1$, the system of equations describes fluid flow in round slot, that is, in round formation without skeleton and pressure $p = p_2$ is pressure of pore fluid. At $0 < \varphi_1 < 1$ and $0 < \varphi_2 < 1$, one can under φ_1 and φ_2 in equations (4.6.46–4.6.52) understand, for example, concentration φ_1 of movable fluid and concentration φ_2 of immovable medium represented by the sum of concentrations of immovable fluid and formation skeleton.

Interaction of the skeleton and moving fluid is accomplished through layers of immovable fluid adhering to wall surfaces of the formation skeleton channels through adhesive forces and may occupy significant part of the pore space. Thus, compatibility between real flow in the formation and a model described by equations (4.6.12–4.6.16) is set with the help of concentrations φ_1 and φ_2. In general, as for two-phase flows in pipes, concentrations φ_1 and φ_2 are functions to be empirically obtained.

The system of equations ($i = 1, 2, \ldots, N$) describing flow of the formation fluids and rock solids is

average momentum equation

$$\sum \rho_i \varphi_i \left(\frac{\partial v_i}{\partial t} + v_i \frac{\partial v_i}{\partial r} \right) + \frac{\partial p}{\partial r} = -\frac{\lambda_c}{2H} \sum \varphi_i \rho_i v_i |v_i|; \tag{11.1.1}$$

equation of mass conservation

$$\frac{\partial \varphi_i \rho_i}{\partial t} + \frac{1}{r} \frac{\partial r \varphi_i \rho_i v_i}{\partial r} = 0; \tag{11.1.2}$$

thermodynamic equation of state

$$p = p(\rho_i, \bar{T}); \tag{11.1.3}$$

equation of concentrations

$$\varphi_i = \varphi_i(p, \rho_1, \rho_2, \ldots, \rho_N, v_1, v_2, \ldots, v_N, \lambda_c); \tag{11.1.4}$$

equation for hydraulic resistance factor

$$\lambda_c = \lambda_c(p, \rho_1, \rho_2, \ldots, \rho_N, v_1, v_2, \ldots, v_N, \varphi_1, \varphi_2, \ldots, \varphi_N). \tag{11.1.5}$$

11.2 STATIONARY LAMINAR FLOWS OF INCOMPRESSIBLE AND COMPRESSIBLE FLUIDS AND GASES

As it was noted above, the flow in a formation may be of great variety. Flows of fluid can happen both in immovable and in movable formation. Consider fluid flow in immovable formation skeleton (Leonov and Isaev, 1982).

In this case $N=2$ and the momentum equation (11.1.1) at $v_1=0$, $v_2=v$, $\varphi=\varphi_2$ in inertialess approximation is written as

$$\frac{\partial p}{\partial r} = -\frac{\lambda_c}{2H}\varphi\rho v|v|, \tag{11.2.1}$$

where v is true velocity of the fluid expressed through average velocity v_{av} in accordance with (3.7) as $v=v_{av}/\varphi$ (at absorption $v>0$ whereas at inflow (show) $v<0$); φ is concentration of the moving fluid. Immovable fluid layers are connected to solid skeleton. These layers are in particular adsorption layers as well as fluid in dead (stagnation) zones of the flow.

Equation (11.2.1) is the equation of average filtration in the formation. Since values of λ_c under the formation conditions are unavailable, we take λ_c equal to hydraulic resistance factor in laminar flow of single-phase viscous fluid between plates

$$\lambda_c = 24/Re, \tag{11.2.2}$$

where $Re=|v_{av}|h\rho_{av}/\mu$; $v_{av}=\varphi v$; $\rho_{av}=\varphi\rho$; $h=\varphi H$.

Substitution of (11.2.2) in (11.2.1) yields

$$\frac{\partial p}{\partial r} = -\frac{12\mu}{H^2\varphi^3}v_{av}. \tag{11.2.3}$$

If to introduce the permeability factor

$$k_f = H^2\varphi^3/12 \tag{11.2.4}$$

equation (11.2.3) reduces to Darcy law in differential form

$$v_{av} = -\frac{k_f}{\mu}\frac{\partial p}{\partial r}. \tag{11.2.5}$$

Application of formula (11.2.3) is restricted by certain assumptions to be made in this section.

Since the average velocity is $v_{av}=Q/(2\pi rH)$, from (11.2.5) we get

$$Q\frac{dr}{r} = -\frac{2\pi Hk_f}{\mu}dp. \tag{11.2.6}$$

Integration of equation (11.2.6) in limits from r_w (well radius) to r_c (circuit radius) and from p_w to p_f gives for incompressible fluid at $Q=$ constant

$$Q = -\frac{2\pi H k_f}{\mu}(p_f - p_w)/\ln\frac{r_c}{r_w}. \tag{11.2.7}$$

This formula is known as Dupuis formula.

Now let us derive a formula to calculate flow rate Q in filtration of compressible fluid (gas). The equation of mass conservation gives

$$\rho Q = \rho_0 Q_0 = \text{const}, \tag{11.2.8}$$

where ρ_0 and Q_0 are density and flow rate of fluid at normal conditions.

Multiplying (11.2.6) by ρ, we get with regard to (11.2.8)

$$\rho_0 Q_0 \frac{\mathrm{d}r}{r} = -\frac{2\pi H k_f}{\mu}\rho\,\mathrm{d}p. \tag{11.2.9}$$

Replacement of ρ in (11.2.9) through its value from gas state equation

$$p = \bar{z}R\bar{T}\rho g \tag{11.2.10}$$

and integration in the same limits gives the formula for gas inflow from the formation (Pihachev and Isaev, 1973)

$$Q_0 = -\frac{\pi H k_f (p_f^2 - p_w^2)}{\mu \rho_0 g \bar{z} R\bar{T}\ln(r_c/r_w)} = -\frac{\pi H k_f (p_f^2 - p_w^2)}{\mu p_0 \ln(r_c/r_w)}. \tag{11.2.11}$$

Suppose that in laminar inflow of viscous-plastic fluid it is valid (11.2.4) and apply to filtration in the formation the formula (6.3.14)

$$v_{av} = -\frac{k_f}{\eta}\left[\frac{\partial p}{\partial r} - a\frac{\tau_0}{\sqrt{k_f}} + b\frac{\left(\frac{\tau_0}{\sqrt{k_f}}\right)^3}{\left(\frac{\partial p}{\partial r}\right)^2}\right], \tag{11.2.12}$$

where $a = \sqrt{3}/2$; $b = \sqrt{3}/18$.

From (6.3.16) with regard to (11.2.4), we obtain the formula for pressure drop in filtration of viscous-plastic fluid

$$\Delta p = \frac{r_c \tau_0}{\sqrt{3 k_f}\beta}\ln\frac{r_c}{r_w}. \tag{11.2.13}$$

At $\tau_0 \to 0$ ($\beta \to 0$), formulas (11.2.12) and (11.2.13) transform into formulas for viscous fluid.

For power fluid under suggestion that the relation (11.2.4) is valid for laminar flow, we get from (6.4.14)

$$v_{\text{av}} = \pm a \left(\frac{\partial p}{\partial r} \right)^{1/n}; \quad a = \left(\sqrt{3k_{\text{f}}} \right)^{1/n+1} \frac{n}{2n+1} \left(\frac{1}{k} \right)^{1/n}. \qquad (11.2.14)$$

In the lost of circulation it is taken sign plus. In the well flow it is taken sign minus.

To calculate pressure, one can use formulas of Section 6.4, taking in them k_{f} in the form of (11.2.4).

11.3 NONSTATIONARY LAMINAR FLOWS OF INCOMPRESSIBLE AND COMPRESSIBLE FLUIDS AND GASES

Flow velocities of fluids with different rheology considered in the previous section are variable along radius, that is, they depend on coordinate r, but are independent of time t. In this section, we shall find the connection of pressure drop in the formation $\Delta p(t) = p_{\text{f}}(t) - p_{\text{c}}(t)$ with the flow rate $Q(t)$ of fluid or formation solid in nonstationary flow when pressure and flow rates vary with time.

Suppose that Darcy law in the form of (11.2.5), which is experimentally confirmed for stationary flow of viscous fluid, is also valid for nonstationary flow of viscous fluids as well as relations (11.2.12) and (11.2.14) for viscous-plastic and power fluids. Each of the expressions (11.2.5), (11.2.12), and (11.2.14) replaces momentum equation (11.1.1). Thus, to determine relations between $\Delta p(t)$ and $Q(t)$ it is needed to solve the system of equations with replacement of momentum equation (11.1.1) by one of the equations (11.2.5), (11.2.12), or (11.2.14), in which equations (11.1.4) and (11.1.5) are always used.

In inflow of viscous slightly compressible fluid, the system of equations has the following form:

momentum equation

$$v_{\text{av}} = -\frac{k_{\text{f}}}{\mu} \frac{\partial p}{\partial r}; \qquad (11.3.1)$$

equation of mass conservation

$$\frac{\partial \rho}{\partial t} + \frac{1}{r} \frac{\partial r \rho v_{\text{av}}}{\partial r} = 0; \qquad (11.3.2)$$

state equation of slightly compressible fluid

$$\rho = \rho_0 [1 + \beta_0 (p - p_0)]. \qquad (11.3.3)$$

Substitution of (11.3.1) and (11.3.3) in (11.3.2) gives the equation for elastic filtration regime of slightly compressible fluid in Schelkachev form (Schelkachev, 1990)

$$\frac{\partial p}{\partial t} = \chi\left(\frac{\partial^2 p}{\partial r^2} + \frac{1}{r}\frac{\partial p}{\partial r}\right), \tag{11.3.4}$$

where $\chi = k_{\mathrm{f}}/(\mu\beta_0)$ is piezoconductivity factor characterizing pressure redistribution in the formation.

Equation (11.3.4) is similar to the heat conduction equation.

In fluid filtration or flow of formation as viscous fluid during well testing, it is necessary to know pressure drop that could be determined by solution of the equation with the following initial and boundary conditions for infinite formation

$$\begin{aligned} p(r,t) &= p_{\mathrm{f}} \quad \text{at} \quad t=0; \\ \frac{2\pi\kappa_{\mathrm{f}}H}{\mu}\left(r\frac{\partial p}{\partial r}\right)_{r=r_{\mathrm{w}}} &= Q = \text{const} < 0 \quad \text{at} \quad t>0; \end{aligned} \tag{11.3.5}$$

$$p(r,t) = p_{\mathrm{f}} \quad \text{at} \quad r \to \infty. \tag{11.3.6}$$

The exact solution (Thomson–Kelvin formula, see Businov and Umrichin (1973)) of this problem at $r_{\mathrm{w}} = 0$ is

$$p_{\mathrm{c}} - p(r,t) = -\frac{Q\mu}{4\pi\kappa_{\mathrm{f}}H}\mathrm{Ei}\left(-\frac{r^2}{4\chi t}\right), \tag{11.3.7}$$

where $\mathrm{Ei}(-r^2/4\chi t)$ is integral exponential function, values of which could be found from special tables (see Tables of the integral exponential function (Gradstein I.S. and Ryszik I.M.)).

By expanding $\mathrm{Ei}(-r^2/4\chi t)$ into a series and restricting by the first term of the series, we obtain

$$\mathrm{Ei}\left(-\frac{r^2}{4\chi t}\right) = 0.5772 - \ln\frac{4\chi t}{r^2}. \tag{11.3.8}$$

Then, the formula (11.3.7) transforms to

$$p_{\mathrm{f}} - p = \frac{Q\mu}{4\pi\kappa_{\mathrm{f}}H}\left[\ln\frac{(t_{\mathrm{f}}+t)4\chi}{r^2} - 0.5772\right]. \tag{11.3.9}$$

The greatest error of Ei at $(4\chi t/r^2) \geq 8.33$ is 1%.

The approximate solution (11.3.9) is used to determine the hydroconductivity factor of the formation $\kappa_{\mathrm{f}}H/\mu$ and consequently permeability factor from results of well testing, in particular, at shutdown of the circulation in the well. Here, it is assumed that in nonstationary flows in elastic regimes superposition of flows takes place. In particular, the pressure sought is equal to the sum of pressure components.

Suppose that the well is put into operation with constant flow rate Q (Businov and Umrichin, 1973). The distribution of pressure p_f in the formation can be obtained with formula (11.3.9). Suppose the well is shutdown in time t_f after well start-up. Since the well shutdown, the pressure in it begins to rise and perturbation caused by sudden shutdown propagates over the whole formation. In such a case, one can accept that the rise of pressure p'' happens owing to the sign change of the flow rate Q.

Thus, the process may be represented as follows: beginning from the instance of time t_f at one and the same place of the formation operation and injection wells as if working jointly and continuously.

From (11.3.9), it follows

$$\begin{aligned} p' &= p_f - p_{t_f+t} = \frac{Q\mu}{4\pi\kappa_f H}\left[\ln\frac{(t_f+t)4\chi}{r^2} - 0.5772\right], \\ p'' &= p_f - p_t = \frac{Q\mu}{4\pi\kappa_f H}\left[\ln\frac{4\chi t}{r^2} - 0.5772\right]. \end{aligned} \quad (11.3.10)$$

Denoting the bottom pressure through p_{bot} and using assumption of pressure superposition, we get Horner formula (Horner, 1951)

$$p_{bot} - p_f \approx p_t - p_{t_f+t} = p' - p'' = \frac{Q\mu}{4\pi\kappa_f H}\ln\frac{t_f+t}{t}. \quad (11.3.11)$$

With the help of formula (11.3.11), one can determine the hydro-conductivity factor $\kappa_f H/\mu$ from the pressure buildup curve.

Consider an example of determination of formation parameters through results obtained by the formation tester lowered into the bottom.

EXAMPLE 11.3.1

It is required to determine hydro-conductivity and permeability factors of the formation on the following initial data: testing interval $\Delta L = 1445$–1477 m, well diameter $d_w = 0.190$ m; testing were conducted with one open and one closed periods; inflow time $t_f = 42$ min $= 2520$ s; recovery time $t_r = 30$ min $= 1800$ s. Pressure diagram is shown in Fig. 11.1.

SOLUTION Rebuild the curve 2 of pressure restoration in coordinates p_{bot} and $\log[(t_f + t)/t]$. To do this, let us read the values of p_{botm} and $t_m = m\Delta t$

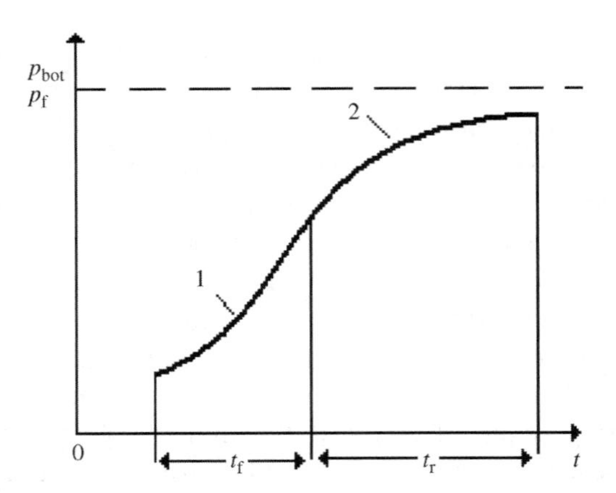

FIGURE 11.1 Pressure variation with time during inflow (1) and pressure recovery (2).

from the pressure restoration curve. Take the number of points $m = 6$. Then $\Delta t = t_r/6 = 1800/6 = 300$ s. Calculated values of $(t_f + t_m)/t_m$ and $\log[(t_f + t_m)/t_m]$ are presented below.

m	1	2	3	4	5	6
p_{botm} (MPa)	14.06	14.59	14.85	15.04	15.18	15.27
t_m, s	300	600	900	1200	1500	1800
$(t_f + t_m)/t_m$	9.4	5.2	3.8	3.1	2.68	2.4
$\log[(t_f + t_m)/t_m]$	0.973	0.716	0.580	0.491	0.428	0.380
Q_{av} (m^3/day)	−1230					

Given data permit to build dependence of p_{bot} on $\log[(t_f + t_m)/t_m]$ (Fig. 11.2). Resolve the formula (11.3.11) with respect to $\kappa_f H/\mu$

$$\frac{\kappa_f H}{\mu} = \frac{Q \ln[(t_f + t)/t]}{4\pi(p_{bot} - p_f)} = 0.183Q \cdot \frac{\lg[(t_f + t)/t]}{p_{bot} - p_f}. \tag{11.3.12}$$

To determine the hydro-conductivity factor it is necessary to know the flow rate Q. Take Q as mean value of the flow rate during inflow time

$$Q_{av} = V/t_1, \tag{11.3.13}$$

where V is fluid volume arrived at a time of the inflow.

The value of formation pressure $p_{av} = 16.1$ MPa is obtained extending the line in Fig. 11.2 up to its intersection with pressure axis.

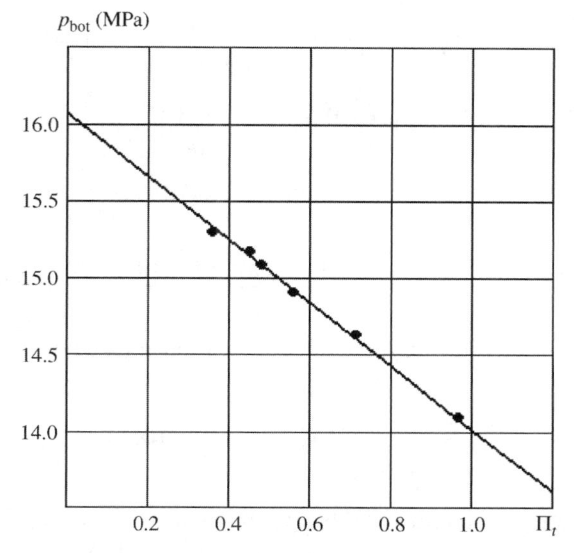

FIGURE 11.2 Experimental dependence of pressure on $\prod_t = \log[(t_f + t)/t]$.

The current value of the hydro-conductivity factor $(\kappa_f H/\mu)_m$ is obtained from formula (11.3.12) by inserting in it p_{botm} and associating with it $\log[(t_f + t_m)/t_m]$ instead of p_b

$$\left(\frac{k_f H}{\mu}\right)_m = -0.183 Q_{av} \log \frac{[(t_f + t_m)/t_m]}{p_f - p_{bm}}. \qquad (11.3.14)$$

Mean value of the hydro-conductivity factor is defined as

$$\left(\frac{k_f H}{\mu}\right)_{av} = \sum \frac{(k_f H/\mu)_m}{m}. \qquad (11.3.15)$$

After determining the viscosity of the obtained fluid sample and knowing the formation thickness H, we get the permeability factor

$$k_f = \frac{\mu}{H}\left(\frac{k_f H}{\mu}\right)_{av}. \qquad (11.3.16)$$

Then with (11.3.14) is found

$$\left(\frac{k_f H}{\mu}\right)_1 = -\frac{-1230 \times 0.183 \times 0.973}{(16.1 - 14.06) \times 10^6} = 1.07 \times 10^{-4}\ \mathrm{m^3/Pa{\cdot}s};$$

$$\left(\frac{k_f H}{\mu}\right)_2 = 1.07 \times 10^{-4}\ \mathrm{m^3/Pa{\cdot}s};$$

$$\left(\frac{k_{\rm f}H}{\mu}\right)_3 = 1.04 \times 10^{-4}\,{\rm m^3/Pa{\cdot}s}; \quad \left(\frac{k_{\rm f}H}{\mu}\right)_4 = 1.04 \times 10^{-4}\,{\rm m^3/Pa{\cdot}s};$$

$$\left(\frac{k_{\rm f}H}{\mu}\right)_5 = 1.05 \times 10^{-4}\,{\rm m^3/Pa{\cdot}s}; \quad \left(\frac{k_{\rm f}H}{\mu}\right)_6 = 1.03 \times 10^{-4}\,{\rm m^3/Pa{\cdot}s}.$$

Mean value of the hydro-conductivity factor in accordance with formula (11.3.15) is

$$\left(\frac{k_{\rm f}H}{\mu}\right)_{\rm av} = \frac{(1.07+1.07+1.04+1.04+1.05+1.03)10^{-4}}{6} = 1.05\times10^{-4}\,{\rm m^3/Pa{\cdot}s}.$$

For given viscosity $\mu = 0.01$ Pa·s and formation thickness $H = 5$ m, we get from (11.3.16)

$$k_{\rm f} = \frac{0.01}{5}1.05 \times 10^{-4} = 2.1 \times 10^{-7}\,{\rm m^2}.$$

The exact solution (Pihachev and Isaev, 1973) of equation (11.3.4) when the second condition (11.3.5) is obeyed at some boundary $r = r_{\rm k}$

$$p_{\rm k} - p_{\rm f} = -\frac{\mu r_{\rm c}}{k_{\rm f}} \frac{{\rm Ei}\left(-\frac{r_{\rm c}}{4\chi t}\right)}{2\exp\left(-\frac{r}{4\chi t}\right)} v|_{r=r_{\rm w}}$$

permits to obtain the velocity of fluid filtration at well walls or well wall narrowing in viscous solids

$$v_{\rm av}|_{r=r_{\rm w}} = -\frac{k_{\rm f}(p_{\rm k}-p_{\rm f})}{\mu r_{\rm c}} \frac{2\exp\left(-\frac{r_{\rm w}}{4\chi t}\right)}{{\rm Ei}\left(-\frac{r_{\rm w}}{4\chi t}\right)}. \tag{11.3.17}$$

To get the relationship between $\Delta p(t)$ and $Q(t)$ in the flow of formation solid and fluid contained in it, the rheological model of which is viscous-plastic slightly compressible medium, one should use the following system of equations:

momentum equation (11.2.12) (Leonov and Isaev, 1982; Leonov and Triadski, 1980)

$$v_{\rm av} = -\frac{k_{\rm f}}{\eta}\left[\frac{\partial p}{\partial r} - a\frac{\tau_0}{\sqrt{k_{\rm f}}} + b\frac{\left(\tau_0/\sqrt{k_{\rm f}}\right)^3}{(\partial p/\partial r)^2}\right]; \tag{11.3.18}$$

equation of mass conservation

$$\frac{\partial \rho}{\partial t} + \frac{1}{r}\frac{\partial(r\rho v_{\rm av})}{\partial r} = 0; \tag{11.3.19}$$

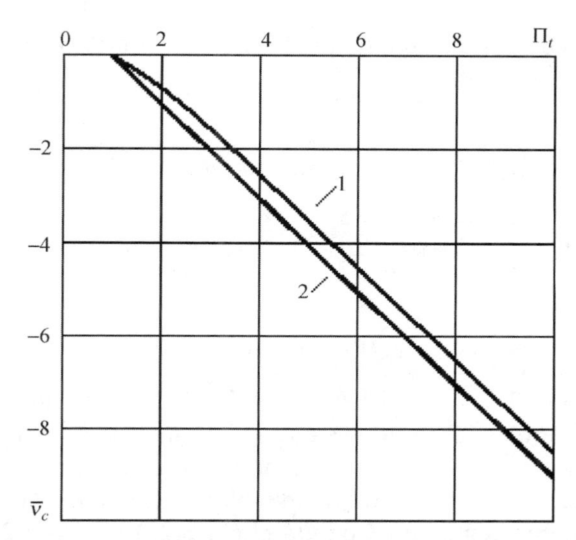

FIGURE 11.3 Graphics of function $\overline{v}_{av} = 6\eta v_{av}/I'\,\tau_0 = \overline{v}_{av}(\Pi_t = (\partial p/\partial r)/(\tau_0/\sqrt{3k_f}))$ (curve 1) and its approximation (curve 2).

state equation

$$\rho = \rho_0(1+\beta_0(\rho-\rho_0)). \tag{11.3.20}$$

Equations (11.3.18)–(11.3.20) describe nonstationary flow of solid formation if to take $k_f = H^2\varphi^3/12$, $\varphi = 1$.

The product of density ρ and velocity v_{av} at $(\partial p/\partial r)/\left(\sqrt{3}/2\right)\times\left(\tau_0/\sqrt{k_f}\right) \geq 2$ and $\beta_0(p-p_0) \leq 1$ can be approximated by dependence (see Fig. 11.3)

$$\rho v_{av} = -\rho\frac{k_f}{\eta}\left[\frac{\partial p}{\partial r} - a\frac{\tau_0}{\sqrt{k_f}} + b\frac{\left(\frac{\tau_0}{\sqrt{k_f}}\right)^3}{\left(\frac{\partial}{\partial r}\right)^2}\right] \approx \rho_0\left(\frac{\partial p}{\partial r} - \frac{\tau_0}{\sqrt{3k_f}}\right). \tag{11.3.21}$$

Substitution of (11.3.21) and (11.3.20) in the left part of (11.3.19) gives the following piezoconduction equation:

$$\frac{\partial p}{\partial t} = \frac{\chi}{r}\frac{\partial}{\partial r}\left[r\left(\frac{\partial p}{\partial r} - \frac{\tau_0}{\sqrt{3k_f}}\right)\right], \tag{11.3.22}$$

where $\chi = H^2/(12\eta\beta_0)$.

Now, let us formulate the boundary value problem on the inflow of slightly compressible viscous-plastic solid to the well.

It is required to solve equation (11.3.22) at the following conditions:

$$p(r,0) = p_{\mathrm{f}} \quad \text{at} \quad t = 0; \tag{11.3.23}$$

$$p(r_{\mathrm{w}},t) = p_{\mathrm{w}} \quad \text{at} \quad t > 0; \tag{11.3.24}$$

$$\frac{\partial p}{\partial r}\Big|_{r=r_{\mathrm{w}}} = \frac{\tau_0}{\sqrt{3k_{\mathrm{f}}}} \quad \text{at} \quad t > 0; \tag{11.3.25}$$

$$p(r_{\mathrm{c}},t) = p_{\mathrm{f}} \quad \text{at} \quad t > 0. \tag{11.3.26}$$

Condition (11.3.23) shows that the solid or fluid is in a quiescent state at constant pressure equal to the formation one p_{f}. The condition (11.3.24) implies that after drilling into the formation, pressure at the well contour (at $r = r_{\mathrm{c}}$) remains constant in the course of the process equal to the pressure p_{w} of fluid column in the well. Conditions (11.3.25) and (11.3.26) show that at certain movable boundary with radius $r = r_{\mathrm{c}}(t)$ separating the formation into undisturbed and disturbed regions the flow rate vanishes, that is, condition (11.3.26) is a consequence of (11.3.18) with regard to (11.3.21) at $\bar{\mathrm{v}} = 0$.

Approximate solution of equation (11.3.22) is sought in the following form:

$$p(r,t) = b_0(t) + b_1(t)\ln\frac{r}{r_{\mathrm{c}}} + b_2(t)\frac{r}{r_{\mathrm{c}}} + \frac{\tau_0}{\sqrt{3k_{\mathrm{f}}}}(r - r_{\mathrm{c}}). \tag{11.3.27}$$

Equation (11.3.27) with boundary condition (11.3.24)–(11.3.26) yields the following solution:

$$p(r,t) = p_{\mathrm{f}} - \frac{\tau_0}{\sqrt{3k_{\mathrm{f}}}}(r_{\mathrm{c}} - r) - \frac{\tau_0}{\sqrt{3k_{\mathrm{f}}}}(R_0 - r_{\mathrm{c}})\frac{\ln(r/r_{\mathrm{c}}) + 1 - (r/r_{\mathrm{c}})}{\ln(r_{\mathrm{w}}/r_{\mathrm{c}}) + 1 - (r_{\mathrm{w}}/r_{\mathrm{c}})}, \tag{11.3.28}$$

where $R_0 = r_{\mathrm{w}} + ((p_{\mathrm{f}} - p_{\mathrm{w}})/\tau_0)\sqrt{3\kappa_{\mathrm{f}}}$ and radius $r_{\mathrm{c}} = r_{\mathrm{c}}(t)$ to be determined.

Substitution of (11.3.28) in equation (11.3.22) leads to the following differential equation for dimensionless radius r_{c}:

$$\mathrm{d}\zeta = \left[2(\xi + 1) + \frac{2\ln\xi - \xi^2 + 1}{\xi\ln\xi - \xi + 1} + \frac{6\xi^2\ln\xi - 7\xi^3 + 6\xi^2 + 3\xi - 2}{(\lambda - \xi)(\xi - 1)}\right]\mathrm{d}\xi, \tag{11.3.29}$$

where $\zeta = 12(\chi/r_{\mathrm{w}}^2)t$; $\xi = r_{\mathrm{c}}/r_{\mathrm{w}}$; and $\xi_0 = R_0/r_{\mathrm{w}}$.

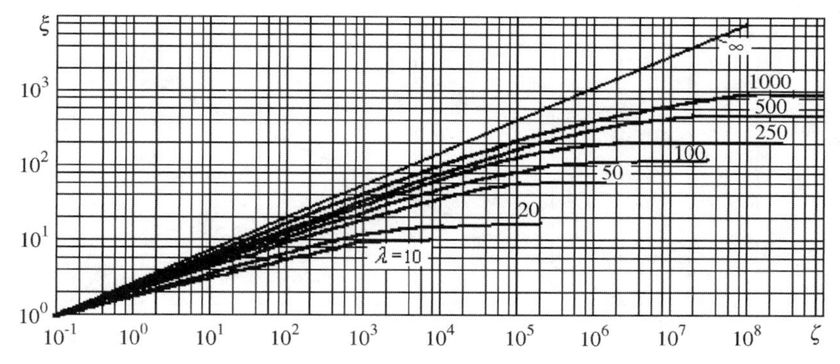

FIGURE 11.4 Graphic of the function $\xi(\zeta, \lambda)$.

It should be noted that at $\xi = 1$, that is, at $r_c = r_w$, there is $\zeta = 0$, at $\xi \to \lambda$, that is, at $p_f \to p_w$, there is $\zeta \to \infty$.

Equation (11.3.29) was solved numerically in the whole range of conditions encountered in drilling. Values of the parameter λ were varied from 10 to 10^3. Calculation was also made for $\lambda = \infty$ (at $\tau_0 = 0$). Values of ξ were changed in the range from 1 to 10^3 for λ from 10 to 10^3 and in the range from 1 to 10^4 for $\lambda = \infty$.

In Fig. 11.4, results of numerical calculation of equation (11.3.29) are plotted in the form of graphics of function $\xi(\zeta, \lambda)$ (Leonov and Triadski, 1980). For values of λ from 10 to 10^3, approximation function is chosen

$$\xi(\zeta, \lambda) = 0.2^{\zeta + \lambda}\left[1 - \exp\left(-\frac{\zeta^{0.4}}{\alpha(\lambda)}\right)\right], \tag{11.3.30}$$

where α is a function of the parameter λ.

The graphic of $\alpha(\lambda)$ is shown in Fig. 11.5. Differentiation of (11.3.28) with respect to r, replacement of r_c in ξr_w, and further substitution of the obtained derivative at $r = r_w$ in (11.3.18) with regard to (11.3.21) yield the formula for inflow rate of the fluid in the well or average rate of well bore narrowing in the case when the solid represents slight compressible viscous-plastic medium (Leonov and Triadski, 1980)

$$v_c\big|_{r=r_w} = \frac{1}{3}\frac{\sqrt{3\kappa_f}\tau_0}{\eta}(\lambda - \xi)\frac{\xi - 1}{\xi - 1 - \xi \ln \xi}. \tag{11.3.31}$$

From (11.3.31), one can estimate the velocity of core equal to maximal velocity $v_{max} = 3/2 v_{av}\big|_{r=r_w}$.

As $\tau_0 \to 0$ and $\zeta = 12\chi t/r_w^2 > 4 \times 10^3$, results of calculations with formulas (11.3.31) and (11.3.17) practically coincide. In the range $10^3 <$ $< 4 \times 10^3$, maximal difference is 2.7% at $\zeta = 10^3$.

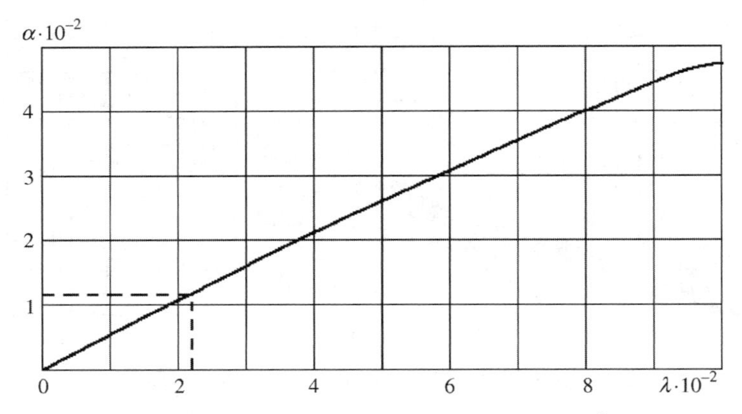

FIGURE 11.5 Graphic of the function $\alpha(\lambda)$.

Obtained regularities permit to calculate mean and maximal constriction rates of well bore walls in viscous-plastic solids as follows. Let the thickness of the formation is $H = 2\sqrt{3\kappa_f}$. With given rheological properties of solid τ_0 and η, elastic contraction factor β_0, geometric sizes r_w, H, and pressure drop $p_f - p_w$, we determine radius $R_0 = r_w + (p_f - p_w)/(2\tau_0/H)$, dimensionless parameter $\lambda = R_0/r_c$, and piezoconductivity factor $\chi = H^2/(12\eta\beta_0)$. Then, with the graphic in Fig. 11.5, we get $\alpha(R_0/r_w)$, with the formula (11.3.30) or from the graphic in Fig. 11.4 the current radius $r_c(t)$ of the disturbance. At last, formula (11.3.31) gives the rate of well bore constriction.

EXAMPLE 11.3.2

It is required to determine mean and maximal rate of well bore constriction at the following initial data: $\tau_0 = 20 \times 10^6$ Pa, $\eta = 2.6 \times 10^{12}$ Pa·s, $\beta_0 = 0.5 \times 10^{-9}$ Pa^{-1}, $r_w = 0.107$ m, $\Delta p = p_f - p_w = 30.2 \times 10^6$ Pa.

SOLUTION Determine

$$\chi = \frac{H^2}{12\eta\beta_0} = \frac{30^2}{12 \times 2.6 \times 10^{12} \times 0.5 \times 10^{-9}} = 5.77 \times 10^{-2}\ \mathrm{m^2/s};$$

$$R_0 = r_w + \frac{p_f - p_w}{2\tau_0/H} = 0.107 + \frac{30.2 \times 10^6}{2 \times 20 \times 10^6} 30 = 22.755\ \mathrm{m};$$

$$\frac{R_0}{r_c} = \frac{22.755}{0.107} = 212.682.$$

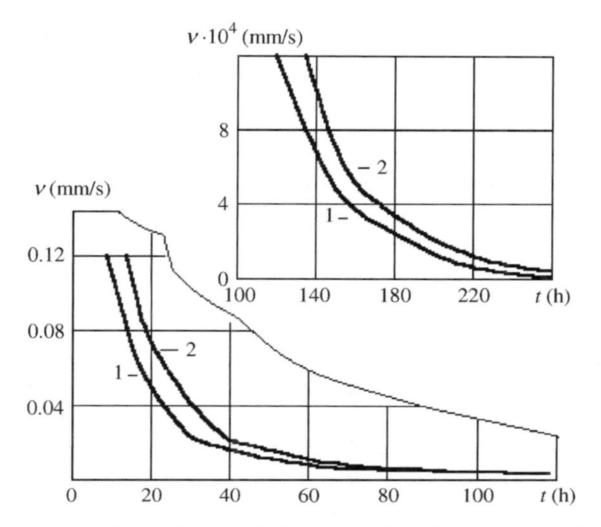

FIGURE 11.6 Variation of constriction rate of well walls with time: 1 – mean rate, 2 – maximal rate.

From Fig. 11.5, we get $\alpha(213) = 122$. The current radius is calculated with formula (11.3.30), mean velocity with (11.3.31). Results of calculations are presented in Fig. 11.6.

Deriving relation between $\Delta p(t)$ and $Q(t)$ in the flow of the formation solid and/or the fluid contained in rheological models that are slightly compressible power medium serves the following system of equation (Leonov and Isaev, 1982; Leonov and Triadski, 1980) where momentum equation is taken (11.2.14):

momentum equation in the form of (11.2.14)

$$v_{\text{av}} = -a\left|\frac{\partial p}{\partial r}\right|^{1/n}; \tag{11.3.32}$$

equation of mass conservation

$$\frac{\partial p}{\partial t} + \frac{1}{r}\frac{\partial r\rho v_{\text{av}}}{\partial r} = 0; \tag{11.3.33}$$

equation of state

$$\rho = \rho_0[1 + \beta_0(p - p_0)]. \tag{11.3.34}$$

Substitution of (11.3.32) and (11.3.34) in (11.3.33) yields

$$\frac{\partial p}{\partial t} = \frac{\chi_0}{r}\frac{\partial}{\partial r}\left[r\sqrt[n]{\frac{\partial p}{\partial r}}\right], \tag{11.3.35}$$

where

$$\chi_0 = \frac{1}{\beta_0}\frac{n}{2n+1}\frac{\left(\sqrt{3\kappa_{\rm f}}\right)^{(n+1)/n}}{\kappa^{1/n}}.$$

Solution of equation (11.3.35) is found with the help of integral relation method

$$p(r,t) = b_0(t) + b_1(t)\left(\frac{r}{r_{\rm k}}\right)^{1-n} + b_2(t)\left(\frac{r}{r_{\rm k}}\right). \tag{11.3.36}$$

Taking into account conditions (11.3.23) and (11.3.24) and condition at the contour

$$\frac{\partial p}{\partial r}\Big|_{r=r_{\rm c}} = 0,$$

we get

$$p(r,t) = p_{\rm f} - (p_{\rm f} - p_{\rm w})\frac{r^{1-n}r_{\rm c}^{n} - nr_{\rm k} - (1-n)r}{{r_{\rm w}}^{1-n}r_{\rm c}^{n} - nr_{\rm k} - (1-n)r_{\rm w}}, \tag{11.3.37}$$

where $r_{\rm c}(t)$ is determined from the differential equation obtained after substitution of (11.3.37) into (11.3.35)

$$\sqrt{\frac{r_{\rm w}^{1-n}r_{\rm c}^{n} - nr_{\rm c} - (1-n)r_{\rm w}}{\left[1-\left(\frac{r_{\rm c}}{r_{\rm w}}\right)^{n}\right](1-n)}}\cdot\frac{d}{{\rm d}t}\left[\frac{\frac{n(1-n)}{6(3-n)}r_{\rm c}^{3} - \frac{nr_{\rm w}^{2}}{2}r_{\rm c} + \frac{r_{\rm w}^{3-n}}{3-n}r_{\rm c}^{n} - (1-n)\frac{r_{\rm w}^{3}}{3}}{r_{\rm w}^{1-n}r_{\rm c}^{n} - nr_{\rm c} - (1-n)r_{\rm w}}\right]$$

$$= -\chi_0 r_{\rm w}\frac{\sqrt[n]{p_{\rm f} - p_{\rm w}}}{p_{\rm f} - p_{\rm w}} \tag{11.3.38}$$

with initial condition $r_{\rm c}(0) = r_{\rm w}$.

Equation (11.3.38) was solved numerically (Leonov and Triadski, 1980) for $0.4 < n \leq 1$ and $10^0 \leq r_{\rm c}/r_{\rm w} \leq 10^4$. Results are shown in Fig. 11.7 as variation of $\xi = r_{\rm c}/r_{\rm w}$ with

$$v = \frac{12\chi_0(p_{\rm f} - p_{\rm w})^{(1-n)/n}}{{r_{\rm w}}^{(n+1)/n}}t$$

for $n = 0.4$; 0.5; 0.6; 0.7; 0.8; 0.9; 1.0 (curves 1–7, respectively).

For $0.4 \leq n \leq 0.9$, approximation function $\xi = (r_{\rm c}/r_{\rm w}) = 1 + \varepsilon v^{\delta}$ is found, where ε and δ are functions of n presented in Fig. 11.8 by curves 1 and 2, respectively.

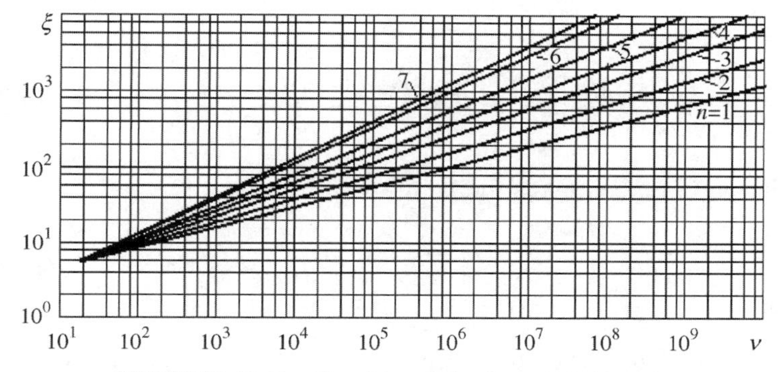

FIGURE 11.7 Graphic of the function $\xi(\nu, n)$.

The constriction rate of well walls is obtained from (11.3.32) at $r = r_{\mathrm{w}}$ and $k_{\mathrm{f}} = H^2/12$

$$v_{\mathrm{av}}\big|_{r=r_{\mathrm{w}}} = -a\sqrt[n]{\frac{\partial p}{\partial r}}\bigg|_{r=r_{\mathrm{w}}}. \tag{11.3.39}$$

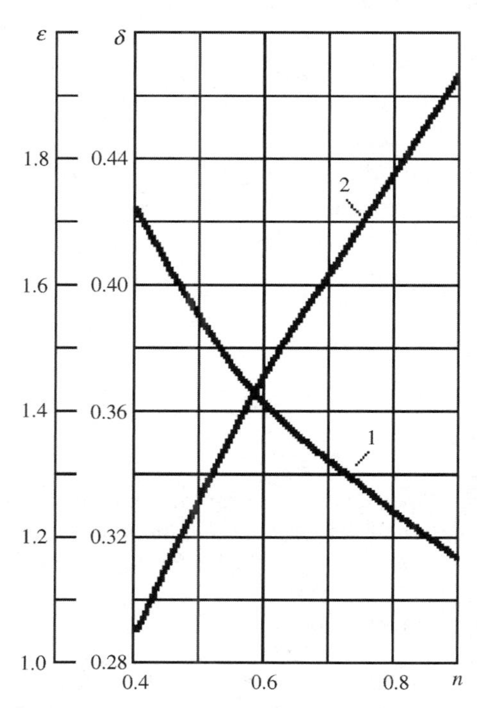

FIGURE 11.8 Graphics of functions $\varepsilon(n)$ and $\delta(n)$.

Since

$$\left.\frac{\partial p}{\partial r}\right|_{r=r_\mathrm{w}} = \frac{(p_\mathrm{f}-p_\mathrm{w})(1-n)(\xi^n-1)}{r_\mathrm{w}(1-n+n\xi-\xi^n)}, \tag{11.3.40}$$

the relation (11.3.39) takes the form

$$v_\mathrm{av}\big|_{r=r_\mathrm{w}} = -a\sqrt[n]{\frac{(p_\mathrm{f}-p_\mathrm{w})(1-n)(\xi^n-1)}{r_\mathrm{w}(1-n+n\xi-\xi^n)}}. \tag{11.3.41}$$

For viscous fluid flow ($n \to 1$ and $k=\mu$), the relation (11.3.41) transforms to (Leonov and Triadski, 1980)

$$v_\mathrm{av}\big|_{r=r_\mathrm{w}} = \frac{-(H/2)^2}{3\mu}\frac{p_\mathrm{f}-p_\mathrm{w}}{r_\mathrm{w}}\frac{1-\xi}{\xi-1-\xi\ln\xi}. \tag{11.3.42}$$

Calculation of the constriction rate of well walls with time is performed as follows. For given rheological properties of the solid k, n, and the formation thickness H, we get a with (11.2.14). Then, for given values of pressure drop p_f-p_w, well radius r_w and elastic compression factor of the solid β_0 factor $\chi_0=\alpha/\beta_0$ and dimensionless time

$$\nu = 12\chi_0\frac{(p_\mathrm{f}-p_\mathrm{w})^{(1-n)/n}}{r_\mathrm{w}^{(n+1)/n}}t$$

are determined.

From the graphic in Fig. 11.8, we get $\varepsilon(n)$ and $\delta(n)$ and with the formula $\xi=1+\varepsilon\nu^\delta$ the current radius of the disturbance $\xi(t)$. Average rate of well diameter constriction is found from (11.3.41). Maximal rate of the constriction is determined by

$$v_\mathrm{max} = \frac{2n+1}{n+1}v_\mathrm{av}. \tag{11.3.43}$$

EXAMPLE 11.3.3

It is required to get dependences of mean and maximal constriction rates of well walls with time at the following initial data: $n=0.8$; $k=1.24\times10^{15}\,\mathrm{Pa\,s}^n$; $H=30\,\mathrm{m}$; $r_\mathrm{c}=0.107\,\mathrm{m}$; $p_\mathrm{f}-p_\mathrm{w}=30.2\times10^6\,\mathrm{Pa}$; $\beta_0=0.5\times10^{-9}\,\mathrm{Pa}^{-1}$.

SOLUTION Calculations give

$$\chi_0 = 3.68\times10^{-8}\,\mathrm{Pa}^{(1-1/n)}\,\mathrm{m}^{(1+n)/n}\,\mathrm{s}^{-1};$$
$$\nu = 5.03\times10^{-3}t.$$

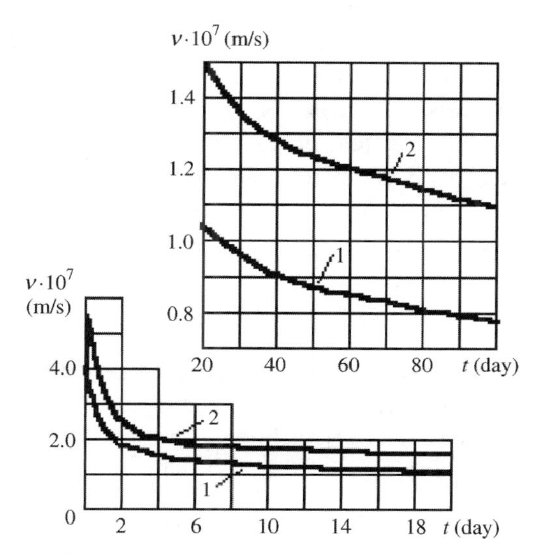

FIGURE 11.9 Fall of mean and maximal constriction rates of well walls with time.

From Fig. 11.8, we find $\varepsilon = 1.24$ and $\delta = 0.437$. The current disturbance radius is $\xi = 1 + 0.1227t^{0.437}$.

With formulas (11.3.41) and (11.3.43), we draw in Fig. 11.9 graphics of dependences of mean (curve 1) and maximal (curve 2) constriction rates of well walls with time.

11.4 FLOWS OF FORMATION FLUIDS AND ROCK SOLIDS IN REGIMES DIFFERENT FROM LAMINAR

In Sections 11.1–11.3, cases of stationary and nonstationary laminar flows of formations, fluids, and rock solids were considered, when regularities hold (11.3.42), (11.3.31), and (11.3.41) expressing relations between average velocities and flow rates, pressure gradients, and parameters of fluids and rock solids in formations. In many cases, flows in real formations are governed by these laws.

However, there are sometimes deviations from these laws caused by the following reasons: (1) turbulent flow resulting from high velocity of flow and chaotic arrangement of elementary channels in the formation, for example, in a solid with different number, sizes, and orientation of cracks; (2) nonuniform filtration; (3) movable formation boundaries; (4) rheological equations of state used above describe real fluids and rock solids inadequately; (5) fluid inertia, temperature field, phase transitions, and so on are not taken into account.

If to take into account these factors, derivation of theoretical formulas for speed of filtration would represent very complex problem because of the absence of closure equations for concentration φ and stress τ. Nevertheless, in some practical cases, the derivation from laminar flow in real formation can be taken into account with replacement of equation (11.1.1) by equation of the type (Businov and Umrichin, 1973; Pihachev and Isaev, 1973)

$$\frac{\partial p}{\partial r} = -a_1 v_{av} - b_1 \rho v_{av} |v_{av}|, \qquad (11.4.1)$$

where a_1 and b_1 are determined with the help of experiments and in some cases proved to be constant for given formation-well system.

At $b_1 = 0$, equation (11.4.1) gives linear filtration law (11.2.5) if to insert $a_1 = \mu/k_f$ in (11.4.1). Taking $a_1 = 0$ in (11.4.1), we get

$$\frac{\partial p}{\partial r} = -b_1 \rho v_{av} |v_{av}|. \qquad (11.4.2)$$

This quadratic filtration law would coincide with (11.2.1) for laminar and turbulent flow if to take b_1 as

$$b_1 = \lambda_c/(2H\varphi), \qquad (11.4.3)$$

where $\lambda_c = 24/Re$ is hydraulic resistance factor for laminar flow while for turbulent flow it is in the first approximation constant $\lambda_c = \text{const}$.

Resolving (11.4.2) relative $Q = v_{av}S$ with regard to (11.4.3), we obtain

$$Q = v_{av}S = C \times \sqrt{\frac{\partial p}{\partial r}} \times S, \qquad (11.4.4)$$

where $C = \sqrt{2H\varphi/(\lambda_c \rho)}$ is filtration resistance factor.

With the help of equation (11.4.1) at $a_1 \neq 0$ and $b_1 \neq 0$, one can get formulas connecting fluid flow rate in the formation with pressure under assumption that coefficients α_1 and b_1 are constant and may be experimentally determined.

Using (11.4.1), we get equations for stationary flow of incompressible fluid in circular formation

$$\frac{\partial p}{\partial r} = -a_1 v_{av} - b_1 \rho v_{av} |v_{av}|; \qquad (11.4.5)$$

$$\rho v_{av} S = \rho Q = \text{const}; \qquad (11.4.6)$$

$$\rho = \text{const}; \qquad (11.4.7)$$

$$\varphi = \text{const}. \qquad (11.4.8)$$

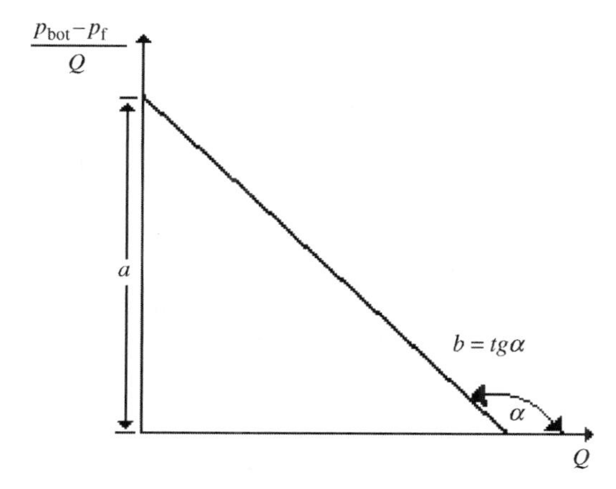

FIGURE 11.10 Indicator curve.

Here, φ is the content of moving fluid in the whole cross section of the circular formation; $v_{av} = Q/S$ is average velocity of the fluid related to the area of formation cross section $S = 2\pi rH$; and Q is fluid flow rate.

Integration of (11.4.5) in limits from $r = r_w$ (well radius) to $r = r_c$ (external boundary of the reservoir) and from $p = p_{bot}$ (bottom pressure) to $p = p_f$ (formation pressure) with regard to gives

$$p_f - p_{bot} = -aQ - bQ|Q|, \tag{11.4.9}$$

where $a = a_1/2\pi H \ln(r_c/r_w)$;

$$b = -\frac{b_1\rho\left(\frac{1}{r_c} - \frac{1}{r_w}\right)}{(2\pi H)^2}.$$

Quantities a and b are called filtration factors. They depend on properties of the critical area of formation.

Curves built up with equation (11.4.9) in coordinates $p_f - p_{bot}$, Q, or $(p_f - p_{bot})/Q$, Q are called indicator curves. If indicator curve for formation is known, one can easily obtain with it coefficients a and b (Fig. 11.10).

EXAMPLE 11.4.1

It is required to determine reservoir characteristic a using (11.4.9) and assuming the filtration law to be linear, that is, $b = 0$. Results of well tests in pumping down incompressible fluid are identical to the formation one with flow rate Q. Formation pressure is $p_f = 11$ MPa.

Results of well tests

n	1	2	3	4
p_{bot} (MPa)	12.32	13.6	14.9	16.1
Q (m^3/day)	25	50	75	100

SOLUTION Determine difference $p_{bot} - p_f$ and factor $a_n = (p_{bot} - p_f)/Q$, which in given case may be called absorption factor.

n	1	2	3	4
$p_{bot} - p_f$ (MPa)	1.32	2.6	3.9	5.1
a_n (MPa day/m^3)	0.053	0.052	0.052	0.051

Build up the indicator curve in coordinates $p_{bot} - p_f$, Q (Fig. 11.11). In given case, the linear with respect to flow rate filtration law is justified. Average factor a is $a = (0.053 + 0.052 + 0.052 + 0.051)/4 = 0.052$ MPa day/m^3.

Found values of the factor a for different wells of one and the same field can be unequal and depend on some factors: concentration of the filtered fluid φ, formation thickness H, external boundary of the formation r_c, well radius r_w, and coefficient a_1, so that $a = f(\varphi, H, r_c, r_w, a_1)$.

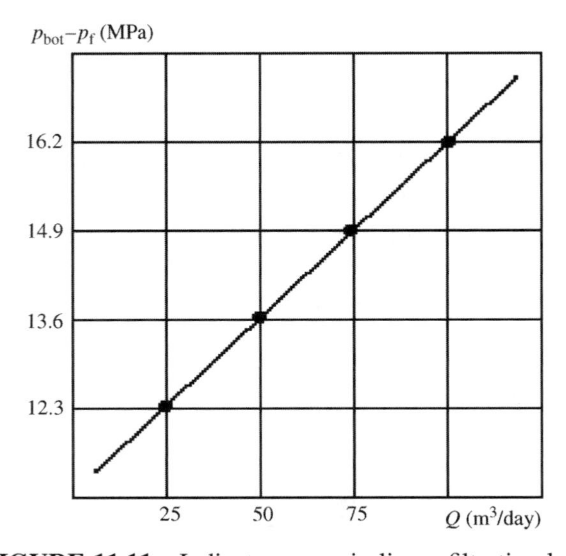

FIGURE 11.11 Indicator curve in linear filtration law.

From (11.2.4) and (11.2.7), it follows

$$a = \frac{\pi\varphi^3 H^3}{(6\mu \ln(r_c/r_w))}. \quad (11.4.10)$$

Quantities entering in the coefficient a are at present difficult to determine owing to the absence of reliable methods. In practice, when testing wells on oil field one can build up prognostic map with distribution of the absorption factor $a = (p_{bot} - p_f)/Q$ and introduce known factors into technological calculations in designing new wells. Knowledge of the factor a allows to estimate possible intensity of absorption.

Mathematical model of compressible fluid flow differs from the incompressible one only by the form of thermodynamic equation of state

$$\frac{\partial p}{\partial r} = -a_1 v_{av} - b_1 \rho v_{av} |v_{av}|; \quad (11.4.11)$$

$$\rho v_{av} S = \rho_0 v_0 S = \rho_0 Q_0 = \text{const}; \quad (11.4.12)$$

$$p = \rho g \times \bar{z} R \times \bar{T}; \quad (11.4.13)$$

$$\varphi = \text{const}. \quad (11.4.14)$$

Integration of equation (11.4.11) in the same limits as it was done with equation (11.4.5) gives with regard to (11.4.12)–(11.4.14) (in first approximation for gas it can be taken $\varphi \approx 1$)

$$p_f^2 - p_{bot}^2 = -aQ_0 - bQ_0|Q_0|, \quad (11.4.15)$$

where

$$a = \frac{a_1 \rho_0 g \bar{z} R \bar{T}}{\pi H} \ln\frac{r_c}{r_w}; \quad b = -\frac{b_1 \rho_0^2 g \bar{z} R \bar{T}}{2(\pi H)^2} \left(\frac{1}{r_c} - \frac{1}{r_w}\right).$$

One can also build up here an indicator curve in coordinates $(p_f^2 - p_{bot}^2)/Q_0$ and Q_0 (Businov and Umrichin, 1973; Pihachev and Isaev, 1973).

In both cases, factor a is length of a segment cutoff by the indicator curve on the ordinate axis and factor b is tangent of the angle of inclination to the abscissa axis. In testing the formation, there are commonly a lot of experimental points. In order to get factors a and b and consequently to build up the indicator curve, one should use the least squares method. In the absence of some data, factors a and b may be estimated by formulas

$$a = \mu/k_{\mathrm{f}}; \quad b = 63\rho \times 10^6/(k_{\mathrm{f}}/m)^{3/2}$$

proposed by Shirkovski (see Evdokimova and Kochina, 1979).

To apply the formula (11.4.9) or (11.4.15) at $b=0$, one should determine previously critical Reynolds number with Millionschikov formula (see Evdokimova and Kochina, 1979)

$$Re_{\mathrm{cr}} = \frac{|v_{\mathrm{av}}|\sqrt{k_{\mathrm{f}}}\rho}{m^{3/2}\mu}.$$

If Re_{cr} is greater than Re for fluid, one should use (11.4.9) at $b=0$ or (11.4.15).

EXAMPLE 11.4.2

It is required to determine formation characteristics a and b using (11.4.15) and results of gas well tests with inflow ($Q<0$) at $p_{\mathrm{f}}=17.5$ MPa.

Results of well tests

n	1	2	3	4	5
p_{bot} (MPa)	17.4	16.9	16	14.6	12.4
$Q_0 \times 10^3$ (m^3/day)	−20	−40	−60	−80	−100

SOLUTION To build up the indicator curve, we should get factors a and b using the least squares method. Equation (11.4.15) may be rewritten as

$$\frac{p_{\mathrm{f}}^2 - p_{\mathrm{bot}}^2}{Q_0} = -a - b|Q_0|. \quad (11.4.16)$$

In coordinates $(p_{\mathrm{bot}}^2 - p_{\mathrm{f}}^2)/Q_0 = y$, $|Q_0| = x$, this equation represents a straight line $y = a + bx$, where coefficients a and b are determined by formulas

$$a = \frac{\sum y_n \sum x_n^2 - \sum x_n \sum x_n y_n}{n\sum x_n^2 - (\sum x_n)^2}; \quad (11.4.17)$$

$$b = \frac{n\sum x_n y_n - \sum x_n \sum y_n}{n\sum x_n^2 - (\sum x_n)^2}. \quad (11.4.18)$$

Thus, to determine a and b, it is necessary to calculate x_n, y_n, $\sum x_n$, $\sum x_n^2$, $\sum (x_n)^2$, $\sum x_n y_n$ with given initial data of the example.

n	1	2	3	4	5	Σ
$x_n = Q_{0n}$	-20×10^3	-40×10^3	-60×10^3	-80×10^3	-100×10^3	-300×10^3
$y_n = \left(\frac{p_{bot}^2 - p_f^2}{Q_0}\right)_n$	1.745×10^{-4}	5.16×10^{-4}	8.37×10^{-4}	11.64×10^{-4}	15.2×10^{-4}	42.12×10^{-4}
$x_n^2 = Q_{0n}^2$	4×10^8	16×10^8	36×10^8	64×10^8	100×10^8	2.2×10^{10}
$x_n y_n = (p_{bot}^2 - p_f^2)_n$	3.49	20.64	50.25	93.09	152	319.47

Formulas (11.4.17) and (11.4.18) give

$$a = \frac{\sum\left(\frac{p_{\text{bot}}^2 - p_{\text{f}}^2}{Q_0}\right)_n \sum Q_{0n}^2 - \sum |Q_{0n}| \sum (p_{\text{bot}}^2 - p_{\text{f}}^2)_n \frac{|Q_0|}{Q_0}}{n \sum Q_{\text{on}}^2 - (\sum Q_{\text{on}})^2}.$$

$$= -\frac{-42.120 \times 10^{-4} \times 2.2 \times 10^{10} + 3 \times 10^5 \times 319.47}{5 \times 2.2 \times 10^{10} - 9 \times 10^{10}}$$

$$= 1.59 \times 10^{-4}\ \text{MPa}^2\ \text{day/m}^3;$$

$$b = \frac{n \sum (p_{\text{bot}}^2 - p_{\text{f}}^2)_n \frac{|Q_0|}{Q_0} - \sum |Q_{0n}| \sum \left(\frac{p_{\text{bot}}^2 - p_{\text{f}}^2}{Q_0}\right)_n}{n \sum Q_{0n}^2 - (\sum Q_{0n})^2}$$

$$= \frac{5 \times 319.47 - 3 \times 10^5 \times 42.120 \times 10^{-4}}{5 \times 2.2 \times 10^{10} - 9 \times 10^{10}} = 1.67 \times 10^{-8}\ (\text{MPa day})^2/\text{m}^6.$$

Hence, the equation of the indicator curve is

$$p_{\text{bot}}^2 - p_{\text{f}}^2 = -1.59 \times 10^{-4} Q_0 - 1.67 \times 10^{-8} Q_0 |Q_0|, \quad \text{where} \quad Q_0 < 0.$$

CHAPTER 12

NONSTATIONARY FLOWS OF GAS–LIQUID MIXTURES IN WELL-FORMATION SYSTEM

One-dimensional equations (4.6.1)–(4.6.5) describing nonstationary flow of gas–liquid mixtures in pipes (well) together with equations (11.1.1)–(11.1.5) describing nonstationary flow of fluid in formation represents combined system of equations for nonstationary flow of gas–liquid mixtures in well-formation system. Matching of solutions of these equations is performed at the bottom. At given initial and boundary conditions the solution of combined system of equations in general case can be obtained only by numerical methods.

Consider most important nonstationary flows of gas–liquid mixtures taking place in drilling as a result of interaction between well and formation when formation gas enters the well. Mathematical formulation of such problems being special cases of equations (4.6.1)–(4.6.5) and (11.1.1)–(11.1.5) at given assumptions may be brought to clear analytical solutions and numerical algorithms.

There are three types of formation fluid inflow: show, outburst, and blowout.

Applied Hydro-Aeromechanics in Oil and Gas Drilling. By Leonov and Isaev

12.1 ESTIMATION OF BOTTOM-HOLE DECOMPRESSION IN REMOVAL OF GAS BENCH FROM A WELL

One of the main reasons of emergency outbursts in well drilling turned often into gas blowout is decompression with time of the column of washing fluid owing to its aeration.

Assume that gas enters the critical area of formation against the production horizon in form of single bench with volume equal to the volume of fluid displaced from the well. As compared to any other form of gas distribution over height of the annular channel, such assumption allows to get estimation of pressures increasing safety of works on outburst liquidation.

A scheme of a well with gas bench is shown in Fig. 12.1 (Sheberstov et al., 1968). In washing the well with open mouth, this bench in ascending over the annular channel in plug flow regime is augmented in volume. The bench being at the bottom has minimal volume and pressure equal to the bottom one p_{bot}. As the bench is lifting the pressure in it falls, the bench volume increases, and thereby the height of the fluid column in the well decreases. The bottom pressure reduces and when the upper boundary of the bench reaches the open mouth it becomes minimal. In the absence of the counter-pressure a portion of gas almost immediately flows out from the well. Pressure in the remained part of the bench drops to atmospheric pressure, and bottom pressure can become even less than that of reservoir that leads to intensification of gas inflow up to gas blowout.

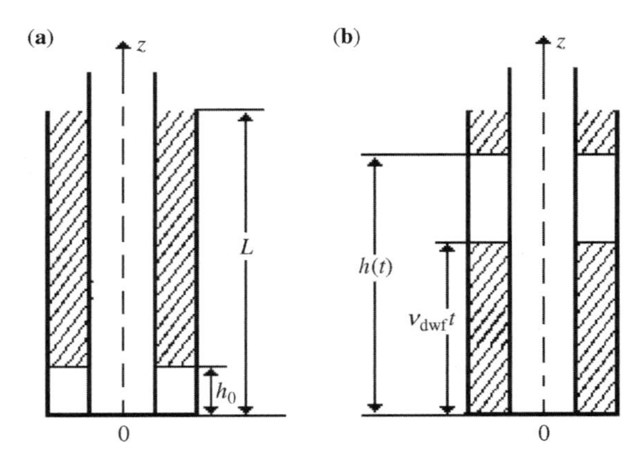

FIGURE 12.1 Scheme of a well with gas bench: *a*—gas bench at the bottom ($t = 0$), b—gas bench at the instance of time t; L—well depth; $h(t)$—distance from the bottom to the upper boundary of the bench at the instance of time t.

Let us estimate the reduction of the bottom pressure as a result of gas bench lift from the bottom up to the well mouth in turbulent flow of washing fluid ($\lambda = \text{const}$) without regard for plastic properties of fluid at given volume $V = h_0 \times S_{as}$ of the bench at the bottom, where S_{as} is the area of the annular space cross section.

The flow of the drilling fluid above the bench, when considering the solution as incompressible single-phase fluid, can be described by equations (Sheberstov et al., 1968) following from (4.6.33)–(4.6.39) at $\varphi_2 = 1$

$$-\frac{\partial p}{\partial z} = \rho g + \frac{\lambda \rho v^2}{2d_h} + \rho \frac{\partial v}{\partial t}, \tag{12.1.1}$$

$$\frac{\partial \rho v}{\partial z} = 0, \tag{12.1.2}$$

$$\rho = \text{const}. \tag{12.1.3}$$

In accordance with equations (12.1.2) and (12.1.3), the fluid velocity v does not depend on z. Consequently, $v = v(t)$ for all $z \geq h(t)$. Then,

$$v = \frac{dh}{dt}, \quad \frac{dv}{dt} = \frac{d^2h}{dt^2}. \tag{12.1.4}$$

Substitution of (12.1.4) in (12.1.1) and integration of the result from $z = h(t)$ to $z = L$ yields

$$p(h(t)) = \left[\rho \frac{d^2h}{dt^2} + \frac{\lambda \rho}{2d_h}\left(\frac{dh}{dt}\right)^2 + \rho g\right][L - h(t)] + p_{bean}(t), \tag{12.1.5}$$

where p_{bean} is pressure at the bean; $p(h(t))$ is the pressure at instance of time t on the upper boundary of the gas bench as viewed from the above moving fluid.

The washing fluid displaces gas bench from the well bottom with average velocity $v_{dwf} = Q_{pd}/S_{as}$, where Q_{pd} is constant delivery of pumps. We accept that gas bench is initially ($t = 0$) at the bottom, and fluid column above the gas bench is at rest (circulation is absent). Then, the pressure acting on the gas bench will be equal to hydrostatic pressure plus pressure at the bean

$$p_{bot\,0} = p_{bean} + [L - h_0]\rho g. \tag{12.1.6}$$

Assume that the pressure at the upper bench boundary is not distinct from the pressure averaged over gas volume $p(h(t))$. The equation of gas state is taken in form of Boyle–Mariotte law

$$pV = \text{const}, \tag{12.1.7}$$

where V is variable bench volume.

Then, with regard to (12.1.6) and (12.1.7), there is (see Fig. 12.1b)

$$p(h)[h - v_{\text{dwf}}t]S_{\text{as}} = [p_{\text{bean}} + (L - h_0)\rho g]h_0 S_{\text{as}}. \tag{12.1.8}$$

Resolving (12.1.8) with respect to pressure $p(h)$ and inserting the result in (12.1.5), we receive differential equation for motion of the upper boundary of the gas bench

$$\rho\frac{\mathrm{d}^2h}{\mathrm{d}t^2} + \frac{\lambda\rho}{2d_{\text{h}}}\left(\frac{\mathrm{d}h}{\mathrm{d}t}\right)^2 = \frac{[p_{\text{bean}} + (L - h_0)\rho g]h_0}{[h(t) - v_{\text{dwf}}t][L - h(t)]} - \frac{p_{\text{bean}}}{[L - h(t)]} - \rho g. \tag{12.1.9}$$

Since earlier the gas bench at $t = 0$ was taken to be at the bottom and fluid above it to be at rest, the solution of the equation (12.1.9) should be sought at following initial conditions:

$$h = h_0, \quad \mathrm{d}h/\mathrm{d}t = 0 \quad \text{at} \quad t = 0. \tag{12.1.10}$$

If the solution $h = h(t)$ of the equation (12.1.9) is known, one can find from

$$L = h(t_{\text{L}}) \tag{12.1.11}$$

the instant of time t_{L} at which the upper boundary of the bench arrived the well mouth. The bottom pressure at this instant of time would be

$$p_{\text{bot}} = p_{\text{bean}} + v_{\text{dwf}}t_{\text{L}}\rho g + \frac{\lambda t_{\text{L}}\rho}{2d_{\text{h}}}(v_{\text{dwf}})^3, \tag{12.1.12}$$

where v_{dwf} is the velocity of the lower bench boundary. This formula is valid at $t \geq t_{\text{L}}$. When the circulation at $t = t_{\text{L}}$ comes to an end, one should set $\lambda = 0$. At this, the bottom pressure p_{bot} would be minimal (p_{botm}). The ratio of p_{botm} obtained from (12.1.12) at $t = t_{\text{L}}$ to initial pressure (12.1.6) at $t = 0$ would be

$$\frac{p_{\text{botm}}}{p_{\text{bot}\,0}} = \frac{p_{\text{bean}} + v_{\text{pdf}}t_{\text{L}}\rho g}{p_{\text{bean}} + (L - h_0)\rho g}. \tag{12.1.13}$$

Solving the system of equations (12.1.9) and (12.1.11) with given data ρ, L, λ, d_{h}, v_{dwf}, and initial conditions (12.1.10), one can determine time t_{L}. Substitution of this time in (12.1.13) gives minimal pressure expected at the bottom p_{botm}.

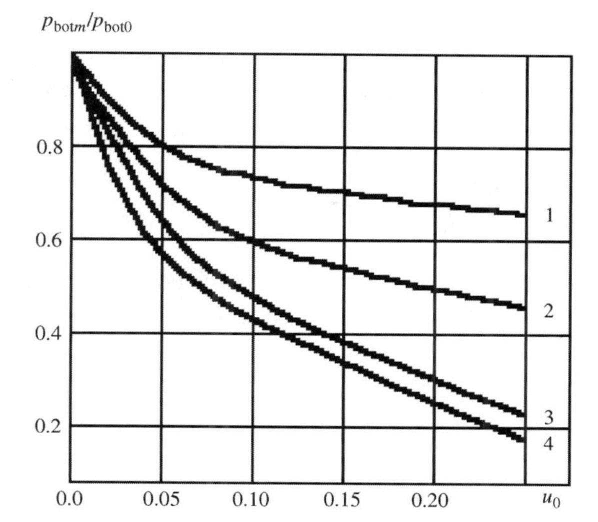

FIGURE 12.2 Dependence of relative value of bottom pressure decrease on the ratio of gas bench height to well depth.

To estimate approximately p_{botm}/p_{bot0}, one can use graphics in Fig. 12.2 plotted by results of numerical solution of the system (12.1.9)–(12.1.11) at atmospheric pressure at the bean ($p_{bean} = p_0$) and $L = (1\text{–}5) \times 10^3$ m, $v = 0.3\text{–}3$ m/s, $\rho = (1\text{–}2) \times 10^3$ kg/m^3, $d_h = 0.0533\text{–}0.286$ m (Sheberstov et al., 1968).

Calculations show that the ratio p_{botm}/p_{bot0} at $t = t_L$ depends mainly on two parameters $u_0 = h_0/L$ and $k^2 = \lambda v_{dwf}^2/(2gd_h)$. Figure 12.2 represents graphics of p_{botm}/p_{bot0} at different values of k^2. Curves 1, 2, 3, and 4 are obtained at k^2 equal to 0.303, 0.0578, 0.0002, and 0, respectively.

In washing out the gas bench up to its arrival at the well mouth, the pressure at the bottom decreases continuously whereas pressures at the boundary of the gas bench as viewed from liquid and gas are equal. By one or another technological reason, for example, caused by failure of surface pipeline system, the process of the gas bench washing out can be stopped at $t < t_L$. If at this instant of time the lower bench boundary were at a height $h_1 \geq h_1^*$ (h_1^* is critical height), outburst of fluid above the bench from the well would happen, since the pressure in bench p_h would exceed the pressure p_f of the fluid column above the bench.

The critical height h_1^* and associated time $t^* = h_1^*/v_{dwf}$ can be obtained from the condition of gas bench equilibrium in the well in the absence of circulation

$$p_h = p_f, \tag{12.1.14}$$

where $p_{\mathrm{f}} = p_0 + \rho g[L - h_2]$ is the hydrostatic pressure of the fluid above the bench, h_2 is the distance from the bottom to the upper boundary of the bench, and p_0 is the atmospheric pressure.

We get p_{h} from (12.1.8) at $h = h_2$ and substitute it in (12.1.14)

$$\frac{[p_0 + \rho g(L - h_0)]h_0}{(h_2 - h_1)} = p_0 + \rho g(L - h_2). \tag{12.1.15}$$

Here, h_1 is distance from the bottom to lower boundary of the bench.

At given values of L, h_0, and value of h_1 obtained from (12.1.15), one can determine the coordinate h_2 of the upper boundary of the bench. At $h_2 < L$, that is, when the upper boundary is in the well, the equation (12.1.15) reduces to

$$h_2^2 - h_2\left(L + \frac{p_0}{\rho g} + h_1\right) + \left[\frac{p_0}{\rho g} + (L - h_0)\right]h_0 + \left(L + \frac{p_0}{\rho g}\right)h_1 = 0. \tag{12.1.16}$$

At ascent of the lower boundary to the height $h_1 < h_1^*$, for example, owing to rise of the fluid column above the bench, a portion of fluid would displace from the well. To each new position of h_1 corresponds h_2 determined from the equation (12.1.16). At $h_1 > h_1^*$ complete displacement of fluid occurs above the bench from the well, that is outburst. In such a case, the solution of the equation (12.1.16) with respect to h_2 does not exist, since the discriminant of this equation becomes negative. Therefore, the critical height $h_1 = h_1^*$ can be obtained by equating the discriminant to zero

$$\left(L + \frac{p_0}{\rho g} + h_1\right)^2 - 4\left\{\left[\frac{p_0}{\rho g} + (L - h_0)\right]h_0 + \left(L + \frac{p_0}{\rho g}\right)h_1\right\} = 0.$$

This equation gives

$$h_1 = h_1^* = L + \frac{p_0}{\rho g} - 2\sqrt{\left[\frac{p_0}{\rho g} + (L - h_0)\right]h_0}.$$

From (12.1.16) at zero, discriminant yields the coordinate of the upper bench boundary

$$h_2^* = L + \frac{p_0}{\rho g} - \sqrt{\left[\frac{p_0}{\rho g} + (L - h_0)\right]h_0}.$$

Thus, if at $t = t^*$ the circulation ceases, from the well outburst of a fluid column with height takes place as

$$l = L + \frac{p_0}{\rho g} - h_2^* = \sqrt{\left[\frac{p_0}{\rho g} + (L - h_0)\right] h_0}$$

and the well would be emptied up to the depth $2l$. At this, the pressure at the bottom p_{bot} is determined by the height h_1^*

$$p_{\mathrm{botm}} = \rho g h_1^* = \rho g \left[L + \frac{p_0}{\rho g} - 2\sqrt{\left[\frac{p_0}{\rho g} + (L - h_0)\right] h_0}\right].$$

The ratio of this pressure to the initial one $p_{\mathrm{bot0}} = p_{\mathrm{dwf}}((L - h_0)\rho g$ is

$$\frac{p_{\mathrm{bot}}}{p_{\mathrm{bot\,0}}} = \frac{1 + \delta_0 - 2\sqrt{\delta_0 + (1 - u_0)u_0}}{\delta_0 + (1 - u_0)}, \qquad (12.1.17)$$

where $\delta_0 = p_0/\rho g L$.

The curve 4 in Fig. 12.2 corresponds to the formula (12.1.17).

12.2 RECOGNITION OF THE GAS OUTBURST AND SELECTION OF REGIMES OF ITS LIQUIDATION

Gas is carried out from the well by washing fluid supporting the pressure at the bottom equal or a little bit greater than the formation pressure in order to exclude further inflow of gas from the formation. At this, maximal pressure acting on the weak formation at the height H of uncased part of the well and on the cased column appears in course of washing out the bottom gas bench from the annular space as the upper boundary of the bench approaches, respectively, weak formation and well mouth (Fig. 12.3). In this connection when the gas enters the well, it is required to recognize or rather to calculate previously the pressure p_{as} acting on the weak formation and the pressure p_{bean} at well mouth (bean). If they do not exceed restrictions imposed on the strength of the formation and cased column, outburst otherwise blowout happens.

Hence, the type of failure is determined as follows (Kipunov et al., 1983): outburst

$$\begin{cases} p_{\mathrm{as}} \leq p_{\mathrm{hf}}, \\ p_{\mathrm{bean}} \leq p_{\mathrm{bp}}. \end{cases} \qquad (12.2.1)$$

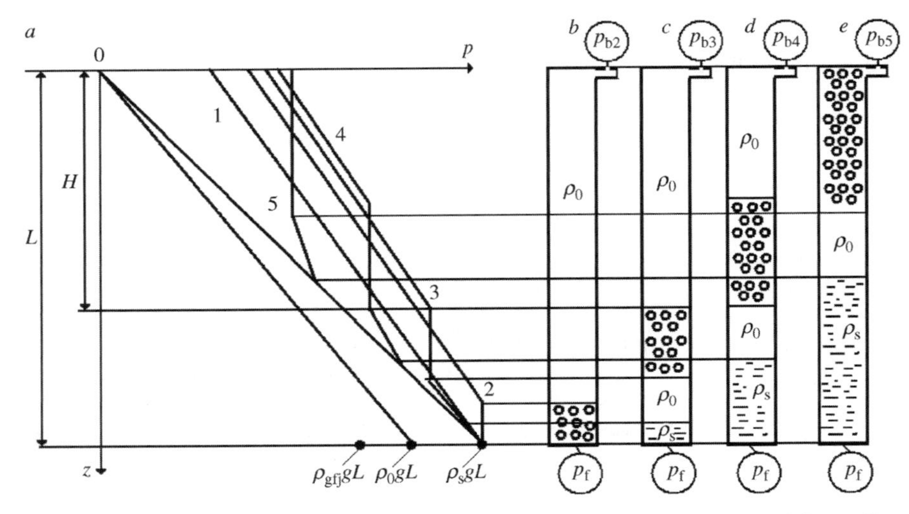

FIGURE 12.3 Graphics of pressure distribution (a) in annular space of the well at different positions (b–e) of gas bench washing out by solution with density ρ_b.

blowout

$$\begin{cases} p_{as} > p_{hf}, \\ p_{bean} > p_{bp}. \end{cases} \quad (12.2.2)$$

Here, p_{as} is the pressure in the annular space at the instant of time when the top of the gas bench in course of washing out reaches the depth H, p_{hf} is the pressure of the hydraulic fracturing of the weakest formation unblocked by columns, p_{bean} is the pressure at the bean when the top of gas bench reaches the well mouth, and p_{bp} is the bursting pressure of the last casing column caused by internal pressure at the well mouth.

Figure 12.3 shows pressure distribution in the annular space of the well at different times of gas bench washing out by solution with density p_s. The curve 1 characterizes pressure in the well in opening up of horizon with formation pressure $p_f = \rho_s gL$, exceeding pressure of the solution column with density ρ_0 chosen in accordance with expected pressure $p_{gtj} = \rho_{gtj} gL$, indicated in geological–technical job; curves 2, 3, 4, 5 correspond to pressure distribution in well at positions b, c, d, e of the gas bench.

Let us get formulas to calculate p_{as} and p_b in advance, accepting that at the well mouth operates ideal controller (adjustable choke) providing each instance of time during gas bench washing out the pressure at the bottom equal to bottom pressure p_b equal to formation one p_f. One can perform the gas bench washing out in two ways: pumping of solution with initial

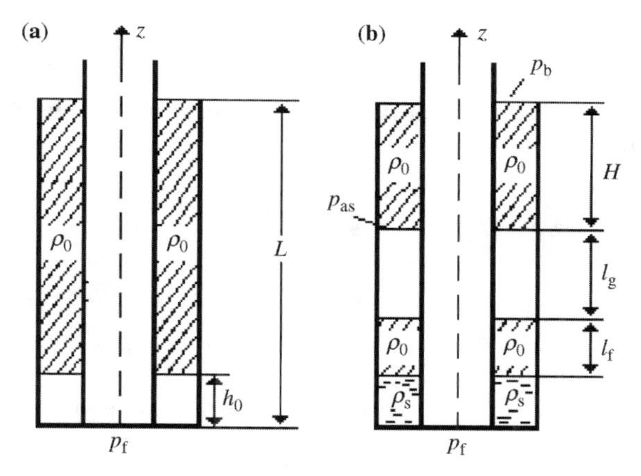

FIGURE 12.4 Scheme of gas bench motion in annular space: (a) at the bottom ($t = 0$); (b) at time t.

density ρ_0 (method of driller) and weighted drilling solution with density ρ_s (method of waiting and weighting).

In accordance with simplified scheme of Fig. 12.4 and using separately gravity force and friction losses, one can write the following relations

$$\Delta p_1 = p_f - p_{as} = \rho_0 g l_g + \frac{\lambda \rho_0 v_{as}^2}{2d_h} l_f + \rho_b g(L - H - l_f - l_g) + \frac{\lambda \rho_s v_{as}^2}{2d_{hf}}(L - H - l_f - l_g); \quad (12.2.3)$$

$$\Delta p_2 = p_{as} - p_b = \rho_0 g H + \frac{\lambda \rho_0 v^2}{2d_{hf}} H.$$

As in the previous section, we use designation $k^2 = \lambda v_{as}^2/(2d_h g)$. Then, equation (12.2.3) takes the form

$$p_f = p_{as} + \rho_0 g(1 + k^2) l_f + \rho_s g(L - H - l_f)(1 + k^2) - \rho_s g(1 + k^2) l_g;$$

$$p_{as} = p_b + \rho_0 g \left[1 + k^2 \left(\frac{v}{v_{as}} \right)^2 \right] H. \quad (12.2.4)$$

From (12.1.7) it follows

$$p_f h_0 S_{as} = p_{as} l_g S_{as}, \quad (12.2.5)$$

where $h_0 = V/S_{as}$, S_{as} the is cross section area of the annular channel and V is the volume of gas entering the bottom calculated by increase of the fluid level in receiving tanks.

In the term of the second equation (12.2.4) describing pressure losses owing to friction, we take $v = v_{as}$. Then, obtaining from (12.2.5) the bench height l_g and substituting it in (12.2.4), we get quadratic equation for p_{as}, whose root having physical meaning is (Kipunov et al., 1983)

$$p_{as} = 0.5\left\{\sqrt{p_1^2 + 4\rho_s g(1+k^2)p_f h_0} + p_1\right\}, \tag{12.2.6}$$

where $p_1 = p_f - \rho_0 g(1 + k^2)l_g - \rho_s g[L - H - l_f](1 + k^2)$; p_f is formation pressure determined in closed well by formula $p_f = p_{ap} + \rho_0 gL$ (p_{ap} is the pressure in ascending pipe).

The pressure in the bean is found from (12.2.4) at given pressure p_{as}

$$p_b = p_{as} + \rho_0 gH(1+k^2). \tag{12.2.7}$$

As it is seen from (12.2.6) and (12.2.7), the pressure in the bean is determined by pressure distribution in the whole well-formation system.

In gas washing out by solution with initial density ρ_0 ($\rho_0 = \rho_s$), the pressure p_1 in (12.2.6) is

$$p_1 = p_{ap} + \rho_0 gH(1+k^2) - \rho_0 gLk^2. \tag{12.2.8}$$

At $\rho_0 \neq \rho_s$, the pressure p_1 is calculated with

$$p_1 = \left[(\rho_s - \rho_0)g\frac{V_f}{S_{as}} + \rho_s gH\right](1+k^2) - \rho_s gLk^2, \tag{12.2.9}$$

where V_{ds} is internal volume of the drill-stem needed to calculate pressure in the annular space developed by fluid being displaced from pipes.

The inflow of the fluid can be found from the enhancement of washing fluid volume V in receiving tanks at $0 < V \leq V_{sh} = S_{as}(p_b - p_{gtc})/g\rho_0$. Here, p_{gtc} is bottom pressure indicated in geotechnical conditions. At $V_{sh} < V \leq V_{blow} = [p/(\rho_s g) + \rho_0 l_f/\rho_s + L - l_T - p_f/(\rho_s g)]pS_{as}/p_f$, outburst takes place ($p = \min[(p_g - \rho_0 gH), p_{bp}]$) while at $V \geq V_{blow}$, it is blowout. Density of the weighted drilling solution is selected to balance the bottom pressure at well vertical depth L

$$\rho_s = \frac{p_f}{gL}. \tag{12.2.10}$$

Accuracy of pressure p_b calculations with formula (12.2.7) is verified under field conditions. Relative discrepancy between predicted and experimental curves does not exceed 20%. Having established with (12.2.1) that predicted fluid inflow at the bottom corresponds to the outburst, one can turn to calculate regime of its liquidation.

In the course of showings of oil and gas liquidation, it is required to wash out the fluid from the annular space at the bottom-hole pressure

$$p_{bot} = p_f + S, \qquad \rho_s = \frac{p_f + S}{gL} \tag{12.2.11}$$

exceeding slightly the formation pressure p_f, where S is given excess of the bottom pressure (safety margin) over the formation. In accordance with (7.5.4), it is commonly taken that $0 \le S \le \Delta p_p$. If the pressure p_{bot} acting on the formation happens to be less than formation pressure, it would cause additional inflow of fluid from the formation, which makes the process of outburst liquidation longer or the outburst may convert into blowout. On the other hand, if the bottom pressure p_{bot} becomes significantly higher than that of formation, hydraulic fracturing of the formation and absorption of the fluid become difficult to be eliminated.

In practice, the bottom pressure can be judged from the pressure in the ascending pipe

$$p_{ap} = p_{bot} - \rho_{av} gL + k_c Q^2 \frac{\rho_{av}}{\rho_0}, \tag{12.2.12}$$

where

$$\rho_{av} = \frac{V_s \rho_s + \rho_0 (V_{ds} - V_b)}{V_{ds}} \tag{12.2.13}$$

is the average density of the solution in the drill-stem, V_s is the volume of the weighted fluid with density ρ_s pumped into the drill-stem, V_{ds} is the internal volume of the drill-stem, and g is the gravity acceleration;

$$k_c = \frac{(p_{ap} - p_{bot})}{Q_{pd}^2} \tag{12.2.14}$$

is the proportionality factor determined before arrival of the formation fluid; Q_{pd}, p_{as}, ρ_0 are, respectively, delivery of pumps, pressure in the ascending pipe, and density of the solution in the well at delivery of the fluid.

The factor k_c is obtained assuming equality of pressure losses and flow rates in circulation system during drilling and outburst liquidation. The distinction between values of k_c to be minimal, parameters p_p, p_{ib}, and Q_{pd} have to be measured at the beginning of each bit drill with open well mouth when $p_{ib} \approx p_0$.

Control of the bottom pressure p_{bot} can be performed by varying flow cross section area of the bean, density, and feed of the washing fluid delivered into drill-stem. Method of outburst liquidation depends on parameters and quantity of washing fluid being at hand on the drilling site at the instance of outburst as well as on technical characteristic of the circulation system.

To exclude possible blowing, two most common liquidation methods of gas–oil–water shows were obtained: method of driller and method of waiting and weighting.

Driller method consists of two stages. Washing out of gas bench at the first stage is performed by old drilling solution, whereas at the second stage weighted drilling solution is pumped to restore equality (or a trifle more) of bottom pressure to that of formation. In use of the second method, needed weighted solution with density ρ_s is pumped at once to restore equality (or a trifle more) of bottom pressure to that of formation. Use of both methods requires controlling pressure in the ascending pipe or in the pump or at a certain stage of driller method in the annular space to adjust the degree of expansion or constriction of the bean flow orifice.

Consider well operation in washing out of gas bench with method of waiting and weighting. At given blowout, killing delivery Q and density of the weighted fluid ρ_s, the bottom pressure may be supported equal to $\bar{p}_{bot}$ with the help of pressure drop adjusting in the bean $\bar{p}_{ib}$so that the pressure in the ascending pipe would obey equation (12.2.12). At this, in the beginning of fluid washing out from the well until the pump goes to given delivery Q to provide needed bottom pressure $\bar{p}_{bot}$, it is required to support in the annular space the pressure obtained by well sealing. After the pump yields needed delivery, it would be necessary to support initial pressure of pumping p_{init}

$$p_{init} = p_{ap}|_{V_b=0} = \Delta p_p + p_{apcw} + S \tag{12.2.15}$$

even in the ascending pipe, where $\Delta p_p = k_c Q^2$ is pumping pressure of the old solution, p_{apcw} is the pressure in the ascending pipe of closed well when determining bottom pressure, and S is the reserve of safety provided by (12.2.11).

When the whole drill-stem will be filled by weighted solution with density $p_s = (p_f + S)/(gL)$, the pressure in the ascending pipe becomes equal to the end pressure

$$p_{end} = p_{ap}|_{V_s=V_f} = k_c Q^2 \frac{\rho_s}{\rho_0} = \Delta p_p \frac{\rho_s}{\rho_0}. \tag{12.2.16}$$

Adjusting the bean, one should support the pressure in the ascending pipe up to final displacement of the whole inflow volume, that is, of total washing out the bottom fluid from the annular space characterized by equality of densities of fluids pumping into and coming out from the well.

The pressure p_{ap} in equation (12.2.12) is linear function of the density ρ_{av} that in its turn is related to the volume V_s of the weighted solution or number of pump strokes N needed to pump the volume V_s. One can plot a graphic of pressure needed to support pressure in the ascending pipe versus volume of pumped solution (Fig. 12.5a).

Coordinates of points B and C in Fig. 12.5a are determined by pressures in the ascending pipe $p_{init} = p_{ap}|_{V_s=0}$ and $p = p_{ap}|_{V_s=V_f}$ (see formulas (12.2.15) and (12.2.16)) at the beginning and the end of drill-stem filling with weighted solution having density $\rho_{ws} = \rho_s$.

Current values of $p_{ap}(N_p)$ are determined from intersection of the pointer 2 with the ordinate axis as a result of successive motion along pointers 1 and 2. Graphics in Fig. 12.5a should be used in prompt regulation of current pressure in the ascending pipe and consequently of the bottom pressure in the well during liquidation of gas–water–oil shows.

EXAMPLE 12.2.1

It is required to build graphic $p_{ap}(N_p)$ determining regulation of the bean as the weighted solution is pumping into the drill-stem in washing out gas bench by method of waiting and weighting. To fill the drill-stem with internal volume 24 m^3, the pump needs 857 strokes.

Initial data are $p_{ap} = 40 \times 10^5$ Pa, $\rho_0 = 1200\,kg/m^3$, $\rho_s = 1400\,kg/m^3$, $V_f = 24\,m^3$, $N_p = 857$, $Q = 0.014\,m^3/s$, $k_c = 4 \times 10^{10}\,kg/m^7$, and $S = 10 \times 10^5$ Pa.

SOLUTION We receive with formulas (12.2.15) and (2.2.16)

$$p_{init} = p_{ap}|_{V_s=0} = 4 \times 10^{10}(14 \times 10^{-3})^2 + 40 \times 10^5 + 10^6 \text{Па} = 128 \times 10^5 \text{ Pa},$$

$$p_{end} = p_{ap}|_{V_s=V_f} = 4 \times 10^{10}(14 \times 10^{-3})^2 1400/1200 = 91 \times 10^5 \text{ Pa}.$$

Intermediate values of $p_{ap}(N_p)$ presented in Fig. 12.5a (Gabolde and Nguyen, 1991) are beforehand tabled. These values permit to build graphics with which one can easily and quickly control pressure in the ascending pipe during blowout liquidation.

Operation of well pressures in washing out the gas bench by driller method differs from waiting and weighting method in which the liquidation of the gas–water–oil show is performed by old drilling solution with density ρ_0. In course of its drive into the well, one should support initial pressure of circulation determined by (12.2.15) until solution volume equal to the

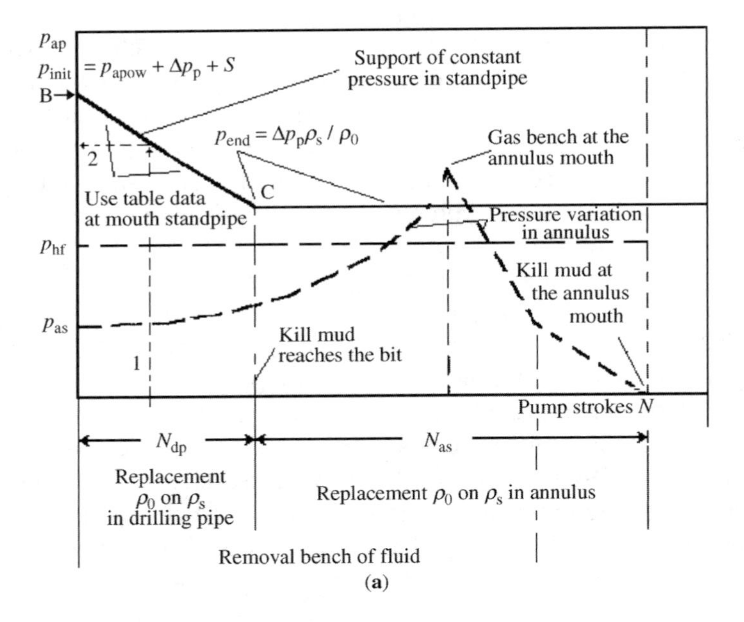

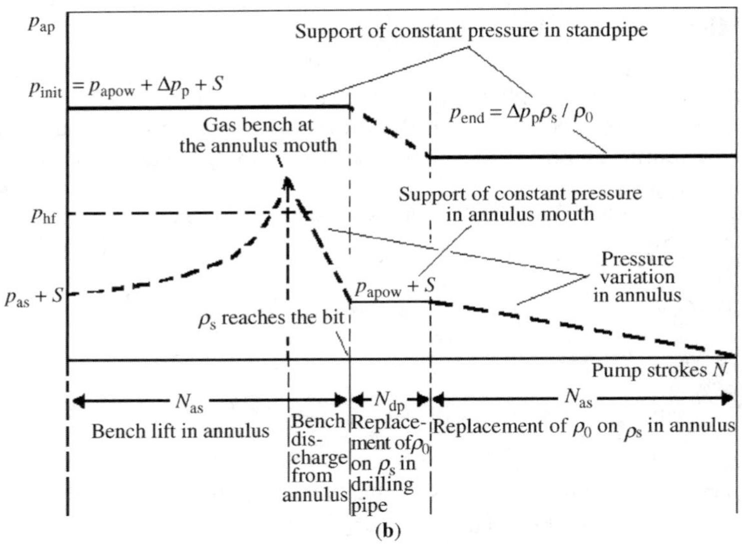

FIGURE 12.5 (a) Graphic of needed pressure variation in ascending pipe (solid line) during well filling with killing fluid. By dotted line, pressure variation is shown in annular space when during operation of well killing the pressure in the ascending pipe is supported in accordance with the solid line. (b) Graphics of needed pressure variation in ascending pipe and annular space (solid line) in well operation during gas–water–oil show by driller method.

volume of the whole annular space would be pumped, that is all formation fluid would be displaced. After this, the well is sealed. If pressures in the ascending pipe and in the annular space coincide and are equal to the pressure in the ascending pipe at the end of its stabilization, that is, after show detection, one can begin to pump the weighting solution whose density is enough to balance the bottom pressure. At this the operation of the well should be performed with the help of manometer in the annular space supporting in it pressure equal to the pressure of well closing after first stage of well killing. When the weighed solution reaches the bottom, one should turn to well operation with manometer in the ascending pipe supporting in it pressure equal to the end pressure of circulation. Graphics of dependences to operate pressures at the well mouth and in annular space are shown in Fig. 12.5b (Gabolde and Nguyen, 1991) by solid lines.

The advantage of driller method consists that one should know only three values of pressure: initial pressure, pressure of initial well sealing at the mouth, and end pressure. However, in using driller method pressures in the annular space and time of sealing exceed those in method of waiting and weighting. Therefore, in cases when there are zones of absorption or great length of open well bore, it is better to use method of waiting and weighting in liquidation of gas–water–oil show. The disadvantage of this method is that one has to operate pressure in the ascending pipe with intermediate values of pressure during pumping of weighted solution in drill pipes, while its advantage is lesser time required for well killing.

12.3 CALCULATION OF AMOUNT, DENSITY, AND DELIVERY OF FLUID NEEDED TO KILL OPEN GAS BLOWOUT

One of main methods to liquidate open gas blowout is pumping of killing fluid into the flow of blowing gas through pipes provided in emergency well or through specially inclined wells connected directly or through hydraulic fracturing crack with blowing well. The rate of pumping and the amount of killing fluid are most important parameters needed for proper choice of surface pump equipment, fluid reserve, and designing of inclined well constructions.

In Fig. 12.6, a scheme of blowout killing is shown. Consider the process of blowout killing (Sheberstov et al., 1969). Take as origin ($t = 0$) the beginning of killing fluid delivery with density ρ_k and constant flow rate Q_3 into the bore of blowing well at the point $z = L$ with gas flow rate Q_0. In killing blowout in section h of the well flows gas, while in section L flows

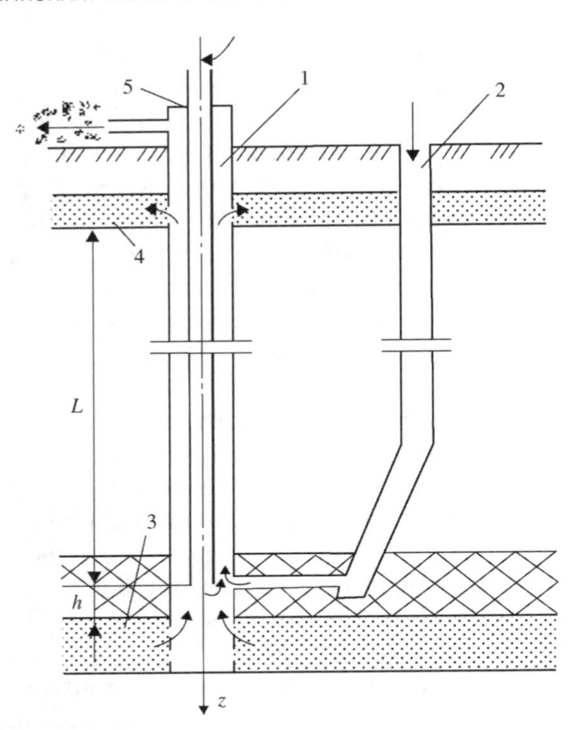

FIGURE 12.6 A scheme of "well-formation" system in process of blowout killing (pointers denote possible directions of gas, fluid, and mixture flows): 1—blowout well; 2—inclined well; 3, 4—productive and absorptive formations, 5—tubing and drill pipe.

gas–liquid mixture that goes to intake formation. If the well is not communicated with the formation, gas–liquid mixture can flow out through delivery lines of blowout equipment and in the case of broken mouth in the crater.

In what follows, we accept $h = 0$. The fluid moving along the well to the mouth increases the bottom pressure $p_{\mathrm{bot}} = p(L, t)$, and thereby decreases gas inflow. Blowout killing happens when the bottom pressure becomes equal to or greater than the reservoir pressure

$$p_{\mathrm{bot}} = p(L, t) \geq p_{\mathrm{f}} = \mathrm{const}. \tag{12.3.1}$$

Thus, in the well bore nonstationary flow of two-phase mixture takes place. Let us assume that the killing fluid is incompressible, gas obeys the law (4.3.4), phase transitions are absent, fluid and gas move in the well with equal velocities ($v_1 = v_2 = v$), that is, volume and true gas contents coincide ($\varphi = \beta$). Accept also that the flow is turbulent ($\lambda = \text{const}$) and inertial forces are ignored.

At given assumptions, the system of equations (4.6.33)–(4.6.39) describing the flow in pipes may be simplified to

momentum equation of gas–liquid mixture

$$\frac{\partial p}{\partial z} = g[\rho\beta + \rho_k(1-\beta)]\left[1 + \frac{\lambda}{2gd}v^2\right]; \tag{12.3.2}$$

equations of mass conservation for gas and killing fluid

$$\frac{\partial \rho\beta}{\partial t} + \frac{\partial \rho\beta v}{\partial z} = 0, \tag{12.3.3}$$

$$\frac{\partial \rho_{\mathrm{k}}(1-\beta)}{\partial t} + \frac{\partial \rho_{\mathrm{k}}(1-\beta)v}{\partial z} = 0; \tag{12.3.4}$$

equations of state for gas and killing fluid

$$p = \bar{z}\rho g R\bar{T}, \tag{12.3.5}$$

$$\rho_{\mathrm{k}} = \text{const}, \tag{12.3.6}$$

where z and T are, respectively, overcompressibility factor and temperature averaged along well bore;

equation of concentration with volume flow rate depending on time

$$\beta = \frac{Q(t)}{Q(t) + Q_{\mathrm{k}}}. \tag{12.3.7}$$

Resolving the equation (11.4.15) of gas flow in the formation with respect to bottom pressure, we get

$$p_{\mathrm{f}} = p(L,t) = \sqrt{p_{\mathrm{bot}}^2 - aQ_0(t) - bQ_0^2(t)}, \tag{12.3.8}$$

where $Q_0(t)$ is volume flow rate of gas at the bottom at time t reduced to normal conditions. Note that equation (12.3.8) is at the same time boundary condition for the system of equations (12.3.2)–(12.3.7).

Before the beginning of blowout killing, the gas flow in well-reservoir system is stationary, since the well blows with known constant flow rate Q_0.

As initial conditions, we take pressure distribution in formation and well (12.3.8) and blowout flow rate appropriate to stationary flow

$$Q_0 = \text{const} \quad \text{for} \quad 0 \le z \le L \quad \text{at} \quad t = 0.$$

Additional boundary conditions are

$$\begin{aligned} &p = p_{\mathrm{wm}} = \mathrm{const} \quad \text{at} \quad z = 0 \quad \text{and} \quad t \geq 0; \\ &Q_{\mathrm{b}} = \mathrm{const} \quad \text{at} \quad z = L \quad \text{and} \quad t > 0, \end{aligned} \tag{12.3.9}$$

where p_{wm} is pressure at the well mouth.

Density ρ_{m} of the mixture is

$$\rho_{\mathrm{m}} = \rho\beta + \rho_{\mathrm{k}}(1-\beta). \tag{12.3.10}$$

Substitution of β from (12.3.7) and ρ from (12.3.5) into (12.3.10) yields

$$\rho_{\mathrm{m}} = \frac{p}{g\bar{z}R\bar{T}}\frac{Q}{Q+Q_{\mathrm{k}}} + \rho_{\mathrm{k}}\frac{Q_{\mathrm{k}}}{Q+Q_{\mathrm{k}}}. \tag{12.3.11}$$

The system of equations (12.3.2)–(12.3.7) is solved by approximate method. Consider a sequence of times $t = t_0, t_1, \ldots, t_i$ determined as follows: $t_0 = 0$ is the beginning time of the fluid delivery in the blowout well; t_{i+1} is the time at which a layer of gas–liquid mixture beings at $t = t_i$ and at the bottom rises to the height L. At these times, one can estimate density and velocity of the mixture as well as pressure and flow rate of gas. For any intermediate instant of time $t = t'$ and cross section z are true inequalities

$$t_i \leq t' \leq t_{i+1}, \tag{12.3.12}$$

$$Q(t_{i+1}) \leq Q(t') \leq Q(t_i). \tag{12.3.13}$$

Inequality (12.3.13) means that gas flow rate reduces with time.

The flow at time intervals (12.3.12) is assumed to be stationary. Then, the process may be considered as change of stationary states for $i = 0, 1, 2,$ and so on. In these intervals $\partial\rho\beta/\partial t = 0$ and in accordance with (12.3.3), there is

$$Q_0(t')\rho_0 = Q(t')\rho(t'). \tag{12.3.14}$$

Transform equation (12.3.11) with regard to (12.3.5) and (12.3.14)

$$\rho_{\mathrm{m}}(z,t) = \frac{Q_0(t)\rho_0 + Q_{\mathrm{k}}\rho_{\mathrm{k}}}{Q_0(t)p_0 + Q_{\mathrm{k}}p(z,t)}p(z,t). \tag{12.3.15}$$

Due to (12.3.13), we have

$$\rho_{\mathrm{m}}(z,t) \geq \frac{Q_0(t_i)\rho_0 + Q_{\mathrm{k}}\rho_{\mathrm{k}}}{Q_0(t_i)p_0 + Q_{\mathrm{k}}p(z,t)}p(z,t). \tag{12.3.16}$$

The velocity of gas–liquid mixture can be represented as

$$v = \frac{Q_0 + Q_k}{S_{as}} = \frac{Q_0(t) + Q_k}{S_{as}}, \qquad (12.3.17)$$

where S_{as} is cross section area of annular channel of the blowing well.

Using (12.3.14), we get

$$v(z, t) = \frac{Q_0(t)p_0 + p(z, t)Q_k}{S_{as}\, p(z, t)}. \qquad (12.3.18)$$

Now, with (12.3.11) we can estimate the velocity

$$v(z, t) \geq \frac{Q_0(t_{i+1})p_0 + Q_k p(z, t)}{S_{as}\, p(z, t)}. \qquad (12.3.19)$$

Substitution in (12.3.2) instead of density ρ_c and velocity v right parts of (12.3.16) and (12.3.19) yields the following inequality

$$\frac{\partial p}{\partial Z} \geq \frac{[Q_0(t_i)\rho_0 + Q_k\rho_k]g}{Q_0(t_i)p_0 + Q_k p}\, p\left[1 + \frac{\lambda}{2gd_h}\left(\frac{Q(t_{i+1})p_0 + Q_k p}{S_{as}p}\right)^2\right]. \qquad (12.3.20)$$

For convenience of following calculations let us replace $Q_0(t_i)$ with G_{i+1}, $Q(t_{i+1})$ with G_i, and integrate the result with respect to z from $z = L$ to $z = 0$ and with respect to p from $p = p(L, t)$ taken from (12.3.8) at $t = t_{i+1}$ to $p = p_m$

$$\frac{G_{i+1}}{Q_k} = \frac{\frac{Lg\rho_k}{p_0}(1 + k^2) - \frac{p_i}{p_0} + \frac{p_{wm}}{p_0} + \frac{k^2(G_i/Q_k)}{1+k^2}\ln Z + \frac{G_i}{Q_k}k\frac{1-k^2}{1+k^2}\arctan B}{0.5\ln A - k\arctan B - \frac{L\rho_0 g}{p_0}(1 + k^2)}, \qquad (12.3.21)$$

where $p_i = \sqrt{p_f^2 - aG_i - bG_i^2}$,

$$k^2 = \frac{\lambda Q_k^2}{2gd_h S_{as}^2},$$

$$A = \frac{\left(\frac{p_i}{p_0}\right)^2\left[1 + k^2\left(1 + \frac{G_i}{Q_k}\frac{p_0}{p_i}\right)^2\right]}{\left(\frac{p_{wm}}{p_0}\right)^2\left[1 + k^2\left(1 + \frac{G_i}{Q_{3wm}}\frac{p_0}{p_{wmy}}\right)^2\right]},$$

$$B = \frac{\frac{G_i}{Q_k} k(1+k^2)\left(\frac{p_i}{p_0} - \frac{p_{wm}}{p_0}\right)}{k^2\left(\frac{G_i}{Q_3}\right)^2 + \left[(1+k^2)\frac{p_i}{p_0} + k^2\frac{G_i}{Q_k}\right]\left[(1+k^2)\frac{p_{wm}}{p_0} + k^2\frac{G_i}{Q_k}\right]}.$$

It should be noted that in the solution (12.3.21) at given t_i, the flow rate $Q_0(t_i)$ is always greater than $Q_0(t_{i+1})$, whereas G_i less than G_{i+1}. The sequence of calculation of blowout killing time t_k is as follows. Substituting in the right part of equation (12.3.21)$G_0 = 0$, we get G_1. Then after inserting the obtained value G_1 in the right part of equation (12.3.21), we receive G_2, and so on until at some n will be $G_n \geq Q_0(0)$. Since the time interval $t_{i+1} - t_i$ lasts no more than SL/Q_k, the time of blowout killing is restricted by

$$t_k \leq t \leq \frac{nLS}{Q_k}. \quad (12.3.22)$$

As far as all gas is removed from the well, one has to pump in time t_k additionally to the fluid volume equal to well volume V_w. The total fluid volume V may be estimated by

$$V < (n+1)V_w. \quad (12.3.23)$$

Figure 12.7 represents typical dependences of fluid volume needed for well killing on flow rate at different densities of the killing fluid calculated by formulas (12.3.21) and (12.3.23).

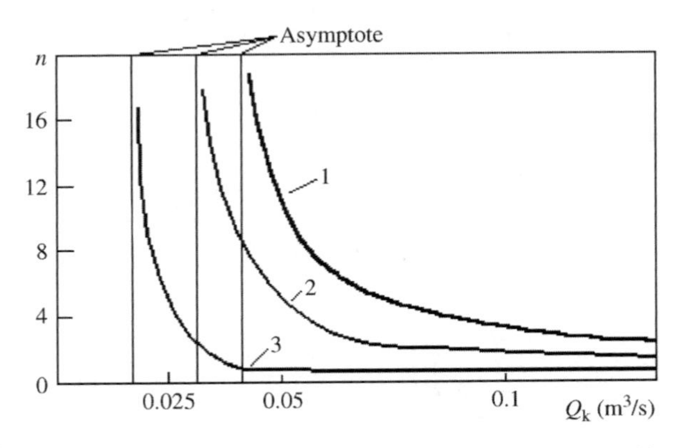

FIGURE 12.7 Dependence of fluid volume versus flow rate at different densities given in Example 12.3.1.

EXAMPLE 12.3.1

It is required to determine parameters of gas crossflow killing in a well at initial data:

$$L = 1200\,\mathrm{m}, \quad p_{\mathrm{wm}} = 34 \times 10^5\,\mathrm{Pa}, \quad p_{\mathrm{f}} = 147 \times 10^5\,\mathrm{Pa}, \quad d = d_{\mathrm{h}} = 0.083\,\mathrm{m},$$
$$S = 0.011\,\mathrm{m}^2, \quad p_0 = 10^5\,\mathrm{Pa}, \quad \rho_0 = 0.8\,\mathrm{kg/m}^3, \quad Q_0 = 23.7\,\mathrm{m}^3/\mathrm{s},$$
$$a = 72.1 \times 10^{10}\Pi a^2\,\mathrm{s/m}^3, \quad b = 8.2 \times 10^{10}\,\mathrm{Pa}^2\,\mathrm{s}^2/\mathrm{m}^6, \lambda_{\mathrm{m}} = 0.02.$$

SOLUTION If the killing fluid has density $\rho_{\mathrm{k}} = 1300\,\mathrm{kg/m}^3$ and delivery $Q_{\mathrm{k}} = 0.05\,\mathrm{m}^3/\mathrm{s}$, successive calculations with (12.3.21) give flow rates

$$G_0 = 0, \quad G_1 = 3\,\mathrm{m}^3/\mathrm{s}, \quad G_2 = 6.73\,\mathrm{m}^3/\mathrm{s}, \quad G_3 = 15.1\,\mathrm{m}^3/\mathrm{s}, \quad G_4 = 74.6\,\mathrm{m}^3/\mathrm{s}.$$

Since $G_4 > Q_0$ from (12.3.23) at $n = 4$, the volume is $V < 5V_{\mathrm{w}} = 66\,\mathrm{m}^3$.

Curves 1, 2, and 3 in Fig. 12.7 represent $n = n(Q_3)$ at $\rho_{\mathrm{k}} = 1000$, 1300, and 2400 $\mathrm{kg/m}^3$.

With formulas (12.3.21) in Petrov et al. (1974), an album of graphics to select blowout killing regime was made. Graphics are represented in dimensionless coordinates as functions of Froude *Fr* and Strouhal *Sh* numbers

$$\sqrt{Fr} = f\left(Sh, \frac{P_{\mathrm{f}}}{P_{\mathrm{b0}}}\right), \tag{12.3.24}$$

where $Fr = Q_{\mathrm{k}}^2/S_{\mathrm{as}}^2 g d_{\mathrm{h}}$, $Sh = Q_{\mathrm{k}} t_{\mathrm{k}}/S_{\mathrm{as}} L$; t_k is the blowout killing time when the bottom pressure p_{bot} becomes equal to the formation pressure p_{f}; p_{bot0} is the bottom pressure in the blowing well before killing.

Each graphic represents a set of curves to determine parameters $p_{\mathrm{f}}/(g\rho_{\mathrm{k}}L)$, $p_{\mathrm{f}}/p_{\mathrm{wm}}, L/d_{\mathrm{h}}$. In Fig. 12.8, it is shown that a part of these graphics at $L/d_{\mathrm{h}} = 1600$. To use graphics, one should first calculate bottom pressure p_{bot0} with initial data for pure gas, then determine parameters $p_{\mathrm{f}}/p_{\mathrm{wm}}$, $p_{\mathrm{f}}/\rho_{\mathrm{bot}} gL, p_{\mathrm{f}}/p_{\mathrm{bot\,0}}$, and with these parameter select appropriate graphic and curve of blowout killing. In accordance with (12.3.22) and (12.3.24), we have

$$Sh = n = \frac{V}{V_{\mathrm{w}}}. \tag{12.3.25}$$

Now setting a certain volume V for killing fluid and calculating the number *Sh* with (12.3.25), one gets the value of $\sqrt{Fr}$ from appropriate curve and determines the delivery of pumps with formula

$$Q_{\mathrm{p}} = \sqrt{Fr}\sqrt{g d_{\mathrm{h}}} S_{\mathrm{as}}. \tag{12.3.26}$$

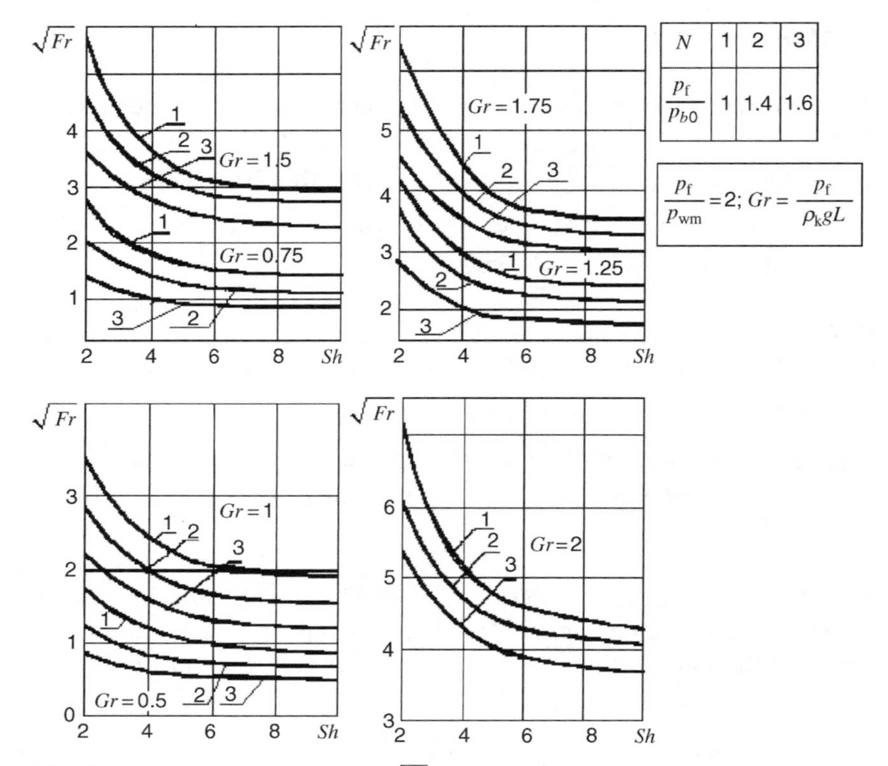

FIGURE 12.8 Dependences of $\sqrt{Fr}$ on Sh at $p_f/p_{wm} = 2$ and different dimensionless parameter $Gr = p_f/\rho_k gl$ in process of gas blowout killing.

12.4 CALCULATION OF PRESSURE AT THE WELL MOUTH IN BLOWOUT KILLING BY DIRECT PUMPING OF KILLING FLUID IN THE WELL

In Section 12.3, it was described that a method to calculate well blowout regime is considered as nonstationary flow of gas cut fluid with gas number tending to zero with time. However, it is impossible to produce such flow in the blowing well because of technological reason, for example, when the drill tool is absent or lowered not deep into the well. In these cases well killing can be performed by direct pumping of the killing fluid into the blowing well, of course if the well strength allows it.

Gas blowing is commonly killed by direct pumping in the well with safe mouth equipped with closed preventers through which delivery lines flow out gas. In blowout killing gas is directed to one or several delivery lines. The killing fluid is pumped through other delivery lines connected with cementing trailers and drill pumps. Immediately, the flow begins to be pumped into the well, valves at outlets for gas are closed, and well killing starts. In some cases, the killing fluid is pumped also through drill pipes, more often through one pipe, lowered not deeply into the well. During fluid pumping the pressure at the mouth increases, tending to the bottom one minus pressure of fluid and gas columns in the well.

Let us adduce calculation of pressure variation with time at well mouth in course of blowout killing at given flow rates of the killing fluid. It is necessary to know the pressure variation during blowout killing to select characteristics and operation regimes of pumps and cementing aggregates in order to kill blowing and conserve safety of cased drill-stem and equipment of the well mouth whose breaking strength is sometimes below the excess pressure of gas in the well closed and fully emptied from the fluid.

Change of pressure with time at the mouth can be determined by combined consideration of descending flow of killing fluid in the well and gas inflow from the formation. In plug delivery of killing fluid with flow rate $Q_{\rm bot}$ at any instant of time t, the difference $\Delta p(t) = p_{\rm bot} - p_{\rm wm}$ between pressure $p_{\rm bot}$ at the interface of gas and fluid and pressure $p_{\rm wm}$ at the well mouth is taken equal to difference of pressure losses caused by friction $\Delta p_{\rm fr}$ and hydrostatic pressure of fluid column $\Delta p_{\rm hyd}$. Then,

$$\Delta p(t) = p_{\rm bot} - p_{\rm wm} = \Delta p_{\rm hyd} - \Delta p_{\rm fr}. \tag{12.4.1}$$

As derived from (6.2.29) or (6.5.53), pressure losses due to the fluid friction are

$$\Delta p_{\rm fr}(t) = \frac{\lambda \rho_{\rm k} v^2}{2d_{\rm h}} L_1(t), \tag{12.4.2}$$

where $v = Q_{\rm k}/S$ and $Q_{\rm k}$ are velocity and flow rate of the killing fluid; S is the cross section area of the well; L_1 is the distance from the mouth to the interface between gas and fluid in the well.

Hydrostatic pressure of killing fluid column is

$$p_{\rm hyd} = \rho_{\rm k} g L_1. \tag{12.4.3}$$

The depth L_1 is determined by formula

$$L_1 = vt, \tag{12.4.4}$$

where t is the time elapsed after beginning of blowout killing.

Then, with regard to (12.4.2)–(12.4.4), from (12.4.1) the pressure at the mouth can be obtained

$$p_{\rm wm} = \frac{\lambda \rho_{\rm k} v^3}{2d_{\rm h}} t - \rho g v t + p_{\rm bot}. \tag{12.4.5}$$

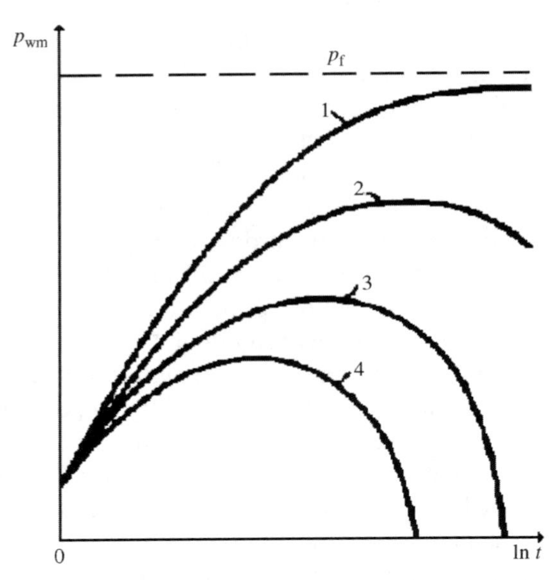

FIGURE 12.9 Variation of pressure at the mouth (curves 1–4) during blowout killing at different flow rates of the killing fluid $Q_{k1} = 0 < Q_{k2} < Q_{k3} < Q_{k4}$.

In nonstationary inflow of gas from the formation into closed well, the pressure p_{bot} is determined by formula (Businov and Umrichin, 1973)

$$p_{bot} = \sqrt{\alpha + \beta \ln t}, \qquad (12.4.6)$$

where $\alpha = p_f^2 + \beta \ln \frac{2.25\chi}{r_w^2} + bQ_0^2$, $\beta = Q_0 \mu p_0 / (2\pi k_f H)$; b is the factor in the formula (11.4.15). Thus, in each concrete case of blowout killing one can plot the dependence of pressure with (12.4.5) and (12.4.6) at the mouth on time of blowout killing for given flow rate Q_k and density ρ_k of the killing fluid. Then, with the use of obtained graphical dependence $p_{wm}(t)$, suitable pumping regime of killing fluid and equipment capable to realize it can be selected. Figure 12.9 gives characteristic dependences $p_{wm}(t)$ built with (12.4.5).

It should be noted that above presented method of gas blowout killing with direct pumping of the killing fluid is most effective in liquidation of gas inflow from formations providing slow pressure restoration in wells after their closing. In these cases, having enough time for blowout killing, it is able to create great column of killing fluid in the well and in doing so to lower the curve $p_{wm}(t)$ of pressure increment at the mouth.

CHAPTER 13

NONSTATIONARY FLOWS OF FLUID MIXTURES IN WELL-FORMATION SYSTEM: CALCULATION OF FLUID–GAS BLOWOUT KILLING

Study of nonstationary flows of fluid mixtures in well-reservoir system or its elements is of great importance for designing and performing technologic processes of fluid blowout or interstratal crossflow killing and well-cementing process. Present chapter contains calculation method of fluid blowout killing and interstratal crossflows based on theoretical grounds of nonstationary mixture flows. Calculation method of single-stage cementing of cased drill-stems will be outlined in the Chapter 14.

Figure 13.1 gives a scheme of well-reservoir system in blowout killing and liquidation of interstratal crossflows. Also, in killing of emergency gas blowout (see Section 12.3), main parameters in selection of surface equipment and designing of inclined wells are volume V, delivery Q_k, and density ρ_2 of the killing fluid. As origin of time ($t = 0$), it is taken at the beginning of the delivery of weighted flushing fluid with flow rate $Q_k = \text{const}$ in borehole of the blowing well. In course of killing process in the formation uncovered by blowing well moves the reservoir fluid, whereas in the section $z_h(t)$ the mixture of killing and reservoir fluids is assumed to be incompressible. Inertial terms in momentum equation are neglected.

Applied Hydro-Aeromechanics in Oil and Gas Drilling. By Leonov and Isaev

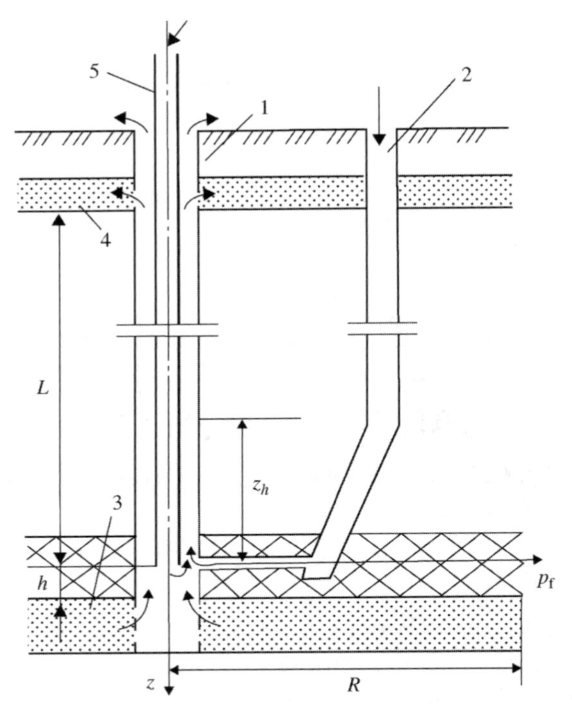

FIGURE 13.1 Scheme of well-formation system in blowout killing (pointers denote possible directions of gas, fluid, and mixture): 1—blowing well; 2—inclined well; 3, 4—productive and absorptive formations; 5—tubing and drill pipe.

The system of equations (4.6.33)–(4.6.39) for combined nonstationary flow in a well and formation has the following form.

1. Flow in the well:
   momentum equation

$$\frac{\partial p}{\partial z} = g[\rho_1\varphi_1 + \rho_2\varphi_2] + \frac{\lambda}{2d}\left[\rho_1\varphi_1 v_1^2 + \rho_2\varphi_2 v_2^2\right]; \tag{13.1}$$

   equations of mass conservation

$$\begin{aligned} \frac{\partial \rho_1\varphi_1}{\partial t} + \frac{\partial \rho_1\varphi_1 v_1}{\partial z} &= 0,\\ \frac{\partial \rho_2\varphi_2}{\partial t} + \frac{\partial \rho_2\varphi_2 v_2}{\partial z} &= 0; \end{aligned} \tag{13.2}$$

equations of state

$$\begin{aligned} \rho_1 &= \text{const}, \\ \rho_2 &= \text{const}; \end{aligned} \tag{13.3}$$

equations of concentrations

$$\begin{aligned} &\varphi_1 = \varphi_1(p, \rho_1, \rho_2, v_1, v_2, \lambda), \\ &\varphi_1 + \varphi_2 = 1; \end{aligned} \tag{13.4}$$

equation for hydraulic resistance factor

$$\lambda = \text{const}. \tag{13.5}$$

2. For flow in the reservoir, the solution in form of indicator curve (10.2.7) at variable flow rate of the reservoir fluid $Q_1(t)$ is assumed to be true

$$p_{\text{bot}} = p_{\text{f}} - bQ_1(t),$$

where

$$b = \frac{\mu \ln(R/r_{\text{w}})}{2\pi k_{\text{f}} H}; \quad R = 1.77\sqrt{\kappa_{\text{pc}} t_{\text{blow}}}; \tag{13.6}$$

κ_{pc} is the factor of piezoconductivity; t_{blow} is the time counted from the beginning of well blowing.

Before the beginning of blowout killing ($t < 0$), the flow in well-reservoir system is taken as stationary. The well is blowing with constant flow rate $Q_1 = Q_0 = \text{const}$ at $Q_{\text{k}} = 0$. Then, from equation (13.1) with $v_1 = Q_0/S$ and $\varphi_2 = 0$, we get pressure at $z = 0$ before killing begins

$$p_{\text{bot}} = p_{\text{wm}} + L\rho_1 g\left[1 + \frac{\lambda Q_0^2}{2gdS^2}\right], \tag{13.7}$$

where p_{wm} is pressure at the well mouth in blowing and opposite the absorptive formation in crossflow.

Equating (13.6) and (13.7) at $Q_1(t) = Q_0$, we obtain flow rate of the blowing well at given factor b of the formation

$$Q_0 = \frac{2(p_{\text{f}} - p_{\text{wm}} - M_0)}{\sqrt{b^2 + \frac{2M_0\lambda}{gdS^2}[p_{\text{f}} - p_{\text{wm}} - M_0]} + b}, \tag{13.8}$$

where $M_0 = \int_0^L \rho_1 g \, \mathrm{d}z = \rho_1 g L$.

Thus, the initial pressure distribution over the well bore is taken equal to (13.7) at Q_0, calculated by (13.8).

It is taken that at $t = 0$, the killing fluid is delivered with flow rate Q_{k} in the well ($z = 0$). The time of blowout killing T_{k} can be estimated from above by equations (13.1)–(13.6).

Since the total derivative of incompressible fluid density with respect to time is zero, the continuity equation at $v_1 = v_2$ gives

$$\frac{\partial v_1}{\partial z} = \frac{\partial v_2}{\partial z} = 0. \tag{13.9}$$

Consequently, velocities v_1 and v_2 are independent of coordinate z

$$v_1 = v_1(t); \qquad v_2 = v_2(t). \tag{13.10}$$

The form of function (13.4) is unknown. Nevertheless, it may be asserted that

$$\varphi_1 = 1, \quad \varphi_2 = 0 \quad \text{at} \quad z_{\text{h}} < z < L, \tag{13.11}$$

where z_{h} is a coordinate of the boundary between mixture and formation fluid displaced from the well by the formation fluid at time t_{h} as a result of killing fluid pumping.

Let us integrate (13.1) over z with regard to (13.10) from $z = 0$ (beginning of ascending flow of the killing fluid) to $z = L$ (well mouth in blowout or absorptive formation in crossflow)

$$p_{\text{bot}} = p_{\text{wm}} + \left(\int_0^L \rho_1 \varphi_1 \mathrm{d}z + \int_0^L \rho_2 \varphi_2 \mathrm{d}z \right) g + \frac{\lambda}{2d} v_1^2 \int_0^L \rho_1 \varphi_1 \mathrm{d}z + \frac{\lambda}{2d} v_2^2 \int_0^L \rho_2 \varphi_2 \mathrm{d}z. \tag{13.12}$$

Integrals in (13.12) can be obtained with the help of (13.11)

$$\int_0^L \rho_2\varphi_2 \mathrm{d}z = \frac{\rho_2}{S}\int_0^{z_h} S_2 \mathrm{d}z,$$

where $S_2(z)$ is cross section area of the well occupied by killing fluid.

Integral in the right part of the last expression is equal to volume V_2 of the killing fluid having been pumped into the well to time t_k. Then,

$$\int_0^L \rho_2\varphi_2 \mathrm{d}z = \rho_2\frac{V_2}{S} = \rho_2\frac{Q_k}{S}t_k. \tag{13.13}$$

The second integral is

$$\int_0^L \rho_1\varphi_1 \mathrm{d}z = \rho_1\int_0^L (1-\varphi_2)\mathrm{d}z = \rho_1 L - \rho_1\frac{Q_k}{S}t_k.$$

Substitution of obtained integrals in (13.12) yields

$$\begin{aligned} p_{bot} = {} & p_{wm} + \rho_1 gL + \frac{Q_k}{S}(\rho_2-\rho_1)gt_k \\ & + k^2 g\left\{(Sv_1)^2\rho_1 L + \frac{Q_k}{S}t_k\left[\rho_2(Sv_1)^2 - \rho_1(Sv_2)^2\right]\right\}, \end{aligned} \tag{13.14}$$

where $k^2 = \lambda/2gd_h S^2$.

The relation (13.14) is true for $z_k < L$ at any t_k.

Thus, one can choose killing regimes, that is, parameters ρ_2 and Q_k of the killing fluid, so that killing of the blowout happens before the boundary z_k would reach L. Otherwise, at

$$z_k = (p_f - p_{wm} - M_0)[1 + (p_f - p_{wm} - M_0)/(2bQ_k)]/[(\rho_2-\rho_1)g] > L$$

the calculations should be performed by several stages.

Assuming $z_k < L$ to be obeyed, we equate right parts of (13.6) and (13.14)

$$\begin{aligned} p_f - bQ(t) = {} & p_{wm} + \rho_1 gL + \frac{Q_k}{S}(\rho_2-\rho_1)gt_k \\ & + k^2\left\{(Sv_1)^2\rho_1 gL + \frac{Q_k}{S}gt_k\left[\rho_2(Sv_1)^2 - \rho_1(Sv_2)^2\right]\right\} \end{aligned} \tag{13.15}$$

to obtain the time t_{k} of mixture boundary advance

$$t_{\mathrm{k}} = \frac{p_{\mathrm{f}} - bQ(t) - p_{\mathrm{f}} - \rho_1 gL[1 + (kSv_1)^2]}{\frac{Q_{\mathrm{f}}}{S} g\left\{\rho_2 - \rho_1 + k^2\left[\rho_2 (Sv_2)^2 - \rho_1 (Sv_1)^2\right]\right\}}. \tag{13.16}$$

The relation (13.16) permits to get the time $t_{\mathrm{k}} = \bar{t}_{\mathrm{k}}$, when $Q_1(\bar{t}_{\mathrm{h}})$ becomes equal to zero. At $\bar{t}_{\mathrm{h}}$, the whole mixture in the well would have velocity $v_1 = v_2 = Q_{\mathrm{k}}/S$. Then,

$$\bar{t}_{\mathrm{k}} = \frac{p_{\mathrm{f}} - p_{\mathrm{wm}} - \rho_1 gL\left[1 + (kQ_{\mathrm{k}})^2\right]}{\frac{Q_{\mathrm{k}}}{S} g(\rho_2 - \rho_1)\left[1 + (kQ_{\mathrm{k}})^2\right]}. \tag{13.17}$$

At $\bar{t}_{\mathrm{k}}$, the inflow of the formation fluid ceases. If at this to bring pumps to a stop ($Q_{\mathrm{k}} = 0$), the inflow of fluid would be recommenced because the bottom pressure p_{bot} falls owing to termination of friction force action. Therefore, pumps should continue for some time t_{p} pumping of killing fluid until the mixture with height z_{h} formed to time $\bar{t}_{\mathrm{k}}$ would be displaced from the well.

The time of pumping $\bar{t}_{\mathrm{p}}$ needed to displace mixture column from the well is

$$\bar{t}_{\mathrm{p}} = LS/Q_{\mathrm{k}}. \tag{13.18}$$

If calculations show that $\bar{t}_{\mathrm{k}} > \bar{t}_{\mathrm{p}}$, blowout killing of crossflow does not happen. Then, one should repeat calculation of $\bar{t}_{\mathrm{k}}$ and $\bar{t}_{\mathrm{p}}$ with formulas (13.17) and (13.18) with varied values of ρ_2 and Q_{k} until fulfillment of inequality $z_{\mathrm{k}} < L$. Substitution of $\bar{t}_{\mathrm{k}} = \bar{t}_{\mathrm{p}} = LS/Q_{\mathrm{k}}$ into (13.17) and resolving the result with respect to ρ_2 yields

$$\rho_2 = \frac{p_{\mathrm{f}} - p_{\mathrm{wm}} - \rho_1 gL\left[1 + (kQ_{\mathrm{k}})^2\right]}{gL\left[1 + (kQ_{\mathrm{k}})^2\right]} + \rho_1. \tag{13.19}$$

This formula allows to get minimal value of density ρ_2 and delivery Q_{k} of the killing fluid with which the killing of blowout is possible in the case $z_{\mathrm{k}} < L$.

Hence, the blowout can be killed in time T_{k}, calculated by formula

$$T_{\mathrm{k}} = \bar{t}_{\mathrm{k}} + \bar{t}_{\mathrm{p}}, \quad z_{\mathrm{k}} < L. \tag{13.20}$$

EXAMPLE 13.1

It is required to estimate the time of blowout killing at the following initial data:

$$L = 1000\,\text{m},\quad p_{\text{wm}} = p_0 = 10^5\,\text{Pa},\quad p_{\text{f}} = 130\times 10^5\,\text{Pa},$$
$$d_{\text{h}} = 0.126\,\text{m},\quad S = 0.0124\,\text{m}^2,\quad \rho_1 = 1000\,\text{kg/m}^3,\ \rho_2 = 1800\,\text{kg/m}^3,$$
$$Q_{\text{k}} = 0.04\,\text{m}^3/\text{s},\quad b = 3\times 10^7\,\text{Pa s/m}^3,\quad \lambda = 0.015.$$

SOLUTION We get

$$k^2 = \frac{\lambda}{2gd_{\text{h}}S^2} = \frac{0.015}{2\times 9.81\times 0.126\times 0.0124^2} = 39.46\,\text{s}^2/\text{m}^6,$$

with formula (13.17)

$$\bar{t}_{\text{k}} = \frac{0.0124}{0.04}\times\frac{13\times 10^6 - 0.1\times 10^6 - 10^3\times 10^3\times 9.81\left[1 + 39.46\times 0.04^2\right]}{9.81\{(1800-1000)[1+39.46\times 0.04^2]\}} = 91.8\,\text{s},$$

with formula (13.18)

$$\bar{t}_{\text{p}} = LS/Q_{\text{k}} = 10^3\times 0.0124/0.04 = 310\,\text{s}.$$

The time of blowout killing is no more than

$$T_{\text{k}} = \bar{t}_{\text{h}} + \bar{t}_{\text{p}} = 91.8 + 310 = 401.8\,\text{s}.$$

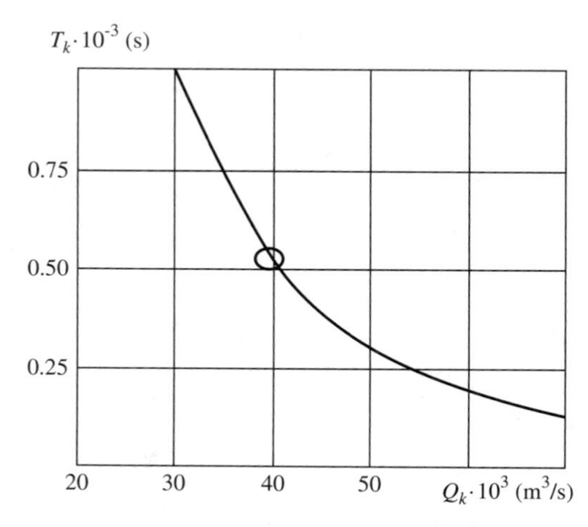

FIGURE 13.2 Dependence of T_{k} versus Q_{k}.

Needed volume of the killing fluid is

$$V = T_k Q_f = 401.8 \times 0.04 = 16.1\ \mathrm{m}^3.$$

With formula (13.8) at $t_0 = 0$, we receive

$$M_0 = \rho_1 g L = 9.81 \times 10^3 \times 10^3 = 9.81 \times 10^6\ \mathrm{Pa};$$

$$z_k = \frac{13 \times 10^6 - 10^5 - 9.81 \times 10^6}{(1800 - 1000)9.81} \times \left(1 + \frac{13 \times 10^6 - 10^5 - 9.81 \times 10^6}{2 \times 3 \times 10^7 \times 0.04}\right) = 901\ \mathrm{m} < L = 1000\ \mathrm{m}.$$

Figure 13.2 shows the curve of blowout killing for $z_k > L$ at $p_f = 16.1 \times 10^6$ MPa. Data for other parameters are taken from Example 13.1. By circle, the point is shown at

$$Q_k = 0.04\ \mathrm{m}^3/\mathrm{s}, \qquad T_k = 517\ \mathrm{s}.$$

CHAPTER 14

DISTRIBUTION OF CONCENTRATION AND PRESSURE IN DISPLACEMENT OF NEWTONIAN AND VISCOUS-PLASTIC FLUIDS FROM CIRCULAR PIPES AND ANNULAR CHANNELS: HYDRAULIC CALCULATION OF CEMENTATION REGIME

The practice of well cementation and experimental investigations testifies that insufficient displacement of washing and flashing fluids from the interval of cementation may be a reason of interstratal crossflows, shows, and bursting of drill-stems under action of rock pressure and other troubles.

14.1 MAIN REASONS OF INCOMPLETE DISPLACEMENT OF FLUIDS

For any combined flows of several fluids, their volume concentrations were defined in Section 3 as

$$\varphi_i = V_i/V. \tag{14.1.1}$$

The best quantitative characteristic of replacement fluids by another fluids in pipes and annular channels is distribution of their local concentrations that can be obtained from (14.1.1) by contraction of a chosen volume V into a point. Local concentration φ_i is equal to 1 when only ith fluid is in the point and 0 when this fluid is absent.

The completeness of replacement of washing and flushing fluids by cement solution is sometimes characterized by displacement factor under

Applied Hydro-Aeromechanics in Oil and Gas Drilling. By Leonov and Isaev

which it is an understood volume concentration of displacing (cement) solution in channel section when first particles have become available at far boundary of the section. Volume concentration may also be calculated with (14.1.1), if to accept in it V_i as volume of displacing ith fluid and V as section volume.

Volume concentration in the section of displacement depends on characteristics of the flow, properties of fluids, and so on but is independent of coordinates of the channel. Therefore, to analyze distribution of fluids over the channel length, it is better to use surface concentration of the fluid equal to the ratio of cross-section area S_i of the channel occupied by fluid to the whole cross-section area S

$$\varphi_i = S_i/S. \tag{14.1.2}$$

Surface concentration (14.1.2) is obtained from (14.1.1) by contraction of volume V containing given cross section to this cross section. The displacement of washing and flushing fluids by cement slurry is commonly incomplete. When pumping a volume of cement slurry equal to or even more than volume of cementation interval, concentrations obtained from (14.1.1) or (14.1.2) are less than 1. Thus, investigation of the function (14.1.2) permits to establish the reasons for incomplete displacement of fluids. In general case, the function (14.1.2) for viscous-plastic fluid can depend on all other flow parameters entering in equations (4.6.26) and (4.6.27), namely channel geometry Γ and flow rate Q

$$\begin{aligned}\varphi_i = \varphi_i(&z, t, v_1, v_2, \ldots, v_N, \rho_1, \rho_2, \ldots, \rho_N,\\ &\tau_{01}, \tau_{02}, \ldots, \tau_{0N}, \eta_1, \eta_2, \ldots, \eta_N, \Gamma, Q)\end{aligned} \tag{14.1.3}$$

Because of different average velocities v_i of fluids, the displacing fluid in some parts of the channel leaves behind the displaced fluid moving with lesser velocity. Lag of displaced fluid is particularly noticeable in the presence of channel cross-section eccentricity e favoring the formation of dead zones.

Figure 14.1 shows a scheme of dead zone in eccentric channel. In the presence of displaced 1 and displacing 2 fluids with different densities ρ_1 and ρ_2 in and around the dead zone, the condition of dead zone existence can be approximately written in form similar to (6.7.15)

$$\frac{4\tau_{01}}{d_h - 2e} \geq \frac{\Delta p}{L} + (\rho_1 - \rho_2) \cdot g \cdot \cos(g, z), \tag{14.1.4}$$

where $\Delta p = |p_2 - p_1|$; $d_h = d_1 - d_2$.

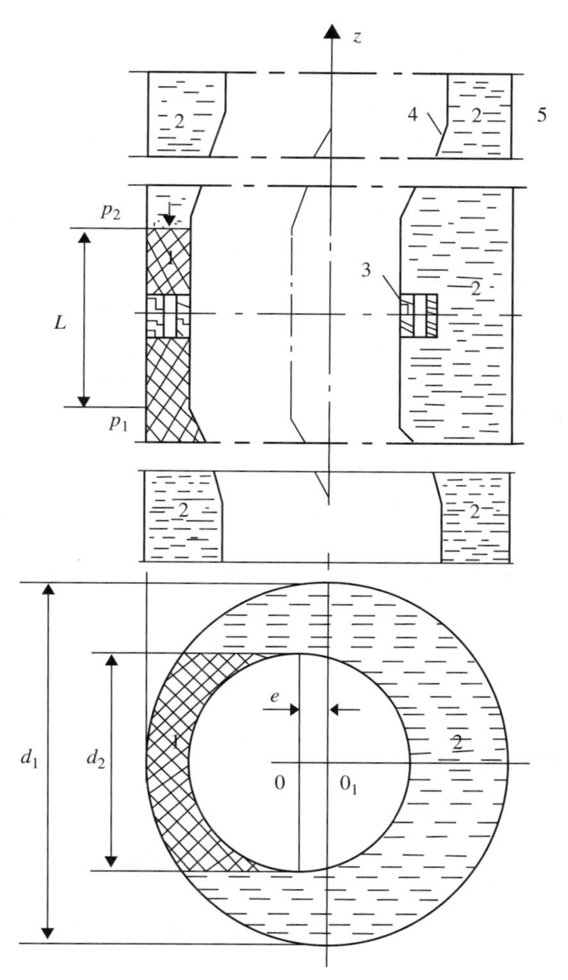

FIGURE 14.1 Scheme of a dead zone in eccentric annular space: (1) dead zone; (2) displaced fluid; (3) rigid-type centralizer; (4) casing; (5) well cavernous section.

The term $(\rho_1 - \rho_2) \cdot g \cos(g, z)$ is a vertical component of buoyancy force vector. In the case when due to thixotropic variation of dead-load shear stress θ_1 of the fluid in dead zone becomes greater than τ_{01}, one should use θ_1 instead of τ_{01} in (14.1.4).

If in (14.1.4) to reverse sign on opposite, necessary condition for fluid to be moved from the dead zone would be expressed by relation, then

$$\frac{4\tau_{01}}{(d_{\mathrm{h}}-2e)\left[\frac{\Delta p}{L} + (\rho_1-\rho_2) \cdot g \cdot \cos(g, z)\right]} < 1. \qquad (14.1.5)$$

In cementing practice, there are more frequent cases with $\rho_1 < \rho_2$ and $\cos(g, z) < 0$. Thus, in order to provide motion in the dead zone, it is sufficient to satisfy the inequality (14.1.5) through enhancement of $(\rho_2 - \rho_1)$, $\Delta p/L$, d_h, $|\cos(g, z)$ (the latter means that the well axis tends to take vertical position) and decrease of e and τ_{01} or θ_1. The replacement becomes worse with the formation of thickened filter clay coating facing conductive absorptive formations. This coating has elevated shear stress and is poorly washed off by displacing fluid.

In turbulent flow washout of the boundary between successively moving fluids takes place. This phenomenon of turbulent diffusion leads to equalizing average velocities v_i, lowers delay of fluids, and consequently, brings to rise of displacement completeness.

Let us find relation between parameters of displacing 2 and displaced 1 viscous-plastic fluids when the flow of displacing fluid in broad part of clearance has pressure gradient sufficient for flow of displaced fluid from narrow part of the clearance.

The condition of limit equilibrium of displacing viscous-plastic fluid in broad clearance is approximately

$$\frac{4\tau_{02}}{d_h + 2e} = \frac{\Delta p}{L}. \tag{14.1.6}$$

Substitution of $\Delta p/L$ in (14.1.5) gives

$$\frac{4\tau_{01}}{(d_h - 2e)\left[\frac{4\tau_{02}}{d_h + 2e} + (\rho_1 - \rho_2) \cdot g \cdot \cos(g, z)\right]} < 1. \tag{14.1.7}$$

In cementation at conditions of abnormal high formation pressure, there is frequently $\rho_1 \approx \rho_2$. Then, (14.1.7) takes form

$$\frac{\tau_{02}}{\tau_{01}} > \frac{d_h + 2e}{d_h - 2e}. \tag{14.1.8}$$

This relation between dynamic shear stresses is recommended to support in cementing of casings eccentrically located in wells.

14.2 DISTRIBUTION OF CONCENTRATIONS IN DISPLACEMENT OF ONE FLUID BY ANOTHER FLUID

In the present section, formulas for calculation concentration distribution along a channel in displacement of one fluid by another will be derived.

Let fluids have equal rheological characteristics and vary in colors. Such statement of the problem is the simplest one and easily permits to study the effect of velocity profile nonuniformity on concentration distribution and to get best displacement regime.

Since fluids differ only in colors, distribution of velocities and pressure drop at constant flow rate Q of displacing fluid at any displacement channel section are determined by formulas for viscous, viscous-plastic, and power fluids presented in Section 6.

Instead of the system of equations (4.6.31), (4.6.32), (4.6.21), (4.6.14), and (4.6.39), consideration of such stated problem allows to invoke only one equation (14.1.2) for concentration φ_2 of displacing fluid with addition of equation $z_h = z_h(r, t)$ describing motion of the interface between fluids

$$\begin{cases} \varphi_2(z_h) = \dfrac{S_2(z_h)}{S} = \dfrac{4r^2(z_h)}{d^2}; \\ \varphi_1 + \varphi_2 = 1; \end{cases} \qquad (14.2.1)$$

$$\frac{dz_h}{dt} = W(r). \qquad (14.2.2)$$

and initial conditions

$$\left. \begin{aligned} \varphi_2 = 0, \quad z > 0 \\ \varphi_2 = 1, \quad z \le 0 \end{aligned} \right\} t = 0. \qquad (14.2.3)$$

Conditions (14.2.3) at $t = 0$ give initial location of the interface as

$$z_h = 0. \qquad (14.2.4)$$

As velocity $w(r)$ in (14.2.2) should be taken: (6.3.23) and (6.3.26) for viscous-plastic fluid, and (6.4.25) for power fluid in pipe; (6.3.37) and (6.3.38) for viscous-plastic fluid, and (6.4.35) and (6.4.36) for power fluid in annular channel.

Turn now to the problem on laminar flow of viscous-plastic fluid in pipes. Consideration of fluids with another rheology can be performed in a similar way. Assume that at $t \ge 0$ stationary velocity profile (6.3.23), (6.3.26) in the whole pipe for both fluids takes place (Fig. 14.2).

Formulas (6.3.23) and (6.3.26) determine velocity at any point of the flow, in particular at points of the interface at any instant of time. Since radius r_h of any interface point does not vary with time, we get from (6.3.23) and (6.3.26)

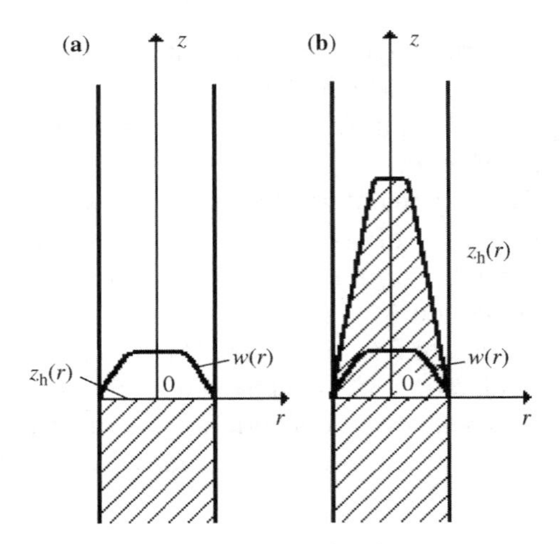

FIGURE 14.2 Position of the interface $z_h(r)$ in displacement of one fluid by another: (a) $t=0$ and (b) $t=t_1>0$.

$$r_h = \frac{d}{2}\left(\beta + \sqrt{(1-\beta^2)-\frac{w(r)4\eta}{|A|R^2}}\right), \tag{14.2.5}$$

where $w(r)$ is determined by formulas (6.3.23) and (6.3.26).

Before the root in (14.2.5) the plus sign is taken because at $r=d/2$ the velocity $w=0$. Points of the core interface with coordinates $r<R_0$ move with constant velocity w_0 expressed by the formula (6.3.26). For any point r_h of the interface from (14.2.2) with regard to (14.2.4), we get

$$\frac{z_h}{t} = w(r_h). \tag{14.2.6}$$

It should be noted that at the instant of time under consideration the coordinate of any interface point obeys inequality $z_h \leq w_{max}t$. Dividing both parts of (14.2.6) by average velocity $v=Q/S$, we receive

$$\frac{z_h}{vt} = \frac{w(r_h)}{v}, \tag{14.2.7}$$

where the left part represents dimensionless coordinate $\bar{z}$ of the interface between fluids

$$\bar{z} = \frac{r_h}{l(t)} \tag{14.2.8}$$

and $l(t)=vt$ is the distance that a particle would come in time t with average velocity v.

Then, (14.2.5) with regard to (14.2.7) and (14.2.8) may be written as (Broon and Leonov, 1981)

$$r = \frac{r_{\mathrm{h}}}{R} = \beta + \sqrt{(1-\beta^2) - 2\bar{A}\bar{z}}, \qquad (14.2.9)$$

where $\bar{A} = 2\eta v/(|A|R^2)$; $\beta = 4\tau_0 l/(\Delta p \cdot d)$ and $\bar{z} \le \bar{z}_{\max} = w_{\max}/v = (1-\beta)^2/(2\bar{A})$.

At insertion into the expression for $\bar{A}$, the average velocity $v = Q/S$ and Q from (6.3.27) yields

$$\bar{A} = 0.25\left(1 - \frac{4}{3}\beta + \frac{1}{3}\beta^4\right). \qquad (14.2.10)$$

Use of (14.2.9) in (14.2.1) gives the following formula for concentration in the cross section z at the instant of time t:

$$\varphi_2 = (\bar{r}^2) = \beta + \left(\sqrt{(1-\beta^2) - 2\bar{A}\bar{z}}\right)^2. \qquad (14.2.11)$$

For viscous fluid $\beta = 0$ and from (14.2.11) it follows $\varphi_2 = 1 - 0.5\bar{z}$, where $\bar{z} \le \bar{z}_{\max} = w_{\max}/v = 2$.

Resolving (14.2.11) with respect to z, we get

$$\bar{z} = \frac{(1-\beta)^2 - (\varphi_2^{0.5} - \beta)^2}{2\bar{A}}. \qquad (14.2.12)$$

Formulas (14.2.11) and (14.2.12) are true for laminar flow.

Obtain now expression for concentration φ_2 in turbulent flow when velocity distribution in (14.2.2) is described by power law (6.5.33)

$$w(r) = w_{\max}\left(1 - \frac{r}{R}\right)^{\frac{1}{N}}. \qquad (14.2.13)$$

At this the average velocity is

$$v = \frac{Q}{S} = \frac{2\pi}{\pi R^2}\int_0^R w_{\max}\left(1 - \frac{r}{R}\right)^{\frac{1}{N}} r\,\mathrm{d}r = \frac{2w_{\max}N^2}{(N+1)(2N+1)}. \qquad (14.2.14)$$

In turbulent flow relations (14.2.6)–(14.2.8) are also valid. Substitution of $w(r)$ and v from (14.2.13) and (14.2.14) into (14.2.7) yields

$$\bar{z} = \frac{(N+1)(2N+1)}{2N^2}(1-\bar{r})^{\frac{1}{N}}. \qquad (14.2.15)$$

Resolve this expression with respect to $\bar{r}$ and insert it in (14.2.1). As a result we obtain concentration φ_2

$$\varphi_2 = \left[1 - \left(\frac{2N^2\bar{z}}{(N+1)(2N+1)}\right)^N\right]^2. \qquad (14.2.16)$$

The concentration φ_2 in turbulent flow vanishes at

$$\bar{z} = \bar{z}_{\max} = (N+1)(2N+1)/(2N^2).$$

At $N = 6$, the formula (14.2.16) takes form

$$\varphi_2 = [1 - 0.245\,\bar{z}^6]^2. \qquad (14.2.17)$$

Resolving (14.2.16) relative $\bar{z}$, we get

$$\bar{z} = (1 - \varphi_2^{0.5})\frac{(N+1)(2N+1)}{2N^2}. \qquad (14.2.18)$$

Consider displacement of fluid with variable average velocities v_i in time intervals Δt_i. Let k, l, m be the number of time intervals Δt_i at which a point with coordinate $\bar{r}$ of the interface moves in turbulent regime, in laminar regime in gradient layer, and in flow core, respectively. Then, this point in all time intervals $t = \Sigma\Delta t_i$ travels the distance $z = \Sigma z_i$. The distance z_i covered by the point in time Δt_i may be found from formulas (14.2.12) and (14.2.18) using in them (14.2.8):

for laminar flow in gradient layer ($\bar{r} > \beta$ or $\varphi_2 > \beta^2$)

$$\bar{z}_i = \frac{(1-\beta)^2 - (\varphi_2^{0.5} - \beta)^2}{2\bar{A}_i} v_i\,\Delta t_i, \quad i = 1, 2, \ldots, l;$$

for laminar flow in flow core ($\bar{r} \leq \beta$ or $\varphi_2 \leq \beta^2$)

$$\bar{z}_i = \frac{(1-\beta)^2}{2\bar{A}_i} v_i\,\Delta t_i, \quad i = 1, 2, \ldots, m;$$

for turbulent flow

$$z_i = \frac{(N_i+1)(2N_i+1)}{2N_i^2}(1 - \varphi_2^{0.5})^{\frac{1}{N_i}} v_i\,\Delta t_i, \quad i = 1, 2, \ldots, k.$$

The total distance $z = \Sigma z_i$ is

$$z = \sum_{i=1}^{k} \frac{(N_i+1)(2N_i+1)}{2N_i^2}(1 - \varphi_2^{0.5})^{\frac{1}{N_i}} v_i\,\Delta t_i + \sum_{i=1}^{l} \frac{(1-\beta_i)^2 - (\varphi_2^{0.5} - \beta_i)^2}{2\bar{A}_i} v_i\,\Delta t_i + \sum_{i=1}^{m} \frac{(1-\beta_i)^2}{2\bar{A}_i} v_i\,\Delta t_i. \qquad (14.2.19)$$

As generalization of (14.2.8) would be a dimensionless quantity $\bar{z}$ equal to the ratio of the distance z calculated by formula (14.2.19) to the distance traveled by interface point in time interval $\Delta t = \Sigma \Delta t_i$ when moving with average velocity v_i at each time interval Δt_i

$$\bar{z} = \frac{z}{\sum l_i} = \frac{z}{\sum\limits_{i=1}^{k} v_i\,\Delta t_i + \sum\limits_{i=1}^{l} v_i\,\Delta t_i + \sum\limits_{i=1}^{m} v_i\,\Delta t_i}. \tag{14.2.20}$$

Substitution of (14.2.19) in (14.2.20) gives

$$\bar{z} = \sum_{i=1}^{k} \frac{(N_i+1)(2N_i+1)}{2N_i^2}\left(1-\varphi_2^{0.5}\right)^{\frac{1}{N_i}} \bar{V}_i + \sum_{i=1}^{l} \frac{(1-\beta_i)^2-(\varphi_2^{0.5}-\beta_i)^2}{2\bar{A}_i}\bar{V}_i + \sum_{i=1}^{m} \frac{(1-\beta_i)^2}{2\bar{A}_i}\bar{V}_i, \tag{14.2.21}$$

where

$$\bar{V}_i = \frac{v_i\,\Delta t_i}{\sum\limits_{i=1}^{k} v_i\,\Delta t_i + \sum\limits_{i=1}^{l} v_i\,\Delta t_i + \sum\limits_{i=1}^{m} v_i\,\Delta t_i} \tag{14.2.22}$$

are parts of displacing fluid volume pumped into the channel at correspondent flow regimes.

$V_h = zS$ is volume of the channel interval in which displacement takes place, $V = (\sum l_i)S$ is the total volume of the displacing volume pumped into the channel in time Δt. Then,

$$V_h/V = \bar{z}. \tag{14.2.23}$$

Formulas (14.2.21) and (14.2.23) are chief formulas in the calculation method of cementation with concentration φ_2 needed in given cross section. Similar formulas can be obtained for displacement of fluids from annular channel. At this in equations (14.2.1) and (14.2.2) as S_2, S and $w(r)$ should be taken areas and velocity distribution, respectively, in annular channel. Typical concentration curves for laminar $\varphi_{2L}(z)$ and turbulent $\varphi_{2T}(z)$ flows built with formulas (14.2.11) and (14.2.17) are represented in Fig. 14.3. It is seen that for both regimes, there are regions of better displacement. There is also an interval $0 \leq \bar{z} \leq \bar{z}_n$ at the upper boundary of

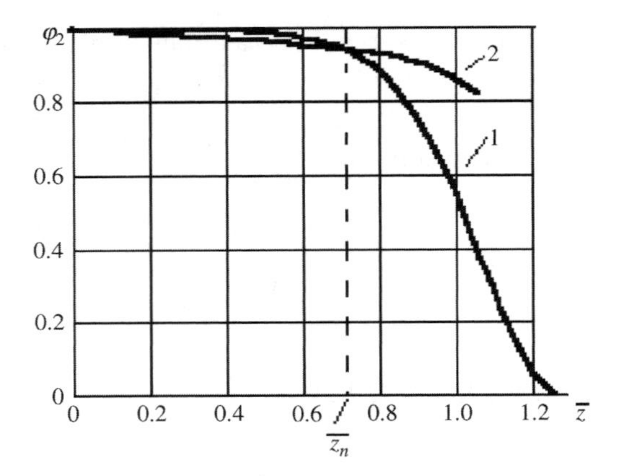

FIGURE 14.3 Dependences of $\varphi_2(\bar{z})$ for hydraulic smooth pipes: (1) turbulent flow ($Re = 4 \times 10^3$, $N = 6$) and (2) laminar flow of viscous-plastic fluid ($\beta = 0.9$; $Se = 384$).

which $\bar{z} \leq \bar{z}_n$ it would be

$$\varphi_{2L} = \varphi_{2T}. \tag{14.2.24}$$

The cross section $\bar{z} = \bar{z}_n$ separates the region of better displacement (plug regime) in laminar flow.

Thus,

$$\begin{aligned} \varphi_{2T} &> \varphi_{2L}, \quad \bar{z} < \bar{z}_n; \\ \varphi_{2T} &\leq \varphi_{2L}, \quad \bar{z} \geq \bar{z}_n. \end{aligned} \tag{14.2.25}$$

Consider an interval $0 \leq \bar{z} \leq 1$, which in accordance with (14.2.23) corresponds to pumping of displacing fluid with volume no less than the volume of displacement interval. At this interval, there is single root of the equation (14.2.24). Solution of the equation (14.2.24) is represented in Fig. 14.4 in form of dependence $\beta = \beta(\bar{z})$ at $N = 6$. It is seen that the region II of values β and $\bar{z}$ characterizing better displacement in laminar regime is less than the region I of better displacement in turbulent regime.

With rise of delivery Q and viscosity η of displacing fluid, the Saint-Venant parameter

$$Se = \frac{\tau_0 d}{\eta v} = \frac{\pi}{4} \frac{\tau_0 d^3}{\eta Q} \tag{14.2.26}$$

declines, and inversely it increases with enhancement of τ_0 and d. Growth of Se in its turn leads to rise of β (see Fig. 6.7) and in accordance with (14.2.11) to build up of φ_{2L}. At $\bar{z} \leq 1$, this causes widening of better displacement

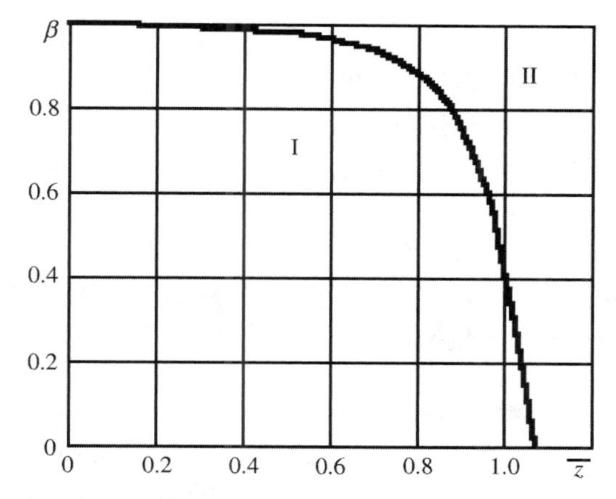

FIGURE 14.4 Regions of better displacement for turbulent (I) and laminar (II) flow of viscous-plastic fluid.

region in laminar flow. Thus, increase of τ_0, d and decrease of η, Q brings to gain in displacement in laminar flow. The strongest influence on the displacement exerts parameter d. The best displacement in laminar flow is achieved when $Se \to \infty$, for example, at $Q \to 0$, which requires infinite time of displacement. In cementing, such process is of course unlikely. Displacement in turbulent flow is preferable to perform with great delivery Q, since with rise of Q or Re, the number N increases (see Section 6.5), which in accordance with (14.2.16) leads to increase of φ_{2T}.

As follows from formulas (14.2.11) or (14.2.16) and (14.2.23), one may increase φ_2 by decreasing $\bar{z}$. The latter can be achieved at given displacement interval of z through enhancement of the displacement fluid volume V or by decrease of the displacement interval z at given V. Hence, in any cross section of the displacement interval, any desired concentration of the displacement fluid can be attained.

It is seen from the preceding discussion that all undertakings to rise in displacement completeness in cementing by controlling parameters τ_0, η, d, Q, $\bar{z}$ are confined only to engineering reasoning.

Variation of fluid discharge in cementing is restricted by maximal Q_{max} and minimal Q_{min} values of pump delivery ($Q_{min} \leq Q \leq Q_{max}$). z_φ is the coordinate of the cross section in which it is required to get maximal permissible concentration φ_2 at given volume V or given concentration φ_2 at minimal allowable volume V of displacing fluid. Then, the concentration φ_2 in the cross section z_φ would be maximal at the following deliveries:

(1) $Q = Q_{max}$ at $Q_{cr} \leq Q_{min}$, where Q_{cr} is the flow rate at which laminar flow transits into turbulent one;

(2) $Q = Q_{\max}$ at $Q_{\min} \leq Q_{cr} \leq Q_{\max}$ and $\varphi_{2T}(z_\varphi) \geq \varphi_{2L}(z_\varphi, \beta)$, where β should be calculated with delivery of $Q_{\min}$;

(3) $Q = Q_{\min}$ at $Q_{cr} \geq Q_{\max}$;

(4) $Q = Q_{\min}$ at $Q_{\min} \leq Q_{cr} \leq Q_{\max}$ and $\varphi_{2T}(z_\varphi) \leq \varphi_{2L}(z_\varphi, \beta)$.

14.3 TAKING INTO ACCOUNT NEEDED DISPLACEMENT COMPLETENESS IN CALCULATION OF CEMENTING

As criterion to estimate displacement completeness in well annular space, concentration φ_2 in given cross section z at the end of well driving is used. Calculation of cementation with regard to given displacement completeness can be performed by solving one of the following basic problems.

1. It is required to select a displacement regime providing concentration in given cross section no less than prearranged one in driving minimal volume of displacing grouting mortar. This problem comes into existence when admissible concentration of washing and flushing fluids in cement solution providing qualitative cementation is known and given, for example, a concentration unaffecting the properties of grouting mortar and providing its setting in the whole cementation interval.
2. It is necessary to select regime of displacement providing maximal concentration in given cross section at fixed volume V of displacing grouting mortar. This problem arises when needed concentration is unknown or the concentration obtained from the solution of the first problem is economically disadvantageous or impossible to apply by technological reason. Below it is accepted that achievement of φ_2 in given range of initial parameters of cementation is justified by gain in quality of well seating.

Let us use formulas derived in the previous section when considering flows of many colored fluids describing well displacement of fluids with like properties. Such fluids are more frequently encountered in well seating under conditions of abnormal high formation pressure and in the presence of absorptive horizons. Formulas are also derived for pipes with circular cross section, but displacement happens in annular channel. Since solution for annular pipe can be obtained in the main by numerical methods to estimate concentration φ_2, we invoke formulas given above for circulate pipes at $d = d_h$. It can be done when the ratio of internal diameter of the annular channel to the external one is far less than unity ($\delta \ll 1$). The parameter β is determined by dependences for annular channels presented in Section 6. For annular space, it should be also obtained volume V_{as}. The displacement in cemented casing is assumed to be complete.

Calculation of $V(\varphi_2)$ is performed with grouting mortar having properties of the drill fluid. It is permissible when properties of fluids as compared to those containing in calculations make displacement better. Such case takes place, when for example, the density ρ and rheological properties (η, τ_0) of the cement solution quantitatively exceed analogous properties of flushing and washing fluids. Difference in geometry of the casing and annular channel can be taken into account by step-by-step calculation of sections with constant diameters. In estimated calculations, real annular space is replaced by a pipe with volume equal to the given annular space.

Cross section z_φ with needed φ_2 can coincide with the roof of a horizon with maximal abnormal high formation pressure most disposed to shows, the boundary of the interval of well bore constriction caused by rock pressure and the end of cementation interval. In these cases calculations should be performed with account for interaction of the well with formations in well-formation system. The problem on needed concentration φ_2 of grouting mortar is not certainly solved. For approximate calculations, one can use experimental data for $\varphi_2 \geq 0.8$ providing setting of grouting mortar mixed with another fluids during expectation of cement solidification.

14.4 METHOD OF HYDRAULIC CALCULATION OF CEMENTATION REGIMES WITH REGARD TO GIVEN CONCENTRATION IN CHANNEL CROSS SECTION

In the present section hydraulic calculation method of single-step cementation dedicated to select concentration of grouting mortar in a certain channel cross section no less than the given one and excluding absorption, shows, and disturbances of casing leak-proof at the mouth is described (Broon and Leonov, 1981).

Results of calculations are volumes of pumped fluids, delivery of pumps (cementation aggregates, drill pumps, and so on), pressure at the casing mouth, and counter-pressure at annulus mouth providing the absence of crossflows during expectation of cement solidification.

Consider successive motion of washing, flushing, displacing fluids, and grouting mortars in the well at constant delivery of pumps. The coordinate system (Fig. 14.5) is directed along fluid flow in pipes and annular space. In

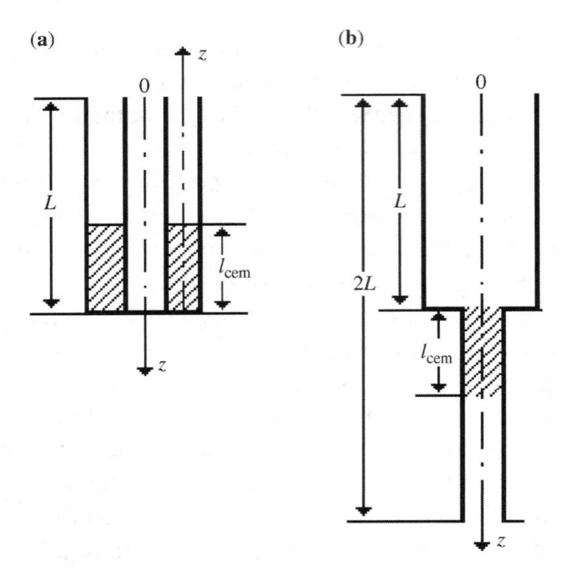

FIGURE 14.5 Well-circulation system (a) and associated calculation scheme (b).

Fig. 14.5b, a sketch of circulation system is shown in which for the sake of simplicity the length of the casing equal to the length of the well is taken. The well is assumed to be initially filled with washing fluid. As other fluids with another rheological properties are being pumped with constant flow rate into the well, the pressure drop $\Delta p = |p_2 - p_1|$ (p_2, p_1 are pressures at the mouth of the annular space and in the casing) in general case varies with time.

At arbitrary instant of time t, one can calculate coordinates of washing z_{wash}, flushing z_{flush}, driving z_{driv} fluids, and grouting mortar z_{gm} boundaries under condition that they are taken as planes, that is, fluid flow is considered as slug flow. Thus, the pressure drop caused by friction of each fluid at the interface at given delivery of pumps Q is determined by formulas of Section 6. For example, if any, from above mentioned fluids appears to be viscous-plastic, the pressure drop in laminar flow of this fluid would be determined by formula (6.3.48):

$$\Delta p = 4\tau_0 l/(\beta d_h), \tag{14.4.1}$$

where β is obtained from the curve 1 for pipe ($d_h = d$) and from the curve 2 for annular space (see Fig. 6.7) in accordance with Saint-Venant number

$$Se = \tau_0 d_h S/(\eta Q) \tag{14.4.2}$$

calculated in advance.

At this, the length l should be less than or equal to the length l_{f} of the circulation system section occupied by given fluid with constant diameter d_{h} ($l \leq l_{\mathrm{f}}$). In laminar flow of viscous fluid, the pressure drop is determined by formulas (6.2.30) and (6.2.31) and for power fluid by (6.4.27) and (6.4.41).

In turbulent flow of any fluid in Section 1, the pressure drop Δp is calculated by Darcy–Weisbach formula (6.5.1):

$$\Delta p = \lambda \rho \frac{Q^2}{2 d_{\mathrm{h}} S^2} l, \qquad (14.4.3)$$

where λ is determined by (6.5.38) or (6.5.58). Pressure losses are calculated with formulas for laminar or turbulent flow depending on critical parameters obtained in advance. In flow of viscous-plastic and viscous fluids to determine critical Reynolds number Re_{cr} and delivery Q_{cr}, one can use the formula (6.6.7)

$$\mathrm{Re}_{\mathrm{cr}} = 2100 + 7.3 He^{0.58}; \qquad (14.4.4)$$

$$Q_{\mathrm{cr}} = \mathrm{Re}_{\mathrm{cr}} \frac{S\eta}{d_{\mathrm{h}} \rho}. \qquad (14.4.5)$$

Sections containing formations most disposed to absorption are marked off with coordinates of these formations. Total pressure drop Δp at the section l is

$$p_2 - p_1 = \rho g \cos(g, z) l - \Delta p, \qquad (14.4.6)$$

where p_1 is pressure in the near end cross section, whereas p_2 is in the far end cross section of the section l.

The first term in the right part (14.4.6) represents hydrostatic component of the pressure drop in which $\cos(g, z) = 1$ for the casing and $\cos(g, z) = -1$ for annular space of vertical wells; for inclined well one should take another values of $\cos(g, z)$.

Total pressure drop in the whole circulation system or in its part is determined by summation over all pressure drops $(p_2 - p_1)$

$$\sum (p_2 - p_1) = \sum (\rho g \cos(g, z) l) - \sum \Delta p_{\mathrm{p}}. \qquad (14.4.7)$$

If the summation is performed over all circulation system, the formula (14.4.7) gives the difference of pressures between the casing and annular space at the well mouth.

Thus, in order to calculate the whole pressure drop $(p_2 - p_1)$ at given instant of time with formula (14.4.6), it is needed to know positions of boundaries z_i to this time and delivery of pumps Q. Location of boundaries z_i should be determined at characteristic instance of time associated with minimal and maximal pressures at the mouth and in well cross sections of interest, for example, opposite show formations disposed to absorption. The choice of instances of time depends on the volume of pumped fluid. Further will be considered: beginning $(t = t_1)$ and end of the flushing fluid volume V_{flush} pumping $(t = t_2)$, beginning $(t = t_3)$ of grouting mortar driving, instances of time $t = t_4$, t_5, t_6 of driving one, two and, three thirds of grouting mortar, respectively.

To selected instances of time are consistent certain fluid volumes delivered into the well:

$$\begin{aligned}
V_1 &= 0 && \text{at} \quad t = t_1 = 0;\\
V_2 &= V_{\text{flush}} && \text{at} \quad t = t_2;\\
V_3 &= V_{\text{p}} + V_{\text{flush}} && \text{at} \quad t = t_3;\\
V_4 &= V_3 + V/3 && \text{at} \quad t = t_4;\\
V_5 &= V_4 + V/3 && \text{at} \quad t = t_5;\\
V_6 &= V_5 + V/3 && \text{at} \quad t = t_6,
\end{aligned} \tag{14.4.8}$$

where V_{as} is volume of the annulus.

Hence, for given volumes at any of the associated instances of times t_i, it is possible to get coordinates $z_i(t)$ of the corresponding boundary.

The method of cementation includes the following stages.

1. The volume of the flushing fluid V_{flush} is taken as known. Permissible density of fluids ρ_{per} is determined by

$$\rho_{\text{per}} = \rho_{\text{w}} k_{\text{a}} k_{\text{s}},$$

where ρ_{w} is density of water; k_{a} is the factor of the abnormality; k_{s} is the factor of the safety taking into account condition of show nonadmission from corresponding formation during cementation.

Density of the washing fluid should obey condition $\rho_{\text{wash}} \geq \rho_{\text{per}}$. Let us verify whether the given volume V_{flush} can be used assuming absence of shows. In order to do this, compare the density of the flushing fluid ρ_{flush} with permissible one ρ_{per}. If $\rho_{\text{flush}} \geq \rho_{\text{per}}$, we leave V_{flush}, otherwise at $\rho_{\text{flush}} < \rho_{\text{per}}$, one should calculate new volume

V_{flush} with formula

$$V_{\text{flush}} = S_k \frac{\rho_{\text{wash}} - \rho_{\text{per}}}{\rho_{\text{wash}} - \rho_{\text{flush}}} l_{\text{sh}}, \tag{14.4.9}$$

where l_{sh} is the depth of the showing formation. In the following calculations we take the lesser volume V_{flush} from calculated with (14.4.9) and given in initial data.

Determine now the volume of driving fluid V_{df} with formula

$$V_{\text{df}} = S_{\text{p}}(L - h_0), \tag{14.4.10}$$

where h_0 is the height of the cement box.

In the first approximation, we take the volume of grouting mortar V driven into annular space to be equal to the volume of the section to be cemented

$$V = S_{\text{as}} l_{\text{cem}}, \tag{14.4.11}$$

where l_{cem} is the length of the section to be cemented.

2. Total volume of the grouting mortar with regard to the volume of cement box is

$$V_{\text{cem}} = V + S_{\text{p}} h_0. \tag{14.4.12}$$

Since volumes are determined, one can get coordinates of fluid boundaries at instances of time t_i with formulas tabulated in Table 14.1.

When filling Table 14.1, if values of boundary coordinates exceed $2L$, they should be taken equal to $2L$, if they are negative they have to be accepted equal to zero. At each instance of time t_i, we divide pipe and annulus spaces by sections l_k along which diameters, slopes to z-axis, and properties of fluids do not vary. Sections containing formations most disposed to absorption are divided by more fine parts separated by coordinates of these formations.

Figure 14.6 demonstrates one of the possible arrangements of fluid boundaries at driving time t_4. In accordance with this scheme, sections l_k are presented by the following six sections ($k = 6$):

l_1:	$z_2 = z_{\text{flush}} \le z < 2L = z_1$;
l_2:	$z_3 = z_{\text{h}} \le z < z_2$;
l_3:	$z_4 = z_{\text{cem}} \le z < z_3$;
l_4:	$z_5 = L \le z < z_4$;
l_5:	$z_6 = z_{\text{driv}} \le z < z_5$;
l_6:	$0 \le z < z_6$.

TABLE 14.1

t_i	V_i	z_{wash}	z_{flush}	z_{cem}	z_{driv}
t_1	$V_1 = 0$	$2L$	0	0	0
t_2	$V_2 = V_{flush}$	$2L$	V_{flush}/S_p; at $V_{flush}/S_p > L$ is $z_{flush} = L + (V_{flush} - V_p)/S_{as}$	0	0
t_3	$V_3 = V_p + V_{flush}$	$2L$	$L = V_{flush}/S_{as}$	L	$L - V_{cem}/S_p$
t_4	$V_4 = V_3 + V/3$	$2L$	$L + V_{flush}/S_{as} + V/(3S_{as})$	$L + V/(3S_{as})$	$L - V_{cem}/S_p + V/(3S_p)$
t_5	$V_5 = V_4 + V/3$	$2L$	$L + V_{flush}/S_{as} + 2V/(3S_{as})$	$L + 2V/(3S_{as})$	$L - V_{cem}/S_p + 2V/(3S_p)$
t_6	$V_6 = V_5 + V/3$	$2L$	$L + (V_{flush} + V)/S_{as}$	$L + V/S_{as}$	$L - h_0$

Hence, nonstationary flow in the process of cementation is approximately replaced by a number of stationary states at t_i for which with formulas of stationary flows can be determined all parameters of interest.

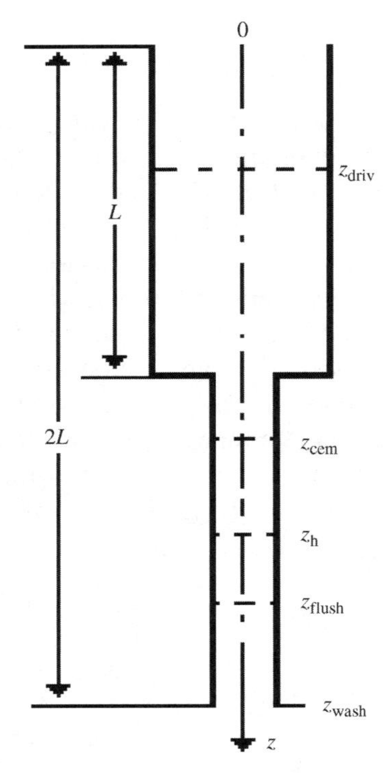

FIGURE 14.6 Possible scheme of arrangement of fluid boundaries in fluid driving.

3. Let at any instance of time are determined $l_k = \Delta z_k$. Opposite formation most disposed to absorption with coordinate $z = z_h$, let us get pressure $p_{st}(z_h)$ less friction losses

$$p_{st}(z_h) = p_{cp} - \sum \rho_1 g \cos(g, z) l_k, \qquad (14.4.13)$$

where p_{cp} is counter-pressure at the annular space mouth, for example, equal to pressure drop in controlled bean. The sum is performed over section from $z = 2L$ to $z = z_h$. In Fig. 14.6 such cross sections are two $l_1 = z_1 - z_2$ and $l_2 = z_2 - z_3$. They are located between $z = 2L$ and $z = z_h$. Pressure p_{st} at the casing mouth is determined with formula similar to (14.4.13)

$$p_{st}(0) = p_{cp} - \sum \rho_l g \cos(g, z) l_k, \qquad (14.4.14)$$

where summing is performed over all sections from cross section $z = 2L$ to $z = 0$. Then the permissible pressures at the casing mouth and opposite absorptive formations minus pressure obtained by formulas (14.4.13) and (14.4.14) are

$$p_{per}(0) = p_{perm} - p_{st}(0); \qquad (14.4.15)$$

$$p_{per}(z_h) = p_h - p_{st}(z_h), \qquad (14.4.16)$$

where p_{perm} is given admissible pressure at the mouth determined by strength characteristics of the casing and drilling wellhead equipment; p_h is the given absorption pressure (hydraulic fracturing pressure). If the pressure p_{per} calculated by (14.4.15) and (14.4.16) proved to be negative, initial data should be changed, for example, density of fluid in order to decrease p_{st}.

4. In pumping volumes V_i at each t_i, we determine delivery of pumps Q_i to obey the relation

$$\sum (\Delta p_{fr}) \leq p_{per}, \qquad (14.4.17)$$

at the casing mouth ($z = 0$) and opposite the absorptive formation ($z = z_h$). Here p_{per} is calculated with formulas (14.4.15) and (14.4.16) and summation is performed over corresponding sections l_k. Friction losses Δp_{fr} are calculated depending on the flow regime in section l_k with formulas (14.4.1) and (14.4.3). In accordance with initial data, the delivery Q_i lies in the range $Q_{min} \leq Q_i \leq Q_{max}$. We begin calculation of Q_i from delivery Q_6 in pumping the volume V_6. First, it is taken $Q_6 = Q_{min}$. Then after calculation of Δp_{fr}, the relation (14.4.17) is verified. If (14.4.17) is not satisfied at $z=0$ or $z=z_h$, one should change initial data. Otherwise we take $Q_6=Q_{max}$ and verify the relation (14.4.17). If it is satisfied at $z=0$ and/or at $z=z_h$, we take $Q_6=Q_{max}$. If (14.4.17) for $Q_6=Q_{min}$, it is satisfied but for $Q_6=Q_{max}$ at $z=0$ and $z=z_h$ it is not, we get Q_6 between Q_{min} and Q_{max} satisfying equation

$$\sum(\Delta p_{fr}) = p_{per}. \tag{14.4.18}$$

The solution of equation (14.4.18) with respect to Q_6 might be reckoned as obtained, when

$$\left|\sum_l(\Delta p_{fr}) - p_{per}\right| < \Delta_{adm},$$

where Δ_{adm} is admissible error in pressure.

Calculations can be reduced if to seek permissible values of Q_6 at $z=0$ between Q_{min} and limit value of Q_6 got at $z=z_h$. If the flow regime in sections [0.2L] and [z_h, 2L] with $Q_6=Q_{max}$ is turbulent and the condition (14.4.17) is not obeyed, one should calculate new Q_6 with formula

$$Q_6(z) = Q_{max}\sqrt{p_{per}/\sum_l(\Delta p_{fr})}, \tag{14.4.19}$$

where $\sum(\Delta p_{fr})$ is received at $Q=Q_{max}$. If the condition (14.4.17) is not again satisfied, the search of new Q_6 ought to be performed between values Q_{min} and Q calculated with (14.4.19). When (14.4.17) is obeyed, the solution of (14.4.18) should be sought between Q obtained with (14.4.19) and Q_{max}.

Once the delivery Q_6 is obtained, one has to get Q_5. The search of Q_5 is conducted between Q_{min} and Q_6, if at $Q_5=Q_6$ (14.4.17) it is not satisfied and between Q_6 and Q_{max} if (14.4.17) is obeyed. Search of the following Q_i can be stopped and taken equal to previous delivery, when it is managed to show that pressures received at delivery Q_{i+1} represent upper estimations of pressures in pumping volumes with the same delivery.

5. In stage 1 with formula (14.4.10), the first approximation was obtained in a volume of grouting mortar V without regard for concentration φ_2. Now, calculate the next approximation of the volume V with regard to concentration in given cross section z_φ in course of fluid pumping. New volume V is found with formula

$$V = S_{as} l_{cem}/\bar{z}, \tag{14.4.20}$$

where $\bar{z}$ is determined by (14.2.21) under condition that all $N_i = N =$ const

$$\bar{z} = \frac{(N+1)(2N+1)}{2N^2}(1-\varphi_2^{1/2})^{1/N}\sum_{i=1}^{k}\bar{V}_i + \sum_{i=1}^{l}\frac{(1-\beta_i)^2-(\varphi_2^{1/2}-\beta_i)^2}{2\bar{A}_i}\bar{V}_i + \sum_{i=1}^{m}\frac{(1-\beta_i)^2}{2\bar{A}_i}\bar{V}_i. \tag{14.4.21}$$

In the formula (14.4.21) enters dimensionless volumes $\bar{V}$

$$\frac{V_4-V_3}{V}, \quad \frac{V_5-V_4}{V}, \quad \frac{V_6-V_5-\Delta V}{V}, \quad \frac{\Delta V}{V}, \tag{14.4.22}$$

where ΔV is the given volume of driving fluid pumped to the end of cementation with minimal delivery of pumps $Q_{\min}$.

Quantities V, V_4, V_5, V_6 in (14.4.22) are taken from previous calculations. Volumes V_4, V_5, V_6, ΔV correspond to deliveries Q_4, Q_5, Q_6, $Q_{\min}$, respectively. Since they are known, regimes (turbulent or laminar) may be found comparing Q_4, Q_5, Q_6, and $Q_{\min}$ with Q_{cr} for washing fluid. Put out from volumes (14.4.22) those which are driven in laminar regime. Calculate at first number Se with (14.4.2) taking τ_0 and η for washing fluid. Then, from Fig. 6.7 curve 2 gets corresponding values of β. Relative volumes $\frac{V_4-V_3}{V}, \frac{V_5-V_4}{V}, \frac{V_6-V_5-\Delta V}{V}$ are driven in turbulent flow while $\Delta V/V$ are driven in laminar flow and $\varphi_2 < \beta^2$. Then, the second term in (14.4.21) is absent, since there are no volumes with $\varphi_2 > \beta^2$. Substitution in (14.4.21) of

$$\sum_{i=1}^{k=3}\bar{V}_i = \frac{V_4-V_3}{V} + \frac{V_5-V_4}{V} + \frac{V_6-V_5-\Delta V}{V} = \frac{V-\Delta V}{V},$$

$$\sum_{i=1}^{l=0}\frac{(1-\beta_i)^2-(\varphi_2^{1/2}-\beta_i)^2}{2\bar{A}_i}\bar{V}_i = 0,$$

$$\sum_{i=1}^{m=1}\frac{(1-\beta_i)^2}{2\bar{A}_i}\bar{V}_i = \frac{(1-\beta_1)^2}{2\bar{A}_1}\bar{V}_1 = \frac{(1-\beta_1)^2}{2\bar{A}_1}\frac{\Delta V}{V}$$

yields

$$\bar{z} = \frac{(N+1)(2N+1)}{2N^2}\frac{V-\Delta V}{V} + \frac{(1-\beta_1)^2}{2\bar{A}_1}\frac{\Delta V}{V},$$

where β_1 is found from Fig. 6.7 curve 2 at given $Se = \tau_0 d_h S_{as}/(Q_{\min}\eta)$.

New volume V is determined from (14.4.20). If the volume V calculated from (14.4.20) appears to be less than $S_{as}l_{cem}$, we take it equal to $S_{as}l_{cem}$.

Now, get the difference between the volume obtained and the previous one. If it is lesser Δ_0 given initially, the calculation of V can be stopped. Otherwise, we return to stage 2, determine V_{cem} with (14.4.12), in which new volume V should be inserted, and turn to the following calculation. If the number of steps would be greater than 3 or on the following step appears to be $Q < Q_{min}$, the search of V is finished. As a result, we take the volume V of grouting mortar and values Q_i associated with its pumping and driving and calculated at the previous step. For all Q_i and Q_{min}, we determine pressure at casing mouth with formula

$$p_{cm} = \sum_k (\Delta p_{as}) - \sum_{l_k} \rho g \cos(g, z) l_k. \tag{14.4.23}$$

Numerical values of terms in the last expression should be obtained if they were not calculated before. If at the beginning of driving $t = t_3$ would be $p_{cm} \leq p_0$ (atmospheric pressure), then in pumping from $t = t_2$, the pressure at the bean $p_{cp} = |p_{cm} - p_0|$ should be given.

Here, and in what follows, it is assumed that pressure p_{cp} in course of cementing and waiting of grouting mortar solidification does not exceed admissible pressure at the mouth of annular space based on the pipe strength.

The coordinate of the upper boundary of cement solution is determined by

$$z_{cem} = L + \frac{V}{S_{as}}. \tag{14.4.24}$$

Needed counter-pressure p_{cp} at the mouth of annular space during solidification of grouting mortar is

$$p_{cp}(t) = \rho_{per} g l_{sh} - g \left[\rho_{wash}(2L - z_{flush}) + \rho_{flush}(z_{flush} - z_{cem}) + \left(\rho_{cem} + \frac{\rho_w - \rho_{cem}}{t_c} t \right) (z_{cem} - z_{sh}) \right], \tag{14.4.25}$$

where t is time from the beginning of grouting mortar solidification; t_c is the time of hydraulic pressure lowering in a column of grouting mortar up to the pressure of water column; $z_{sh} = 2L - l_{sh}$ is the coordinate of the showing formation.

Pressure p_{cp} calculated at $t = t_c$ is supported constant during remaining period of solidification at least up to the end of grouting mortar solidification. If after completion of grouting mortar driving to close preventers, the pressure p_{cm} at the mouth of the annular space would be increased during time t_c in accordance with (14.4.25).

The volume of driving fluid V_{drf} with regard to compressibility factor k_c is

$$V_{drfc} = V_{drf} \cdot k_c. \quad (14.4.26)$$

Times of cementation t_i in pumping and driving volumes V_i with flow rates Q_i are calculated with

$$t_i = V_i / Q_i. \quad (14.4.27)$$

EXAMPLE 14.4.1

It is required to perform hydraulic calculation of single-stage cementation providing to attain given concentration of the grouting mortar at the upper boundary of cementation interval, to determine volumes of flushing fluid, grouting mortar, and driving fluid. Calculate regimes of cementation, pressure at the casing mouth, and additional pressure on the formation best disposed to shows during waiting of grouting mortar solidification.

Initial data	
Needed concentration φ_2	0.9
Depth (m)	
of well L	3000
of casing descent l	3000
well diameter, d_w (m)	0.250
Casing diameter (m)	
external d_{ex}	0.178
internal d	0.158
Height of cementing box in casing h_0 (m)	15
Height of cemented section l_{cem} (m)	1400
Depth of formation bedding (m)	
disposed to show l_{sh}	1800
disposed to absorption l_h	2900
Pressure in showing formation p_{sh} (MPa)	22.85
Pressure of the show beginning (hydraulic fracturing) p_{hf} (MPa)	64
Factor of showing formation abnormality k_a	1.294
Compressibility factor of driving fluid k_c	1.03
Safety factor k_s	1.05
Admissible pressure at casing mouth p_{per} (MPa)	36

Counter-pressure in annular space at the mouth p_{cp} (MPa)	0.1
Factor of roughness in pipes and annular space $k_{rp} = k_{ras}$ (m)	3×10^{-4}
Pipe delivery (m^3/s)	
minimal Q_{min}	0.004
maximal Q_{max}	0.1
Volume of flushing fluid V_{flush} (m^3)	8
Density of fluid (kg/m^3)	
washing ρ_{wash}	1430
flushing ρ_{flush}	1000
grouting mortar ρ_{cem}	1800
driving ρ_{dr}	1430
Dynamic shear stress of fluid τ_0 (Pa)	
washing	4
flushing	0
grouting mortar	8
driving	4
Plastic viscosity factor of fluid η (Pa s)	
washing	0.02
flushing	0.02
grouting mortar	0.05
driving	0.02
Permissible error when determining	
volumes Δ_0 (m^3)	0.5
pressures Δ_p (Pa)	5×10^5
Volume of grouter mortar driven with $Q = Q_{min}$ (m^3)	2
Time of solution hydraulic pressure drop up to pressure of water column t_c (s)	10,800

1. Determine: areas S_p, S_{as}, and volumes V_p, V_{as} of casing and annular space, volume V of cementation interval

$$S_p = \frac{\pi d^2}{4} = \frac{3.14 \times 0.158^2}{4} = 1.96 \times 10^{-2}\ m^2;$$

$$S_{as} = \frac{\pi(d_w^2 - d_p^2)}{4} = \frac{3.14 \times (0.25^2 - 0.178^2)}{4} = 2.42 \times 10^{-2}\ m^2;$$

$$V_p = S_p L = 1.96 \times 10^{-2} \times 3000 = 58.8\ m^3;$$

$$V_{as} = S_{as} L = 2.42 \times 10^{-2} \times 3000 = 72.6\ m^3;$$

$$V = S_{as} l_{cem} = 2.42 \times 10^{-2} \times 1400 = 33.9\ m^3,$$

coordinate of the absorptive formation

$$z_{\mathrm{h}} = 2L - l_{\mathrm{h}} = 6000 - 2900 = 3100\,\mathrm{m},$$

admissible density

$$\rho_{\mathrm{per}} = \rho_{\mathrm{w}} k_{\mathrm{a}} k_{\mathrm{s}} = 1000 \times 1.294 \times 1.05 = 1359\,\mathrm{kg/m^3}.$$

2. Improve the volume of flushing fluid with (14.4.9):

$$V_{\mathrm{flush}} = S_{\mathrm{as}} \frac{\rho_{\mathrm{wash}} - \rho_{\mathrm{per}}}{\rho_{\mathrm{wash}} - \rho_{\mathrm{flush}}} l_{\mathrm{sh}} = 2.42 \times 10^{-2} \frac{1430 - 1359}{1430 - 1000} 1800 = 7.2\,\mathrm{m^3}.$$

Since obtained volume $V_{\mathrm{flush}} = 7.2\,\mathrm{m^3}$ is less than initial volume $V_{\mathrm{flush}} = 8\,\mathrm{m^3}$, we accept $V_{\mathrm{flush}} = 7.2\,\mathrm{m^3}$.

With formula (14.4.10), we get the volume of driving fluid

$$V_{\mathrm{dr}} = S_{\mathrm{p}}(L - h_0) = 1.96 \times 10^{-2}(3000 - 15) = 58.5\,\mathrm{m^3}.$$

3. Determine critical deliveries for all fluids in casing and annular channel with formula (14.4.5). First calculate

$$\left(\frac{d}{d_{\mathrm{h}}}\right)^2 = \left(\frac{0.158}{0.250 - 0.178}\right)^2 = 4.82.$$

Then,

$$He_{\mathrm{wash}} = He_{\mathrm{drive}} = \frac{4 \times 1430 \times 0.158^2}{0.02^2} = 3.57 \times 10^5;$$

$$He_{\mathrm{flush}} = 0; \quad He_{\mathrm{cem}} = \frac{8 \times 1800 \times 0.158^2}{0.05^2} = 1.44 \times 10^5;$$

$$Q_{\mathrm{crwash}} = \left[2100 + 7.3 He_{\mathrm{wash}}^{0.58}\right] \frac{S_{\mathrm{p}} \eta_{\mathrm{wash}}}{d \rho_{\mathrm{wash}}}$$

$$= \left[2100 + 7.3(3.57 \times 10^5)^{0.58}\right]$$

$$\times \frac{1.96 \times 10^{-2} \times 0.02}{0.158 \times 1430} = 0.025\,\mathrm{m^3/s};$$

$Q_{\mathrm{crdrive}} = Q_{\mathrm{crwash}} = 0.025\,\mathrm{m^3/s}$, since parameters of washing and flushing fluids are identical;

$$Q_{\text{crcem}} = \left[2100 + 7.3(1.44 \times 10^5)^{0.58}\right] \times \frac{1.96 \times 10^{-2} \times 0.05}{0.158 \times 1800} = 0.032\,\text{m}^3/\text{s};$$

$$Q_{\text{crdril}} = 2100\frac{1.96 \times 10^{-2} \times 0.02}{0.158 \times 1000} = 0.0052\,\text{m}^3/\text{s}.$$

For annular space, we get He similar to above calculated dividing them by $(d/d_{\text{h}})^2 = 4.82$,

$$He_{\text{wash}} = He_{\text{driv}} = \frac{3.57 \times 10^5}{4.82} = 7.4 \times 10^4;$$

$$He_{\text{dril}} = 0; \quad He_{\text{cem}} = \frac{1.44 \times 10^5}{4.82} = 2.99 \times 10^4;$$

$$Q_{\text{crwash}} = \left[2100 + 7.3(7.4 \times 10^4)^{0.58}\right] \times \frac{2.42 \times 10^{-2} \times 0.02}{(0.25\text{–}0.178) \times 1430} = 0.033\,\text{m}^3/\text{s};$$

$$Q_{\text{crdrive}} = Q_{\text{crwash}} = 0.033\,\text{m}^3/\text{s},$$

$$Q_{\text{crcem}} = 0.046\,\text{m}^3/\text{s}; \qquad Q_{\text{crdril}} = 0.014\,\text{m}^3/\text{s}.$$

4. With formula (14.4.12) we have

$$V_{\text{cem}} = V + S_{\text{p}}h_0 = S_{\text{as}}l_{\text{cem}} + S_{\text{p}}h_0 = 2.42 \times 10^{-2} \times 1400 + 1.96 \times 10^{-2} \times 15 = 34.2\,\text{m}^3$$

Volumes $V_1, V_2, \ldots, V_6$ are

$$V_1 = 0; \quad V_2 = V_{\text{flush}} 7.2\,\text{m}^3, \quad V_3 = V_{\text{p}} + V_{\text{flush}} = 58.8 + 7.2 = 66\,\text{m}^3;$$

$$V_4 = V_3 + V/3 = 66 + 11.3 = 77.3\,\text{m}^3;$$

$$V_5 = V_4 + V/3 = 77.3 + 11.3 = 88.6\,\text{m}^3;$$

$$V_6 = V_5 + V/3 = 88.6 + 11.3 = 99.9\,\text{m}^3.$$

5. Fill out the last row of Table 14.1

t_i	V_i	z_{wash}	z_{flush}	z_{cem}	z_{dr}
t_6	$V_6 = 99.9$	6000	$3000 + (33.9 + 7.2)/2.42 \times 10^{-2} = 4698$	$3000 + 33.9/2.42 \times 10^{-2} = 4401$	2985

6. Calculate permissible delivery Q_6 in pumping volume V_6. In accordance with stage 5 and initial data separate the following sections l_k $(k = 6)$:

$$z_{flush} = 4698 \leq z < 6000 = z_1; \quad \cos(g, z) = -1,$$
$$l_1 = \Delta z_1 = 6000 - 4698 = 1302 \text{ m};$$
$$z_{cem} = 4401 \leq z < 4698 = z_2, \quad \cos(g, z) = -1, \quad l_2 = 297 \text{ m};$$
$$z_h = 3100 \leq z < 4401 = z_3, \quad \cos(g, z) = -1, \quad l_3 = 1301 \text{ m};$$
$$L = 3000 \leq z < 3100 = z_4, \quad \cos(g, z) = -1, \quad l_4 = 100 \text{ m};$$
$$z_{dr} = 2985 \leq z < 3000 = z_5; \quad \cos(g, z) = 1, \quad l_5 = 15 \text{ m};$$
$$0 \leq z < 2985 = z_6, \quad \cos(g, z) = 1, \quad l_6 = 2985 \text{ m}.$$

7. Determine pressures at $z = 0$ and $z = z_h = 3100$ m with formulas (14.4.13) and (14.4.14):

$$p_{st}(0) = p_{cp} - \sum_{k=1}^{6} \rho_k g \cos(g, z) l_k$$
$$= 10^5 - 9.81[1430(-1)1302 + 1000(-1)297 + 1800(-1)1301 + 1800(-1)100 + 1800 \times 1 \times 15 + 1430 \times 1 \times 2985] = 3.88 \times 10^6 \text{ Pa}.$$

$$p_{st}(3100) = p_{cp} - \sum_{k=1}^{3} \rho_k g \cos(g, z) l_k = 44.25 \times 10^5 \text{ Pa}.$$

Get permissible pressures at the mouth of casing and opposite formation disposed to absorption with formulas (14.4.15) and (14.4.16):

$$p_{per}(0) = p_{perm} - p_{st}(0) = 36 \times 10^6 - 3.88 \times 10^6 = 32.1 \times 10^6 \text{ Pa};$$
$$p_{per}(3100) = p_h - p_{st}(z_h) = 64 \times 10^6 - 44.25 \times 10^6 = 19.8 \times 10^6 \text{ Pa}.$$

8. Determine friction losses with $Q_6 = Q_{min} = 0.004 \text{ m}^3/\text{s}$ in the well. Calculate Δp_k in sections l_k. Since $Q_6 = 0.004 < Q_{cr}$ for all fluids in

pipes and annular space, flow with $Q_6 = Q_{\min}$ is laminar. Determine Δp_1 ($k=1$) in section l_1. With formula (14.4.2) calculate number Se

$$Se = \frac{\tau_0 d_{\rm h} S}{\eta Q} = \frac{4 \times (0.25-0.178) \times 2.42 \times 10^{-2}}{0.02 \times 0.004} = 87.1.$$

With Fig. 6.7 curve 2, we obtain $\beta = 0.72$ and with (14.4.1)

$$\Delta p_1 = \frac{4\tau_0 l_1}{\beta d_{\rm h}} = \frac{4 \times 4 \times 1302}{0.72(0.25-0.178)} = 0.402 \times 10^6 \text{ Pa}.$$

Section l_2 is occupied with viscous fluid; therefore, Δp_2 is determined by (6.2.30)

$$\Delta p_2 = \frac{128\mu Q}{\pi d_{\rm h}^3 (d_{\rm c} + d_{\rm H})} f(\delta) \cdot l_2 = \frac{128 \times 0.02 \times 0.004}{3.14(0.25-0.178)^3 (0.25+0.178)} 1.5 \times 297 = 9.09 \times 10^3 \text{ Pa}.$$

In other sections, the fluid is viscous-plastic. Calculate Δp_3 and Δp_4 in annular space and Δp_5, Δp_6 in pipes in the same manner as Δp_1. As a result, we get

$$\Delta p_3 = 0.825 \times 10^6 \text{ Pa}, \qquad \Delta p_4 = 0.063 \times 10^6 \text{ Pa},$$
$$\Delta p_5 = 0.004 \times 10^6 \text{ Pa}, \qquad \Delta p_6 = 0.37 \times 10^6 \text{ Pa}.$$

Pressure drop owing to friction losses at the mouth ($z = 0$) in casing and opposite absorptive formation ($z = 3100$ m) is determined as a sum of corresponding drops:

$$p(0) = \sum_{k=1}^{6} \Delta p_k = (0.402 + 0.00909 + 0.825 + 0.063 + 0.004 + 0.37) \times 10^6 = 1.67 \times 10^6 \text{ Pa};$$

$$p(3100) = \sum_{k=1}^{3} \Delta p_k = (0{,}402 + 0.00909 + 0.825) \times 10^6 = 1.24 \times 10^6 \text{ Pa}.$$

9. Verify fulfillment of the condition (14.4.17) at the mouth and opposite the absorptive formation:

   at the mouth ($z = 0$)

$$\sum_{k=1}^{6} \Delta p_k = 1.67 \times 10^6 \text{ Pa} < p_{\rm per} = 32.1 \times 10^6 \text{ Pa};$$

opposite the absorptive formation ($z = z_h = 3100$ m)

$$\sum_{k=1}^{3} \Delta p_k = 1.24 \times 10^6 \text{ Pa} < 19.8 \times 10^6 \text{ Pa}.$$

Condition (14.4.17) is obeyed.

10. Turn now to verify the condition (14.4.17) taking $Q_6 = Q_{max} = 0.1$ m^3/s.

    Determine flow regime, since $Q_6 = Q_{max} = 0.1$ m^3/s exceeds all before obtained values of Q_{cr}, the regime is turbulent and pressure drop Δp_k can be determined by formula (14.4.3):

$$\Delta p_1 = \lambda_1 \rho_1 \frac{Q_6^2}{2 d_h S_{as}^2} l_1 = 0.034$$

$$\times 1430 \frac{0.1^2}{2(0.25-0.178)(2.42 \times 10^{-2})^2} 1302 = 7.5 \times 10^6 \text{ Pa}.$$

Here, λ_1 is calculated with formula (6.5.58)

$$\lambda_1 = 0.106 \left(\frac{1.46 k_{ras}}{d_h} + \frac{100}{\text{Re}_1} \right)^{0.25}$$

$$= 0.106 \left(\frac{1.46 \times 3 \times 10^{-4}}{0.25-0.178} + \frac{100}{\frac{0.1(0.25-0.178)1430}{2.42 \times 10^{-2} \times 0.02}} \right)^{0.25} = 0.034.$$

Similarly, we have $\Delta p_2 = 1.3 \times 10^6$ Pa, $\Delta p_3 = 10.3 \times 10^6$ Pa, $\Delta p_4 = 0.79 \times 10^6$ Pa.

Calculation of Δp_5 and Δp_6 is performed by formula (14.4.3) with regard to geometric sizes of casing $\Delta p_5 = 0.062 \times 10^6$ Pa, $\Delta p_6 = 9.1 \times 10^6$ Pa.

Get now $\sum \Delta p_k$ in all sections up to the boundary of absorptive formation:

$$p(0) = \sum_{k=1}^{6} \Delta p_k = (7.5 + 1.3 + 10.3 + 0.79 + 0.062 + 9.1)$$

$$\times 10^6 = 29.1 \times 10^6 \text{ Pa};$$

$$p(3100) = \sum_{k=1}^{3} \Delta p_k = (7.5 + 1.3 + 10.3) \times 10^6 = 19.1 \times 10^6 \text{ Pa}.$$

Verify condition (14.4.17)

$$\sum_{k=1}^{6} \Delta p_k = 29.1 \times 10^6 \text{ Pa} < p_{\text{per}} = 32.1 \times 10^6 \text{ Pa};$$

$$\sum_{k=1}^{6} \Delta p_k = 19.1 \times 10^6 \text{ Pa} < p_{\text{per}} = 19.8 \times 10^6 \text{ Pa}.$$

Condition (14.4.7) at the mouth ($z = 0$) and opposite absorptive formation ($z = z_{\text{h}} = 3100$ m) is obeyed. Thus, the delivery $Q_6 = Q_{\max}$ is allowable. Calculations show that

$$\sum_{k=1}^{6} \Delta p_k \quad \text{and} \quad \sum_{k=1}^{3} \Delta p_k$$

exceed analogous sums in pumping other volumes. Therefore, other deliveries Q_i can be accepted equal to $Q_6 = Q_{\max}$:

$$Q_1 = Q_2 = Q_3 = Q_4 = Q_5 = Q_6 = Q_{\max} = 0.1 \text{ m}^3/\text{s}.$$

11. Determine $\bar{z}$ with formula (14.4.21). Parts of grouting mortar volume driven in turbulent regime less than the volume $\Delta V = 2 \text{ m}^3$ driven at the end of cementation in laminar pump delivery $Q_{\min} = 0.004 \text{ m}^3/\text{s}$. Use of formula (14.4.22) gives

$$\sum_{i=1}^{3} \bar{V}_i = \frac{33.9 - 2}{33.9} = 0.941.$$

Also determine

$$\bar{V} = \Delta V / V = 2/33.9 = 0.059.$$

This volume is driven in laminar regime and as it was found before $\beta = 0.72$ at $Q = Q_{\min}$. Then, $\varphi_2 = 0.9 > \beta^2 = 0.52$. Consequently, in (14.4.21), one should discard the last sum.

Let $N = 6$, then

$$\bar{z} = \frac{91}{72}(1 - \varphi_2^{1/2})^{1/6} \sum_{i=1}^{k} \bar{V}_i + \frac{(1-\beta)^2 - (\varphi_2^{1/2} - \beta)^2}{0.5\left(1 - \frac{4}{3}\beta + \frac{1}{3}\beta^4\right)} \bar{V}$$

$$= \frac{91}{72}(1 - 0.9^{1/2})^{1/6} 0.941 + \frac{(1-0.72)^2 - (0.9^{1/2} - 0.72)^2}{0.5\left(1 - \frac{4}{3}0.72 + \frac{1}{3}0.72^4\right)} 0.059$$

$$= 0.749.$$

With formula (14.4.20), we get new driven volume of grouting mortar

$$V = \frac{S_{as} l_{cem}}{\bar{z}} = \frac{2.42 \times 10^{-2} \times 1400}{0.749} = 45.2\,\text{m}^3.$$

Determine the difference between this and previous volumes

$$45.2 - 33.9 = 11.3\,\text{m}^3 > \Delta_0 = 0.5\,\text{m}^3.$$

Since obtained difference exceeds the given one, the calculation to get V should be continued.

12. For the next step we take $V = 45.2\,\text{m}^3$. With formula (14.4.12), we receive

$$V_{cem} = V + S_p h_0 = 45.2 + 1.96 \times 10^{-2} \times 15 = 45.49\,\text{m}^3.$$

13. Fill out Table 14.1

t_i	V_i	z_{wash}	z_{flush}	z_{cem}	z_{dr}
t_1	0	6000	0	0	0
t_2	7.2	6000	367	0	0
t_3	66	6000	3298	3000	679
t_4	81.1	6000	3921	3623	1448
t_5	96.2	6000	4543	4245	2216
t_6	111.3	6000	5165	4868	2985

Find sections l_k in driving the volume $V_6 = 111.3\,\text{m}^3$.

$$z_{flush} = 5165 \le z < 6000 = z_1; \quad \cos(g,z) = -1, \quad l_1 = 835\,\text{m};$$

$$z_{cem} = 4868 \le z < 5165 = z_2, \quad \cos(g,z) = -1, \quad l_2 = 297\,\text{m};$$

$$z_h = 3100 \le z < 4868 = z_3, \quad \cos(g,z) = -1, \quad l_3 = 1768\,\text{m};$$

$$L = 3000 \le z < 3100 = z_4, \quad \cos(g,z) = -1, \quad l_4 = 100\,\text{m};$$

$$z_{dr} = 2985 \le z < 3000 = z_5, \quad \cos(g,z) = 1, \quad l_5 = 15\,\text{m};$$

$$0 \le z < 2985 = z_6, \quad \cos(g,z) = 1, \quad l_6 = 2985\,\text{m}.$$

As well as in the previous stage, we find that the condition (14.4.17) in driving the volume $V_6 = 111.3\,\text{m}^3$ with delivery $Q_6 = Q_{min} = 0.004\,\text{m}^3/\text{s}$ is obeyed. We set $Q_6 = Q_{max} = 0.1\,\text{m}^3/\text{s}$. Omitting similar calculations, we find that condition (14.4.17) at

$z = z_h = 3100$ m is not obeyed. Then take new Q_6 with formula (14.4.19):

$$Q_6 = Q_{max}\sqrt{p_{per}(z_h)/\sum_{i=1}^{3}(\Delta p_k)}$$

$$= 0.1\sqrt{18 \times 10^6/(20.4 \times 10^6)} = 0.0942\ \text{m}^3/\text{s}.$$

The sum $\sum_{i=1}^{3}(\Delta p_k) = 20.4 \times 10^6$ Pa is calculated in the same manner as it was done in stage 10 of this example. Take $Q_6 = 0.0942\ \text{m}^3/\text{s}$. Calculations demonstrate fulfillment of the conditions (14.4.17) and $p_{per}(3100) = p(3100) = \sum_{i=1}^{3}(\Delta p_k)$ with given accuracy.

14. Determine the allowable delivery in driving volume V_5. In accordance with Table 14.1 (raw $t = t_5$), we get sections l_k. Calculate corresponding pressures, verify conditions (14.4.17), and assure ourselves that at $Q_5 = Q_{max} = 0.1\ \text{m}^3/\text{s}$, they are obeyed.
15. Similar to the stage 10 one can show that

$$Q_1 = Q_2 = Q_3 = Q_4 = Q_5 = 0.1\ \text{m}^3/\text{s}.$$

16. Determine volume of the grouting mortar driven in turbulent regime minus volume $\Delta V = 2\ \text{m}^3$ driven in laminar regime with minimal delivery of pumps $Q_{min} = 0.004\ \text{m}^3/\text{s}$. Use of formulas (14.4.22) gives

$$\sum_{i=1}^{3}\Delta\bar{V}_i = \frac{45.2-2}{45.2} = 0.956.$$

Then, $\bar{V} = \Delta V/V = 2/45.2 = 0.044$.

Performing calculations with formula (14.4.21) similar to stage 11 we get $\bar{z} = 0.754$.

With formula (14.4.20), we receive new volume

$$V = \frac{S_{as}l_{cem}}{\bar{z}} = \frac{2.42 \times 10^{-2} \times 1400}{0.754} = 44.9\ \text{m}^3.$$

Since the difference between obtained volume $V = 44.9\ \text{m}^3$ and previous one is less than given error $\Delta_0 = 0.5\ \text{m}^3$, we take $V = 45.2\ \text{m}^3$ as the volume sought.

TABLE 14.2

Time Elapsed from Beginning of Cementation t_i (s)	Total Volume of Pumped Fluids $\sum V$ (m^3)	Maximal Allowable Delivery Q (m^3/s)	Pressure at the Mouth of Casing p_1 (MPa)	Pressure at the Bean p_b (MPa)
$t_1 = 0$	0	0.1	26.4	0.1
$t_2 = 72$	7.2	0.1	28.1	0.1
$t_3 = 660$	66	0.1	18.9	0.1
$t_4 = 811$	81.1	0.1	24.7	0.1
$t_5 = 962$	96.2	0.1	30	0.1
$t_6 = 1101$	109.3	0.0942	32.5	0.1
$t_7 = 1601$	111.3	0.004	7.4	0.1

Calculate with (14.4.27) times t_i corresponding to V_i and Q_i and enter them in Table 14.2. Find pressure at the mouth with formula (14.4.27) in driving the volume V_6 with delivery $Q_6 = Q_{\min}$

$$p_{\text{wm}}(Q_{\min}) = \sum(\Delta p_{\text{p}}) + p_{\text{st}} = 1.83 \times 10^6 + 5.6 \times 10^6 = 7.43 \times 10^6 \text{ Pa}.$$

Pressure at the mouth in driving volume V_6 with delivery $Q_6 = 0.0942$ m^3/s is greater than pressure at $Q_6 = 0.0942$ m^3/s to the end of driving volume V_6—ΔV. When driving V_6, we have

$$p_{\text{wm}}(Q_6 = 0.0942) = \sum(\Delta p_{\text{p}}) + p_{\text{st}} = 26.9 \times 10^6 + 5.6 \times 10^6 = 32.5 \times 10^6 \text{ Pa},$$

where Δp_{p} in the section with flushing fluid is replaced by calculated greater value of Δp_{p} for washing fluid. In calculation of p_{wm} we also replace rheological parameters τ_0 and η of the flushing fluid in the casing by the same parameters for washing fluid

$$p_{\text{wm}}(Q_5 = 0.1) = \sum(\Delta p_{\text{p}}) + p_{\text{c}} = (29.5 + 0.52) \times 10^6 = 30 \times 10^6 \text{ Pa};$$

$$p_{\text{C}}(Q_4 = 0.1) = (29.1 - 4.4) \times 10^6 = 24.7 \times 10^6 \text{ Pa};$$

$$p_{\text{C}}(Q_3 = 0.1) = (28.5 - 9.6) \times 10^6 = 18.9 \times 10^6 \text{ Pa};$$

$$p_{\text{C}}(Q_2 = 0.1) = (26.4 + 1.65) \times 10^6 = 28.1 \times 10^6 \text{ Pa};$$

$$p_{\text{C}}(Q_1 = 0.1) = (26.4 + 0) \times 10^6 = 26.4 \times 10^6 \text{ Pa}.$$

Determine with formula (14.4.25) maximal counter-pressure in the bean excluding show during waiting of grouting mortar solidification. Since

$t_c = 3\,\text{h} = 10{,}800\,\text{s}$, we get

$$p_{cp}(t = t_c) = 10^5 + \rho_{per} g l_{sh} - g \left[\rho_{wash}(2L - z_{flush}) + \rho_{flush}(z_{flush} - z_{cem}) + \left(\rho_{cem} + \frac{\rho_w - \rho_{cem}}{t_c} t \right)(z_{cem} - z_{sh}) \right]$$
$$= 10^5 + 1359 \times 9.81 \times 1800 - 9.81 \left[1430(6000 - 5165) + 10^3 \times (5165 - 4868) + \left(1800 + \frac{1000 - 1800}{10,800} 10,800 \right) \times (4868 - 4200) \right] = 2.9 \times 10^6\,\text{Pa}.$$

Find volume of the driven fluid with regard to compressibility

$$V_{drc} = V_{dr} \cdot k_c = 58.5 \times 1.03 = 60.3\,\text{m}^3.$$

Determine the ratio

$$\frac{V + h_0 S_p}{S_{as} l_{cem} + h_0 S_p} = \frac{45.2 + 15 \times 1.96 \times 10^{-2}}{2.42 \times 10^{-2} \times 1400 + 15 \times 1.96 \times 10^{-2}} = 1.33.$$

Hence, to receive given concentration $\varphi_2 = 0.9$ at the upper boundary of cementation interval, it is necessary to use volume of the grouting mortar 1.33 greater than that in calculation with plug displacement scheme.

Key results of calculations:

volume of flushing fluid $V_{flush} = 7.2\,\text{m}^3$,
volume of grouting mortar $V_{cem} = 45.5\,\text{m}^3$,
volume of driven fluid $V_{dr} = 60.3\,\text{m}^3$.

Counter-pressure at the mouth in annular space during waiting of grouting mortar solidification should be evenly raised during time $t = t_c$ from 10^5 to $p_{cp} = 2.9 \times 10^6\,\text{Pa}$ and further supported constant.

Cementation time is $t = t_7 = 1601\,\text{s} \approx 26.7\,\text{min}$.

Factors of cementation regime are shown in Table 14.2.

With data of Table 14.2 in Fig. 14.7, curve 1 characterizing variation of pressure in cementing casing head depending on total volume pumped fluids is plotted. Curve 2 illustrates the delivery variation of pumps with total fluid volume delivery.

14.5 CALCULATION OF SINGLE-STAGE WELL CEMENTATION: METHOD AND CALCULATION OF CEMENTATION WITH FOAM–CEMENT SLURRY

Calculation method of well cementation with foam–cement slurry permits to determine main operating parameters at constant and variable aeration

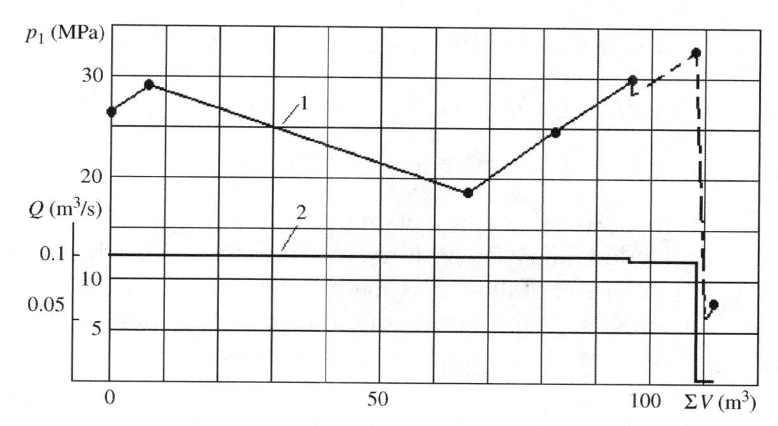

FIGURE 14.7 Dependence of pressure in cementing casing head and delivery of pumps on total volume of pumped fluids: (1) $p_1 = p_1(\Sigma V)$ and (2) $Q = Q(\Sigma V)$.

degree of the grouting mortar. As the basis of the method, the condition to maintain required properties of gas–liquid mixture in the annular space of the well from the mouth or given cross section up to the weakest absorbing formation is used as a result of which the formation pressure is balanced by bottom-hole pressure of the gas–liquid mixture column and the absorption is excluded. It is also accepted that the foam–cement solution represents two-phase liquid–gas system, gas bubbles are distributed in the solution uniformly and continuously, velocities of gas and liquid phases are identical, gas solubility in rigid and liquid phases as well as temperature exothermicity is absent.

14.5.1 Calculation of Cementing Parameters

1. Cementing with variable aeration degree.
   (A) Without a column of nonaerated (clean) cement solution in the upper part of the well annular space.

   Condition at which the absorption of the gas–liquid mixture by the weakest formation is excluded looks like

$$p_{ab} = \rho_m \cdot g \cdot H + p_0. \tag{14.5.1}$$

   From here it follows

$$\rho_m = (p_{ab} - p_0)/g \cdot H,$$

   where p_{ab} is absorption pressure, Pa; H is the vertical depth of the relevant absorption formation roof, m; p_0 is the atmospheric pressure.

The degree of cement solution aeration with depth is

$$a(H)=Q_0/Q_c=(\rho_c-\rho_m)/[\rho_{ix}-(\rho_m\cdot g\cdot H+p_0)/\bar{z}\cdot R\cdot\bar{T}\cdot g] \cdot(\rho_m\cdot g\cdot H+p_0)/\bar{z}\cdot R\cdot\bar{T}\cdot g\cdot\rho_0, \quad (14.5.2)$$

where Q_0 and Q_c are volume flow rate of gas (air) at normal conditions and cement slurry, m^3/s; $\bar{z}$ is the average factor of gas overcompressibility; $\bar{T}$ is the average temperature in the well bore, K; ρ_0 is the gas density at normal conditions, kg/m^3.

Volume of gas V_G in cement slurry is

$$V_G=\{[F_{as}\bar{z}R\bar{T}\cdot(\rho_c-\rho_m)]/\rho_m\} \cdot\ln|(p_0-\bar{z}\cdot R\cdot\bar{T}\cdot g\cdot r_c)/(p_{ab}-\bar{z}\cdot R\cdot\bar{T}\cdot g\cdot r_c)|, \quad (14.5.3)$$

where F_{as} is area of well annular space cross section, m^2.

The volume of the cement slurry in the well annular space is

$$V_c=V_{as}-V_G, \quad (14.5.4)$$

where V_G and V_{as} are volumes of gas and annular space of the well, respectively, m^3.

Saving of the oil-well cement Sav as a result of use of the foam–cement slurry instead of cement slurry is

$$Sav=(V_{as}-V_c)\cdot\rho_c. \quad (14.5.5)$$

The time t of the foam–cement pumping in the well at constant delivery of compressors and variable delivery of pumps of cementing aggregates is

$$t=\{F_{as}\cdot(\rho_c-\rho_m)/(Q_0\cdot\rho_m\cdot\rho_0)\} \cdot\{-\rho_m H-\rho_c\cdot\bar{z}\cdot R\cdot\bar{T}\cdot\ln[(\bar{z}\cdot R\cdot\bar{T}\cdot g\cdot\rho_c-(\rho_m\cdot g\cdot H+p_0))/ (\bar{z}\cdot R\cdot\bar{T}\cdot g\cdot\rho_c-p_0)]\}. \quad (14.5.6)$$

Delivery variation of pumps Q_c of cementing aggregates with well depth is

$$Q_c(H)=\{Q_0\cdot\rho_0\cdot\bar{z}\cdot R\cdot\bar{T}\cdot g\cdot[\rho_m-(\rho_m\cdot g\cdot H+p_0)/ \bar{z}\cdot R\cdot\bar{T}\cdot g]/(\rho_c-\rho_m)\cdot(\rho_m\cdot g\cdot H+p_0)\}. \quad (14.5.7)$$

(B) At the presence in the upper part of the well annular space of clean cement slurry column providing stability of the foam–cement slurry and required physical parameters of foam–cement stone.

The height H^* of the clean cement slurry is

$$H^* = (p_a - p_0)/(\rho_c \cdot g). \qquad (14.5.8)$$

The pressure p_a (Fig. 14.8) in the cross section A–A between columns of clean cement and foam–cement slurries is

$$p_a = \left(x - (x^2 - 4y)^{1/2}\right)/2, \qquad (14.5.9)$$

where
$x = \rho_c \cdot g \cdot H + p_0 + \bar{z} \cdot R \cdot \bar{T} \cdot g \cdot \rho_c - (\rho_c/(a \cdot p_0) \cdot (p_f - \rho_c \cdot g \cdot H - p_0)$,
$y = p_f \cdot \bar{z} \cdot R \cdot \bar{T} \cdot g \cdot \rho_c$.

Needed density of the foam–cement slurry is

$$\rho_m = (p_f - p_a)/(g \cdot (H - H^*)). \qquad (14.5.10)$$

The degree of cement slurry aeration with a height reduced to normal conditions yielding stability of the foam–cement slurry and required physical parameters of the foam–cement stone is

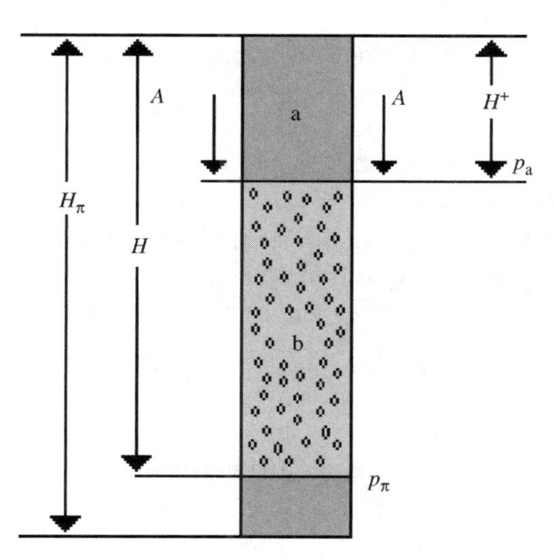

FIGURE 14.8 For determination of the pressure p_a in the cross section A–A between columns of clean (a) and foam–cement (b) slurries.

$$a(H) = \{(\rho_c - \rho_m)/[\rho_{mix} - (\rho_m \cdot g(H - H^*) + p_a)/\bar{z} \cdot R \cdot \bar{T} \cdot g]\} \cdot \{[\rho_m \cdot g(H - H^*) + p_a]/(\bar{z} \cdot R \cdot \bar{T} \cdot g \cdot \rho_0)\}. \quad (14.5.11)$$

The degree of grouting mortar aeration reduced to the pressure under clean cement slurry column providing stability of the foam–cement slurry and required parameters of the foam–cement stone is

$$a_r = (\rho_c - \rho_m)/[\rho_m - (p_a/(\bar{z} \cdot R \cdot \bar{T} \cdot g))]. \quad (14.5.12)$$

The volume of gas in grouting mortar in the interval of aeration is

$$V_G = \{[F_{as}\bar{z} \cdot R \cdot \bar{T} \cdot (\rho_c - \rho_m)]/\rho_m\} \cdot \ln|(p_a - \bar{z} \cdot R \cdot \bar{T} \cdot g \cdot \rho_c)/(p_f - \bar{z} \cdot R \cdot \bar{T} \cdot g \cdot \rho_c)|. \quad (14.5.13)$$

The time of foam–cement slurry pumping in the well is

$$t = \{S_{as}(\rho_c - \rho_m)/(Q_0 \rho_m \rho_0)\} \cdot \{(-\rho_m(H - H^*) - \rho_c \bar{z} \cdot R\bar{T} \cdot \ln[(\bar{z} \cdot R\bar{T} g \rho_c - (\rho_m g(H - H^*) + p_a))/(\bar{z} \cdot R\bar{T} \cdot g \cdot \rho_c - p_a)]\}. \quad (14.5.14)$$

The delivery variation of pumps of cementing aggregates with well depth is

$$Q_c(H) = \{Q_0 \cdot \rho_0 \bar{z} \cdot R \cdot \bar{T} \cdot g[\rho_m - (\rho_m \cdot g(H - H^*) + p_a)/(\bar{z} \cdot R \cdot \bar{T} \cdot g)]/((\rho_c - \rho_m)(\rho_m \cdot g(H - H^*) + p_a)\}. \quad (14.5.15)$$

2. Cementing with constant aeration degree.
   (A) Without a column of nonaerated (clean) cement solution in the upper part of the well annular space.

   Aeration degree of the grouting mortar is

$$a = (\rho_c/\rho_0)[(p - p_0)/\rho_c g - H]/[H - \bar{z} \cdot R \cdot \bar{T} \cdot \ln(p/p_0)]. \quad (14.5.16)$$

The pressure p of the foam–cement slurry column with the well depth H is

$$H = 1/[1 + (\rho_c/a \cdot \rho_0)] \cdot z \cdot R \cdot \bar{T} \cdot \ln(p/p_0) + \{1/[(a \cdot \rho_0 + \rho_c) \cdot g]\} \cdot (p - p_0). \quad (14.5.17)$$

The fall H_{fc} of foam–cement slurry column level in the cementing interval owing to degassing of the unstable part of the solution is

$$H_{fc} = 1/(1+\rho_c/a\cdot\rho_0)\cdot\bar{z}\cdot R\cdot\bar{T}\cdot\ln\{a_r\cdot\bar{z}\cdot\bar{T}/(a_r\cdot z_0\cdot T_0)\} + (1/g)\cdot\{[(a_r\bar{z}\cdot\bar{T})/(a_r\bar{z}_0\cdot\bar{T}_0)]-p_0\} \cdot\{1/((a\cdot\rho_0)+\rho_c)-(1/\rho_c)\}, \quad (14.5.18)$$

where z_0 is overcompressibility factor of gas at normal conditions; T_0 is the normal temperature, K

The volume of gas in the cement slurry is

$$V_G = S_{as}\bar{z}\,R\bar{T}\cdot(\eta/(\eta+1))\cdot\ln(p/p_0), \quad (14.5.19)$$

where $\eta = a\rho_0/\rho_c$ is aeration mass factor.

The time of foam–cement slurry pumping at constant delivery of pumps of cementing aggregates and compressors is

$$t = \{S_{as}\cdot[H-\bar{z}\cdot R\bar{T}\cdot[\eta/(\eta+1)]\cdot\ln(p/p_0)]\}/Q_c. \quad (14.5.20)$$

(B) In the presence of the clean cement slurry in the upper part of the annular space providing stability of the foam–cement slurry and physical parameters of the foam–cements stone.

The degree of cement slurry aeration providing stability of the foam–cement slurry and physical parameters of the foam–cements stone is

$$a = (\rho_c/\rho_0)\cdot[(p_f-p_a)/(\rho_c\cdot g)-(H-H^*)]/[(H-H^*)\cdot\bar{z}\cdot R\cdot\bar{T}\cdot\ln(p_f/p_a)]. \quad (14.5.21)$$

The pressure p_a in the cross section A–A (see Fig. 4.8) is determined by simultaneous solution of equations (14.5.22) and (14.5.23)

$$(p_f-p_0-H\cdot\rho_c\cdot g)/(p_0-p_a+H\cdot\rho_c\cdot g-\rho_c\cdot g\cdot\bar{z}\cdot R\cdot\bar{T}\cdot\ln(p_f/p_a)) = a_r^{min}(p_a/(\bar{z}\cdot R\bar{T}\cdot g\rho_c)), \quad (14.5.22)$$

$$\rho_c\cdot g\cdot\bar{z}\cdot R\cdot\bar{T}\cdot\ln(p_a)-2\cdot p_a+p_0^+\rho_c g\cdot H -\rho_c\cdot g\cdot\bar{z}\cdot R\cdot\bar{T}\cdot\ln(p_f)+\rho_c\cdot g\cdot\bar{z}\cdot R\cdot\bar{T} = 0, \quad (14.5.23)$$

where a_r^{min} is minimal aeration degree of the grouting mortar reduced to the pressure p_a providing stability of the foam–cement slurry and physical parameters of the foam–cement stone.

The density of the foam–cement slurry under the column of cement slurry is
(a) at given mass factor of aeration η

$$\rho_m = \eta/[\eta + (p_a/(\bar{z}\cdot R\cdot \bar{T}\cdot g\cdot \rho_c))]\cdot[p_a/(\bar{z}\cdot R\cdot \bar{T}\cdot g) - \rho_c] + \rho_c; \quad (14.5.24)$$

(b) when calculating a_r^{min}

$$\rho_m = [\rho_c + (a_r^{min}\cdot p_a/(\bar{z}\cdot R\cdot \bar{T}\cdot g))]/[a_r^{min} + 1]. \quad (14.5.25)$$

Volume of gas in the grouting mortar in aeration interval is

$$V_G = S_{as}\cdot \bar{z}\cdot R\cdot \bar{T}\cdot \eta/(\eta + 1)\cdot \ln(p_f/p_a). \quad (14.5.26)$$

The time of the foam–cement slurry pumping in the well is

$$t = \{S_{sp}\cdot[(H - H^*) - z\cdot R\cdot \bar{T}\cdot \eta/(\eta\cdot 1)\cdot \ln(p_f/p_a)]\}/Q_c. \quad (14.5.27)$$

Examples of calculation

It is required to cement with foam–cement slurry up to the mouth a 219 mm intermediate casing with length 1569 m in a well drilled by a bit with diameter 270 mm using washing drilling mud with density $\rho_{dm} = 1200\ kg/m^3$. Absorption pressure at the casing shoe is $p_{ap} = 211.5\times 10^5$ Pa. Driving fluid is drilling mud. Initial data are given below.

Well construction	
casing	
diameter external d_{ex} (m)	0.299
diameter internal D_{in} (m)	0.276
landing depth H_{ld} (m)	550
intermediate casing	
diameter external d_{ex} (m)	0.219
diameter internal D_{in} (m)	0.199
landing depth H_{ld} (m)	1560
Cementing interval H (m)	0–1540
Height of cement box in a column H_c (m)	20
Bit diameter in drilling under intermediate column D_g (m)	0.270

Average factor of well bore vugular porosity k	1.2
Well bore temperature T (K) at depth	
200 m	280
500 m	285
1000 m	290
1500 m	295
Pressure at the well mouth p_0 (Pa)	1×10^5
Absorption pressure at the intermediate column shoe p_{sh} (Pa)	211.5×10^5
Density (kg/m^3)	
cement powder ρ_{cp}	3100
fluid cement mixing ρ_{fl}	1000
mud powder ρ_1	2500
asbestos ρ_2	2500
air ρ_0	1.29
initial cement slurry ρ_c	1730
Air overcompressibility factor $\bar{z}$	~1.0
Gas constant R (m/K)	29.27
Gravity acceleration g (m/s^2)	9.81
Annular space cross-section area (m^2)	
in the interval under casing shoe S'_{as}	0.045
in the interval of casing carry S''_{as}	0.022
weighted average in interval 0–1540 m S_{as}	0.039
Compressor delivery СД 9/101 Q_0 (m^3/s)	0.15

Calculation of the cementing regime parameters is performed for two cases when in the upper part of the well annular space there is a column of clean "cement" slurry.

(A) Cementing with varied aeration degree.

1. We determine with the formula (14.5.9) the pressure p_a between columns of "clean" cement and foam–cement slurries at given aeration degree ($a = 10$)

$$p_a = (x - (x^2 - 4y)^{1/2})/2;$$

$$\begin{aligned} x &= \rho_c \cdot g \cdot H + p_0 + \bar{z} \cdot R \cdot \bar{T} \cdot g \cdot \rho_c - (\rho_c / a \cdot \rho_0) \cdot (p_f - \rho_c \cdot g \cdot H - p_0) \\ &= 1540 \times 1730 \times 9.81 + 10^5 + 1 \times 29.27 \times 280 \times 9.81 \times 1730 \\ &\quad - [1730/(10 \times 1.29)] \times (211.5 \times 10^5 - 1540 \\ &\quad \times 1730 \times 9.81 - 10^5) = 8.52 \times 10^8 \text{ Pa}; \end{aligned}$$

$$y = p_f \cdot \bar{z} \cdot R \cdot \bar{T} \cdot g \cdot \rho_c = 211.5 \times 10^5 \times 1.0$$

$$\times 29.27 \times 280 \times 9.81 \times 1730 = 3.047 \times 10^{15}\,\text{Pa}^2.$$

$$p_a = \{8.52 \times 10^8 - [(8.52 \times 10^8)^2 - 4 \times 3.047 \times 10^{15}]^{1/2}\}/2$$

$$= 35.9 \times 10^5\,\text{Pa}.$$

2. With the formula (14.5.1) we get the height of "clean" cement slurry column

$$H^* = (p_a - p_0)/\rho_c \cdot g = (35.9 \times 10^5 - 10^5)/(1730 \times 9.81) = 205.6\,\text{m}.$$

3. Calculate the density of the foam–cement slurry with the formula (14.5.10).

$$\rho_m = (p_f - p_a)/(g \cdot (H - H^*))$$

$$= 211.5 \times 10^5 - 35.9 \times 10^5)/(9.81 \times (1540 - 205.6))$$

$$= 1341\,\text{kg/m}^3.$$

4. With the formula (14.5.12), we calculate the reduced aeration level under the column of "clean" cement slurry

$$a_r = (\rho_c - \rho_m)/[\rho_m - (p_a/(\bar{z}R\bar{T}g))]$$

$$= (1730 - 1341)/[1341 - (35.9 \times 10^5/(1 \times 29.27 \times 280 \times 9.81))]$$

$$= 0.3.$$

5. Determine aeration degree of the cement slurry with depth using the formula (14.5.11)

at $H = 205.6$ m (testing)

$$a = \{(\rho_c - \rho_m)/[\rho_m - (\rho_m \cdot g(H - H^*) + p_a)/\bar{z} \cdot R\bar{T} \cdot g]\}$$

$$\cdot \rho_m \cdot g(H - H^*) + p_a]/\bar{z} \cdot R\bar{T} \cdot g \cdot \rho_0\}$$

$$= \{(1730 - 1341)/[1341 - [(1341 \times 9.81$$

$$\times (205.6 - 205.6) + 35.9 \times 10^5)/(1 \times 29.27 \times 280 \times 9.81)]\}$$

$$\times [1341 \times 9.81 \times (205.6 - 205.6) + 35.9 \times 10^5]/(1.0$$

$$\times 29.27 \times 280 \times 9.81 \times 1.29) = 10;$$

at $H = 500\,\mathrm{m}$

$$a = \{(1730-1341)/[1341-[(1341\times9.81\times(500-205.6)$$
$$+35.9\times10^5)/1.0\times29.27\times285\times9.81]]\}$$
$$\times[1341\times9.81\times(500-205.6)+35.9\times10^5]/1.0\times29.27$$
$$\times285\times9.81\times1.29=22.$$

Similarly, at $H = 1000\,\mathrm{m}$, $a = 43.4$; at $H = 1540\,\mathrm{m}$, $a = 69$.

6. With the formula (14.5.13), we obtain the volume of gas in cement slurry in aeration interval

$$V_G = \{[S_{as}\bar{z}R\bar{T}\cdot(\rho_c-\rho_m)]/\rho_m\}$$
$$\cdot\ln|(p_a-\bar{z}\cdot R\cdot\bar{T}\cdot g\cdot\rho_c)/(p_f-\bar{z}\cdot R\bar{T}\cdot g\cdot\rho_c)|$$
$$=0.039\times1.0\times29.27\times290(1730-1341)/1341$$
$$\ln|(35.9\times10^5-1.0\times29.27\times290\times9.81\times1730)/$$
$$(211.5\times10^5-1.0\times29.27\times290\times9.81\times1730)|=12.8\,\mathrm{m}^3.$$

7. The volume of the grouting mortar in the aeration interval is obtained from the formula

$$V_c = V_{as}-V_G = F_{as}\cdot(H-H^*)-V_G$$
$$= 0.039\times(1540-205.6)-12.8 = 39.2\,\mathrm{m}^2.$$

8. With the formula (14.5.15), the delivery of cement slurry by pumps of cement aggregates with well height is obtained. At $Q_0 = 0.3\,\mathrm{m}^3/\mathrm{s}$, there is

at $H = 205.6\,\mathrm{m}$

$$Q_c(H)=Q_0\cdot r_0\cdot\bar{z}\cdot R\cdot\bar{T}\cdot g[\rho_m-(\rho_m\cdot g\cdot(H-H^*)+p_a)/\bar{z}\cdot R\bar{T}\cdot g]/$$
$$\{(\rho_c-\rho_m)(\rho_m\cdot g(H-H^*)+p_a)\}$$
$$=0.3\times1\times29.27\times280\times9.81\times1.29/(1730-1341)$$
$$\times\{[1341-(1341\times9.81\times(205.6-205.6)$$
$$+35.9\times10^5)/1\times29.27\times280\times9.81]\}/(1341\times9.81$$
$$\times(205.6-205.6)+35.9\times10^5)=3.0\times10^{-3}\,\mathrm{m}^3/\mathrm{s}=30\mathrm{L/s}.$$

Similarly, we have $Q_c = 13.6$ l/s at $H = 500$ m; $Q_c = 6.9$ l/s at $H = 1000$ m and $Q_c = 4.3$ l/s at $H = 1540$ m.

9. Calculate the time of foam–cement slurry pumping in a well with depth from $H = H^*$ ($t = 0$) up to $H = 500$ m with the formula (14.5.14)

$$t = \{S_{as}(\rho_c - \rho_m)/(Q_0 \cdot \rho_m \cdot \rho_0)\} \cdot \{-\rho_m \cdot H - H^*) - \rho_c \cdot \bar{z} \cdot R\bar{T} \cdot \ln[(\bar{z} \cdot R\bar{T} \cdot g \cdot \rho_c - (\rho_m \cdot g \cdot (H - H^*) + p_a))/(\bar{z} \cdot R \cdot \bar{T} \cdot g \cdot \rho_c - p_a)]\}$$
$$= [0.039 \times (1730 - 1341)/(0.3 \times 1341 \times 1.29)] \times \{-1341 \times (500 - 205.6) - 1730 \times 1.0 \times 29.27 \times 285 \times \ln[(1730 \times 1.0 \times 29.27 \times 285 \times 9.81 - 1341 \times 9.81 \times (500 - 205.6) + 35.9 \times 10^5)/ (1730 \times 1.0 \times 29.27 \times 285 \times 9.81 \times 35.9 \times 10^5)]\}$$
$$= 440.4 \text{ s} = 7.34 \text{ min}.$$

In a similar way we get $t = 35.4$ min up to the depth $H = 1000$ m and $t = 1.38$ h up to $H = 1540$ m. At $t = 1.38$ h, the compressors are cut and "clean" cement slurry is pumped with $Q_c = 4.3$ l/s to block off the casing shoe and form cement box.

10. Determine now the volume of cement slurry needed to create in the upper part of the annular space of a column of "clean" cement slurry to insulate the casing shoe and perform a cement box:
    to create a column of "clean" cement slurry

$$V''_c = S''_{as} \cdot H^* = 0.022 \times 205.6 = 4.5 \text{ m}^3;$$

to insulate the casing shoe and perform a cement box

$$V'_c = S_{as} \cdot (H_f - H) + 0 \cdot 785 \cdot D_f^2 \cdot H_c = 0.045 \times (1560 - 1540) + 0.785 \times 0.199^2 \times 20 = 0.9 + 0.62 = 1.52 \text{ m}^3.$$

11. The total volume of the cement slurry spent for cementing is

$$V_0 = V_c + V'_c + V''_c = 39.2 + 1.52 + 4.5 = 45.2 \text{ m}^3.$$

12. The time to pump calculated volumes of "clean" cement slurry
    (a) in intervals 0–205.6 m, at $Q_c = 30$ l/s, $t_1 = V''_c/Q_c = 4.5/0.03 = 150$ s $= 3.5$ min;
    (b) in isolating the column shoe and creating the cement box at $Q_c = 4.8$ l/s,
    $t_f = V'_c/Q_c = 1.52/0.0048 = 317$ s $= 5.3$ min;

driving fluid at $Q_f = 20$ l/s

$$t_f = V_f/Q_f = 0.785 \cdot D_f^2 \cdot (H_f - H_c)/Q_f$$
$$= 0.785 \times 0.1992 \times (1560 - 20)/0.02 = 2394\text{ s} = 39.6\text{ min}.$$

13. The total time needed to pump and drive the cement slurry in the annulus is

$$t_0 = t_1 + t_2 + t_c + t_f = 2.5 + 82.7 + 5.3 + 39.9 = 130\text{ min} = 2.2\text{ h}.$$

14. The pressure in the cementing head at the end of driving is

$$\begin{aligned} p_c &= \rho_c \cdot H^* \cdot g \cdot \rho_m \cdot (H - H^*) \cdot g + \rho_c \cdot (H_f - H) \\ &\quad \cdot g - \rho_c \cdot (H_f - H_c) \cdot g - \rho_c \cdot H_c \cdot g \\ &= 1730 \times 205.6 \times 9.81 + 1341 \times (1540 - 205.6) \times 9.81 \\ &\quad + 1730 \times (1760 - 1740) \times 9.81 - 1200 \times (1760 - 20) \\ &\quad \times 9.81 - 1730 \times 20 \times 9.81 = 5.6 \times 10^5\text{ Pa}. \end{aligned}$$

As a result of calculation, an operation graphic of cementing aggregate pumps and compressors through step approximation of calculated curves for delivery of pumps 9T ЦА-32OA at different operation regimes is built (Fig. 14.9). The step approximation in pump delivery is chosen in accordance with certificate characteristics of pumps 9T. Data about operation regimes of pumps 9T are represented in Table 14.3.

(B) Cementing with constant aeration rate.

1. Determine the pressure p_a and minimal degree of aeration a_r^{min} reduced to pressure p_a in cross section A–A (Fig. 14.6) between columns of "pure" cement and foam–cement slurry using formulas (14.5.22) and (14.5.23)

$$(p_f - p_0 - H \cdot \rho_c \cdot g)/(p_0 - p_a - H \cdot \rho_c \cdot g - \rho_c \cdot g \cdot \bar{z} \cdot R \cdot \bar{T} \cdot \ln(p_f/p_a))$$
$$= a_r^{min} \cdot (p_a/\bar{z} \cdot R\bar{T} \cdot g \cdot \rho_c)$$

and

$$\rho_c \cdot g \cdot \bar{z} \cdot R \cdot \bar{T} \cdot \ln p_a - 2 \cdot p_a + p_0$$
$$+ H \cdot \rho_c \cdot g - \rho_c \cdot g \cdot \bar{z} \cdot R \cdot \bar{T} \cdot \ln p_f + \rho_c \cdot g \cdot \bar{z} \cdot R\bar{T} = 0.$$

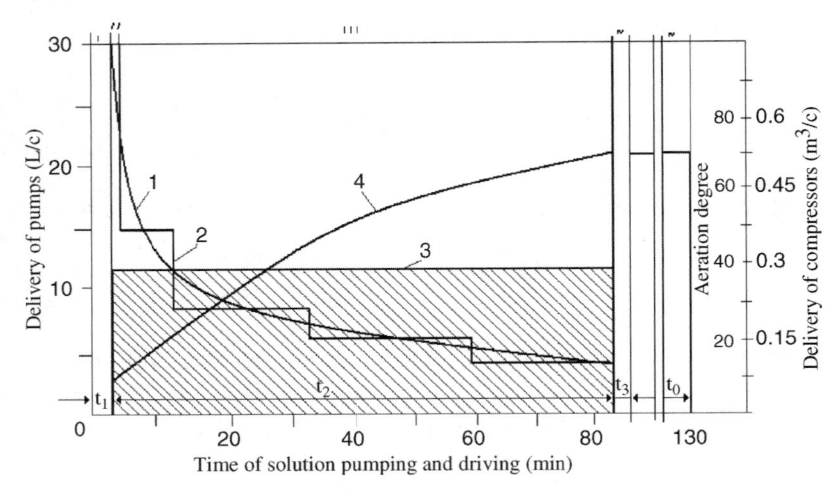

FIGURE 14.9 Operating conditions of cement aggregate pumps and compressors СД 9/101 with time. (1 and 2) calculated and actual delivery of pumps ЦА; (3) delivery of compressors; (4) aeration degree.

Introduction of numerical values in these relations gives

$$(211.5 \times 10^5 - 10^5 - 1540 \times 1730 \times 9.81)/(10^5 - p_a + 1540 \times 1730 \times 9.81 - 1730 \times 9.81 \times 1.0 \times 29.27 \times 285 \times \ln(211.5 \times 10^5/p_a)) = a_r^{\min}(p_a/(1.0 \times 29.27 \times 285 \times 9.81 \times 1730))$$

TABLE 14.3

	Cementing Time (min)								
Factor	0	2.5	4	14	35	61	83	88.5	130
Number of aggregates	2	2	2	1	1	1	1	1	2
Connected velocity	V	V	V	V	IV	III	III	III	IV
Rotating speed of the engine shaft ($\min^{-1}$)	–	1440	1440	1440	1300	1320	1060	1060	1500
Delivery of pumps Q_c (l/s)	–	2 × 15	2 × 15	15	9	6	4.8	4.8	20
Aeration ("+"– there is, "–"– there is not)	–	–	+	+	+	+	+	–	–

Note. The check of the cement slurry volume: $V_c^t = \sum Q_c \cdot \Delta t = (1.5 \times 60 \times 30) + (10 \times 60 \times 15) + (21 \times 60 \times 9) + (26 \times 60 \times 6) + (22 \times 60 \times 4.8) = 38.7\ \text{m}^3$; $V_c = 39.2\ \text{m}^3$; $V_c^t \approx V_c$.

and

$$1730 \times 9.81 \times 1.0 \times 29.27 \times 285 \times \ln(p_a) - 2 \times p_a + 10^5 + 1540 \times 1730 \times 9.81 - 1730 \times 9.81 \times 1.0 \times 29.27 \times 285 \times \ln(211.5 \times 10^5) + 1730 \times 9.81 \times 1.0 \times 29.27 \times 285 = 0.$$

Solution of these equations yields

$$p_a = 71.4 \times 10^5 \text{ Pa}; \qquad a_r^{min} = 0.84.$$

2. With the formula (14.5.8), we calculate the height of "pure" cement slurry column

$$H^* = (p_a - p_0)/\rho_c \cdot g = (71.4 \times 10^5 - 10^5)/(1730 \times 9.81) = 415\,\text{m}.$$

3. With the formula (14.5.21), we obtain the degree of cement slurry aeration

$$\begin{aligned} a &= (\rho_c/\rho_0) \cdot [(p_c - p_a)/\rho_c \cdot g - (H - H^*)]/ [(H - H^*) - \bar{z} \cdot R \cdot \bar{T} \cdot \ln(p_f/p_a)] \\ &= (1730/1.29) \times [(211.5 \times 10^5 - 71.4 \times 10^5)/ 1730 \times 9.81 - (1540 - 415)]/[(1540 - 415) \\ &\quad - 1.0 \times 29.27 \times 285 \times \ln(211.5 \times 10^5/71.4 \times 10^5)] = 50.6. \end{aligned}$$

4. Determine the density of foam–cement slurry under the column of "pure" cement slurry with the formula (14.5.25)

$$\begin{aligned} \rho_m &= [\rho_c + (a_r^{min} p_a/\bar{z}R\bar{T} \cdot g)]/[a_r^{min} + 1] \\ &= (1730 + (0.84 \times 71.4 \times 10^5/(1.0 \times 29.27 \times 285 \times 9.81)))/(0.84 + 1) = 980\,\text{kg/m}^3. \end{aligned}$$

5. With the formula (14.5.26), we calculate the volume of gas in cement slurry in aeration interval determining first $\eta = a\rho_0/\rho_c = (50.6 \times 1.29/1730) = 0.038$. Then,

$$\begin{aligned} V_G &= S_{as} \cdot \bar{z} \cdot R \cdot \bar{T} \cdot \eta/(\eta + 1) \cdot \ln(p_f/p_a) \\ &= 0.039 \times 1.0 \times 29.27 \times 285 \times (0.038/(0.038 + 1)) \times \ln(211.5 \times 10^5/71.4 \times 10^5) = 12.9\,\text{m}^3. \end{aligned}$$

6. Find with the formula (14.5.4), the volume of the cement slurry in the aeration interval

$$V_c = V_{as} - V_G = F_{as}(H - H^*) - V_G$$
$$= 0.039 \times (1540 - 415) - 12.9 = 31\ \text{m}^3.$$

7. Calculate the delivery of cement aggregate pumps

$$Q_c = (Q_0/a) = 0.3/50.6 = 0.0059\ \text{m}^3/\text{c} = 5.9\ \text{L/s}.$$

Such delivery can provide one pump 9T ЦА 32-OA with connected III velocity and rotating speed 1300 min^{-1} of the engine shaft.

8. Determine the pumping time of the foam–cement slurry with the formula (14.5.27)

$$t_2 = \{S_{as} \cdot [(H - H^*) - \bar{z} \cdot R \cdot \bar{T} \cdot \eta/(\eta + 1) \times \ln(p_f/p_a)]\}/Q_c$$
$$= \{0.39 \times [(1540 - 415) - 1.0 \times 29.27 \times 285$$
$$\times (0.038/(0.038 + 1)$$
$$\times \ln(211.5 \times 10^5/71.4 \times 10^5)]\}/0.0059 = 5244\ \text{c} = 1.46\ \text{h}.$$

9. Obtain the volume of the cement slurry spent to create a column of "pure" cement slurry in the upper part of the annular space, to isolate the column shoe and to form a cement box:
to create a column of "pure" cement slurry

$$V''_c = S''_{as} \cdot H^* = 0.022 \times 415 = 9.13\ \text{m}^3;$$

to isolate the column shoe and to form a cement box

$$V'_c = 1.52\ \text{m}^3.$$

10. Total volume of the cement slurry is

$$V_0 = V_c + V'_c + V''_c = 31.0 + 1.52 + 9.13 = 41.7\ \text{m}^3.$$

11. Determine the time of calculated volumes of "pure" cement slurry:
(a) in the interval 0–415 m at $Q_c = 9.0\ l/\text{s}$,

$$t_1 = (V''_c/Q_c) = 9.13/0.009 = 1014\ \text{s} \approx 17\ \text{min};$$

TABLE 14.4

Cementing Interval (m)	Cement Slurry Volume (m^3)	Delivery of Pumps (l/s)	Cementing Time (min)	Degree of Aeration at $Q_0 = 0.3$ m^3/s
0–415	9.13	9.0	17.0	–
415–1540	31.0	5.9	87.4	50.6
1540–1560	1.52	5.9	4.3	–
0–1560[a]	–	20.0	39.9	–

[a] Driving.

(b) when isolating the column shoe and forming the cement box for cement slurry delivery $Q_c = 5.9\ l/\text{s}$, we have $t_{is} = V'_c / Q_c = 1.52/0.0059 = 257.6\ \text{s} = 4.3\ \text{min}$ and for driving fluid delivery $Q_f = 20\ l/\text{s}$, the time will be $t_f = 2394\ \text{s} = 39.9\ \text{min}$.

12. The total time of cement slurry pumping and driving is $t_0 = t_1\ t_2 + t_{is} + t_f = 17 + 87.4 + 4.3 + 39.9 = 148.6\ \text{min} = 2.5\ \text{h}$.

Obtained results are summarized in Table 14.4.

14.5.2 Selection and Calculation of Cement Slurry Composition

The composition of the foam–cement slurry is selected individually for each method of cementing using date on temperature in the well bore, estimated time of cement slurry pumping, and driving obeying condition $a_{aer} \geq a_r$, where a_{aer} is aeration degree of the grouting mortar reduced to atmospheric pressure at 100% stability of the foam–cement slurry measured in laboratory

$$a_c = \frac{r_c - r_{mix}}{r_{mix} - p_0/(\bar{z}RTg)}. \tag{14.5.28}$$

For example, for method of cementing with constant aeration rate at $T = 295$ K, $t_0 = 2.2$ h and $a_r = 0.3$ cement slurry compositions No. 1–5 can be used (Table 14.5), whereas for method of cementing with constant aeration degree at $t_0 = 2.5$ h and $a_r = 0.84$ are suitable compositions No. 1–3 and 5.

After laboratory completion of selected composition, the component compound of the foam–cement slurry may be calculated for drilling tests.

Density of given grouting mortar is

$$\rho_c = \frac{(1+n)\rho_{mix}\rho_{fl}}{\rho_{fl} + n\rho_{cp}}, \tag{14.5.29}$$

TABLE 14.5

	Type—Mass Content of Components					Factors of grouting mortars and stone						
								Solidification Time (h min)				
Composition No.	Cementing Agents	Cement Mixing Fluid	Deflocculant Fluid	Solid Deflocculant	Another Additives	Spreadabilit[a] 10^{-2} (m)	Solidification Temperature (°C)	Beginning	End	Density of Stone, (kg/m^3)	Bending Strength in 2 days (MPa)	a_c
1	Pokrovski ТПЦХ—100	Fresh water—50	CMC Aina—1.0	Aerosil—0.1	Na_2CO_3—5.0	18	22	5–30	8–30	1340	1.6	1.06
2	Pokrovski ТПЦХ—100	Fresh water—70	CMC Aina—0.7	Asbestos—1.4	Liquid glass—10.0	18	22	6–30	10–00	1250	1.2	0.85
3	Pokrovski ТПЦХ—100	Fresh water—50	Kubocsalim—0.5	–	Liquid glass—5.0	19	22	7–30	9–00	1100	1.0	0.98
4	Pokrovski ТПЦХ—100	Солегель of magnesium—65	Afrocs 200—0.7	–	–	17	22	6–00	8–30	1250	1.1	0.81
5	Sdolbunovski ТПЦХ—100	Water of the Barents Sea—55	CMC Aina—0.55	–	Liquid glass—3.0	18	22	4–00	5–30	1220	1.4	1.12
6	Novotroizkiy ТПЦГ—100	Water of the Caspian Sea—50	CMC Aina—1.0	Filtroperlit – 0.75	Liquid glass—0.75	18	75	5–30	6–10	1100	1.2	0.82
7	Karadagskiy ТПЦГ—100	Water of the Caspian Sea—50	CMC Aina—1.0	Bentonitic clay—3.0	Casein—1.0	18	40	2–40	3–40	1200	1.5	0.93

					NaOH—1.0		90	2–10	3–15	1200		0.93
8	Novotroiz-kiy ТПЦГ—100	Water of the Caspian Sea—50	CMC Aina—1.0	Asbestos —3.0	Liquid glass—0.75	18	75	5–40	6–55	1000	1.6	1.30
9	Novotroiz-kiy ТПЦГ—100	Water of the Caspian Sea—50	CMC Aina—0.65	–	SAA—0.22	18	75	4–10	5–35	1330	1.8	0.71
10	ОЦГ—100	Brine water—90	CMC Lotos—2.0	Neftebadskaya clay—7.0	Liquid glass—2.5	17	90	2–50	4–00	850	1.0	0.96
					KOH—0.5		120	2–30	4–00	850	1.1	0.96
11	ШПЦС—120	Water of the Caspian Sea—65	ВРП—31	Neftebadskaya clay—10.0	Liquid glass—3.0	17	120	4–00	6–20	1000	1.6	1.23
					KOH—2.5		160	3–40	5–30	1000	1.8	1.23

[a] It is indicated nonfoamed solution.

where n is mass water mixture ratio; ρ_{cp} is the density of the dry rigid phase (cement powder + rigid additives), kg/m^3; $\rho_{fl}$is the density of liquid cement slurry, kg/m^3 or

$$V_c^t \approx V_c \tag{14.5.30}$$

where $\rho_{cp}$is the density of the dry rigid phase, kg/m^3; $\rho_1, \rho_2, \ldots, \rho_k$ are the densities of additives to the cement, kg/m^3; $n_1, n_2, \ldots, n_k$ are the ratios of additive masses to cement powder mass in a unit volume of the cement slurry.

Mass of the rigid phase to prepare a unit volume of a slurry

$$m_{rc} = \frac{\rho_c}{1+n}. \tag{14.5.31}$$

Mass of the cement powder needed to prepare a unit volume of cement slurry is

$$m_{cp} = \frac{m_{rc}}{1+n_1+n_2+\cdots+n_k}. \tag{14.5.32}$$

Delivery of substance needed to prepare grouting mortar consists of cement powder, kg

$$C = k_{cp} m_c V_c; \tag{14.5.33}$$

first additive, kg

$$M_1 = n_1 C; \tag{14.5.34}$$

second (kth) additive, kg

$$M_{2(k)} = n_{2(k)} C; \tag{14.5.35}$$

solidification fluid, m^3

$$V_{fl} = \frac{n(C+M_1+M_2+\cdots+M_k)}{k_{cp}\rho_{fl}}, \tag{14.5.36}$$

where $k_{rc}$is the reserve factor in cementing taking into account loss of the rigid phase during batching mixers and preparing grouting mortar, $k_{cp} = 1.05$–1.1.

Calculation of required grouting mortar volume and its components is given below.

The number of cement mixers needed to perform cementing operation is

$$i_c = \frac{C+M_1+M_2+\cdots+M_k}{m_{bc}V_{cm}}, \tag{14.5.37}$$

where m_{bc} is the bulk density of grouting mortar mixture (cement + rigid additives), kg/m^3; V_{cm} is the capacity of one cement mixer, m^3.

Calculate the composition of a mixture for cementing with constant degree of aeration using the following data on mass content components obtained in a laboratory: cementing agent–ТПЦХ—100; solidification liquid–fresh technical water—60; liquid deflocculant–surface active agent (SAA) Cubocsalim—0.2; rigid deflocculant–paligorskiy clay powder—5 and asbestos of the seventh sort—2; soda ash—5.

1. Determine the density of dry rigid phase with the formula (14.5.30)

$$\begin{aligned}\rho_c &= (\rho_{cp} + n_1 \cdot \rho_1 + n_2 \cdot \rho_2)/(1 + n_1 + n_2)\\ &= (3100 + 0.05 \times 2500 + 0.02 \times 2500)/(1 + 0.05 + 0.02)\\ &= 3061\ \mathrm{kg/m^3}.\end{aligned}$$

2. Calculate the density of initial cement slurry at W/C = 0.6 with the formula (14.5.29)

$$\begin{aligned}\rho_c &= [(1+n) \cdot \rho_c \cdot \rho_{fl}]/(\rho_{fl} + n \cdot \rho_c)\\ &= [(1+0.6) \times 3061 \times 1000]/(1000 + 0.6 \times 3061) = 1730\,\mathrm{kg/m^3}.\end{aligned}$$

3. Obtain the mass of rigid phase in a unit volume of cement slurry with the formula (14.5.11)

$$m_{rc} = \rho_c/(1+n) = 1730/(1+0.6) = 1081\ \mathrm{kg}.$$

4. Determine the mass of cement powder spent to prepare a unit volume of cement slurry with the formula (14.5.32)

$$m_{cp} = m_{rc}/(1 + n_1 + n_2) = 1081/(1 + 0.05 + 0.02) = 1010\,\mathrm{kg}.$$

5. Calculate an amount of material, reagents, and fluid needed to prepare the total volume of grouting mortar with formulas (14.5.33)–(14.5.36): cement powder

$$C = k_{cp} \cdot m_c \cdot V_c = 1.05 \times 1010 \times 41.7 = 44,200\,\mathrm{kg};$$

mud powder

$$M_1 = n_1 \cdot C = 0.05 \times 44,200 = 2200\,\mathrm{kg};$$

asbestos

$$M_2 = n_2 \cdot C = 0.02 \times 44,200 = 885\,\mathrm{kg};$$

SAA Cubocsalim

$$M_3 = n_3 \cdot C = 0.002 \times 44,200 = 88\,\text{kg};$$

soda ash

$$M_4 = n_4 \cdot C = 0.05 \times 44,200 = 2200\,\text{kg};$$

solidification fluid

$$\begin{aligned} V_{\text{fl}} &= n \cdot (C + M_1 + M_2)/(k_{\text{cp}} \cdot \rho_{\text{fl}}) \\ &= 0.6 \times (44,200 + 2200 + 885)/(1.05 \times 1000) = 27\,\text{m}^3. \end{aligned}$$

6. We determine the number of cement mixers i_c needed to perform cementing operation at $V_{\text{mix}} = 14.5\,\text{m}^3$ (2CMH-20)

$$\begin{aligned} i_c &= (C + M_1 + M_2)/(m_{\text{bc}} \cdot V_{\text{cm}}) \\ &= (44,200 + 2200 + 885)/(1081 \times 14.5) = 3. \end{aligned}$$

CHAPTER 15

SEDIMENTATION OF RIGID PHASE IN DRILLING FLUID AFTER DEADLOCK OF MIXING

One of the reasons of undesirable lowering of hydraulic pressure in a well is pressure drop of solution after cessation of its circulation in the well. Such pressure drop is observed in mud solution as well as in cement solution at initial period of its solidification.

Experiments have shown that after termination of mixing cement or mud suspension is not left in rest. A column of suspension may be considered as if it contains deformable skeleton (rigid phase) whose pore space is filled by fluid.

Heavier rigid phase or the suspension settles under action of gravity force descending relative fluid phase and well walls. Pure water is collected in the upper part of the well. Gravity force of the settling phase is imparted to pore water and through this it is imparted to the fluid in uncovered formations. Thus, the hydraulic pressure of fluid being lowered with time caused by setting of rigid phase on walls and bottom of the well counteracts the pressure of the formation fluid.

In the present section, dependences of hydraulic pressure distribution in a well on time in sedimentation of rigid phase of the drilling fluid after cessation of its mixing will be received (Leonov, 1975).

15.1 ONE-DIMENSIONAL EQUATION FOR HYDRAULIC PRESSURE IN SEDIMENTATION OF RIGID PHASE OF SUSPENSION

The process of rigid phase sedimentation in drilling fluid after cessation of mixing can be described by equations (4.6.33) – (4.6.39) under condition of incompressibility of rigid and fluid phases. We have

momentum equation

$$\rho_1\varphi_1\frac{\mathrm{d}v_1}{\mathrm{d}t}+\rho_2\varphi_2\frac{\mathrm{d}v_2}{\mathrm{d}t}+\frac{\partial p}{\partial z}=[\rho_1\varphi_1+\rho_2\varphi_2]g+\frac{\lambda_c}{2d}\left[\rho_1\varphi_1v_1^2+\rho_2\varphi_2v_2^2\right]; \tag{15.1.1}$$

equations of mass conservation

$$\frac{\partial\varphi_1}{\partial t}+\frac{\partial\varphi_1v_1}{\partial z}=0, \tag{15.1.2}$$

$$\frac{\partial\varphi_2}{\partial t}+\frac{\partial\varphi_2v_2}{\partial z}=0; \tag{15.1.3}$$

thermodynamic equations of state

$$\begin{aligned}\rho_1&=\text{const},\\ \rho_2&=\text{const};\end{aligned} \tag{15.1.4}$$

equations of concentrations

$$\varphi_1+\varphi_2=1, \tag{15.1.5}$$

$$\varphi_1=\varphi_1(p,\rho_1,\rho_2,v_1,v_2,d); \tag{15.1.6}$$

equation for hydraulic resistance factor

$$\lambda_c=\lambda_c(p,\rho_1,\rho_2,v_1,v_2,d,\mu). \tag{15.1.7}$$

On the assumption that sedimentation velocity of rigid particles in drilling fluid is very small, the first term in the right part of the equation (15.1.1) is much more than the friction term, and the term $\partial p/\partial z$ in the left part is much more than inertial terms. Then momentum equation takes form

$$\frac{\partial p}{\partial z}=[\rho_1\varphi_1+\rho_2\varphi_2]g. \tag{15.1.8}$$

In accordance with (15.1.5), we have $\varphi_1 = 1 - \varphi_2$. Substitution of this relation in (15.1.8) yields

$$\frac{\partial(p-\rho_1 gz)}{\partial z} = [\rho_2-\rho_1]\varphi_2 g. \tag{15.1.9}$$

The quantity $p' = p-\rho_1 gz$ defines pressure exerted on the fluid by moving rigid phase.

Differentiation of (15.1.9) with respect to z gives

$$\frac{\partial^2 p'}{\partial z^2} = (\rho_2-\rho_1)g\frac{\partial\varphi_2}{\partial z}. \tag{15.1.10}$$

Accept that the velocity of particle sedimentation v_2 depends only on concentration φ_2. Then, products $\varphi_2 v_2$ and $\varphi_1 v_1$ would be also dependent only on concentrations φ_2 or φ_1, and the equation (15.1.3) could be rewritten as

$$\frac{\partial\varphi_2}{\partial t} + \frac{\partial\varphi_2 v_2}{\partial\varphi_2}\frac{\partial\varphi_2}{\partial z} = 0. \tag{15.1.11}$$

Adding together equations (15.1.2) and (15.1.3), we get

$$\varphi_1 v_1 = -\varphi_2 v_2. \tag{15.1.12}$$

Then, from (15.1.11) with regard to (15.1.12), it follows

$$\frac{\partial\varphi_2}{\partial t} - \frac{1}{\partial\varphi_1 v_1/\partial\varphi_2}\frac{\partial\varphi_2}{\partial t} = 0. \tag{15.1.13}$$

Introduce the quantity $\varepsilon = \varphi_1/\varphi_2$ that is analog of porosity factor in soil mechanics. Then, with regard to (15.1.5), we receive

$$\varphi_2 = 1/(1+\varepsilon). \tag{15.1.14}$$

Consider the quantity $\partial(\varphi_1 v_1)/\partial\varphi_2$ entering in (15.1.13).

The product $\varphi_1 v_1$ may be represented as

$$\varphi_1 v_1 = S_1 v_1/S = Q_1/S, \tag{15.1.15}$$

where Q_1 is flow rate of fluid displaced by rigid phase.

Suppose that fluid flow rate is governed by Darcy filtration law

$$\varphi_1 v_1 = \frac{Q_1}{S} = -\frac{k}{\mu}\frac{\partial p'}{\partial z}, \tag{15.1.16}$$

where k is permeability factor of rigid phase skeleton taken equal to a certain average value k_{av} for the whole sedimentation process.

Insertion of the term $\partial p'/\partial z$ from (15.1.9) into (15.1.16) gives

$$\frac{\partial \varphi_1 v_1}{\partial \varphi_2} = -\frac{\partial}{\partial \varphi_2}\left(\frac{k}{\mu}\frac{\partial p'}{\partial z}\right) = -\frac{k}{\mu}\frac{\partial}{\partial \varphi_2}[(\rho_2-\rho_1)\varphi_2 g] = -\frac{k}{\mu}(\rho_2-\rho_1)g. \tag{15.1.17}$$

Substitution of (15.1.17) and (15.1.14) into (15.1.13) yields

$$\frac{\partial \varphi_2}{\partial z} = \frac{1}{(k/\mu)(\rho_2-\rho_1)g}\frac{1}{(1+\varepsilon)^2}\frac{\partial \varepsilon}{\partial t} \tag{15.1.18}$$

with which the equation (15.1.10) transforms to

$$\frac{\partial^2 p'}{\partial z^2} = \frac{1}{(k/\mu)(1+\varepsilon)^2}\frac{\partial \varepsilon}{\partial t}. \tag{15.1.19}$$

Let for drilling fluid exist a compression curve $\varepsilon = \varepsilon(p')$ similar to that of a soil saturated with water. Then,

$$\frac{\partial \varepsilon}{\partial t} = a\frac{\partial p'}{\partial t}, \tag{15.1.20}$$

where $a = \partial\varepsilon(p')/\partial p'$.

From (15.1.20) and (15.1.19) follows one-dimensional equation for p' of heat conduction type

$$\frac{\partial p'}{\partial t} = \alpha\frac{\partial^2 p'}{\partial z^2}, \tag{15.1.21}$$

where $\alpha = k(1 + \varepsilon^2)/(\mu a)$.

Replace permeability factor in equation (15.1.21) by filtration factor k_f

$$k = k_f\frac{\mu}{\rho_1 g}. \tag{15.1.22}$$

Then, the factor α in (15.1.21) takes form

$$\alpha = \frac{k_f(1+\varepsilon)^2}{\rho_1 g a}. \tag{15.1.23}$$

This factor differs somewhat from consolidation factor (Gercevanov and Polshin, 1948)

$$\alpha = \frac{k_f(1-\varepsilon)}{\rho_1 g a} \tag{15.1.24}$$

used in the consolidation theory of soils. On the assumption that a section of the compression curve may be straightened, when solving practical problems, parameters of the consolidation factor are believed to be $\varepsilon = \varepsilon_{av}$ and $a = \text{const}$.

In making stable drilling fluids and grouting mortars, one should tend to decrease factor α through measures intended to reduce k_f and ε and to enhance a. For this purpose, it would be useful to elevate dispersion of rigid phase and to impart Newtonian properties to fluid phase. Grachev (see Grachev et al., 1980) has first experimentally determined the factor α for grouting mortars, and then showed ways of its control.

15.2 LOWERING OF HYDRAULIC PRESSURE IN WELL AFTER DEADLOCK OF SOLUTION CIRCULATION

Consider decline of solution pressure $p'(z, t)$ in a well assuming well walls and bottom to be impenetrable. In this case, boundary conditions for equations (15.1.21) are

at the mouth

$$p'(z,t)|_{z=0} = 0; \tag{15.2.1}$$

at the bottom

$$\left.\frac{\partial p'}{\partial z}\right|_{z=L} = 0. \tag{15.2.2}$$

Initial pressure distribution p' in the well is characterized by difference of hydraulic pressures of the solution p_h and pure fluid p_{fl}

$$p'(z,t)|_{t=0} = \rho_{mix} g z - \rho_1 g z = (p_h - p_{fl})\frac{z}{L}. \tag{15.2.3}$$

The solution of the equation (15.1.21) is sought as a series

$$p'(z,t) = \sum_{n=1}^{\infty} p_n(t) \sin\frac{\pi(2n-1)z}{2L}, \tag{15.2.4}$$

where $p_n(t)$ is a function to be sought. Obviously the solution obeys boundary conditions (15.2.1) and (15.2.2).

Substitution of (15.2.4) on (15.1.21) yields

$$p'(z,t) = \sum_{n=1}^{\infty} \left(\frac{dp_n}{dt} + \frac{\alpha\pi^2(2n-1)^2}{4L^2} p_n \right) \sin\frac{\pi(2n-1)z}{2L}. \tag{15.2.5}$$

From here follows

$$\frac{\mathrm{d}p_n}{\mathrm{d}t} + \frac{\alpha\pi^2(2n-1)^2}{4L^2} p_n = 0. \tag{15.2.6}$$

Solution of this equation is

$$p_n = c_n \exp\left(-\frac{\alpha\pi^2(2n-1)^2}{4L^2} t\right). \tag{15.2.7}$$

Use of initial conditions (15.2.3) gives

$$c_n = \frac{8L}{\pi^2}(p_{\mathrm{h}} - p_{\mathrm{fl}}) \frac{(-1)^{n+1}}{(2n-1)^2} \exp\left(-\frac{\alpha\pi^2(2n-1)^2}{4L^2} t\right). \tag{15.2.8}$$

Substitution of p_{h} and c_n in (15.2.4) yields

$$p' = \frac{8gL}{\pi^2}(\rho_{\mathrm{mix}} - \rho_1) \sum_{n=1}^{\infty} \frac{(-1)^{n+1}}{(2n-1)^2} \sin\frac{(2n-1)\pi z}{2L} \exp\left(-\frac{\alpha\pi^2(2n-1)^2}{4L^2} t\right). \tag{15.2.9}$$

The hydraulic pressure p' is

$$p = p' + \rho_1 g z. \tag{15.2.10}$$

Thus, after deadlock of circulation and leaving the solution in quiescent state, we obtain from (15.2.9) that pressure $p' \to 0$ at $t \to \infty$ and hydraulic pressure in accordance with (15.2.10) tends to be hydrostatic $p = \rho_1 g z$.

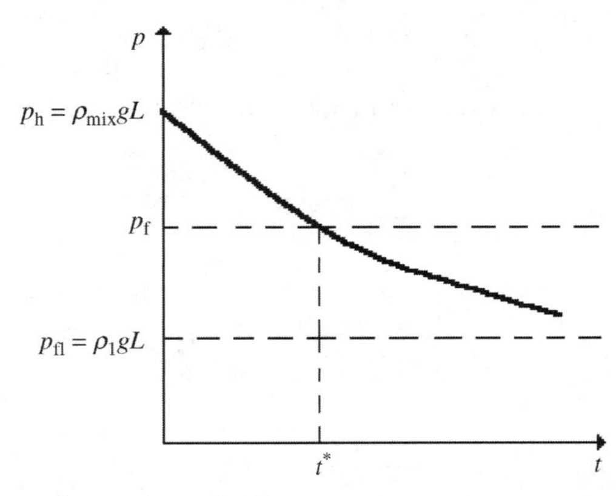

FIGURE 15.1 Drop of hydraulic pressure with time at the well bottom.

Figure 15.1 shows characteristic variation of hydraulic pressure with time at the bottom ($z = L$) obtained from the formula (15.2.10). The point of intersection between this curve and the straight line $p = p_f$ (formation pressure) gives the instant of time t^*, at which the pressure at the bottom falls up to the formation. If to accept that from $t = t^*$ begins inflow of formation fluid into the well, one can consider t^* as characteristic time in which the suspension hydraulic pressure would be reduced to the formation pressure. This estimation enables to forecast the beginning of undesirable inflow of formation fluid into the well and to take precautions against it.

CHAPTER 16

EXPERIMENTAL DETERMINATION OF RHEOLOGICAL CHARACTERISTICS

To get concrete results of the preceding chapters, one needs to know numerical values of rheological parameters of media under study. These parameters are determined with the help of special measuring instruments, among which are rotary and capillary viscometers serving to find rheological characteristics of fluids (Esman, 1982; Filatov, 1973; Shischenko et al., 1976). The method of determining rheological characteristics depends on the type of fluid. So, to determine viscosity factor μ of Newtonian fluid, it is sufficient to carry out one experiment on one of the viscometers. To get dynamic shear stress τ_0 and plastic viscosity factor η of viscous-plastic fluid or parameters k and n of power fluid, one needs to know minimum two experiments. Satisfactory methods to determine the characteristics of rheological nonstationary fluids are not yet elaborated.

Rock solids behave under certain conditions as fluids. For example, at high, slowly varying or permanent loads, many solids flow, in particular salts and muds, that is, they become deformed at certain rates typical for fluids. These rates are small, but values of rheological characteristics are great compared to those of analogues, for example, mud solutions or drilling muds. Therefore, to determine rheological characteristics of rock solids, there are commonly used experiments on uniaxial compression.

Applied Hydro-Aeromechanics in Oil and Gas Drilling. By Leonov and Isaev

16.1 DETERMINATION OF RHEOLOGICAL CHARACTERISTICS WITH ROTARY VISCOMETER

Rotary viscometer represents a measuring instrument consisting of two vertically arranged coaxial cylinders (Fig. 16.1). One of the cylinders, commonly external, can rotate with certain angular velocity. A fluid under test is poured in the gap between cylinders. When the external cylinder rotates with given angular velocity ω_2, the moment M that may be measured is imparted through the fluid in the gap. Assume that ends of the cylinders do not exert a significant effect on the flow in the gap.

By knowing angular velocity ω_2 and measuring moment M, it is possible to calculate mean shear rate $\dot{\gamma}_c$ and stress in fluid τ_c with theoretical formulas being results of momentum equation solution (6.1.6) and formula (4.4.9)

$$\frac{\partial \tau}{\partial r} + \frac{2\tau}{r} = 0; \tag{16.1.1}$$

$$\dot{\gamma} = \frac{\partial w_\varphi}{\partial r} - \frac{w_\varphi}{r}. \tag{16.1.2}$$

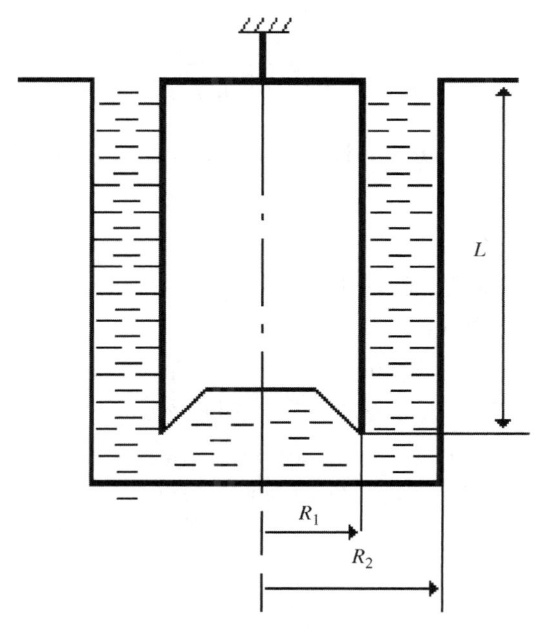

FIGURE 16.1 Scheme of the rotary viscometer.

From (16.1.1) it follows

$$\tau = C/r^2. \tag{16.1.3}$$

The stress τ at $r = R_1$ is equal to the stress τ_1 at the internal cylinder. By definition of moment, it is

$$M = \tau_1 2\pi R_1 L R_1, \tag{16.1.4}$$

where L is cylinder height.

Substituting in (16.1.3) $r = R_1$ and $\tau = \tau_1$, we get from (16.1.4)

$$C = M/(2\pi L) \tag{16.1.5}$$

The formula (16.1.3) with regard to (16.1.5) takes the form

$$\tau = M/(2\pi L r^2). \tag{16.1.6}$$

Rheological equation of the fluid to be tested should have functional form (6.1.9)

$$\tau = \tau(\dot{\gamma}) \tag{16.1.7}$$

Assume that the relation (16.1.7) can be single-valued resolved with respect to $\dot{\gamma}$, that is, to get single dependence $\dot{\gamma} = \dot{\gamma}(\tau)$. For the majority of fluids encountered in drilling, this assumption is true.

Then from (16.1.7) with regard to (16.1.6), we receive

$$\dot{\gamma} = f(\tau) = f\left(\frac{M}{2\pi L r^2}\right). \tag{16.1.8}$$

Linear velocity w_φ in (16.1.2) is connected with angular velocity by relation

$$\omega = w_\varphi / r. \tag{16.1.9}$$

Substitution of ω from (16.1.9) in (16.1.2) and the resulted expression for $\dot{\gamma}$ in (16.1.8) yields

$$\frac{\partial \omega}{\partial r} = \frac{1}{r} f\left(\frac{M}{2\pi L r^2}\right). \tag{16.1.10}$$

Since the internal cylinder is motionless ($\omega_1 = 0$) and the external one rotates with angular velocity ω_2, integration of (16.1.10) in limits from $r = R_1$ to $r = R_2$ gives

$$\omega_2 = \int_{R_1}^{R_2} \frac{1}{r} f\left(\frac{M}{2\pi L r^2}\right) \mathrm{d}r. \tag{16.1.11}$$

Pass now in (16.1.11) from variable r to τ through (16.1.6)

$$\omega_2 = \int_{\tau_1\delta^2}^{\tau_1} \frac{f(\tau)}{2\tau}\,\mathrm{d}\tau, \tag{16.1.12}$$

where $\delta = R_1/R_2$.

Viscometers are generally made with narrow gap between cylinders, so that $(R_2 - R_1)/\mathrm{R}_2 \ll 1$. Therefore, the expression of integrand in (16.1.12) may be replaced by its value at mean point

$$\tau_c = (\tau_1\delta^2 + \tau_1)/2. \tag{16.1.13}$$

Then,

$$\omega_2 \approx \int_{\tau_1\delta^2}^{\tau_1} \frac{f(\tau_c)}{2\tau_c}\,\mathrm{d}\tau = \frac{f(\tau_c)}{2\tau_c}(\tau_1 - \tau_1\delta^2) = \frac{(1-\delta^2)f(\tau_c)}{1+\delta^2} \tag{16.1.14}$$

and with regard to (16.1.8), we get approximate relation

$$\dot{\gamma}_c = f(\tau_c) = \omega_2(1+\delta^2)/(1-\delta^2). \tag{16.1.15}$$

Take into account (16.1.6) in (16.1.13). Then,

$$\tau_c = \tau_1(1+\delta^2)/2 = M(1+\delta^2)/(4\pi R_1^2 L). \tag{16.1.16}$$

Thus, giving angular velocities ω_2 and measuring with (16.1.16) and (16.1.15) moments M, one can obtain relevant set of τ_c and $\dot{\gamma}_c$ with which it is able to get dependence $\tau_c = \tau_c(\dot{\gamma}_c)$ taken as rheological equation (16.1.7).

16.2 DETERMINATION OF RHEOLOGICAL CHARACTERISTICS WITH CAPILLARY VISCOMETER

The basis of capillary viscometer is a tube with internal diameter d (Fig. 16.2). Through the tube flows with given flow rate Q a fluid under study. The pressure drop $\Delta p = p_1 - p_2$ is measured in the operation section of the tube. The operation section should be located at certain distance from tube edges to exclude their influence on the flow. The distance is commonly taken $L > 100d$. The operation section l has to be spaced at such distance from the edges to obey condition $\partial p/\partial z = \mathrm{const}$.

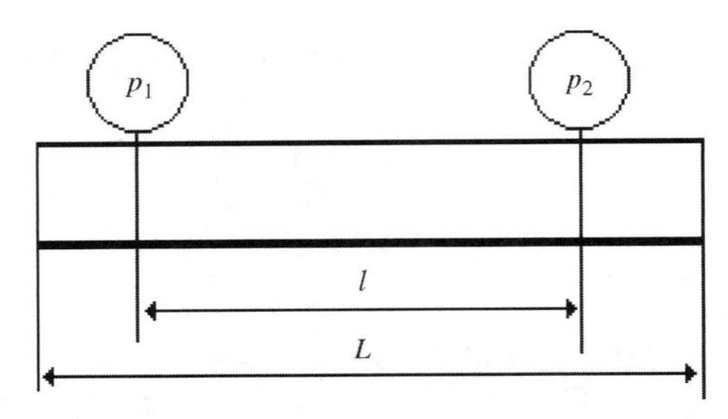

FIGURE 16.2 Scheme of a capillary viscometer.

Flows resulting in a tube were in some detail described in Section 6 for viscous, viscous-plastic and power fluids. In such flows, the following dependence takes place between tangential stresses τ and pressure drop Δp:

$$\tau = -r\,\Delta p/2l. \tag{16.2.1}$$

Expression (16.2.1) is the result of momentum equation and is determined by the second equation (6.2.23). At $r = R$, we get from (16.2.1) the stress at the tube wall τ_w

$$\tau_w = -R\,\Delta p/2l. \tag{16.2.2}$$

From (16.2.1) and (16.2.2), it follows

$$\tau = \tau_w r/R. \tag{16.2.3}$$

Then, the rheological equation

$$\tau = \tau(\dot{\gamma}) \tag{16.2.4}$$

can be single valued resolved with respect to $\dot{\gamma}$, namely,

$$\dot{\gamma} = \dot{\gamma}(\tau) = F(\tau). \tag{16.2.5}$$

For flow in a tube, we have

$$\dot{\gamma} = \partial w/\partial r. \tag{16.2.6}$$

The flow rate through the tube cross section is

$$Q = 2\pi \int_0^R w(r) r\,\mathrm{d}r. \tag{16.2.7}$$

Integration by parts gives

$$Q = \frac{2\pi r^2}{2} w(r)|_0^R - \frac{2\pi}{2} \int_0^R \frac{\partial w}{\partial r} r^2 \, dr. \qquad (16.2.8)$$

Since fluid velocity at tube walls vanishes $w(R) = 0$, from (16.2.7) with regard to (16.2.5) we get

$$Q = -\pi \int_0^R F(\tau) r^2 \, dr. \qquad (16.2.9)$$

In order to operate with positive quantities, we shall take τ and τ_w by absolute value and omit minus sign in formula (16.2.9). Replace r in (16.2.9) by its expression (16.2.3) through τ. As a result, we get

$$Q = \frac{\pi R^3}{\tau_w^3} \int_0^{\tau_w} \tau^2 F(\tau) \, d\tau. \qquad (16.2.10)$$

Taking derivative of τ_w with respect to Q, we obtain Mooney–Rabinovitch relation (Wilkenson, 1960)

$$\frac{dQ}{d\tau_w} = -\frac{3Q}{\tau_w} + \frac{\pi R^3}{\tau_w} F(\tau_w). \qquad (16.2.11)$$

From here with regard to (16.2.5), it yields

$$\dot{\gamma}_w = F(\tau_w) = \frac{1}{\tau_w^2} \frac{d}{d\tau_w} \left(\tau_w^3 \frac{Q}{\pi R^3} \right). \qquad (16.2.12)$$

Thus, by measuring Δp at given Q, we obtain with (16.2.2) appropriate values of τ_w that permit to plot the curve

$$Q/(\pi R^3) = f(\tau_w). \qquad (16.2.13)$$

Approximating it by suitable function and inserting in (16.2.12), we get gradient of the shear rate at the wall $\dot{\gamma}_w$. with τ_w and $\dot{\gamma}_w$, we plot rheological curve $\tau_w = \tau_w(\dot{\gamma}_w)$. Such curve can be obtained for any rheological stationary fluid. When the type of fluid (viscous, viscous-plastic, or power fluid) is beforehand known, there is no need to build curve $\tau_w = \tau_w(\dot{\gamma}_w)$. Really, in the case of viscous fluid, the rheological equation has the form (6.2.21) or $\dot{\gamma} = \tau/\mu$, and in accordance with (16.2.5)$F(\tau) = \tau/\mu$. Then, from (16.2.10) ensues Poiseuille equation (6.2.31) for the pipe flow

$$Q = \frac{\tau_w \pi R^3}{4\mu} = \frac{\pi R^4 \Delta p}{8\mu l}. \tag{16.2.14}$$

In the case of Bingham fluid from (6.3.19) with regard to (16.2.5), it follows

$$\begin{aligned} F(\tau) &= \dot{\gamma} = \frac{\tau - \tau_0}{\eta} \quad \text{at} \quad \tau > \tau_0, \\ F(\tau) &= \dot{\gamma} = 0 \quad \text{at} \quad 0 \le \tau \le \tau_0. \end{aligned} \tag{16.2.15}$$

Substitution of (16.2.15) into (16.2.10) yields Buckingham formula (6.3.27)

$$Q = \frac{\pi R^4 \Delta p}{8\eta l}\left[1 - \frac{4}{3}\frac{2\tau_0 l}{R\,\Delta p} + \frac{1}{3}\left(\frac{2\tau_0 l}{R\,\Delta p}\right)^4\right]. \tag{16.2.16}$$

For power fluid from (6.4.20) with regard to (16.2.5), we get

$$F(\tau) = (\tau/k)^{1/n}. \tag{16.2.17}$$

Insertion of this formula into (16.2.10) gives formula (6.4.26) for power fluid

$$Q = \frac{\pi n}{3n+1}\left(\frac{d}{2}\right)^{\frac{1}{n}+3}\left(\frac{\Delta p}{2lk}\right)^{\frac{1}{n}}. \tag{16.2.18}$$

Consider methods to determine rheological constants in solutions (16.2.14), (16.2.16), and (16.2.18). If the fluid is Newtonian, it is sufficient to perform one measurement of Q and Δp to get viscosity. Then, in accordance with (16.2.14) there is

$$\mu = \pi R^4 \Delta p/(8lQ). \tag{16.2.19}$$

In the case of viscous-plastic fluid, it is sufficient to carry out two measurements of Q and Δp to determine dynamic shear stress τ_0 and plastic viscosity η. Then in accordance with (16.2.16), we obtain two equations for η and τ_0 resulting from measurements $(Q_1, \Delta p_1)$ and $(Q_2, \Delta p_2)$

$$\begin{aligned} \eta &= \frac{\pi R^4 \Delta p_1}{8lQ_1}\left[1 - \frac{4}{3}\frac{2\tau_0 l}{R\,\Delta p_1} + \frac{1}{3}\left(\frac{2\tau_0 l}{R\,\Delta p_1}\right)^4\right]; \\ \eta &= \frac{\pi R^4 \Delta p_2}{8lQ_2}\left[1 - \frac{4}{3}\frac{2\tau_0 l}{R\,\Delta p_2} + \frac{1}{3}\left(\frac{2\tau_0 l}{R\,\Delta p_2}\right)^4\right]. \end{aligned} \tag{16.2.20}$$

Values of η and τ_0 can be obtained graphically as intersection of two curves (16.2.20). If to perform three measurements of parameters η and τ_0, they can be determined in explicit form

$$\tau_0 = \frac{3}{4}\frac{R}{2l}\frac{Q_1 a_1 + Q_2 b_1 + Q_3 c_1}{Q_1 + bQ_2 + cQ_3};\qquad \eta = \frac{\pi R^4}{8l}\frac{\Delta p_1 + b\,\Delta p_2 + c\,\Delta p_3}{Q_1 + bQ_2 + cQ_3}, \tag{16.2.21}$$

where

$$a_1 = \Delta p_3 \frac{1-(\Delta p_2/\Delta p_3)^4}{1-(\Delta p_2/\Delta p_3)^3};\qquad b_1 = \Delta p_1 \frac{(\Delta p_3/\Delta p_1)^4 - 1}{1-(\Delta p_3/\Delta p_2)^3};$$

$$c_1 = \Delta p_1 \frac{1-(\Delta p_2/\Delta p_1)^4}{(\Delta p_2/\Delta p_3)^3 - 1};$$

$$b = \frac{(\Delta p_3/\Delta p_1)^3 - 1}{1-(\Delta p_3/\Delta p_2)^3};\qquad c = \frac{(\Delta p_2/\Delta p_1)^3 - 1}{1-(\Delta p_2/\Delta p_3)^3}.$$

For power fluid, it is sufficient to perform two measurements to determine parameters n and k

$$n = \frac{\log(\Delta p_1/\Delta p_2)}{\log(Q_1/Q_2)};\qquad k = \frac{\left(\frac{\pi n}{3n+1}\right)^n \left(\frac{d}{2}\right)^{1+3n} \Delta p_1}{2lQ_1^n}.$$

16.3 DETERMINATION OF RHEOLOGICAL CHARACTERISTICS OF ROCK SOLIDS

In considering rock solid flows, one also needs to know rheological equations of rock solids. The use of rotary or capillary viscometers in this case is difficult for high yield stress of rock solids. Therefore, they are commonly studied by uniaxial compression (Fig. 16.3). At given constant load at the end of the cylindrical sample with height generally equal to two diameters, one measures normal stress τ_n and deformation rate $\dot{\varepsilon} = \partial w/\partial z$.

Let us consider rock solids as rheologic stationary fluids. It means that in experiments on compression with constant load, the deformation rate $\dot{\varepsilon}$ remains constant and different for different τ_n, that is,

$$\tau_\mathrm{n} = \tau_\mathrm{n}(\dot{\varepsilon}). \tag{16.3.1}$$

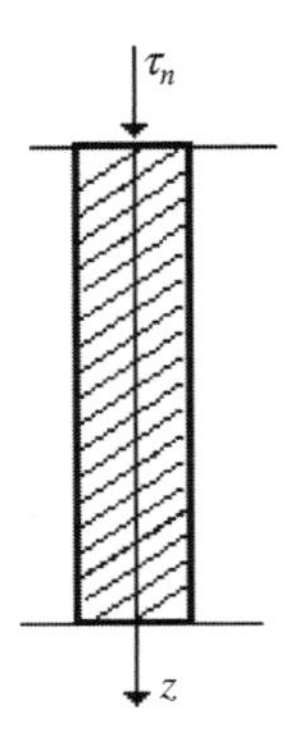

FIGURE 16.3 Scheme of solid sample test on uniaxial compression.

Thus, it is supposed that in experiments on uniaxial compression other stresses are absent. The form of concrete dependence (16.3.1) allows to get rheological equation for tangential stress τ (simple shift) in one-dimensional flows in pipes and slots of the same rock solid

$$\tau = \tau(\dot{\gamma}), \tag{16.3.2}$$

Calculations of tangential stresses with normal components is performed with formulas (Astarita and Marucci, 1974; Sedov, 1983)

$$\tau(\dot{\gamma}) = \frac{\tau_{\mathrm{n}}(\dot{\varepsilon})}{\sqrt{3}}; \tag{16.3.3}$$

$$\dot{\gamma} = \sqrt{3}\dot{\varepsilon}.$$

For example, if in uniaxial compression linear dependence was obtained

$$\tau_{\mathrm{n}} = \tau_{\mathrm{n}0} + \eta'\dot{\varepsilon}, \tag{16.3.4}$$

the appropriate rheological equation for tangential stresses of viscous-plastic fluid would be

$$\tau = \tau_0 + \eta\dot{\gamma}, \tag{16.3.5}$$

where in accordance with (16.3.3) the dynamic shear stress and plastic velocity is determined by formulas

$$\tau_0 = \tau_{\mathrm{n}0}/\sqrt{3}; \qquad \eta = \eta'/3. \tag{16.3.6}$$

On the other hand, if in uniaxial compression nonlinear dependence of the form was obtained

$$\tau_{\mathrm{n}} = k'(\dot{\varepsilon})^{n'}, \qquad (16.3.7)$$

the appropriate rheological equation for tangential stresses of power fluid would be

$$\tau_{\mathrm{n}} = k(\dot{\gamma})^{n}, \qquad (16.3.8)$$

where

$$k = k'/3^{(n+1)2}; \quad n = n'. \qquad (16.3.9)$$

16.4 EXAMPLES OF APPLICATIONS OF RHEOLOGICAL CHARACTERISTICS

To approximate experimental data with analytical dependence, the least squares method is commonly used. Consider this method with the example of rheological curve approximation. Suppose for a set of $\dot{\gamma}_i$, τ_{i} $(i = 1, 2, \ldots, m)$ are obtained. Suppose that the dependence τ on $\dot{\gamma}$ is linear function $\tau = \tau_0 + \eta\dot{\gamma}$, try to obtain the values of τ_0 and η so that the sum of deviation squares of this function in experimental points would be minimal.

Denote root mean square deviation as

$$\sigma = \sum (\tau_i - \tau_0 - \eta\dot{\gamma}_i)^2 \qquad (16.4.1)$$

and get $\min(\sigma)$ considering σ as function of τ_0 and η. This leads to a system of equations

$$\frac{\partial\sigma}{\partial\tau_0} = 0;$$

$$\frac{\partial\sigma}{\partial\eta} = 0.$$

Substitution in this equations expression for σ and resolving them with respect to τ_0 and η gives

$$\tau_0 = \frac{\sum\tau_i \sum\dot{\gamma}_i^2 - \sum\tau_i\dot{\gamma}_i \sum\dot{\gamma}_i}{m\sum\dot{\gamma}_i^2 - (\sum\dot{\gamma}_i)^2}; \qquad (16.4.2)$$

$$\eta = \frac{m\sum\tau_i\dot{\gamma}_i - \sum\dot{\gamma}_i \sum\tau_i}{m\sum\dot{\gamma}_i^2 - (\sum\dot{\gamma}_i)^2}.$$

To approximate experimental data of a power function, it is convenient to use the following expedient. Taking logarithm of $\tau = k\dot{\gamma}^n$, we obtain

$$\log \tau = \log k + n \log \dot{\gamma} \tag{16.4.3}$$

or

$$T = k_1 + n\Gamma,$$

where $T = \log \tau$; $k_1 = \log k$; $\Gamma = \log \dot{\gamma}$.

Let there are m measured values $T_i = \log \tau_i$ and $\Gamma_i = \log \dot{\gamma}_i$. Then in accordance with (16.4.1) and (16.4.2), we have

$$k_1 = \frac{\sum T_i \sum \Gamma_i^2 - \sum T_i \Gamma_i \sum \Gamma_i}{m \sum \Gamma_i^2 - (\sum \Gamma_i)^2}; \tag{16.4.4}$$

$$n = \frac{m \sum T_i \Gamma_i - \sum T_i \sum \Gamma_i}{m \sum \Gamma_i^2 - (\sum \Gamma_i)^2}; \quad k = 10^{K_1}. \tag{16.4.5}$$

EXAMPLE 16.4.1

It is required to get rheological curve $\tau = \tau(\dot{\gamma})$ of a solution with the following data obtained by rotary viscometer with $\delta = R_1/R_2 = 0.9$:

ω_2 (1/s)	1	2	4	8
τ_1 (Pa)	5.5	9.5	18	28

SOLUTION In accordance with (16.1.13) and (16.1.15), there are

$$\dot{\gamma}_c = \omega_2 \frac{1+\delta^2}{1-\delta^2} = \omega_2 \frac{1+0.9^2}{1-0.9^2} = 9.53\omega_2$$

$$\tau_c = \frac{1+\delta^2}{2}\tau_1 = \frac{1+0.9^2}{2}\tau_1 = 0.905\tau_1$$

Readings of the viscometer give

$\dot{\gamma}_c$ (1/s)	9.53	19.06	38.12	76.24
τ_c (Pa)	4.98	8.6	16.3	25.3

These data are represented in Fig. 16.4 by points.

We approximate experimental data first by linear dependence $\tau = \tau_0 + \eta\dot{\gamma}$ and obtain τ_0 and η with the help of least square method. To do it, calculate the following sums:

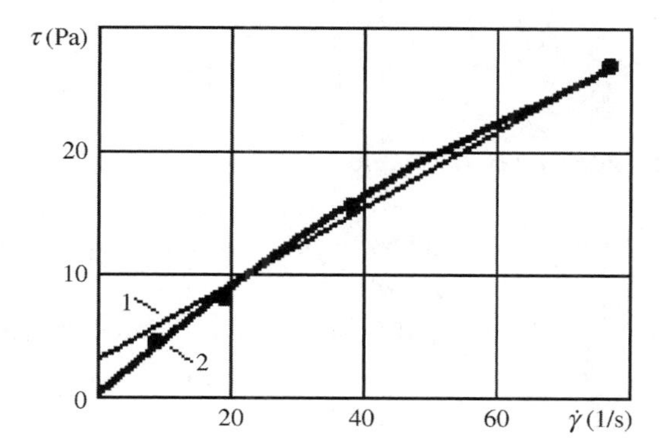

FIGURE 16.4 Dependence of tangential stress on shear rate obtained by rotary viscometer: (1) approximation by linear dependence (model of viscous-plastic fluid) and (2) approximation by power dependence (model of power fluid).

$$\sum \dot{\gamma}_i = 9.53 + 19.06 + 38.12 + 76.24 = 142.3;$$

$$\sum \tau_i = 4.98 + 8.6 + 16.3 + 25.3 = 55.2;$$

$$\sum \tau_i \dot{\gamma}_i = 4.98 \times 9.53 + 8.6 \times 19.06 + 16.3 \times 38.12 + 25.3 \times 76.24 = 2762;$$

$$\sum \dot{\gamma}_i^2 = 9.53^2 + 19.06^2 + 38.12^2 + 76.24^2 = 7720.$$

With formulas (16.4.2), we receive

$$\eta = \frac{4 \times 2762 - 55.2 \times 142.3}{4 \times 7720 - 142.3^2} = 0.3; \qquad \tau_0 = \frac{55.2 \times 7720 - 2762 \times 142.3}{10,631} = 3.11.$$

Hence, there is

$$\tau = 3.11 + 0.3\dot{\gamma}. \tag{16.4.6}$$

Fig. 16.4 shows resulting curve l.

Approximate experimental data by power dependence $\tau = k\dot{\gamma}^n$. Passing from τ_i and $\dot{\gamma}_i$ to $T_i = \log \tau_i$ and $\Gamma_i = \log \dot{\gamma}_i$, we get $T_i = 0.697; 0.934; 1.21; 1.4$; $\Gamma_i = 0.98; 1.28; 1.58; 1.88$. Then calculate $\sum \Gamma_i = 5.72$; $\sum \Gamma_i = 4.24$; $\sum \Gamma_i^2 = 8.63$; $\sum T_i \Gamma_i = 6.42$. With formulas (16.4.4) and (16.4.5), we obtain $n = 0.79$; $k_1 = -0.0697$; and $k = 10^{-0.0697} = 0.85$. Finally, we obtain

$$\tau = 0.85\dot{\gamma}^{0.79}. \tag{16.4.7}$$

This dependence is represented by curve 2 in Fig. 16.4.

Determine now which curve approximates experimental data best. The dependence (16.4.6) gives

$$\sigma = \sum (\tau(\dot{\gamma}_i) - \tau_i)^2 = 0.99^2 + 0.23^2 + 1.75^2 + 0.68^2 = 4.56,$$

whereas the dependence (16.4.7) yields

$$\sigma = \sum (\tau(\dot{\gamma}_i) - \tau_i)^2 = 0.065^2 + 0.124^2 + 1.22^2 + 0.78^2 = 2.12.$$

Hence, power dependence (16.4.7) performs better approximation than linear approximation (16.4.6).

EXAMPLE 16.4.2

It is required to get rheological curve $\tau = \tau(\dot{\gamma})$ with data of Q and ΔP obtained by capillary viscometer and calculated with (16.2.2) and (16.2.13).

Experimental data are				
$Q/\pi R^3$ (1/s)	0.98	5.9	11.8	18.2
τ_w (Pa)	2.5	10.5	18.3	26

SOLUTION With the use of Lagrange interpolation, we get relation between $Q/\pi R^3$ and τ_w

$$Q/\pi R^3 = -0.00013\tau_w^3 + 0.00128\tau_w^2 + 0.47\tau_w - 0.267.$$

Then with expression (16.2.12), we obtain

$$\dot{\gamma}_w = \frac{1}{\tau_w^2}\frac{d}{d\tau_w}\left(\tau_w^3 \frac{Q}{\pi R^3}\right) = -0.00078\tau^3 + 0.064\tau_w^2 + 1.88\tau_w - 0.8.$$

Substitution of viscometer data gives

$\dot{\gamma}_w$ (1/s)	4.29	25.1	50.3	77.6

Values of τ_w relevant to $\dot{\gamma}_w$ are represented in Fig. 16.5 by points.

As in Example 16.4.1, let us get parameters τ_0 and η or k and n in cases of approximation suing models of viscous-plastic and power fluids.

For the model of viscous-plastic fluid, we obtain with the help of least square method $\tau_0 = 1.82$ Pa, $\eta = 0.318$ Pa s. Then, $\tau = 1.82 + 0.318\dot{\gamma}$. With formula (16.4.1), we receive $\sigma = 1.48$.

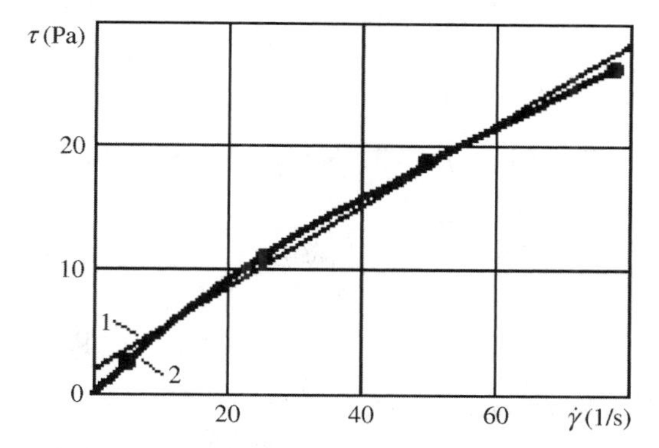

FIGURE 16.5 Dependence of tangential stress on shear rate obtained by capillary viscometer: (1) approximation by linear dependence (model of viscous-plastic fluid) and (2) approximation by power dependence (model of power fluid).

For model of power fluid there is $k = 0.77$; $n = 0.81$.Then $\tau = k\dot{\gamma}^n = 0.77\dot{\gamma}^{0,81}$. The formula (16.4.1) gives $\sigma = 0.0204$.

Since $\sigma = 0.0204$ is less than $\sigma = 1.48$, the model of power fluid approximates experimental data best.

EXAMPLE 16.4.3

It is required to get rheological curve $\tau = \tau(\dot{\gamma})$ of the rock solid from experimental data on uniaxial compression.

Experimental data are				
$\dot{\varepsilon} \times 10^8$ (1/s)	1.2	4.1	13	25
$\tau_n \times 10^{-5}$ (Pa)	1.8	2.5	3.5	5.2

First, let us calculate sums:

$$\sum \tau_{ni} = 13 \times 10^5; \qquad \sum \dot{\varepsilon}_i = 43.3 \times 10^{-8}; \qquad \sum \tau_{ni}\dot{\varepsilon}_i = 187.91 \times 10^{-3};$$
$$\sum \dot{\varepsilon}_i^2 = 812.25 \times 10^{-16}.$$

Then get in (16.3.4) τ_{n0} and η' with formulas similar to (16.4.2):

$$\tau_{n0} = \frac{13 \times 10^5 \times 812.25 \times 10^{-16} - 187.91 \times 10^{-3} \times 43.3 \times 10^{-8}}{4 \times 812.25 \times 10^{-16} - (43.3)^2 \times 10^{-16}} = 1.76 \times 10^5;$$

$$\eta' = \frac{4 \times 187.91 \times 10^{-3} - 43.3 \times 10^{-8} \times 13 \times 10^{5}}{4 \times 812.25 \times 10^{-16} - (43.3)^2 \times 10^{-16}} = 1.37 \times 10^{12};$$

$$\tau_n = \tau_{n0} + \eta' \dot{\varepsilon} = 1.76 \times 10^5 + 1.37 \times 10^{12} \dot{\varepsilon} \tag{16.4.8}$$

From (16.3.6), we obtain

$$\tau_0 = \tau_{n0}/\sqrt{3} = 1.76 \times 10^5/\sqrt{3} = 1.02 \times 10^5 \text{ Pa},$$

$$\eta = \eta'/3 = 1.37 \times 10^{12}/3 = 4.6 \times 10^{11} \text{ Pa s}.$$

The rheological equation (16.3.5) is

$$\tau = 1.02 \times 10^5 + 4.6 \times 10^{11} \dot{\gamma}.$$

Calculate $\Gamma_i = \log \dot{\varepsilon}_i$ and $T_i = \log \tau_{ni}$

Γ_i	−7.921	−7.387	−6.886	−6.602
T_i	5.255	5.3985	5.5441	5.716

From (16.4.4) and (16.4.5), there are $k_1 = 7.89$; $n' = 0.335$; $k' = 7.75 \times 10^7$. Then

$$\dot{\varepsilon}^{n'} = 7.75 \times 10^7 \tau_n = k' \dot{\varepsilon}^{0.335}. \tag{16.4.9}$$

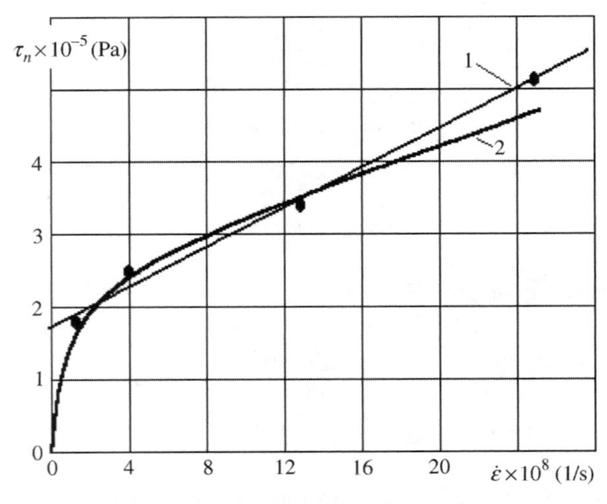

FIGURE 16.6 Dependences of tangential stress on shear rate with experiments on uniaxial compression: (1) approximation by linear dependence (model of viscous- plastic fluid) and (2) approximation by power dependence (model of power fluid).

Figure 16.6 demonstrates experimental data and curves 1 and 2 relevant to formulas (16.4.8) and (16.4.9). In accordance with (16.3.9), we have

$$k = \frac{7.75 \times 10^7}{3^{(1+0.335)/2}} = 3.72 \times 10^7; \qquad n = n' = 0.335.$$

Rheological equation is

$$\tau = 3.72 \times 10^7 \dot{\gamma}^{0.335}.$$

In the given example, the model of viscous-plastic fluid approximates the rheology of rock solid better than the model of power fluid because

$$\sigma = \sum (\tau_{ni} - \tau_{n0} - \eta' \dot{\varepsilon})^2 = 4.98 < \sigma = \sum (\tau_{ni} - k'^{n'})^2 = 30.9.$$

REFERENCES

Altshul, A.V. and Kiselev, P.G. (1975) *Hydraulics and Aerodynamics*. Stroyizsdat, Moscow (in Russian).

Astarita, G. and Marucci, G. (1974) *Principles of Non-Newtonian Fluid Mechanics*. McGraw-Hill, London.

Basarov, I.P. (1991) *Thermodynamics*. High School, Moscow (in Russian).

Bergeron, L. (1950) *De Coup de Belier en Hydraulique au Coup de Foudre en Electricite. Methode Graphique Generale*. Dunod, Paris.

Bingham, E. (1922) *Fluidity and Plasticity*. McGraw-Hill, New York.

Broon, V.G. and Leonov, E.G. (1981) Calculation method of casing cementing in well drilling. *High School Proc., Ser. Geol. Exploration*, (8), 16.

Buckingham, E. (1921) On plastic flow through capillary tubes. *Proc. Am. Soc. Testing Mater.*, **21**, 1154–1161.

Businov, S.N. and Umrichin, I.D. (1973) *Hydrodynamic Researching Methods of Wells and Formations*. Nedra, Moscow (in Russian).

Charniy, I.A. (1963) *Underground Hydro-Gas Dynamics*. Gostoptechizdat, Moscow (in Russian).

Charniy, I.A. (1975) *Non-Stationary Motion of Real Fluid in Pipes*, 2nd ed. Nedra, Moscow (in Russian).

Applied Hydro-Aeromechanics in Oil and Gas Drilling. By Leonov and Isaev

Digalev, V.Z., Malevanski, V.D., Grachev, V.V., and Leonov, E.G. (1987) Method of calculation of pressure losses due to friction in round trip operations. *Oil Industry*, (8), 18–22 (in Russian).

Esman, B.I. (1982) *Thermohydraulics in Well Drilling*. Nedra, Moscow (in Russian).

Evdokimova, V.A. and Kochina, I.N. (1979) *A Workbook on the Underground Hydraulics*. Nedra, Moscow (in Russian).

Filatov, B.S. (1973) Hydrodynamics of drilling fluids and grouting mortars. In: Mischenko, V.I. and Sidorov, N.A. (eds), *Reference Book for Drilling Engineer*, Vol. **1**. Nedra, Moscow (in Russian).

Fredrickson, A.G. and Bird, R.B. (1958) Non-Newtonian flow in annuli. *Ind. Eng. Chem.*, **50** (3), 347–352.

Gabolde, G. and Nguyen, J.-P. (1991) *Drilling Data Handbook*, 6th ed. Editions Technip/Institut Francais du Petrole, Paris/Rueil-Malmaison.

Gercevanov, N.M. and Polshin, D.E. (1948) *Theoretical Fundamentals of Soil Mechanics and Its Practical Applications*. Stroiizdat, Moscow (in Russian).

Gimatudinov, Sh.K. (ed.) (1983) *Reference Book on Designing, Development and Exploitation of Oil Fields*. Nedra, Moscow (in Russian).

Goins, W.C. and Sheffield, R. (1983) *Blowout Prevention. Vol. 1. Practical Drilling Technology*. 2nd ed. Gulf Publication Co., Houston, TX.

Golubev, D.A. (1979) Exact and approximate solutions of general stationary problem on the flow of viscous-plastic systems in annular space. *Problems Tumen Oil Gas*, (42), 22–45 (in Russian).

Grachev, V.V., Leonov, E.G., Malevanski, V.D., and Digalev, V.Z. (1980) Hydrodynamic pressure in a well in round trip operations. *Gas Industry*, (10), 35–38.

Gradstein, I.S. and Ryszik, I.M. (1971) *Tables of Integrals, Sums, Series and Products*. Nauka, Moscow (in Russian).

Grodde, K.H. (1960) Rheologie kolloider Suspension insbesondere der Bohrspülungen. *Erdöl Kohle*, **13** (1), 11–18;13 (2), 79–83.

Gukasov, N.A. (1954) About a case of a motion of a cylindrical body in viscous-plastic disperse medium. *Colloid J.*, **16** (1), 10–12 (in Russian).

Gukasov, N.A. (1976) *Hydrodynamics in Well Setting*. Nedra, Moscow (in Russian).

Gukasov, N.A. and Kochnev, A.M. (1991) *Hydraulics in Exploratory Drilling, Handbook*. Nedra, Moscow (in Russian).

Hanks, R.W. (1963) The laminar-turbulent transition for fluids with a yield stress. *AIChE J.*, **9** (3), 306–309.

Hedström, O.A. (1952) Flow of plastics materials in pipes. *Ind. Eng. Chem.*, **44** (3), 651–656.

Herrick, H.N. (1932) Data on flow of rotary drilling mud. *Oil Gas J.*, Feb. 25, pp. 16, 90.

Horner, D.R. (1951) Pressure build-up in wells. *Proceedings of the 3rd World Petroleum Congress, Section II*, Leiden, p. 503.

Isaev, V.I. (2006) Generalized hydrostatics of single-phase fluids and multi-phase mixtures in gravity field. *J. Oil Gas Eng. Land Sea*, (1) 25–29 (in Russian).

Isaev, V.I. and Leonov, E.G. (1976) Hydraulic calculation in well washing with aerated fluid. In: *Hydrodynamic Problems of Viscous and Viscous-Plastic Aerated Fluid.* Riasan, pp. 46–91 (in Russian).

Isaev, V.I., Leonov, E.G., and Raikevich, S.I. (2001) Hydrostatic pressure of two-phase media in drilling and exploitation of wells. *J. Oil Gas Eng. Land Sea*, (1), 21–22 (in Russian).

Isaev, V.I. and Markov, O.A. (2007) *Well Control. Prevention and Liquidation of Gas–Oil–Water Seepages. Textbook.* Fasis, Moscow (in Russian).

Kipunov, G.N., Leonov, E.G., and Musin, V.V. (1983) Simplified method of blowout and outburst recognition in drilling process. *Oil Industry*, (2), 21–23 (in Russian).

Krilow, A.N. (1936) *Collected Works*, Vol. **7**. Academy of Sciences USSR, Moscow-Leningrad (in Russian).

Lamb, H. (1945) *Hydrodynamics.* Dover, New York.

Leibenson, L.S. (1934) *Oil Field Mechanics.* Part II. ONTI, Moscow (in Russian).

Leonov, E.G. (1973) Calculation method of capacity of pumps and compressors in drilling of wells with aerated fluid washing. *Drilling*, (8), 5–9 (in Russian).

Leonov, E.G. (1975) Estimation of measurements of mud solution pressure in a well with bottomhole manometer. *Chem. Oil Eng.*, (1), 20–21 (in Russian).

Leonov, E.G., Filatov, B.S., and Hahaev, B.N. (1972) Calculation of ball drop time in descending flow of washing fluid in a drillstem. *Oil Industry*, (3), 22–26.

Leonov, E.G. and Isaev, V.I. (1978) *Hydraulic calculations of well washing in drilling.* Gubkin Oil and Gas Institute, Moscow (in Russian).

Leonov, E.G. and Isaev, V.I. (1980a) *Hydraulic Calculations of Well Washing with Aerated Fluid in Drilling.* Gubkin Oil and Gas Institute, Moscow (in Russian).

Leonov, E.G. and Isaev, V.I. (1980b) *Calculation of Hydrodynamic Pressure in Round Trip of Drillstems in Well Drilling. Textbook.* Gubkin Oil and Gas Institute, Moscow (in Russian).

Leonov, E.G. and Isaev, V.I. (1982) On filtration of washing fluid and grouting mortar. In: *Investigation of Backfills. Proceedings of the Gubkin Oil Gas Institute*, p. 162 (in Russian).

Leonov, E.G., Isaev, V.I., and Fisher, V.A. (1984) *Hydraulic Calculation of Circulation System in Well Drilling.* Gubkin Oil and Gas Institute, Moscow (in Russian).

Leonov, E.G., Isaev, V.I., and Lukyanov, I.P. (2001) Theory and method of flow rate calculation of grouting mortar with different rheology to clean inclined directed well bores. *J. Oil Gas Eng. Land Sea*, (8), 24–31 (in Russian).

Leonov, E.G., Isaev, V.I., and Ponomarev, Ju.N. (1980) Application of a computer to hydraulic calculations of well washing in drilling. In: *Technology of Drilling of Oil and Gas Wells. Proceedings of the Gubkin Oil Gas Institute*, Vol. 152, pp. 137–160 (in Russian).

Leonov, E.G. and Triadski, V.M. (1980) Deformation of well walls in argillaceous solids and salts. In: *Technology of Drilling of Oil and Gas Wells. Proceedings of the Gubkin Oil Gas Institute*, Vol. 152, pp. 62–70 (in Russian).

Loitsyansky, L.G. (1987) *Mechanics of Fluid and Gas*. Nauka, Moscow (in Russian).

Macovei, N. (1982) *Hidraulica forajulica*. Editura Tehnica, Bucaresti.

McLean, R.H., Manry, C.W., and Whitaker, W.W. (1967) Displacement mechanics in primary cementing. *J. Petrol. Technol.*, **19** (2), 251–260.

Mezshlumov, A.O. (1976) *Use of Aerated Fluids in Well Drilling*. Nedra, Moscow (in Russian).

Mezshlumov, A.O. and Makurin, N.S. (1967) *Drilling of Wells with Use of Air, Gas and Aerated Fluid*. Nedra, Moscow (in Russian).

Mezshlumov, A.O. and Makurin, N.S. (1976) *Use of Aerated fluid in Well Drilling*. Nedra, Moscow (in Russian).

Mirzadjanzadeh, A.H. (1959) *Problems on Hydrodynamics of Viscous-Plastic and Viscous Fluids in Oil Production*. Azerneftnershr, Baku (in Russian).

Mirzadjanzadeh, A.H. and Entov, V.M. (1985) *Hydrodynamics in Drilling*. Nedra, Moscow (in Russian).

Mittelman, B.I. (1963) *Reference Book on Hydraulic Calculations in Drilling*. Gostoptechizdat, Moscow (in Russian).

Mooney, M.J. (1931) Explicit formulas for slip and fluidity. *J. Rheol.*, **2**, 210–222.

Mosesyan, A.A. and Leonov, E.G. (2002) Calculation of operation characteristics of rotary downhole motors. *J. Oil Gas Eng. Land Sea*, (7–8), 10–15 (in Russian).

Muskat, M. (1937) *The Flow of Homogeneous Fluids Through Porous Media*. McGraw-Hill, New York.

Newton, I. (1871) *Philosophiae Naturalis Principa Mathematica*, V ed. Posthumous.

Nigmatullin, R.I. (1987) *Dynamics of Multiphase Media*, Vols. **I and II**. Nauka, Moscow (in Russian).

Ostwald, W. (1925) Über die Geschwindigkeitsfunktion der Viscosität disperser Systeme. *Kolloid Z.*, **36**, 99.

Pavlowski, N.N. (1922) *The Theory of Motion of Ground Waters Under Hydraulic Engineering Constructions*. Lithograph, Petrograd (in Russian).

Petrov, V.A., Leonov, V.G., Filatov, B.S., and Isaev, B.I. (1974) Investigation of gas outburst and graphic calculation method of its killing. *Gas Industry*, (8), 53–54 (in Russian).

Pihachev, G.B. and Isaev, V.I. (1973) *Underground Hydraulics*. Nedra, Moscow (in Russian).

Poiseuille, J. (1840) Recherches Experimentelles sur le Mouvement des Liquides dans les Tubes de trés petits Diaméters. *Comptes Rendus*, **11**, 961–967, 1041–1048; 1841, **12**, 112–115.

Prandtl, L. and Tietjens, O. (1929) *Hydro- und Aeromechanik nach Vorlesungen von L. Prandtl*, B.1, 2. Berlin.

Rabinowitch, B. (1929) Über die Viscosität von Solen. *J. Phys. Chem.*, **145A**, 1–27.

Reiner, M. (1960) *Deformation, Strain and Flow. An Elementary Introduction to Rheology*. H.K. Lewis & Co. Ltd., London.

Reynolds, O. (1883) An experimental investigation of the circumstances which determine whether the motion of water shall be direct or sinuous, and of the law of resistance in parallel channels. *Philos. Trans. R. Soc. Lond.*, **174**, 935–982.

Samochvalov, S.Ju. and Leonov, E.G. (1989) Estimation of cutting settling velocity in drilling mud solution. *Collection of Scientific Works. Problems of Well Drilling and Development of Oil and Gas Fields*, Vol. **214**, Gubkin Oil and Gas Institute, Moscow, pp. 16–22 (in Russian).

Schelkachev, V.N. (1931) *Motion of Viscous Fluid in a Pipe in the Interior of Which Lies a Pipe of Lesser Diameter*. Soyuzneft, Moscow-Leningrad (in Russian).

Schelkachev, V.N. (1990) *Selected Works*, Vols. **I** and **II**. Nedra, Moscow (in Russian).

Schlichting, H. (1964) *Grenzschicht-Theorie*. G. Braun Verlag, Karlsruhe.

Sedov, L.I. (1983) *Mechanics of Continuum*. Vols. **I** and **II**. Nauka (in Russian).

Sereda, N.G. and Solov'ev, E.M. (1974) *Drilling of Oil and Gas Wells*. Nedra, Moscow (in Russian).

Shazov, N.I. (1938) *Deep Rotary Drillung*. ONTI NKTP USSR, Moscow-Leningrad (in Russian).

Sheberstov, E.V. and Leonov, E.G. (1968) Pressure calculation in a well in drilling with aerated fluids. *Oil Industry*, (12), 14–17 (in Russian).

Sheberstov, E.V., Leonov, E.G., and Malevanski, V.D. (1968) Appearance of gas blowout with the availability of an aerated mud solution bench. *Gas Industry*, (6), 5–7 (in Russian).

Sheberstov, E.V., Leonov, E.G., and Malevanski, V.D. (1969) Calculation of the amount and pumping rate of fluid to kill gas blowout. *Gas Industry*, (4), 7–11 (in Russian).

Shischenko, R.I. (1951) *Hydraulics of Mud Solutions.* Aznefteizdat, Baku (in Russian).

Shischenko, R.I. and Baklanov, B.D. (1933) *Hydraulic Theory of Mud Solutions and Its Practical Application.* Aznefteeizdat, Baku-Moscow (in Russian).

Shischenko, R.I., Esman, B.I., and Kondratenko, P.I. (1976) *Hydraulics of Washing Fluids.* Nedra, Moscow (in Russian).

Shumilov, P.P. (1943) *Theoretical Foundations of Turbo Drilling.* Gostoptechizdat, Moscow-Leningrad (in Russian).

Shwedoff, Th. (1890) Recherches Experimentals sur la Cohesion des Liquids. *J. Phys.*, **9** (2), 34.

Stokes, G.G. (1845) On the theories of the internal friction of fluids in motion and of the equilibrium and motion of elastic solids. *Trans. Cambr. Philos. Soc.*, **8**, 287–305.

Stokes, G.G. (1850) On the effect of internal friction of fluids on the motion of pendulums. *Trans. Cambr. Philos. Soc.*, **9**, 8 *Math. Phys. Papers*, 1901, **3**, 1–141.

Targ, S.M. (1951) *Main Problems on the Theory of Laminar Flows.* GITTL, Moscow-Leningrad (in Russian).

Teletov, S.G. (1958) Problems on hydrodynamics of two-phase mixtures. *MGU Bull.*, (2), 15–27.

Ustimenko, B.P. (1977) *Processes of Turbulent Transport in Rotary Flows.* Nauka, Alma-Ata (in Russian).

Volarovich, M.P. and Gutkin, A.M. (1946) Flow of plastic-viscous body between two parallel plane walls and in annular space between two coaxial cylinders. *J. Tech. Phys.*, **16** (3), 321–328.

Wilkenson, U.L. (1960) *Non-Newtonian Fluids.* Pergamon Press, London.

Zhukowski, N.E. (1948) *Collected Works*, Vol. **II**. Gostechtheorizdat, Moscow (in Russian).

AUTHOR INDEX

Applied Hydro-Aeromechanics in Oil and Gas Drilling. By Leonov and Isaev

SUBJECT INDEX

ABOUT THE AUTHORS

Eugeniy G. Leonov is a Professor in the Department of Oil and Gas Underground Hydrodynamics at the Moscow Gubkin State University of Oil and Gas. He has a doctorate in engineering science. He has authored more than 150 publications, including 6 books and 10 authorship certificates and patents. A prominent authority in the field of accident prevention and liquidation in well drilling, he delivers lectures on problems and failures in oil and gas well drilling.

Valeriy I. Isaev is Associate Professor in the Department of Oil and Gas Underground Hydrodynamics at the Moscow Gubkin State University of Oil and Gas. Possessing a doctorate in engineering science, he has 125 publications to his credit, including 6 books and 3 authorship certificates. He is a leading authority on the mechanics of fluid and oil and gas well drilling and delivers lectures on hydro-aeromechanics in drilling, petrophysics, mechanics of non-Newtonian fluids, and technical hydromechanics.